Bioengineered
NANOMATERIALS

Bioengineered
NANOMATERIALS

Edited by **Atul Tiwari** • **Ashutosh Tiwari**

CRC Press
Taylor & Francis Group
Boca Raton   London   New York

CRC Press is an imprint of the
Taylor & Francis Group, an **informa** business

CRC Press
Taylor & Francis Group
6000 Broken Sound Parkway NW, Suite 300
Boca Raton, FL 33487-2742

First issued in paperback 2017

Version Date: 20130509

ISBN 13: 978-1-138-07675-4 (pbk)
ISBN 13: 978-1-4665-8595-9 (hbk)

Library of Congress Cataloging-in-Publication Data

Bioengineered nanomaterials / edited by Atul Tiwari and Ashutosh Tiwari.
  p. ; cm.
  Includes bibliographical references and index.
  ISBN 978-1-4665-8595-9 (hardcover : alk. paper)
  I. Tiwari, Atul, editor of compilation. II. Tiwari, Ashutosh, 1945- editor of compilation.
  [DNLM: 1. Nanomedicine--trends. 2. Biocompatible Materials. 3. Bioengineering--trends. 4. Nanostructures. QT 36.5]

  R857.M3
  610.28--dc23   2013018422

Visit the Taylor & Francis Web site at
http://www.taylorandfrancis.com

and the CRC Press Web site at
http://www.crcpress.com

Contents

Preface

Many varieties of new, complex diseases are constantly being discovered, which leaves scientists with little choice but to embrace innovative methods for controlling the invasion of life-threatening problems. The use of nanotechnology has given scientists an opportunity to create nanomaterials that could help medical professionals in diagnosing and treating problems quickly and effectively. This book focuses on the novel methodologies and strategies adopted in the research and development of bioengineered nanomaterials and technology. It has been written to provide comprehensive and up-to-date information on cutting-edge research in the area of nanomaterials of biotechnological importance.

Chapter 1 introduces the applications of nanoparticles that have clinical utility in cancer treatment. Chapter 2 provides a brief overview of research related to aptamers. The development of aptamer-modified nanoparticles and their use in medical applications are discussed. Chapter 3 describes five immobilization techniques, that is, adsorption, entrapment, encapsulation, covalent coupling, and cross-linking, that are used in the development of biomaterial-based electrodes. Likewise, Chapter 4 reviews research related to the processing of functional nanostructures with a focus on the crucial role of nanofibers and nanoparticle systems. Chapter 5 examines the historical perspective of nanoemulsion vaccine adjuvants, the latest advancements in nanoemulsion adjuvant research, and the direction of future development. Chapter 6 describes the fascinating properties of biocompatible, inorganic nanoparticles of carbonate apatite, including electrostatic interactions with a variety of charged molecules for capturing and subsequent cellular delivery of potential therapeutics. Chapter 7 reviews the synthesis, applications, and biological properties of various ceramic platforms used in medical applications and their potential application as therapeutic nanoparticles in anticancer protein delivery. Chapter 8 discusses the essential aspects in designing the colloidally stable iron oxide nanoparticles. The encapsulation of transplanted cells in nanoporous semipermeable membranes to facilitate adequate diffusion while minimizing immune and foreign cell responses is described in Chapter 9. This chapter also outlines the structural and chemical attributes of various encapsulants. Chapter 10 describes the release of actives from nanoparticles for the application related to topical drug delivery as transdermal patches. A detailed description of toxicity mechanisms, as well as limitations and future prospects of nanotoxicity testing with potential new applications of silver nanocomposites as antibacterial support systems, is provided in Chapter 11. Chapter 12 is devoted to the laser ablation of solid targets to produce a myriad of morphologically distinct nanostructures. This will help readers in understanding the unique one-pot green synthesis method, which gives a precise control over the size, composition, and biofunctionalization, and their results into multifunctional nanoparticles for use in theranostics. The principal applications of nanomedicine in brain tumor treatment, followed by a report on various preclinical and/or clinical studies conducted in brain tumor treatment, are covered in Chapter 13. Chapter 14 provides detailed information on the development of nanomaterials, especially mesoporous silica nanoparticles, that can incorporate high volumes of therapeutic drugs. Such mesoporous nanoparticles can be promising in a wide range of applications, such as information storage systems, magnetic nanodevices, drug delivery, magnetic hyperthermia, medical diagnostics, and ferrofluids. Chapter 15 demonstrates the use of near-infrared-resonant gold nanoshells and carbon nanotubes in tumor imaging and nanodelivery systems for theranostics of carcinomas.

Finally, Chapter 16 has critically upended the requirements for bone regeneration and how biological inspiration is being coupled with the nanoscale engineering of biomaterials in order to create innovative biomimetic scaffolds. The chapter also describes innovative approaches to improve bioactive properties and molecular signaling in cells to stimulate bone repair.

This book contains in-depth information on bioengineered nanomaterials being developed in leading research laboratories around the world. Although the primary focus of attended nanomaterials is on biomedical applications, these technologies would be interesting if explored in multidisciplinary passage. The comprehensively written chapters are targeted to a broad readership, including students and researchers from diverse backgrounds such as chemistry, physics, materials science and engineering, medical science, pharmacy, biotechnology, and biomedical engineering. This book can be used as a textbook by students of bioengineering courses as well as a reference book for researchers.

We are confident that the most recent and detailed information in this book will be useful for students, researchers, scientists, engineers, and professors.

Atul Tiwari
Ashutosh Tiwari

Editors

Atul Tiwari, PhD, is faculty in the Department of Mechanical Engineering at the University of Hawaii, Honolulu, Hawaii. He earned an MS in chemistry from the University of Kanpur, India, a PhD in polymer science from Macromolecular Research Centre, RD University, Jabalpur, India, and an MS in mechanical engineering from the University of Hawaii at Manoa in the United States. He earned chartered chemist and chartered scientist status from the Royal Society of Chemistry, United Kingdom, and is a member of several professional bodies in the United Kingdom, the United States, and India.

Dr. Tiwari has invented and patented several technologies and has published more than 60 research articles, book chapters, and books in the area of materials science and engineering. He has been actively engaged as a consultant in various fields of materials science and engineering. Dr. Tiwari also serves as a reviewer and coeditor for international publication houses and as a board member in many prestigious institutions worldwide.

Ashutosh Tiwari, PhD, is faculty at the Biosensors and Bioelectronics Centre, IFM-Linkoping University and editor in chief of *Advanced Materials Letters*. He is a materials chemist and a graduate of the University of Allahabad, India. He currently serves as adjunct/visiting professor at many prestigious institutions worldwide. He is also actively engaged as a reviewer, editor, and member of scientific bodies worldwide.

Dr. Tiwari obtained various prestigious fellowships, including JSPS, Japan; SI, Sweden; and Marie Curie, England/Sweden. In his academic career, he has published over 175 articles, acquired several patents, and participated in conference proceedings in the field of materials science and technology. He has also edited or authored ten books on the state of the art of materials science.

Contributors

Nana M. Agyei
Department of Chemistry
University of Limpopo
Medunsa, South Africa

Hossein Abbastabar Ahangar
Department of Materials Science and
  Engineering
Islamic Azad University
Isfahan, Iran

Eilis Ahern
School of Pharmacy
University College Cork
Cork, Ireland

Toshihiro Akaike
Department of Biomolecular Engineering
Graduate School of Bioscience and
  Biotechnology
Tokyo Institute of Technology
Yokohama, Japan

Ricard Armengol
Provital Group
Barcelona, Spain

Ana Baptista
Faculdade de Ciências e Tecnologia
(Faculty of Sciences and Technology)
Departamento de Ciência dos Materiais
(Materials Science Department and
  CENIMAT/I3N)
Universidade Nova de Lisboa
(New University of Lisbon)
Caparica, Portugal

Juana Benavente
Facultad de Ciencias
Departamento de Física Aplicada
Universidad de Málaga
Málaga, Spain

João Paulo Borges
Faculdade de Ciências e Tecnologia
(Faculty of Sciences and Technology)
Departamento de Ciência dos Materiais
(Materials Science Department and
  CENIMAT/I3N)
Universidade Nova de Lisboa
(New University of Lisbon)
Caparica, Portugal

Levon A. Bostanian
College of Pharmacy
Xavier University of Louisiana
New Orleans, Louisiana

Maria Caffo
Department of Neurosciences Psychiatry
  and Anesthesiology
Neurosurgical Clinic
School of Medicine
University of Messina
Messina, Italy

Gerardo Caruso
Department of Neurosciences
Neurosurgical Clinic
School of Medicine
University of Messina
Messina, Italy

C.J. Chen
Center of Nanomagnetics and
  Biotechnology
Florida State University
Tallahassee, Florida

Ezharul Hoque Chowdhury
Jeffrey Cheah School of Medicine and
  Health Sciences
Monash University
Selangor, Malaysia

Timothy Doody
School of Pharmacy
University College Cork
Cork, Ireland

Gregor P.C. Drummen
Bionanoscience and Bio-Imaging
    Program
Cellular Stress and Ageing
    Program
Bio&Nano-Solutions
Düsseldorf, Germany

P.B.A. Fechine
Group of Advanced Chemistry
    Materials
Department of Analytical
    Chemistry and
    Physical-Chemistry
Federal University of Ceará
Ceará, Brazil

Isabel Ferreira
Faculdade de Ciências e Tecnologia
(Faculty of Sciences and Technology)
Departamento de Ciência dos Materiais
(Materials Science Department and
    CENIMAT/I3N)
Universidade Nova de Lisboa
(New University of Lisbon)
Caparica, Portugal

R.M. Freire
Group of Advanced Chemistry
    Materials
Department of Analytical
    Chemistry and
    Physical-Chemistry
Federal University of Ceará
Ceará, Brazil

David H. Gracias
Department of Chemical and Biomolecular
    Engineering
and
Department of Chemistry
Johns Hopkins University
Baltimore, Maryland

Jesús Hierrezuelo
Facultad de Ciencias
Deptemento de Química Orgánica
Universidad de Málaga
Málaga, Spain

Sharif Hossain
Department of Biomolecular Engineering
Graduate School of Bioscience and
    Biotechnology
Tokyo Institute of Technology
Yokohama, Japan

Sachin Kadam
Department of Chemical and Biomolecular
    Engineering
Johns Hopkins University
Baltimore, Maryland

Jagat R. Kanwar
Laboratory of Immunology and Molecular
    Biomedical Research
Institute for Frontier Materials
Deakin University
Waurn Ponds, Victoria, Australia

Rupinder K. Kanwar
Laboratory of Immunology and Molecular
    Biomedical Research
Institute for Frontier Materials
Deakin University
Waurn Ponds, Victoria, Australia

S. Kwon
Department of Biological Engineering
Utah State University
Logan, Utah

Giuseppe La Fata
Department of Neurosciences Psychiatry
    and Anesthesiology
Neurosurgical Clinic
School of Medicine
University of Messina
Messina, Italy

José M. Lanao
Department of Pharmacy and
  Pharmaceutical Technology
Salamanca Institute for Biomedical
  Research (IBSAL)
University of Salamanca
Salamanca, Spain

Grace Ledet
College of Pharmacy
Xavier University of Louisiana
New Orleans, Louisiana

E. Longhinotti
Group of Bioinorganic Chemistry
Department of Organic and Inorganic
  Chemistry
Federal University of Ceará
Ceará, Brazil

Juan Manuel López-Romero
Facultad de Ciencias
Departamento de Química Orgánica
Universidad de Málaga
Málaga, Spain

Cristina Maderuelo
Faculty of Pharmacy
Pharmaceutical R&D Laboratory
Department of Pharmacy and
  Pharmaceutical Technology
University of Salamanca
Salamanca, Spain

Ganesh Mahidhara
Laboratory of Immunology and Molecular
  Biomedical Research
Institute for Frontier Materials
Deakin University
Waurn Ponds, Victoria, Australia

Tarun K. Mandal
College of Pharmacy
Xavier University of Louisiana
New Orleans, Louisiana

Laura Martínez-Marcos
Faculty of Pharmacy
Pharmaceutical R&D Laboratory
University of Salamanca
Salamanca, Spain

Lucia Merlo
Pharmaceutical R&D Laboratory
Department of Neurosciences Psychiatry
  and Anesthesiology
Neurosurgical Clinic
School of Medicine
University of Messina
Messina, Italy

Mambo Moyo
Department of Environmental, Water,
  and Earth Sciences
Tshwane University of Technology
Pretoria, South Africa

Jonathan O. Okonkwo
Department of Environmental, Water,
  and Earth Sciences
Tshwane University of Technology
Pretoria, South Africa

Marcello Passalacqua
Department of Neurosciences Psychiatry
  and Anesthesiology
Neurosurgical Clinic
School of Medicine
University of Messina
Messina, Italy

Laura Peláez
Facultad de Ciencias
Deptmento de Física Aplicada
Universidad de Málaga
Málaga, Spain

Erik Reimhult
Department of Nanobiotechnology
Institute for Biologically Inspired Materials
University of Natural Resources and Life
  Sciences, Vienna
Vienna, Austria

Christine Reinemann
Centre for Environmental Biotechnology
Helmholtz Centre for Environmental
  Research—UFZ
Leipzig, Germany

Rodrigo Rico
Facultad de Ciencias
Departamento de Química Orgánica
Universidad de Málaga
Málaga, Spain

Kislay Roy
Laboratory of Immunology and Molecular
    Biomedical Research
Institute for Frontier Materials
Deakin University
Waurn Ponds, Victoria, Australia

Katie B. Ryan
School of Pharmacy
University College Cork
Cork, Ireland

Daniela Schmid
Molecular Therapeutics
School of Pharmacy
Queen's University Belfast
Belfast, United Kingdom

Christopher J. Scott
Molecular Therapeutics
School of Pharmacy
Queen's University Belfast
Belfast, United Kingdom

R. Sharma
Center of Nanomagnetics and
    Biotechnology
Florida State University
Tallahassee, Florida

Paula Soares
Faculdade de Ciências e Tecnologia
(Faculty of Sciences and Technology)
Departamento de Ciência dos Materiais
(Materials Science Department and
    CENIMAT/I3N)
Universidade Nova de Lisboa
(New University of Lisbon)
Caparica, Portugal

E.H.S. Sousa
Group of Bioinorganic Chemistry
Department of Organic and Inorganic
    Chemistry
Federal University of Ceará
Ceará, Brazil

Beate Strehlitz
Department of Environmental and
    Biotechnology Centre
Helmholtz Centre for Environmental
    Research—UFZ
Leipzig, Germany

Francesco Tomasello
Department of Neurosciences Psychiatry
    and Anesthesiology
Neurosurgical Clinic
School of Medicine
University of Messina
Messina, Italy

Reza Zamiri
Department of Materials Engineering
    and Ceramics
University of Aveiro
Aveiro, Portugal

1

Emerging Potential of Nanoparticles for the Treatment of Solid Tumors and Metastasis

Daniela Schmid and Christopher J. Scott

CONTENTS

1.1 Introduction

Cancer is the third leading cause of death worldwide, and mortality rates are continuously rising due to a growing population, longer life expectancy, and changing lifestyle. Although radiation therapy, surgery, and the development of a range of new chemotherapeutic drugs have proven to be effective control for certain primary tumors, efficient treatment for other cancer types and in particular metastatic cancer is still lacking [1]. Since Paul Ehrlich, inspired by an opera, postulated the theory of *Zauberkugeln*—"magic bullets"—over 100 years ago, generations of researchers have been inspired to pursue the development of molecular therapeutics that target cancer cells directly and exclusively [2]. The application of nanotechnology holds promise in the realization of this goal, and in this chapter, advances in nanomedicine for the treatment of cancer will be reviewed.

1.2 Cancer Therapy and Drug Development

The first reports to the application of chemotherapeutic drugs date back to the early 1940s when the use of folic acid analogues resulted in remission of acute lymphocytic leukemia. Several years later, methotrexate was discovered and approved, which is still in use for several types of cancer today [3]. Other advances at this time included the development of plant-derived anticancer drugs such as vinca alkaloids, taxanes, and camptothecins [4] or DNA-interfering agents like cisplatin and doxorubicin [5,6]. Due to poor stability, solubility, and half-life issues, new derivatives of these first-generation drugs were developed with improved pharmacokinetic properties. Despite these advances, side effects like nausea, diarrhea, and neutropenia or more severe toxicity issues such as myocardiotoxicity remain [7–9].

Increasing scientific understanding of the molecular basis for cancer, including the identification of oncogenes and tumor suppressors, in tandem with phenomenal advances in associated technologies has enabled researchers to further develop molecularly targeted therapeutics. The rationale behind this approach is that if the tumor can be targeted specifically, more potent drugs with less off-target side effects may be realized. Therapeutics that has been developed by this approach includes both small molecules and biologics (Figure 1.1).

With the identification of possible therapeutic targets, chemists have used a variety of approaches including structure–activity relationship (SAR) and high-throughput screening to identify potent inhibitors with high specificity [10]. Small molecules have been developed toward a range of new targets revealed by molecular medicine including cytoplasmic tyrosine kinases, which play fundamental roles in cellular processes such as cell proliferation and survival. Highly specific molecules have been rationally designed toward these enzymes to inhibit cell signaling upon binding [11]. The first small-molecule inhibitor of this class, the tyrosine kinase inhibitor imatinib (Glivec®), was approved in 2001 for the treatment of chronic myelogenous leukemia (CML) and inhibits the transfer

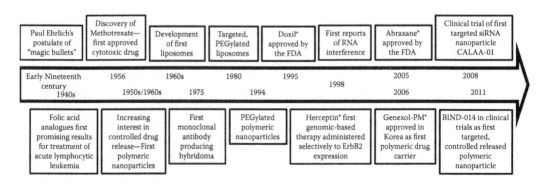

FIGURE 1.1
Milestones in cancer drug development—from the theory of magic bullets to the clinical evaluation of long-circulating therapeutic nanoparticles targeted to cancer cells.

of phosphate from adenosine triphosphate (ATP) to a tyrosine residue of target proteins [12,13]. Remarkably, the response rate to imatinib is over 90%, reducing the white blood cell count in patients to normal. However, resistance in patients has been observed due to gene mutations in the target kinase domain [14]. Several other small-molecule kinase inhibitors have been approved since imatinib became available on the market. Thus, inhibitors like sunitinib (Sutent®) or pazopanib (Votrient®) have broadened the scope of treatment options for renal cell carcinoma where other treatment options are limited [15]. Despite the success that has been realized with small-molecule drugs, many fail to reach the clinic due to issues such as poor bioavailability and lack of specificity.

Another new technology that holds much promise for developing synthetic drug molecules with excellent specificity is RNA interference (RNAi). Fire and Mello first showed the potential of RNAi using double-stranded RNA in *Caenorhabditis elegans* [16]. This principle turned out to be transferrable to mammalian cells and potential RNAi-based therapeutics have been rapidly developed. The main disadvantage of RNAi is its *in vivo* bioavailability, as they do not easily penetrate cell membranes and have very short serum half-lives due to plasma degradation. Thus, efficient delivery remains a challenge for the clinical translation of RNAi-based therapy [17–19].

The major group of molecularly targeted drugs now entering clinical evaluation is biologics, dominated by antibodies. The publication on monoclonal antibody hybridoma generation by Köhler and Milstein in 1975 set a milestone in personalized medicine for targeting specific antigens, in providing a methodology to scalable produce homogenous antibodies [20]. Since then, atleast 12 therapeutic antibodies, targeting proteins overexpressed in tumors, have been approved by the Food and Drug Administration (FDA). These antibodies can either block receptor activation by blocking ligand binding or induce antiproliferative effects [21]. Trastuzumab (Herceptin®) received FDA approval in 1998 for the treatment of patients with human epidermal growth factor receptor 2 (HER2)-positive metastatic breast cancers and remains one of the most successful monoclonal antibodies in clinical use. The ErbB2 gene coding for HER2 plays an important role in the development of breast cancer and is expressed in 20%–30% of patients with invasive breast carcinoma [22]. In 2010, the FDA granted the clinical use of trastuzumab for the treatment of HER2-overexpressing metastatic gastric or gastroesophageal (GE) junction adenocarcinoma in combination with chemotherapy [23]. Another leading therapeutic antibody is bevacizumab (Avastin®), which was approved in 2004 for the treatment of metastatic colorectal cancer. Today, this inhibitor of the pro-angiogenic vascular endothelial growth factor (VEGF) is approved for a range of solid tumors [24].

Despite the successful drugs that have been brought to the clinic, many are far from ideal in terms of their efficacy and side effects that are still evident in healthy tissue. Indeed the inability to reach a therapeutic concentration of a drug at the disease site is often a reason that many exciting novel compounds do not progress to the clinic. Targeted delivery strategies aim to address these limitations, by reducing off-target effects, improving drug half-life, and therefore increasing drug efficiency at the disease site. The most advanced targeting strategies being examined currently use antibodies. As well as their proven efficacy in their own right as therapeutic agents, antibodies can be exploited to carry cytotoxins or imaging reagents to disease sites, targeting disease-specific biomarkers. Recently, an increasing number of antibody-drug conjugates (ADCs) have been developed that deliver cytotoxic drugs specifically to target cells [25–27]. Gemtuzumab ozogamicin (Mylotarg®) was the first ADC approved by the FDA in 2000 for the treatment of acute myeloid leukemia. The drug cargo in this ADC, calicheamicin, is a cytotoxic DNA-binding agent, conjugated to an antibody targeted to the CD33 receptor, an antigen that is presented on leukemic cells in most patients [28,29]. However, Mylotarg was recently withdrawn from the market due to safety issues and lack of clinical benefit [30]. Nonetheless, interest in this field continues and this is exemplified by brentuximab vedotin, a CD30-targeted antibody conjugated to the potent antimicrotubule agent monomethyl auristatin E. It showed antitumor activity in 85% of the patients inducing mainly mild or moderate side effects. This ADC has just received FDA approval for the treatment of Hodgkin's lymphoma and systemic anaplastic large cell lymphoma [31,32]. Trastuzumab emtansine (T-DM1) uses the highly successful breast cancer treatment trastuzumab as targeting antibody and delivers an antimicrotubule agent mertansine selectively to HER2-positive cells. This ADC is currently in phase 3 clinical studies after it has proven to be well-tolerated and effective with a clinical benefit rate of 48% in phase 2 studies on patients with HER2-positive metastatic breast cancer who were extensively pretreated (median of seven therapeutic agents). T-DM1 has the potential to replace trastuzumab and increase therapeutic efficacy of refractory metastatic breast cancer [33,34].

In addition to these current targeting strategies, nanoscale devices represent an exciting emerging platform that have enormous diversity and could revolutionize cancer therapy through the following key parameters [35]:

- Improved drug pharmacokinetics especially of hydrophobic drugs
- Drug targeting to defined tissue and cellular components
- Drug delivery beyond biological barriers
- Co-delivery of drugs
- Therapy and imaging in a combined system

Strategies for this new era of cancer therapy and diagnosis using nanoparticles reflecting 35 years of research including applications of nanoparticles for drug delivery, photothermal therapy, and imaging for early diagnosis and imaging of disease progression are now discussed.

1.3 Therapeutic Nanoparticles for Tumor Treatment

The advances in pharmaceutics in the 1950s and 1960s initiated interest in the concept of controlled drug release. Peter Paul Speiser and his group, pioneers in the field of nanotechnology, developed polyacrylic beads for oral administration and sustained release of chloramphenicol [36]. Subsequently, his group also generated the first nanoparticles for vaccination purposes using micelles with promising immune responses [37]. Since then, an explosion of interest in nanomedicine has occurred ranging from albumin-based to biodegradable nanoparticles and liposomes for a plethora of disease conditions, particularly cancer [38]. Encapsulation in a nanocarrier can protect the cargo drug from premature clearance, and moreover, exposure of normal healthy tissue to the drug can be reduced or even avoided. Thus, the increased bioavailability along with passive and active targeting (discussed later) can increase the levels of a therapeutic at the tumor site. The fact that nanoparticles can be internalized by cancer cells to reach certain cell compartments or bypass biological barriers by appropriate functionalization has the potential to further increase the efficacy of the cargo molecule if its site of action is intracellular [39,40].

1.3.1 Pharmacokinetics and Biodistribution of Nanoparticles

The pharmacokinetics and biodistribution of anticancer drugs in nanoparticle formulations are characterized by the interplay of plasma clearance rate, drug release during circulation, and the accumulation at the tumor site. All these parameters can be influenced by nanoparticle type, size, and physiochemical properties, which have been investigated in an effort to improve the biodistribution of drugs to increase their efficacy.

1.3.1.1 Plasma Clearance

The size, shape, and surface characteristics of nanoparticles all influence circulation time in the bloodstream. Small nanoparticles are cleared mainly by the renal system, whereas larger particles are normally removed from the circulation by the mononuclear phagocyte system (MPS). A rapid clearance rate is desirable for delivery of contrast agents for bioimaging with low long-term toxicity. For example, quantum dots with organic coating are cleared rapidly by renal filtration, but just a small change in their diameter can markedly alter their circulation time from 48 min up to 20 h with increasing size from 4.4 to 8.7 nm. The increase in size also altered significantly the accumulation of the particles in different organs with higher uptake into the liver, lung, and spleen with proportionally lower fluorescence in the bladder in comparison to the smaller particles [41]. For cancer therapy, however, a longer circulation time is preferential to allow a reasonable dose to reach the disease site. Drug-loaded nanoparticulate carriers tend to be much larger in size than quantum dots and can interact with the MPS within seconds of administration. Once in circulation, nanoparticles become covered with serum proteins, also called opsonins, such as immunoglobulins or complement proteins. This innate immune response is a mechanism to clear unknown objects in the circulation, and once opsonized, phagocytic MPS cells of mainly the spleen, liver, and bone marrow will then engulf and clear the nanoparticles [42]. This unwanted opsonization of nanoparticles and clearance by the MPS is influenced by several factors, which are schematically summarized in Figure 1.2.

For nanoparticles with a size below 1 μm, the activation of the MPS was shown to be size dependent, with larger nanoparticles cleared faster than smaller ones [43]. Using polymeric nanoparticles with varying sizes of 80, 171, and 243 nm, it was found that the larger the particle, the more protein is adsorbed. These observations correlated with other studies

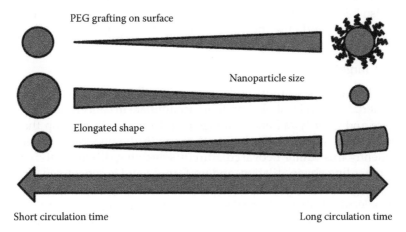

FIGURE 1.2
Schematic illustration of nanoparticle characteristics that influence clearance rate.

examining macrophage uptake where it was found that larger particles were more rapidly taken up by the phagocytic cells [44]. The majority of investigated nanoscale drug delivery systems or devices for imaging have been of spherical shape. Interestingly, however, recent studies suggest that elongated structures alter *in vitro* and *in vivo* performance. It has been observed that gold rod-shaped nanocarriers showed a slower cellular uptake by HeLa cells [45]. This is in agreement with findings examining self-assembled filomicelles compared to spherical micelles, where the filomicelles had longer circulation times and when loaded with chemotherapy resulted in improved antitumor effects [46]. Furthermore, the elongated structure nearly doubled the maximal tolerated dose of paclitaxel with fewer off-target effects. There was 15%–30% reduction in unwanted apoptosis in healthy organs of mice, which were treated with the filomicelles compared to the spherical micelles [47]. However, another study using magnetic iron oxide nanoworms did not observe prolonged circulation time compared to the nanospheres, but did find increased accumulation and extended retention at the tumor site [48].

The hydrophilicity of the particle surface is also crucial to its speed of removal by the blood circulation. Conjugation of hydrophilic polymers such as poloxamer, polysaccharides, and in particular polyethylene glycol (PEG) to the nanoparticle surface improves the half-life significantly [49]. PEG is an FDA-approved polymer and had its major breakthrough in the 1970s when attachment of PEG to proteins showed decreased immunogenicity *in vivo* while retaining their activity [50]. This hydrophilic polymer holds great advantages for nanoparticles as it acts like a barrier between plasma opsonins and the hydrophobic surface of many particle formulations and therefore prevents protein adsorption. Its attachment can also effectively neutralize surface charge of particles and prevent surface interaction with blood components through electrostatic interactions and furthermore provide increased colloidal stability in serum conditions [51]. The levels of PEGylation have been shown to correlate proportionally with reduced protein absorption. With increasing PEG density, adsorption of proteins to gold nanoparticles was decreased by up to 99% with the highest PEG density of around 1 PEG molecule/nm². These observations were complemented by reduced macrophage uptake with the higher PEG density [52]. Moreover, it has been shown using radioactively labeled poly(lactic-*co*-glycolic acid) (PLGA) nanoparticles that nonfunctionalized carriers were removed to the liver within minutes after intravenous administration, whereas the PEGylated nanoparticles remained in circulation for over 24 h and were only gradually removed within this time [53]. The effect of increased circulatory half-life has also been established in numerous studies using PEGylated carriers such as liposomes, dendrimers and other polymeric nanoparticles [54,55].

1.3.1.2 Passive Targeting

The effect of enhanced permeability and retention (EPR) was reported first in 1986, when it was shown that radioactively labeled proteins accumulate in cancerous tissue due to leaky neovasculature and poor lymphatic drainage [56]. Tumors will develop their own blood supply in order to provide the developing tissue with the necessary nutrients and oxygen, but frequently this tumor neovasculature is somewhat different from normal developed vessels regarding shape, dilation, and uniformity and is considered more permeable due to poorly aligned endothelial cell-to-cell junctions [57]. This defective architecture allows large molecules (greater than 40 kDa in size) and nanoparticles to intrude directly into tumor tissue and remain due to reduced clearance [58,59]. This unique characteristic of tumor tissue makes resident blood vessels more permeable for colloidal drug delivery systems than the vasculature in normal tissue. Liposomes with a size of up to 400 nm have been reported to extravasate across tumor microvessels by passing through the cell–cell junctions in the leaky vasculature where they are able to release their payload at the tumor site [60]. The tumor cutoff size can vary from 100 nm to 1.2 μm depending on tumor type and microenvironment. The majority of subcutaneous tumors exhibit endothelial gaps between 200 and 900 nm, whereas gaps in tumors growing in cranial microenvironment can be smaller [61]. The consequences of EPR have been observed in various nanocarrier studies. Examples include doxorubicin-loaded micelles, which reached increased drug concentration in tumor tissue within 96 h compared to free drug that after 24 h was detectable neither in the tumor tissue nor in other organs [62]. Similarly, albumin-based nanoparticles can accumulate in subcutaneously growing tumors within hours and have been used to deliver chemotherapeutic drugs by the EPR effect [63].

1.3.1.3 Active Targeting of Nanoparticles with Surface Ligands

Drug accumulation at the tumor site through the EPR effect does not always correlate with therapeutic outcome, as the drug needs to reach its site of action, frequently to specific cell compartments, for its biological effect to be realized. Surface engineering of nanoparticles using targeting moieties increases the selectivity and efficiency by internalization through receptor-mediated endocytosis [64]. These targeting agents include proteins, mainly antibodies, nucleic acid, or other cancer-specific receptor ligands such as carbohydrates or vitamins [65]. Ligands can be adsorbed or covalently bound using a cross-linker [66]. Figure 1.3 represents schematically the mechanisms of both passive and active targeting by nanoparticles.

Antibodies are probably the most extensively researched targeting agent. The family of epithelial growth factor receptors (EGFRs) is an interesting target for cancer therapy as its signaling pathway triggers cell proliferation. Monoclonal antibodies like trastuzumab (Herceptin) directed toward the extracellular domain of the tyrosine kinase have proven successful for the treatment of HER2-positive carcinomas [67]. Different types of nanoparticles targeting HER2 receptors have been evaluated. Polyamidoamine (PAMAM) dendrimers and liposomes conjugated to monoclonal anti-HER2 antibodies showed increased

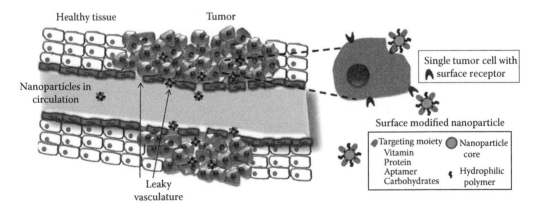

FIGURE 1.3
(See color insert.) Schematic representation of passive and active targeting. Nanoparticles are able to extravasate into tumor tissue through the leaky vasculature in tumor tissue and accumulate due to poor lymphatic drainage—also known as the effect of EPR. Once accumulated in the tumor tissue surface, modified nanoparticles can specifically target cancer cells by different ligands and can be internalized by receptor-mediated endocytosis where payload can be released.

cellular uptake into HER2 expressing cells *in vitro*. The targeted liposomes were also investigated in animal models and were internalized more efficiently by cancer cells, but did not localize more in tumors in comparison to untargeted liposomes [68,69]. In previous studies, these long-circulating HER2-targeted liposomes containing doxorubicin had superior antitumor effects toward HER2-positive xenografts compared to untargeted doxorubicin-loaded liposomes [70].

Other promising cancer targets include the death receptors (DRs) of the tumor necrosis factor (TNF) receptor superfamily, whose members can induce activation of the extrinsic pathway of apoptosis. Nanoparticles targeting these receptors can be simultaneously used as targeting and therapeutic agents as shown with polymeric nanocarriers targeting camptothecin to FAS receptor expressing tumor cells [71]. Some DRs such as TNF-related apoptosis-inducing ligand receptors (TRAILRs) require receptor clustering to activate apoptosis via caspase 8 [72]. Interestingly, polymeric nanoparticles conjugated to antibodies were revealed to present sufficient antibody density on the surface to cross-link receptors. In this study, DR5-targeted nanoparticles exploited receptor-mediated endocytosis to increase uptake, initiate extrinsic apoptosis, and induce cytotoxicity through delivery of camptothecin cargo inside the cell [73].

Nucleic acids can also be used as tumor-targeting strategies. Aptamers are RNA or DNA oligonucleotides, which bind to surface proteins selectively by their unique tertiary conformation. There are several studies using aptamer-functionalized nanoparticles reported using *in vitro* models, but the usefulness of these ligands *in vivo* remains to be determined with factors such as uncertainty around conformational change of aptamers in physiological conditions, negative surface charge (which can increase opsonization rate), and high potential production costs represents challenges to their clinical development in the near future [74]. However, RNA aptamer PLGA-PEG nanoparticles containing docetaxel targeted to prostate-specific membrane antigen induced complete tumor regression with one single injection highlights the potential of this targeting strategy and the necessity of further development [75].

Other targeting strategies involve transferrin and folate. Transferrin is a glycoprotein, which binds and facilitates transfer of iron from the circulation into cells through transferrin receptors. Upregulated expression of these receptors is found in malignant cells due to an increased iron requirement and correlates with tumor stage and prognosis [76,77]. Transferrin-targeted cyclodextrin-based nanoparticles passively accumulate in the tumor microenvironment through EPR to the same extent as untargeted control particles but then facilitate the improved delivery of encapsulated siRNA more efficiently into the tumor cells [78,79]. Similar enhanced antitumor effects were observed with heparin-based paclitaxel-loaded nanoparticles targeted to the folate receptor, which is highly overexpressed on cancer cells but with limited expression on healthy cells [80,81]. There are many more ligands available that may be suitable for nanoparticle targeting depending on type and state of the tumor, and it is anticipated that an increasing volume of evaluative studies will appear in this area [82].

1.3.1.4 Multidrug Resistance

Active targeting of nanoparticles to tumor cells has the potential to improve antitumor activity of the drug molecule while also overcoming multidrug resistance (MDR), which some cancer cells develop upon longer nontoxic exposure to certain types of drugs. MDR can be the result of either the upregulated expression of different transporters to limit the possible maximal concentration of drug possible in the cytoplasm or a genetic alteration to bypass drug-induced cytotoxicity. The ATP-binding cassette transporter

P-glycoproteins (PGPs) is a multidrug efflux pump and can release hydrophobic drugs to the extracellular space upon receptor binding [83]. By targeting receptors expressed on the surface of cancer cells, hydrophobic drugs encapsulated in nanoparticles are internalized by receptor-induced endocytosis rather than simple passive uptake that can overwhelm concentration-dependent limitations caused by this mechanism of MDR. For example, targeting folate receptors on doxorubicin-resistant cells leads to rapid internalization of folate-conjugated liposomes, resulting in drug accumulation in the cytosol and nucleus within 60 and 90 min, respectively. PGP-mediated resistance was bypassed by targeted liposomal delivery reaching a higher drug concentration than with nontargeted liposomes [84]. Another study investigated the potential to overcome PGP-mediated MDR using nanocarriers *in vivo*. Silica-based nanoparticles containing doxorubicin have been injected into doxorubicin-resistant tumors in mice effectively overwhelming the resistance issues that had been associated with the drug [85].

1.3.1.5 Blood–Brain Barrier

The cerebral endothelium is different to peripheral endothelium as it has complex tight junctions between the resident endothelial cells that make transfer of materials even more difficult to the brain than other organs. This phenomenon is known as the blood–brain barrier (BBB) and separates the blood from cerebral tissue, only allowing molecules mainly to pass transcellularly compared to other epithelia that allow passage between cells. Gaseous molecules or small lipid-soluble molecules like barbiturates are able to diffuse through the cell membrane, whereas hydrophilic molecules including nutrients are transported transcellularly by a specific carrier-mediated transport system. Bigger hydrophilic molecules like proteins rely on endocytotic mechanisms such as receptor-mediated or adsorptive transcytosis [86]. Active transport across the barrier using targeted nanoparticles represents an interesting approach to circumvent invasive or disruptive methods of brain tumor treatment. Liposomes were conjugated to mannose analogues and transferrin to deliver daunorubicin to brain tumor cells by passing the BBB. Both the mannose-specific glucose transporter GLUT1 and the transferrin receptor are expressed on the epithelium of the BBB and have been investigated for the suitability for targeting and inducing transcytosis. *In vitro* models indicate that there was increased accumulation of daunorubicin across the synthetic BBB using dual-targeted liposomes compared to nontargeted or single-targeted liposomes. The increased cytotoxicity on brain glioma cells correlated with the results *in vivo* where slower tumor progression and prolonged survival was realized by this approach [87].

1.3.2 Materials Used for Nanoparticle Fabrication

The choice of the selected nanocarrier is dependent on its intended application and the properties of drug to be adsorbed or entrapped. The ideal drug delivery system will protect the drug from early clearance and degradation in the circulation, target specifically the tumor tissue, and provide sustained or controlled release of the drug payload to provide an extended therapeutic window. This has led to a broad range of materials and formulation strategies being explored extensively in recent years. Generally nanoparticles used for cancer therapy can be classified into polymer-/protein-based nanocarriers such as nanospheres or polymer–drug conjugates, dendrimers, metal-based nanoparticles, and self-assembling carriers including liposomes and micelles (see Figure 1.4) [40]. A combination of different materials is very common to optimize delivery and tailor

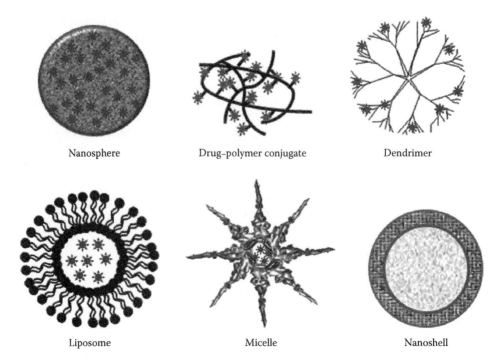

Nanosphere   Drug–polymer conjugate   Dendrimer

Liposome   Micelle   Nanoshell

FIGURE 1.4
Types of different therapeutic nanoparticles including polymer-based nanospheres, drug conjugates, dendrimers, liposomes, micelles, and metal-based nanoparticles such as nanoshells.

TABLE 1.1

Clinically Approved Nanoparticle-Based Therapeutics for Cancer Treatment

Trade Name	Active Drug	NP Material	Cancer Type
Doxil®	Doxorubicin	PEGylated liposome	Ovarian cancer
			HIV-associated Kaposi's sarcoma
			Multiple myeloma
Myocet®	Doxorubicin	Liposome	Metastatic breast cancer
DaunoXome®	Daunorubicin	Liposome	HIV-associated Kaposi's sarcoma
Abraxane®	Paclitaxel	Albumin-based nanoparticle	Metastatic breast cancer
Genexol-PM®	Paclitaxel	PEG–PLA micelle	Metastatic breast cancer

release of cargo drugs. There are several colloidal nanocarriers approved for clinical use (Table 1.1) with a steadily increasing number of nanoparticle-based therapeutics in clinical development with promising recent development for polymer-based nanostructures [88].

Liposomes are probably the most studied and therefore developed vehicles for drug delivery. This spherical type of carrier consists of one or more lipid bilayers of natural or synthetic phospholipids. Amphiphilic phospholipids self-assemble spontaneously with a central aqueous core where both hydrophobic and hydrophilic drugs can be entrapped successfully. These features, along with proven biocompatibility and biodegradability, make liposomes ideal drug carriers [89]. Doxil®, also known as Caelyx®, is a PEGylated liposomal formulation of doxorubicin and was the first nanoparticle-based drug delivery system approved by the FDA in 1995. These stealth nanoparticles with a size of 80–90 nm showed prolonged circulation time and increased drug accumulation at the tumor site resulting in improved efficacy. Interestingly, although clinical studies showed that the liposomal preparation reduced cardiotoxicity, mucosal and skin toxicity as dose-limiting side effects were still evident [90,91]. Similar clinically approved non-PEGylated liposomal formulations for cancer treatment are Myocet® and DaunoXome® for delivery of doxorubicin and daunorubicin, respectively. Several other formulations are currently in clinical trials, including the oxaliplatin formulation MBP-426, a transferrin-targeted liposome currently in phase II trials for gastrointestinal adenocarcinoma [92–94].

1.3.2.1 Polymer- and Protein-Based Nanocarriers

Polymeric nanoparticles are among the most versatile materials for the formulation of nanoparticles. Drugs can be physically entrapped into polymeric nanoparticles using materials such as poly(lactic acid) (PLA) and PLGA that are extensively researched synthetic polymers for the entrapment of hydrophobic drugs. However, naturally occurring polymers are also used such as alginate or chitosan—particularly for the encapsulation of hydrophilic drug molecules [95–97]. Each of these polymers is fully biodegradable and releases the drug over time due firstly to initial drug diffusion and then more prolonged polymer erosion at later stages through polymer hydrolysis [98,99]. BIND-014, an optimized PEGylated PLA nanoparticle formulation, was selected and developed after a screen of 100 targeted nanoparticles with variations in parameters such as polymer composition, size, ligand density, hydrophilicity, and drug loading. These nanoparticles encapsulate docetaxel and also exhibit active targeting using a small-molecule ligand toward prostate-specific membrane antigen. The optimized formulation reached half-lives of up to 20 h with minimal accumulation in liver. Docetaxel was released gradually, which facilitated drug serum concentrations 1000-fold higher than the conventional drug formulation, resulting in superior antitumor effects for the treatment of several solid tumor types in xenograft models. The first clinical results from patients with solid tumors reveal similar pharmacokinetics to the results observed in animal studies with an increased serum concentration compared to the original formulation of docetaxel at the same dose. Also early signs of tumor shrinkage were reported with lower doses of docetaxel compared to doses that are usually administrated [100,101].

Abraxane® is an albumin-based nanoparticle with a size of around 130 nm that has been developed to replace the Cremophor EL-based formulation of paclitaxel (due to the poor water solubility of the drug). This emulsifier can induce hypersensitivity in patients and may also contribute to other toxic side effects of paclitaxel, and therefore, the formulation of the drug in a nanocarrier may circumvent these concerns. Clinical studies showed that albumin-bound paclitaxel can be administrated rapidly with no occurring hypersensitivity [102,103]. Compared to Cremophor EL-based formulations, Abraxane allows the administration of higher drug amounts without increased side effects and higher response rates with increased time to tumor progression in patients with metastatic breast cancer [103]. In preclinical studies, Abraxane had superior antitumor activity in several murine xenograft models with up to a 33% increase in intratumor concentrations. Enhanced uptake is due to the EPR effect, and it is postulated that binding of albumin to 60 kDa glycoprotein receptor (gp60) increases transcytosis of the nanoparticles through epithelial cells into the cancer tissue [104,105].

1.3.2.2 Polymeric Micelles and Dendrimers

Micelles are composed of an amphiphilic diblock polymer and assemble spontaneously in aqueous solution to nano-sized carriers in the size range of 5–100 nm. Hydrophobic drugs can be entrapped in the core, whereas highly soluble drugs can be adsorbed onto the surface. Most micelles are composed of biodegradable polymers, mainly PEG as the hydrophilic component and a variety of polymers for the core including polypeptides (poly-L-lysine) and PLA [106]. Genexol-PM®, a paclitaxel-loaded mPEG-PLA micelle, is approved for metastatic breast cancer in Korea and is currently in clinical studies for approval in United States. In preclinical studies, delivery using micelles enabled the administration of higher doses of drug, which increased tissue concentration of paclitaxel by two- to three-fold and thus improving antitumor activity [107].

Dendrimers are multiple branching 3D molecules. These complex branching polymer molecules form globular bodies, allowing for their exploitation as therapeutic nanoparticles. The stepwise synthesis grants the well-defined structure and low variance in reproduction of the material. Due to the multivalency and number of peripheral groups, different molecules can be conjugated to the polymer including targeting agents, drugs, and hydrophilic surface molecules. Although there are some current issues with lack of controlled release, the ability to control drug loading means these particles will be more frequently examined in the future [108].

1.3.2.3 Cyclodextrin-Based Polymeric Nanocarriers

This relatively new type of nanoparticle is composed of the carbohydrate cyclical β- cyclodextrin. Two PEG molecules give the nanoparticles their hydrophilic character and facilitate the conjugation of drugs [109]. CRLX101 is a polymeric formulation of camptothecin covalently conjugated to the cyclodextrin backbone. In an aqueous environment, the polymer chains form nanoparticle structures due to intramolecular and intermolecular interactions with the hydrophobic drug. The nanoparticles are internalized by endosomal uptake and release camptothecin in its active form with an intact lactone ring. After the release, the nanoparticles fall apart leaving polymer chains behind with a defined size of about 10 nm initiating renal clearance. Early data from clinical studies showed particle half-life of 40 h and promising tumor activity of stable disease in some patients with low side effects and no unexpected effects deriving from the polymer formulation. Clinical phase II studies on solid tumor patients are currently ongoing [110,111].

As described earlier, RNAi has exceptional therapeutic potential, but the major challenge for RNAi is the delivery to the tumor site *in vivo*. CALAA-01 is the first targeted nanocarrier for siRNA in clinical trials. The formulation is based on the RNAi/oligonucleotide nanoparticle delivery technology (RONDEL™), and CALAA-01 is the first candidate for clinical trial emerging from it. The novel nanocarrier delivers siRNA targeting ribonucleotide reductase M2 (RRM2), a clinically validated target to inhibit DNA replication, exploiting active targeting with transferrin. The first results from initial clinical studies did not reveal any serious side effects and showed accumulation of siRNA in biopsies from melanoma as well as decreased mRNA levels for RRM2 [112–114].

1.3.2.4 Gold Nanoparticles and Nanoshells

Gold nanoparticles have been of interest for therapeutic applications since the 1930s. These metal particles can be used for a range of application such as a drug carriers, contrast agents, and radiosensitizers as well as agents for thermal ablation therapy [115]. Upon exposure to light, free electrons on the metal surface oscillate that reaches a maximum at a certain wavelengths, called the localized surface plasmon resonance (LSPR). The oscillation then results in light scattering and conversion into heat. Both effects can be used for nanoparticle tracking, tumor imaging, and thermal ablation, respectively [116]. The administration of EGFR-targeted gold nanoparticles leads to an increased uptake in tumor cells compared to benign cells, analyzed by light scattering imaging. Following laser irradiation at 514 nm, photothermal destruction occurs in cancer cells at low laser intensities with very little effect on healthy cells that required twofold more energy for a similar cytotoxic effect [117]. However, the penetration into deeper tissue requires the use of near-infrared light [118].

Metallic nanoshells are composed of a dielectric core surrounded by a thin metallic layer, typically gold, giving them the physical properties of metal nanoparticles but the potential to be tuned to a desired wavelength. Surface plasmon resonance depends on the core size and shell thickness and can strongly absorb and scatter light [119]. Through suitable fabrication, nanoshells can be designed for light scattering imaging with efficient photothermal effects at the same wavelength. This has recently been shown for HER2-targeted nanoshells [120]. Photothermal effects on cancer tissue were explored in more detail using nanoshells with a 110 nm core and a 10 nm thick shell resulting in an absorbance maximum at 820 nm. Exposure to nanoshells at this wavelength for 4–6 min induced a temperature increase of around 37°C, resulting in thermal tissue damage at a depth of 4–6 mm [121]. Gold nanoshells (AuroLase™) are currently in clinical investigations for treatment of patients with tumors of the head and neck [122].

Gold nanoparticles as drug carriers are being examined for the delivery of TNF-alpha (TNFα) to tumor tissue. At the target site, TNFα alters the permeability of neovasculature, which allows increased influx of chemotherapeutic drugs and therefore higher therapeutic efficacy [123]. Colloidal gold bound to TNFα was investigated in phase I clinical trials and proven to successfully accumulate in tumors by active and passive targeting. Transmission electron microscopy revealed that the colloidal carriers also accumulated in the liver but spared other healthy tissue, especially healthy endothelium, minimizing side effects [124]. These TNFα-targeted gold nanocarriers are now in discovery or preclinical trials in combination with chemotherapeutic drugs such as paclitaxel or doxorubicin [125].

1.4 Nanoparticles for Treatment and Imaging of Metastases

Metastasis is the spread of cancer cells from a primary tumor site to another distant organ. For many cancer patients, metastasis has often occurred even before the primary disease site is detected. These metastases are the main cause of mortality in over 90% of cancer patients [126], highlighting urgent need for new technologies to allow both the early detection of metastases and the effective therapeutics for targeting these secondary tumors.

Metastasis is subdivided into several steps (Figure 1.5). First, cancer cells have to invade and migrate through their localized microenvironment extracellular matrix (ECM) and stromal cell layer of the surrounding tissue at the primary site and intravasate into adjacent blood vessels. Once in circulation, a proportion of the cells will survive the mechanical stresses endured and evade immune cells, in particular natural killer cells. Upon arrival at a distant organ, cells arrest and extravasate into the tissue and adapt to the new environment in order to form micrometastases and proliferate to macroscopic metastases [127].

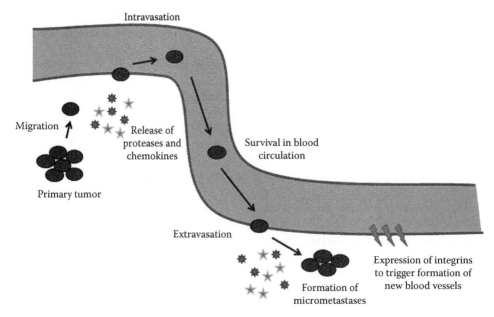

FIGURE 1.5
Schematic presentation of the steps of metastasis—migration, intravasation, and survival in the blood circulation, and extravasation and metastatic colonization with molecular targets used for treatment development of
nanoparticles.

Research of metastasis is dated back to 1889 when Stephan Paget proposed the theory
of the "seed and soil." Analyzing hundreds of autopsies, he discovered that metastases
do not occur randomly but only when the cancer cell—"the seed"—is compatible with the
environment of the distant organ, "the soil." Later in 1929, James Ewing refuted this theory
and proposed that the occurrence of metastasis depends purely on mechanical factors due
to the structure of the vascular system. Over 100 years later, we know that both theories
contribute to metastatic colonization in certain organs. Cancer cells become more likely
trapped in tight capillaries at nearby organs following the blood flow, but certain cell types
selectively metastasize at specific organs compatible with the environmental requirements
of the cells [128,129].

1.4.1 Targeting the "Soil" of Metastases

For many cancer types, an organ-specific pattern of metastasis is observed that can even
occur years after successful treatment of the primary site due to the remnant presence
of dormant malignant cells. Liver colonization is strongly connected to colorectal cancer derived cells as the hepatic portal system directs the blood to the liver. In contrast,
the lung is metastasized by a range of cancer types due to its central role in the venous
system. Other common metastatic sites are brain and bone. Lung or breast carcinomas
preferentially spread to brain, whereas prostate cancer cells mainly metastasize to the
bone [130]. Through selective engineering of nanoparticle properties such as the size,
charge, or surface modification, nanocarriers can be passively delivered to metastatic
sites. As described earlier, the delivery to the brain can be problematic due to the BBB, but
nanocarriers with neutral charge and BBB-specific targeting ligands are more likely to
overcome this barrier. It is thought that the size of the particles for targeting of the brain

should not exceed 100 nm as the uptake decreases rapidly with size. A small particle size is also required for accumulation in the liver, facilitating transfer through small fenestrae in its endothelium. In contrast, for passive targeting to the lung, a positive surface charge and a particle size ≥ 200 nm are beneficial to enable entrapment in the tight pulmonary capillaries [131].

These passive-targeting abilities can be further supplemented with active targeting to specific organs using surface ligands. One strategy has explored the glycyrrhetinic acid (GA) receptors expressed on the hepatocyte membrane that can be used for selective delivery to the liver. GA-targeted nanoparticles doubled the drug concentrations in the liver with superior antitumor effects [132]. Chlorotoxin is another molecule that has been evaluated as a targeting agent as it binds to a cell-surface complex found on the surface of cancer types of neuroectodermal origin, namely, glioma and medulloblastoma [133]. Chlorotoxin-targeted iron oxide nanoparticles coated with chitosan–PEG copolymer with a size of about 30 nm and a small positive surface charge has proven to pass the BBB. An increased signal for the targeted particles was observed for up to 120 h using magnetic resonance imaging (MRI) or near-infrared fluorescence (NIRF) imaging for fluorophore-labeled nanoparticles with minimal accumulation in healthy tissue. This probe could be used as a tool for surgical tumor removal and for early detection of brain metastasis [134]. Approaches to target bone metastases have evaluated bisphosphonate drugs, which are mainly used for treatment of osteoporosis or other bone-related diseases. These molecules contain two phosphate groups, linked by an ester bond, which have high affinity for the calcium phosphate mineral composition in bone by chelation of Ca^{2+} ions [135,136]. Zoledronic acid elicits potent antitumor activity and a high affinity to bone mass and is considered one of the most potent bisphosphonates [135]. Targeted PEGylated nanoparticles with zoledronic acid accumulated 50-fold more at the bone after 24 h than its untargeted formulation and also had up to three times increased affinity toward diseased bone compared to healthy bone [137].

In addition to the development of potential treatments, nanoparticle-based imaging technologies to detect early metastases have increasingly attracted research interest. MRI is clinically used for tumor imaging and often used in combination with contrast agents such as gadolinium III chelates [138]. Superparamagnetic iron nanoparticles (SPION) can replace these contrast agents, and clinically approved examples are Feridex I.V.® for the detection of liver lesions and Feraheme™ for the treatment of iron-deficiency anemia [139]. Another example that has been in clinical evaluation for detection of lymph node metastases is Combidex® (ferumoxtran-10). These monocrystalline SPIONs have a dense coating of dextrans on the surface that allows a rapid uptake to lymphatic tissue [140]. Malignant lymph nodes could be distinguished from benign lymph nodes and were detected with high sensitivity of 100% and specificity of 95.7% in clinical studies on renal cell cancer patients [141]. The potential for SPIONs in diagnostic imaging was further highlighted in a following study on patients with prostate cancer–derived lymph node metastases [142].

1.4.2 Targeting "Seed"-Specific Molecules in the New Organ

Metastatic cells, which seeded into a distant organ, might still express certain surface markers from the site of origin, which differ from the new microenvironment. Active targeting of tumor-specific surface receptors was described earlier for the treatment of primary tumors but can also be applied for the treatment at metastatic sites. HER2-positive breast cancer cells that migrate to secondary sites also express this proto-oncogene at the metastatic site. Fluorophore-tagged SPIONs conjugated to anti-HER2 antibody can detect metastases in transgenic mice that spread to lung, liver, and bone marrow, whereas control IgG–conjugated control SPIONs accumulated nonspecifically in the nondiseased organ tissues. With further development, these nanoparticles could be used for the detection of early micrometastasis using MRI [143].

Sigma receptors are another class of receptors upregulated up to 10-fold in various tumors compared to healthy tissue [144]. Anisamide-targeted liposomes bind these sigma receptors and increase therapeutic effect of chemotherapy by active targeting of prostate cancer cells [145]. In a metastatic model, the therapeutic cargo, in this case siRNA, leads to significant gene silencing in lung metastases as the malignant melanoma cells expressed the receptors also at the metastatic site [146]. However, metastatic diseases are often resistant to conventional therapies, which might have positive effects on primary tumors. As cancer cells undergo genetic mutations in order to survive in the new environment, phenotypic changes occur that can be distinct from primary tumor cells [147]. However, in some instances, such resistance mechanisms may be overwhelmed by nanoparticle-delivered drug cargos as discussed earlier.

1.4.3 Targeting the Process of Metastasis

It has been estimated that only 0.02% of all circulating tumor cells (CTCs) successfully metastasize and colonize in distal sites. This highlights the inefficiency of the metastatic process where cells survive and facilitate formation of micrometastases and proliferate at the new organ [148]. Nonetheless, if the process of metastasis could be prevented with therapy, the clinical benefits would be highly significant. Approaches to target these metastatic cells in the early stages before cells spread to a new organ and are able to form microscopic metastases are being currently investigated [149]. The number of cancer cells that intravasate into the blood circulation becoming CTCs correlates with the state of tumor progression. Elevated levels of CTCs are associated with high risk of metastasis and poor prognosis [150]. The event of CTCs is rare compared to other cells in the blood circulation, at approximately one CTC per 10^{6-7} leukocytes, which requires exquisite test sensitivity [151]. CellSearch® is a diagnostic test approved by the FDA for clinical use to monitor CTCs in patients with metastatic breast cancer and represents a new technology with potential to influence therapy of patients [152]. The system analyzes and distinguishes CTCs from other blood cells using a ferrofluid containing magnetic nanoparticles coated with antibodies recognizing epithelial cell adhesion molecules. This enables the enrichment of CTCs magnetically and additional antibody staining allows the differentiation from contaminating normal leukocytes [153]. To overcome potential limitations of this technology arising from blood volume and sample preparation, new approaches to analyze CTCs *in situ* are currently being developed. *In vivo* flow cytometry using a magnetic multicolor detection strategy makes it possible to enrich and detect rare CTCs, which potentially could improve sensitivity with a threshold of several CTCs in the whole blood circulation. A cocktail of gold carbon nanotubes and magnetic nanoparticles conjugated with folic acid or breast cancer–specific ligands, respectively, can monitor disease progression in a breast cancer xenograft model over 4 weeks by analyzing the increase of CTC [154].

There are several therapeutic agents developed that affect cancer cell invasion and angiogenesis and may hold promise as strategies to block metastasis. The agents aim for a variety of tumor microenvironment targets including cell adhesion proteins such as integrins and proteases. A number of proteases have been implicated in cancer invasion and angiogenesis, inducing dissolution of cell adhesions, degradation of basement membranes, and remodeling of ECM during intra- and extravasation to allow the migration of cells through these otherwise impenetrable barriers. One family of proteases that are extensively researched for their role in tumorigenesis are the matrix metalloproteinases (MMPs) consisting of 23 members with different domains and physiological function [155]. MMP-2 is upregulated on the surface of gliomas in a complex together with other proteins where MMP-2 represents the receptor for chlorotoxin. Studies suggest that chlorotoxin inhibits the proteolytic activity of MMP-2 and reduces its cell-surface expression [133,156]. Chlorotoxin-bound SPIONs, described earlier for targeting nanoparticles to the brain, elicited enhanced anti-invasive potential with increased therapeutic outcome compared to free drug [157].

Exploitation of MMP activity associated with invasive tumors has been explored for imaging cancer invasion and metastasis using fluorescence-activatable cell-penetrating peptides (ACPPs) consisting of a polycation attached to a polyanion via cleavable peptide linker specific to MMP-2 and MMP-9, which are both involved in metastasis [158]. For increased sensitivity, these noninvasive probes were conjugated to dendrimers. The linker is cleaved by proteases in the tumor environment, leaving positively-charged dendrimers with cyanine5 or gadolinium payload behind that can instantly stick to the cells. Delivery by dendrimers increased the half-life and tumor uptake of the probe significantly with superior fluorescent signals and MRI contrast. These probes can be used for the detection of small tumors and sites of metastasis, whereas MRI can be used for surgical planning and fluorescent imaging for intraoperative guidance [159].

Cysteine cathepsins are another family of proteases mainly found in the endolysosomal compartments where they are responsible for proteolysis but also have other physiological roles. Several members of this family have been associated with tumorigenesis due to increased expression and activity in tumors and are known to degrade proteins in the ECM that promotes migration and invasion [160]. Magnetic ferri-liposomes have been developed with iron oxide nanoparticles and a small-molecule cathepsin inhibitor in the core of the bilipid layer. In combination with a magnetic field exposed to the tumor after the injection, nanoparticles accumulated at the tumor site and cathepsin activity were significantly downregulated with inhibition of tumor growth [161].

Targeting the source of these proteases demonstrates another interesting antimetastatic approach. Bone-marrow-derived myeloid cells including tumor-associated macrophages are involved in intravasation and invasion through releasing different factors such as proteases but also growth factors [162]. Clodronate, a bisphosphonate with antitumor activity, has been shown to suppress tumor growth and metastasis. Loaded liposomes (clodrolip) eliminated tumor-associated macrophages and inhibited tumor growth more efficiently than free bisphosphonate [163]. Later studies with similar liposomes showed significant reduction of occurring metastases in bone as well as muscle due to the reduced number of macrophages [164]. Another target with impact on the ECM is lysyl oxidase (LOX). LOX is an enzyme, which plays an important role in the stabilization and repair of the ECM by cross-linkage of collagen and elastin [165]. Clinical evaluation revealed that patients with high LOX levels have poor prognosis as well as reduced metastasis-free survival [166,167]. Monoclonal antibodies directed against LOX were conjugated to polymeric nanoparticle and reduced tumor growth by over 50%, whereas free antibody had only minor effects [168].

The later stages of the metastatic process can also present promising therapeutic strategies and the potential for an early detection of metastatic sites. Micrometastases are not able to extend beyond a certain size without the formation of new blood vessels to supply the cancer cells sufficiently with nutrients and oxygen. Integrins play an essential role during the process of metastasis and are highly involved in angiogenesis [169]. They are expressed in both cancer tissue and angiogenic epithelial cells, making them interesting targets for the inhibition of tumorigenesis [170]. Integrin-$\alpha_v\beta_3$ has attracted attention for the development of antiangiogenic nanoparticles. Cyclic RGD peptides targeting $\alpha_v\beta_3$ were conjugated to the nanoparticle surface to analyze effects of doxorubicin-loaded nanoparticles on primary tumors and spontaneously occurring metastases. Targeting of the nanoparticles to the integrin reduced formation of metastasis significantly by inhibition of new blood vessel formation with only minor effect on the primary tumor due to an established vasculature [171]. SPIONs have also been surface functionalized using RGD peptides. These targeted particles are able to enhance MRI, detecting early stages of metastasis and tumor development. Furthermore, it can give insight into the molecular process of angiogenesis and possible changes of the newly developing blood vessels after treatment [172,173].

1.5 Theranostics

John Funkhouser is credited with the first use of the term "theranostics," while he presented the company's business model to develop diagnostic tests that offer valuable information about suitability of drugs [174]. According to this definition, Herceptin is classified as theranostic. Clinical trials were conducted in combination with HercepTest™, which was simultaneously approved by the FDA in 1998 for the diagnosis of the HER2 status of cancer patients, the marker for therapeutic efficiency. Only patients with HER2-positive tumors are therefore eligible for Herceptin treatment [175]. In relation to nanotechnology, "theranostic" was redefined and denotes the combination of therapeutic and imaging agents in one device [176]. Offspring of such kinds of theranostics can be dated back to 1974 when tumor-targeted IgG was radiolabeled and permitted visualization of diseased tissue using photoscans with increased radioactivity at the diseased site [177]. Ibritumomab tiuxetan (Zevalin®) is a clinically approved theranostic for the treatment of patients with non-Hodgkin's lymphoma. On day one of the treatment regimen, indium-111 (^{111}In) replaces ^{90}Y at the monoclonal antibody ibritumomab tiuxetan allowing imaging for the identification of organ-specific accumulation, which decides over treatment continuation or abortion [178]. However, the field of theranostic nanoparticles is relatively young but is advancing rapidly, and today, multiple applications can be found in preclinical evaluation. Like Zevalin, theranostic nanoparticles can be used for monitoring the biodistribution and treatment efficiency at the diseased site. Furthermore, theranostic agents can be used as guidance for radiation-based therapy and surgery [179].

Several examples for theranostic nanoparticles combining therapy and imaging have been discussed throughout Section 1.3 for early diagnosis and/or treatment of metastases, but the repertoire of multifunctional nanocarriers including gadolinium-labeled liposomes, MRI-sensitive micelles, and fluorescent quantum dots are all tumor targeted and loaded with chemotherapeutic agents [180–182]. There are many more novel nanodevices in development, but so far, no multifunctional theranostic nanocarrier has reached clinical evaluation, probably due to intricacy of the combination of different components in one device at the right dose in a suitable carrier.

1.6 Conclusions and Perspective

Nanocarriers have the potential to enhance the effectiveness of many chemotherapies and imaging tools that we currently use for the treatment of cancer. In addition to advancing current therapies, these nano-strategies also have the potential to bring molecules forward into the clinic, which have so far failed as a consequence of unfavorable biodistribution. The ability to tailor these systems to alter their biodistribution and drug release, depending on the disease further, enhances their capabilities. In this chapter, the recent advances in targeting tumors and metastases for therapy and imaging applications have been reviewed. Clearly, systems that combine these roles will offer much potential in the future, particularly for personalized therapy in cancer treatment, and bring research one step closer to fulfill the desire for "magic bullets" in the near future.

References

1. World Health Organization, Cancer fact sheet No 297. http://www.who.int/mediacentre/factsheets/fs297/en/ (accessed November 2012).
2. Strebhardt, K. and A. Ullrich. 2008. Paul Ehrlich's magic bullet concept: 100 years of progress. *Nature Reviews. Cancer* 8: 473–480. doi:10.1038/nrc2394.
3. Cronstein, B. N. and J. R. Bertino. 2000. *Methotrexate (Milestones in Drug Therapy)*. Birkhauser Verlag AG, Berlin, Germany.
4. Cragg, G. M. and D. J. Newman. 2005. Plants as a source of anti-cancer agents. *Journal of Ethnopharmacology* 100: 72–79. doi:10.1016/j.jep.2005.05.011.
5. Kelland, L. 2007. The resurgence of platinum-based cancer chemotherapy. *Nature Reviews. Cancer* 7: 573–584. doi:10.1038/nrc2167.
6. Unger, C. et al. 2007. Phase I and pharmacokinetic study of the (6-maleimidocaproyl) hydrazone derivative of doxorubicin. *Clinical Cancer Research: An Official Journal of the American Association for Cancer Research* 13: 4858–4866. doi:10.1158/1078-0432.CCR-06-2776.
7. Bennouna, J., J. P. Delord, M. Campone, and L. Nguyen. 2008. Vinflunine: A new microtubule inhibitor agent. *Clinical Cancer Research: An Official Journal of the American Association for Cancer Research* 14: 1625–1632. doi:10.1158/1078-0432.CCR-07-2219.
8. Mita, A. C. et al. 2009. Phase I and pharmacokinetic study of XRP6258 (RPR 116258A), a novel taxane, administered as a 1-hour infusion every 3 weeks in patients with advanced solid tumors. *Clinical Cancer Research: An Official Journal of the American Association for Cancer Research* 15: 723–730. doi:10.1158/1078-0432.CCR-08-0596.
9. Venditto, V. J. and E. E. Simanek. 2010. Cancer therapies utilizing the camptothecins: A review of the in vivo literature. *Molecular Pharmaceutics* 7: 307–349. doi:10.1021/mp900243b.
10. Hoelder, S., P. A. Clarke, and P. Workman. 2012. Discovery of small molecule cancer drugs: Successes, challenges and opportunities. *Molecular Oncology* 6: 155–176. doi:10.1016/j.molonc.2012.02.004.
11. Takeuchi, K. and F. Ito. 2011. Receptor tyrosine kinases and targeted cancer therapeutics. *Biological & Pharmaceutical Bulletin* 34: 1774–1780.
12. Capdeville, R., E. Buchdunger, J. Zimmermann, and A. Matter. 2002. Glivec (STI571, imatinib), a rationally developed, targeted anticancer drug. *Nature Reviews. Drug Discovery* 1: 493–502. doi:10.1038/nrd839.
13. Zhang, J., P. L. Yang, and N. S. Gray. 2009. Targeting cancer with small molecule kinase inhibitors. *Nature Reviews. Cancer* 9: 28–39. doi:10.1038/nrc2559.

14. Kenefick, K. Small-molecule kinase inhibitors: From lab bench to clinic. http://www.promega. com/resources/articles/pubhub/cellnotes/small-molecule-kinase-inhibitors-from-lab-bench-to-clinic/ (accessed November 2012).

15. Bukowski, R. M., U. Yasothan, and P. Kirkpatrick. 2010. Pazopanib. *Nature Reviews. Drug Discovery* 9: 17–18. doi:10.1038/nrd3073.

16. Fire, A. et al. 1998. Potent and specific genetic interference by double-stranded RNA in Caenorhabditis elegans. *Nature* 391: 806–811. doi:10.1038/35888.

17. Castanotto, D. and J. J. Rossi. 2009. The promises and pitfalls of RNA-interference-based therapeutics. *Nature* 457: 426–433. doi:10.1038/nature07758.

18. Bumcrot, D., M. Manoharan, V. Koteliansky, and D. W. Sah. 2006. RNAi therapeutics: A potential new class of pharmaceutical drugs. *Nature Chemical Biology* 2: 711–719. doi:10.1038/nchembio839.

19. Zamore, P. D. 2006. RNA interference: Big applause for silencing in Stockholm. *Cell* 127: 1083–1086. doi:10.1016/j.cell.2006.12.001.

20. Kohler, G. and C. Milstein. 1975. Continuous cultures of fused cells secreting antibody of predefined specificity. *Nature* 256: 495–497.

21. Burden, R. E., J. Caswell, F. Fay, and C. J. Scott. 2012. Recent advances in the application of antibodies as therapeutics. *Future Medicinal Chemistry* 4: 73–86. doi:10.4155/fmc.11.165.

22. Hudis, C. A. 2007. Trastuzumab—Mechanism of action and use in clinical practice. *The New England Journal of Medicine* 357: 39–51. doi:10.1056/NEJMra043186.

23. Pazdur, R. *FDA Approval for Trastuzumab.* http://www.cancer.gov/cancertopics/druginfo/fda-trastuzumab (accessed November 2012).

24. Pazdur, R. *FDA Approval for Bevacizumab.* http://www.cancer.gov/cancertopics/druginfo/fda-bevacizumab (accessed November 2012).

25. Okeley, N. M. et al. 2010. Intracellular activation of SGN-35, a potent anti-CD30 antibody-drug conjugate. *Clinical Cancer Research: An Official Journal of the American Association for Cancer Research* 16: 888–897. doi:10.1158/1078–0432.CCR-09–2069.

26. Petrul, H. M. et al. 2012. Therapeutic mechanism and efficacy of the antibody-drug conjugate BAY 79–4620 targeting human carbonic anhydrase 9. *Molecular Cancer Therapeutics* 11: 340–349. doi:10.1158/1535–7163.MCT-11–0523.

27. Sharkey, R. M., S. V. Govindan, T. M. Cardillo, and D. M. Goldenberg. 2012. Epratuzumab-SN-38: A new antibody-drug conjugate for the therapy of hematologic malignancies. *Molecular Cancer Therapeutics* 11: 224–234. doi:10.1158/1535–7163.MCT-11–0632.

28. Ricart, A. D. 2011. Antibody-drug conjugates of calicheamicin derivative: Gemtuzumab ozogamicin and inotuzumab ozogamicin. *Clinical Cancer Research: An Official Journal of the American Association for Cancer Research* 17: 6417–6427. doi:10.1158/1078–0432.CCR-11–0486.

29. Alley, S. C., N. M. Okeley, and P. D. Senter. 2010. Antibody-drug conjugates: Targeted drug delivery for cancer. *Current Opinion in Chemical Biology* 14: 529–537. doi:10.1016/j.cbpa.2010.06.170.

30. FDA: Pfizer voluntarily withdraws cancer treatment mylotarg from U.S. Market. http://www.fda.gov/NewsEvents/Newsroom/PressAnnouncements/ucm216448.htm (accessed November 2012).

31. Younes, A. et al. 2012. Results of a pivotal phase II study of brentuximab vedotin for patients with relapsed or refractory Hodgkin's lymphoma. *Journal of Clinical Oncology: Official Journal of the American Society of Clinical Oncology* 30: 2183–2189. doi:10.1200/JCO.2011.38.0410.

32. Fanale, M. A. et al. 2012. A phase I weekly dosing study of brentuximab vedotin in patients with relapsed/refractory CD30-positive hematologic malignancies. *Clinical Cancer Research: An Official Journal of the American Association for Cancer Research* 18: 248–255. doi:10.1158/1078–0432. CCR-11–1425.

33. Reichert, J. M. 2011. Antibody-based therapeutics to watch in 2011. *mAbs* 3: 76–99.

34. Krop, I. E. et al. 2012. A phase II study of trastuzumab emtansine in patients with human epidermal growth factor receptor 2-positive metastatic breast cancer who were previously treated with trastuzumab, lapatinib, an anthracycline, a taxane, and capecitabine. *Journal of Clinical Oncology: Official Journal of the American Society of Clinical Oncology* 30: 3234–3241. doi:10.1200/JCO.2011.40.5902.

35. Farokhzad, O. C. and R. Langer. 2009. Impact of nanotechnology on drug delivery. *ACS Nano* 3: 16–20. doi:10.1021/nn900002m.
36. Khanna, S. C. and P. Speiser. 1970. In vitro release of chloramphenicol from polymer beads of alpha-methacrylic acid and methylmethacrylate. *Journal of Pharmaceutical Sciences* 59: 1398–1401.
37. Khanna, S. C., T. Jecklin, and P. Speiser. 1970. Bead polymerization technique for sustained-release dosage form. *Journal of Pharmaceutical Sciences* 59: 614–618.
38. Kreuter, J. 2007. Nanoparticles—A historical perspective. *International Journal of Pharmaceutics* 331: 1–10. doi:10.1016/j.ijpharm.2006.10.021.
39. Kamaly, N., Z. Xiao, P. M. Valencia, A. F. Radovic-Moreno, and O. C. Farokhzad. 2012. Targeted polymeric therapeutic nanoparticles: Design, development and clinical translation. *Chemical Society Reviews* 41: 2971–3010. doi:10.1039/c2cs15344k.
40. Peer, D. et al. 2007. Nanocarriers as an emerging platform for cancer therapy. *Nature Nanotechnology* 2: 751–760. doi:10.1038/nnano.2007.387.
41. Choi, H. S. et al. 2007. Renal clearance of quantum dots. *Nature Biotechnology* 25: 1165–1170. doi:10.1038/nbt1340.
42. Owens, D. E., 3rd and N. A. Peppas. 2006. Opsonization, biodistribution, and pharmacokinetics of polymeric nanoparticles. *International Journal of Pharmaceutics* 307: 93–102. doi:10.1016/j.ijpharm.2005.10.010.
43. Vonarbourg, A., C. Passirani, P. Saulnier, and J. P. Benoit. 2006. Parameters influencing the stealthiness of colloidal drug delivery systems. *Biomaterials* 27: 4356–4373. doi:10.1016/j.biomaterials.2006.03.039.
44. Fang, C. et al. 2006. In vivo tumor targeting of tumor necrosis factor-alpha-loaded stealth nanoparticles: Effect of MePEG molecular weight and particle size. *European Journal of Pharmaceutical Sciences: Official Journal of the European Federation for Pharmaceutical Sciences* 27: 27–36. doi:10.1016/j.ejps.2005.08.002.
45. Chithrani, B. D., A. A. Ghazani, and W. C. Chan. 2006. Determining the size and shape dependence of gold nanoparticle uptake into mammalian cells. *Nano Letters* 6: 662–668. doi:10.1021/nl052396o.
46. Geng, Y. et al. 2007. Shape effects of filaments versus spherical particles in flow and drug delivery. *Nature Nanotechnology* 2: 249–255. doi:10.1038/nnano.2007.70.
47. Christian, D. A. et al. 2009. Flexible filaments for in vivo imaging and delivery: Persistent circulation of filomicelles opens the dosage window for sustained tumor shrinkage. *Molecular Pharmaceutics* 6: 1343–1352. doi:10.1021/mp900022m.
48. Park, J. H. et al. 2008. Magnetic iron oxide nanoworms for tumor targeting and imaging. *Advanced Materials* 20: 1630–1635. doi:10.1002/adma.200800004.
49. Brigger, I., C. Dubernet, and P. Couvreur. 2002. Nanoparticles in cancer therapy and diagnosis. *Advanced Drug Delivery Reviews* 54: 631–651.
50. Abuchowski, A., J. R. McCoy, N. C. Palczuk, T. van Es, and F. F. Davis. 1977. Effect of covalent attachment of polyethylene glycol on immunogenicity and circulating life of bovine liver catalase. *The Journal of Biological Chemistry* 252: 3582–3586.
51. Fang, C., N. Bhattarai, C. Sun, and M. Zhang. 2009. Functionalized nanoparticles with long-term stability in biological media. *Small* 5: 1637–1641. doi:10.1002/smll.200801647.
52. Walkey, C. D., J. B. Olsen, H. Guo, A. Emili, and W. C. Chan. 2012. Nanoparticle size and surface chemistry determine serum protein adsorption and macrophage uptake. *Journal of the American Chemical Society* 134: 2139–2147. doi:10.1021/ja2084338.
53. Avgoustakis, K. et al. 2003. Effect of copolymer composition on the physicochemical characteristics, in vitro stability, and biodistribution of PLGA-mPEG nanoparticles. *International Journal of Pharmaceutics* 259: 115–127.
54. Choi, K. Y. et al. 2011. PEGylation of hyaluronic acid nanoparticles improves tumor targetability in vivo. *Biomaterials* 32: 1880–1889. doi:10.1016/j.biomaterials.2010.11.010.
55. Zhu, S. et al. 2010. PEGylated PAMAM dendrimer-doxorubicin conjugates: In vitro evaluation and in vivo tumor accumulation. *Pharmaceutical Research* 27: 161–174. doi:10.1007/s11095-009-9992-1.

56. Matsumura, Y. and H. Maeda. 1986. A new concept for macromolecular therapeutics in cancer chemotherapy: Mechanism of tumoritropic accumulation of proteins and the antitumor agent smancs. *Cancer Research* 46: 6387–6392.
57. Fang, J., H. Nakamura, and H. Maeda. 2011. The EPR effect: Unique features of tumor blood vessels for drug delivery, factors involved, and limitations and augmentation of the effect. *Advanced Drug Delivery Reviews* 63: 136–151. doi:10.1016/j.addr.2010.04.009.
58. Seymour, L. W. et al. 1995. Influence of molecular weight on passive tumour accumulation of a soluble macromolecular drug carrier. *European Journal of Cancer* 31A: 766–770.
59. Iyer, A. K., G. Khaled, J. Fang, and H. Maeda. 2006. Exploiting the enhanced permeability and retention effect for tumor targeting. *Drug Discovery Today* 11: 812–818. doi:10.1016/j.drudis. 2006.07.005.
60. Yuan, F. et al. 1995. Vascular permeability in a human tumor xenograft: molecular size dependence and cutoff size. *Cancer Research* 55: 3752–3756.
61. Hobbs, S. K. et al. 1998. Regulation of transport pathways in tumor vessels: Role of tumor type and microenvironment. *Proceedings of the National Academy of Sciences of the United States of America* 95: 4607–4612.
62. Yoo, H. S. and T. G. Park. 2004. Folate-receptor-targeted delivery of doxorubicin nano-aggregates stabilized by doxorubicin-PEG-folate conjugate. *Journal of Controlled Release: Official Journal of the Controlled Release Society* 100: 247–256. doi:10.1016/j.jconrel.2004.08.017.
63. Kratz, F. 2008. Albumin as a drug carrier: Design of prodrugs, drug conjugates and nanoparticles. *Journal of Controlled Release: Official Journal of the Controlled Release Society* 132: 171–183. doi:10.1016/j.jconrel.2008.05.010.
64. Wang, M. and M. Thanou. 2010. Targeting nanoparticles to cancer. *Pharmacological Research: The Official Journal of the Italian Pharmacological Society* 62: 90–99. doi:10.1016/j.phrs.2010.03.005.
65. Jokerst, J. V. and S. S. Gambhir. 2011. Molecular imaging with theranostic nanoparticles. *Accounts of Chemical Research* 44: 1050–1060. doi:10.1021/ar200106e.
66. Nobs, L., F. Buchegger, R. Gurny, and E. Allemann. 2004. Current methods for attaching targeting ligands to liposomes and nanoparticles. *Journal of Pharmaceutical Sciences* 93: 1980–1992. doi:10.1002/jps.20098.
67. Pecorino, L. 2008. *Molecular Biology of Cancer*. Oxford University Press Inc., Oxford, U.K.
68. Shukla, R. et al. 2006. HER2 specific tumor targeting with dendrimer conjugated anti-HER2 mAb. *Bioconjugate Chemistry* 17: 1109–1115. doi:10.1021/bc050348p.
69. Kirpotin, D. B. et al. 2006. Antibody targeting of long-circulating lipidic nanoparticles does not increase tumor localization but does increase internalization in animal models. *Cancer Research* 66: 6732–6740. doi:10.1158/0008-5472.CAN-05-4199.
70. Park, J. W. et al. 2002. Anti-HER2 immunoliposomes: Enhanced efficacy attributable to targeted delivery. *Clinical Cancer Research: An Official Journal of the American Association for Cancer Research* 8: 1172–1181.
71. McCarron, P. A. et al. 2008. Antibody targeting of camptothecin-loaded PLGA nanoparticles to tumor cells. *Bioconjugate Chemistry* 19: 1561–1569. doi:10.1021/bc800057g.
72. Kaplan-Lefko, P. J. et al. 2010. Conatumumab, a fully human agonist antibody to death receptor 5, induces apoptosis via caspase activation in multiple tumor types. *Cancer Biology & Therapy* 9: 618–631.
73. Fay, F. et al. 2011. Conatumumab (AMG 655) coated nanoparticles for targeted pro-apoptotic drug delivery. *Biomaterials* 32: 8645–8653. doi:10.1016/j.biomaterials.2011.07.065.
74. Xiao, Z. and O. C. Farokhzad. 2012. Aptamer-functionalized nanoparticles for medical applications: Challenges and opportunities. *ACS Nano* 6: 3670–3676. doi:10.1021/nn301869z.
75. Farokhzad, O. C. et al. 2006. Targeted nanoparticle-aptamer bioconjugates for cancer chemotherapy in vivo. *Proceedings of the National Academy of Sciences of the United States of America* 103: 6315–6320. doi:10.1073/pnas.0601755103.
76. Daniels, T. R., T. Delgado, G. Helguera, and M. L. Penichet. 2006. The transferrin receptor part II: Targeted delivery of therapeutic agents into cancer cells. *Clinical Immunology* 121: 159–176. doi:10.1016/j.clim.2006.06.006.

77. Daniels, T. R., T. Delgado, J. A. Rodriguez, G. Helguera, and M. L. Penichet. 2006. The transferrin receptor part I: Biology and targeting with cytotoxic antibodies for the treatment of cancer. *Clinical Immunology* 121: 144–158. doi:10.1016/j.clim.2006.06.010.

78. Bartlett, D. W., H. Su, I. J. Hildebrandt, W. A. Weber, and M. E. Davis. 2007. Impact of tumor-specific targeting on the biodistribution and efficacy of siRNA nanoparticles measured by multimodality in vivo imaging. *Proceedings of the National Academy of Sciences of the United States of America* 104: 15549–15554. doi:10.1073/pnas.0707461104.

79. Bartlett, D. W. and M. E. Davis. 2008. Impact of tumor-specific targeting and dosing schedule on tumor growth inhibition after intravenous administration of siRNA-containing nanoparticles. *Biotechnology and Bioengineering* 99: 975–985. doi:10.1002/bit.21668.

80. Wang, X. et al. 2009. HFT-T, a targeting nanoparticle, enhances specific delivery of paclitaxel to folate receptor-positive tumors. *ACS Nano* 3: 3165–3174. doi:10.1021/nn900649v.

81. Hilgenbrink, A. R. and P. S. Low. 2005. Folate receptor-mediated drug targeting: from therapeutics to diagnostics. *Journal of Pharmaceutical Sciences* 94: 2135–2146. doi:10.1002/jps.20457.

82. Byrne, J. D., T. Betancourt, and L. Brannon-Peppas. 2008. Active targeting schemes for nanoparticle systems in cancer therapeutics. *Advanced Drug Delivery Reviews* 60: 1615–1626. doi:10.1016/j.addr. 2008.08.005.

83. Gottesman, M. M., T. Fojo, and S. E. Bates. 2002. Multidrug resistance in cancer: Role of ATP-dependent transporters. *Nature Reviews. Cancer* 2: 48–58. doi:10.1038/nrc706.

84. Goren, D. et al. 2000. Nuclear delivery of doxorubicin via folate-targeted liposomes with bypass of multidrug-resistance efflux pump. *Clinical Cancer Research: An Official Journal of the American Association for Cancer Research* 6: 1949–1957.

85. Huang, I. P. et al. 2011. Enhanced chemotherapy of cancer using pH-sensitive mesoporous silica nanoparticles to antagonize P-glycoprotein-mediated drug resistance. *Molecular Cancer Therapeutics* 10: 761–769. doi:10.1158/1535-7163.MCT-10-0884.

86. Abbott, N. J., L. Ronnback, and E. Hansson. 2006. Astrocyte-endothelial interactions at the blood-brain barrier. *Nature Reviews. Neuroscience* 7: 41–53. doi:10.1038/nrn1824.

87. Ying, X. et al. 2010. Dual-targeting daunorubicin liposomes improve the therapeutic efficacy of brain glioma in animals. *Journal of Controlled Release: Official Journal of the Controlled Release Society* 141: 183–192. doi:10.1016/j.jconrel.2009.09.020.

88. Zhang, L. et al. 2008. Nanoparticles in medicine: Therapeutic applications and developments. *Clinical Pharmacology and Therapeutics* 83: 761–769. doi:10.1038/sj.clpt.6100400.

89. Malam, Y., M. Loizidou, and A. M. Seifalian. 2009. Liposomes and nanoparticles: Nanosized vehicles for drug delivery in cancer. *Trends in Pharmacological Sciences* 30: 592–599. doi:10.1016/j.tips.2009.08.004.

90. Gabizon, A., D. Goren, R. Cohen, and Y. Barenholz. 1998. Development of liposomal anthracyclines: From basics to clinical applications. *Journal of Controlled Release: Official Journal of the Controlled Release Society* 53: 275–279.

91. Gabizon, A., H. Shmeeda, and Y. Barenholz. 2003. Pharmacokinetics of pegylated liposomal Doxorubicin: Review of animal and human studies. *Clinical Pharmacokinetics* 42: 419–436.

92. Torchilin, V. P. 2005. Recent advances with liposomes as pharmaceutical carriers. *Nature Reviews. Drug Discovery* 4: 145–160. doi:10.1038/nrd1632.

93. *Study of MBP-426 in Patients with Second Line Gastric, Gastroesophageal, or Esophageal Adenocarcinoma.* http://clinicaltrials.gov/ct2/show/NCT00964080 (accessed November 2012).

94. Mebiopharm products/pipeline. http://www.mebiopharm.com/english/pro.html (accessed November 2012).

95. Khdair, A. et al. 2010. Nanoparticle-mediated combination chemotherapy and photodynamic therapy overcomes tumor drug resistance. *Journal of Controlled Release: Official Journal of the Controlled Release Society* 141: 137–144. doi:10.1016/j.jconrel.2009.09.004.

96. Parveen, S., M. Mitra, S. Krishnakumar, and S. K. Sahoo. 2010. Enhanced antiproliferative activity of carboplatin-loaded chitosan-alginate nanoparticles in a retinoblastoma cell line. *Acta Biomaterialia* 6: 3120–3131. doi:10.1016/j.actbio.2010.02.010.

97. Xu, Q. et al. 2012. Prevention of colorectal cancer liver metastasis by exploiting liver immunity via chitosan-TPP/nanoparticles formulated with IL-12. *Biomaterials* 33: 3909–3918. doi:10.1016/j.biomaterials.2012.02.014.

98. Dinarvand, R., N. Sepehri, S. Manoochehri, H. Rouhani, and F. Atyabi. 2011. Polylactide-co-glycolide nanoparticles for controlled delivery of anticancer agents. *International Journal of Nanomedicine* 6: 877–895. doi:10.2147/IJN.S18905.

99. Xiao, R. Z. et al. 2010. Recent advances in PEG-PLA block copolymer nanoparticles. *International Journal of Nanomedicine* 5: 1057–1065. doi:10.2147/IJN.S14912.

100. Hrkach, J. et al. 2012. Preclinical development and clinical translation of a PSMA-targeted docetaxel nanoparticle with a differentiated pharmacological profile. *Science Translational Medicine* 4: 128ra139. doi:10.1126/scitranslmed.3003651.

101. *A Study of BIND-014 Given to Patients with Advanced or Metastatic Cancer*. http://clinicaltrials.gov/ct2/show/NCT01300533 (accessed November 2012).

102. Ibrahim, N. K. et al. 2002. Phase I and pharmacokinetic study of ABI-007, a Cremophor-free, protein-stabilized, nanoparticle formulation of paclitaxel. *Clinical Cancer Research: An Official Journal of the American Association for Cancer Research* 8: 1038–1044.

103. Gradishar, W. J. et al. 2005. Phase III trial of nanoparticle albumin-bound paclitaxel compared with polyethylated castor oil-based paclitaxel in women with breast cancer. *Journal of Clinical Oncology: Official Journal of the American Society of Clinical Oncology* 23: 7794–7803. doi:10.1200/JCO.2005.04.937.

104. Desai, N. et al. 2006. Increased antitumor activity, intratumor paclitaxel concentrations, and endothelial cell transport of cremophor-free, albumin-bound paclitaxel, ABI-007, compared with cremophor-based paclitaxel. *Clinical Cancer Research: An Official Journal of the American Association for Cancer Research* 12: 1317–1324. doi:10.1158/1078-0432.CCR-05-1634.

105. John, T. A., S. M. Vogel, C. Tiruppathi, A. B. Malik, and R. D. Minshall. 2003. Quantitative analysis of albumin uptake and transport in the rat microvessel endothelial monolayer. *American Journal of Physiology. Lung Cellular and Molecular Physiology* 284: L187–L196. doi:10.1152/ajplung.00152.2002.

106. Torchilin, V. P. 2007. Micellar nanocarriers: Pharmaceutical perspectives. *Pharmaceutical Research* 24: 1–16. doi:10.1007/s11095-006-9132-0.

107. Kim, S. C. et al. 2001. In vivo evaluation of polymeric micellar paclitaxel formulation: toxicity and efficacy. *Journal of Controlled Release: Official Journal of the Controlled Release Society* 72: 191–202.

108. Gillies, E. R., E. Dy, J. M. Frechet, and F. C. Szoka. 2005. Biological evaluation of polyester dendrimer: Poly(ethylene oxide) "bow-tie" hybrids with tunable molecular weight and architecture. *Molecular Pharmaceutics* 2: 129–138. doi:10.1021/mp049886u.

109. Heidel, J. D. and T. Schluep. 2012. Cyclodextrin-containing polymers: Versatile platforms of drug delivery materials. *Journal of Drug Delivery* 2012: 262731. doi:10.1155/2012/262731.

110. Young, C., T. Schluep, J. Hwang, and S. Eliasof. 2011. CRLX101 (formerly IT-101)-A novel nanopharmaceutical of camptothecin in clinical development. *Current Bioactive Compounds* 7: 8–14. doi:10.2174/157340711795163866.

111. Davis, M. E. 2009. Design and development of IT-101, a cyclodextrin-containing polymer conjugate of camptothecin. *Advanced Drug Delivery Reviews* 61: 1189–1192. doi:10.1016/j.addr.2009.05.005.

112. Davis, M. E. 2009. The first targeted delivery of siRNA in humans via a self-assembling, cyclodextrin polymer-based nanoparticle: From concept to clinic. *Molecular Pharmaceutics* 6: 659–668. doi:10.1021/mp900015y.

113. Davis, M. E. et al. 2010. Evidence of RNAi in humans from systemically administered siRNA via targeted nanoparticles. *Nature* 464: 1067–1070. doi:10.1038/nature08956.

114. *Safety Study of CALAA-01 to Treat Solid Tumor Cancers*. http://clinicaltrials.gov/ct2/show/NCT00689065 (accessed November 2012).

115. Jain, S., D. G. Hirst, and J. M. O'Sullivan. 2012. Gold nanoparticles as novel agents for cancer therapy. *The British Journal of Radiology* 85: 101–113. doi:10.1259/bjr/59448833.

116. Jain, P. K., X. Huang, I. H. El-Sayed, and M. A. El-Sayed. 2008. Noble metals on the nanoscale: optical and photothermal properties and some applications in imaging, sensing, biology, and medicine. *Accounts of Chemical Research* 41: 1578–1586. doi:10.1021/ar7002804.

117. El-Sayed, I. H., X. Huang, and M. A. El-Sayed. 2006. Selective laser photo-thermal therapy of epithelial carcinoma using anti-EGFR antibody conjugated gold nanoparticles. *Cancer Letters* 239: 129–135. doi:10.1016/j.canlet.2005.07.035.

118. Weissleder, R. 2001. A clearer vision for in vivo imaging. *Nature Biotechnology* 19: 316–317. doi:10.1038/86684.

119. Lal, S., S. E. Clare, and N. J. Halas. 2008. Nanoshell-enabled photothermal cancer therapy: impending clinical impact. *Accounts of Chemical Research* 41: 1842–1851. doi:10.1021/ar800150g.

120. Loo, C., A. Lowery, N. Halas, J. West, and R. Drezek. 2005. Immunotargeted nanoshells for integrated cancer imaging and therapy. *Nano Letters* 5: 709–711. doi:10.1021/nl050127s.

121. Hirsch, L. R. et al. 2003. Nanoshell-mediated near-infrared thermal therapy of tumors under magnetic resonance guidance. *Proceedings of the National Academy of Sciences of the United States of America* 100: 13549–13554. doi:10.1073/pnas.2232479100.

122. *Pilot Study of AuroLase(tm) Therapy in Refractory and/or Recurrent Tumors of the Head and Neck.* http://clinicaltrials.gov/ct2/show/NCT00848042 (accessed November 2012).

123. Farma, J. M. et al. 2007. Direct evidence for rapid and selective induction of tumor neovascular permeability by tumor necrosis factor and a novel derivative, colloidal gold bound tumor necrosis factor. *International Journal of Cancer. Journal International du Cancer* 120: 2474–2480. doi:10.1002/ijc.22270.

124. Libutti, S. K. et al. 2010. Phase I and pharmacokinetic studies of CYT-6091, a novel PEGylated colloidal gold-rhTNF nanomedicine. *Clinical Cancer Research: An Official Journal of the American Association for Cancer Research* 16: 6139–6149. doi:10.1158/1078-0432.CCR-10-0978.

125. Cytimmune—Drug pipeline. http://www.cytimmune.com/go.cfm?do=page.view&pid=19 (accessed November 2012).

126. Mehlen, P. and A. Puisieux. 2006. Metastasis: A question of life or death. *Nature Reviews. Cancer* 6: 449–458. doi:10.1038/nrc1886.

127. Valastyan, S. and R. A. Weinberg. 2011. Tumor metastasis: Molecular insights and evolving paradigms. *Cell* 147: 275–292. doi:10.1016/j.cell.2011.09.024.

128. Psaila, B. and D. Lyden. 2009. The metastatic niche: Adapting the foreign soil. *Nature Reviews. Cancer* 9: 285–293. doi:10.1038/nrc2621.

129. Fidler, I. J. 2003. The pathogenesis of cancer metastasis: The 'seed and soil' hypothesis revisited. *Nature Reviews. Cancer* 3: 453–458. doi:10.1038/nrc1098.

130. Steeg, P. S. 2006. Tumor metastasis: Mechanistic insights and clinical challenges. *Nature Medicine* 12: 895–904. doi:10.1038/nm1469.

131. Schroeder, A. et al. 2012. Treating metastatic cancer with nanotechnology. *Nature Reviews. Cancer* 12: 39–50. doi:10.1038/nrc3180.

132. Zhang, C. et al. 2012. Doxorubicin-loaded glycyrrhetinic acid-modified alginate nanoparticles for liver tumor chemotherapy. *Biomaterials* 33: 2187–2196. doi:10.1016/j.biomaterials.2011.11.045.

133. Deshane, J., C. C. Garner, and H. Sontheimer. 2003. Chlorotoxin inhibits glioma cell invasion via matrix metalloproteinase-2. *The Journal of Biological Chemistry* 278: 4135–4144. doi:10.1074/jbc.M205662200.

134. Veiseh, O. et al. 2009. Specific targeting of brain tumors with an optical/magnetic resonance imaging nanoprobe across the blood-brain barrier. *Cancer Research* 69: 6200–6207. doi:10.1158/0008-5472.CAN-09-1157.

135. Drake, M. T., B. L. Clarke, and S. Khosla. 2008. Bisphosphonates: Mechanism of action and role in clinical practice. *Mayo Clinic Proceedings. Mayo Clinic* 83: 1032–1045. doi:10.4065/83.9.1032.

136. Ebetino, F. H. et al. 2011. The relationship between the chemistry and biological activity of the bisphosphonates. *Bone* 49: 20–33. doi:10.1016/j.bone.2011.03.774.

137. Ramanlal Chaudhari, K. et al. 2012. Bone metastasis targeting: A novel approach to reach bone using Zoledronate anchored PLGA nanoparticle as carrier system loaded with Docetaxel. *Journal of Controlled Release: Official Journal of the Controlled Release Society* 158: 470–478. doi:10.1016/j.jconrel.2011.11.020.

138. Lu, Z., Y. Wang, F. Ye, A. Vaidya, and E. Jeong. 2007. Noninvasive visualization of in vivo drug delivery of paramagnetic polymer conjugates with MRI. In *Nanotechnology for Cancer Therapy* M. M. Amiji (ed.), pp. 201–212. CRC Press, Boca Raton, FL.

139. Tassa, C., S. Y. Shaw, and R. Weissleder. 2011. Dextran-coated iron oxide nanoparticles: A versatile platform for targeted molecular imaging, molecular diagnostics, and therapy. *Accounts of Chemical Research* 44: 842–852. doi:10.1021/ar200084x.

140. Harisinghani, M. G. et al. 2003. Noninvasive detection of clinically occult lymph-node metastases in prostate cancer. *The New England Journal of Medicine* 348: 2491–2499. doi:10.1056/NEJMoa022749.

141. Guimaraes, A. R. et al. 2008. Pilot study evaluating use of lymphotropic nanoparticle-enhanced magnetic resonance imaging for assessing lymph nodes in renal cell cancer. *Urology* 71: 708–712. doi:10.1016/j.urology.2007.11.096.

142. Fortuin, A. S. et al. 2012. Value of PET/CT and MR lymphography in treatment of prostate cancer Patients with lymph node metastases. *International Journal of Radiation Oncology, Biology, Physics* 84: 712–718. doi:10.1016/j.ijrobp.2011.12.093.

143. Kievit, F. M. et al. 2012. Targeting of primary breast cancers and metastases in a transgenic mouse model using rationally designed multifunctional SPIONs. *ACS Nano* 6: 2591–2601. doi:10.1021/nn205070h.

144. Aydar, E., C. P. Palmer, and M. B. Djamgoz. 2004. Sigma receptors and cancer: Possible involvement of ion channels. *Cancer Research* 64: 5029–5035. doi:10.1158/0008-5472.CAN-03-2329.

145. Banerjee, R., P. Tyagi, S. Li, and L. Huang. 2004. Anisamide-targeted stealth liposomes: A potent carrier for targeting doxorubicin to human prostate cancer cells. *International Journal of Cancer. Journal International du Cancer* 112: 693–700. doi:10.1002/ijc.20452.

146. Li, S. D., S. Chono, and L. Huang. 2008. Efficient gene silencing in metastatic tumor by siRNA formulated in surface-modified nanoparticles. *Journal of Controlled Release: Official Journal of the Controlled Release Society* 126: 77–84. doi:10.1016/j.jconrel.2007.11.002.

147. Nguyen, D. X., P. D. Bos, and J. Massague. 2009. Metastasis: From dissemination to organ-specific colonization. *Nature Reviews. Cancer* 9: 274–284. doi:10.1038/nrc2622.

148. Chambers, A. F., A. C. Groom, and I. C. MacDonald. 2002. Dissemination and growth of cancer cells in metastatic sites. *Nature Reviews. Cancer* 2: 563–572. doi:10.1038/nrc865.

149. Veiseh, O., F. M. Kievit, R. G. Ellenbogen, and M. Zhang. 2011. Cancer cell invasion: Treatment and monitoring opportunities in nanomedicine. *Advanced Drug Delivery Reviews* 63: 582–596. doi:10.1016/j.addr.2011.01.010.

150. Rice, J. 2012. Metastasis: The rude awakening. *Nature* 485: S55–S57. doi:10.1038/485S55a.

151. Gerges, N., J. Rak, and N. Jabado. 2010. New technologies for the detection of circulating tumour cells. *British Medical Bulletin* 94: 49–64. doi:10.1093/bmb/ldq011.

152. Riethdorf, S. et al. 2007. Detection of circulating tumor cells in peripheral blood of patients with metastatic breast cancer: A validation study of the CellSearch system. *Clinical Cancer Research: An Official Journal of the American Association for Cancer Research* 13: 920–928. doi:10.1158/1078-0432.CCR-06-1695.

153. Veridex-CellSearch® video. http://www.veridex.com/CellSearch/CellSearchHCP.aspx (accessed November 2012).

154. Galanzha, E. I. et al. 2009. In vivo magnetic enrichment and multiplex photoacoustic detection of circulating tumour cells. *Nature Nanotechnology* 4: 855–860. doi:10.1038/nnano.2009.333.

155. Kessenbrock, K., V. Plaks, and Z. Werb. 2010. Matrix metalloproteinases: Regulators of the tumor microenvironment. *Cell* 141: 52–67. doi:10.1016/j.cell.2010.03.015.

156. Egeblad, M. and Z. Werb. 2002. New functions for the matrix metalloproteinases in cancer progression. *Nature Reviews. Cancer* 2: 161–174. doi:10.1038/nrc745.

157. Veiseh, O. et al. 2009. Inhibition of tumor-cell invasion with chlorotoxin-bound superparamagnetic nanoparticles. *Small* 5: 256–264. doi:10.1002/smll.200800646.

158. Olson, E. S. et al. 2009. In vivo characterization of activatable cell penetrating peptides for targeting protease activity in cancer. *Integrative Biology: Quantitative Biosciences from Nano to Macro* 1: 382–393. doi:10.1039/b904890a.

159. Olson, E. S. et al. 2010. Activatable cell penetrating peptides linked to nanoparticles as dual probes for in vivo fluorescence and MR imaging of proteases. *Proceedings of the National Academy of Sciences of the United States of America* 107: 4311–4316. doi:10.1073/pnas.0910283107.

160. Mohamed, M. M. and B. F. Sloane. 2006. Cysteine cathepsins: Multifunctional enzymes in cancer. *Nature Reviews. Cancer* 6: 764–775. doi:10.1038/nrc1949.

161. Mikhaylov, G. et al. 2011. Ferri-liposomes as an MRI-visible drug-delivery system for targeting tumours and their microenvironment. *Nature Nanotechnology* 6: 594–602. doi:10.1038/nnano.2011.112.

162. Joyce, J. A. and J. W. Pollard. 2009. Microenvironmental regulation of metastasis. *Nature Reviews. Cancer* 9: 239–252. doi:10.1038/nrc2618.

163. Zeisberger, S. M. et al. 2006. Clodronate-liposome-mediated depletion of tumour-associated macrophages: A new and highly effective antiangiogenic therapy approach. *British Journal of Cancer* 95: 272–281. doi:10.1038/sj.bjc.6603240.

164. Hiraoka, K. et al. 2008. Inhibition of bone and muscle metastases of lung cancer cells by a decrease in the number of monocytes/macrophages. *Cancer Science* 99: 1595–1602. doi:10.1111/j.1349–7006.2008.00880.x.

165. Kagan, H. M. and W. Li. 2003. Lysyl oxidase: Properties, specificity, and biological roles inside and outside of the cell. *Journal of Cellular Biochemistry* 88: 660–672. doi:10.1002/jcb.10413.

166. Erler, J. T. et al. 2006. Lysyl oxidase is essential for hypoxia-induced metastasis. *Nature* 440: 1222–1226. doi:10.1038/nature04695.

167. Bondareva, A. et al. 2009. The lysyl oxidase inhibitor, beta-aminopropionitrile, diminishes the metastatic colonization potential of circulating breast cancer cells. *PloS One* 4: e5620. doi:10.1371/journal.pone.0005620.

168. Kanapathipillai, M. et al. 2012. Inhibition of mammary tumor growth using lysyl oxidase-targeting nanoparticles to modify extracellular matrix. *Nano Letters* 12: 3213–3217. doi:10.1021/nl301206p.

169. Guo, W. and F. G. Giancotti. 2004. Integrin signalling during tumour progression. *Nature Reviews. Molecular Cell Biology* 5: 816–826. doi:10.1038/nrm1490.

170. Desgrosellier, J. S. and D. A. Cheresh. 2010. Integrins in cancer: Biological implications and therapeutic opportunities. *Nature Reviews. Cancer* 10: 9–22. doi:10.1038/nrc2748.

171. Murphy, E. A. et al. 2008. Nanoparticle-mediated drug delivery to tumor vasculature suppresses metastasis. *Proceedings of the National Academy of Sciences of the United States of America* 105: 9343–9348. doi:10.1073/pnas.0803728105.

172. Lin, R. Y. et al. 2012. Targeted RGD nanoparticles for highly sensitive in vivo integrin receptor imaging. *Contrast Media & Molecular Imaging* 7: 7–18. doi:10.1002/cmmi.457.

173. Zhang, C. et al. 2007. Specific targeting of tumor angiogenesis by RGD-conjugated ultrasmall superparamagnetic iron oxide particles using a clinical 1.5-T magnetic resonance scanner. *Cancer Research* 67: 1555–1562. doi:10.1158/0008–5472.CAN-06–1668.

174. Pene, F., E. Courtine, A. Cariou, and J. P. Mira. 2009. Toward theragnostics. *Critical Care Medicine* 37: S50–S58. doi:10.1097/CCM.0b013e3181921349.

175. Ross, J. S. et al. 2009. The HER-2 receptor and breast cancer: Ten years of targeted anti-HER-2 therapy and personalized medicine. *The Oncologist* 14: 320–368. doi:10.1634/theoncologist.2008–0230.

176. Janib, S. M., A. S. Moses, and J. A. MacKay. 2010. Imaging and drug delivery using theranostic nanoparticles. *Advanced Drug Delivery Reviews* 62: 1052–1063. doi:10.1016/j.addr.2010.08.004.

177. Goldenberg, D. M., D. F. Preston, F. J. Primus, and H. J. Hansen. 1974. Photoscan localization of GW-39 tumors in hamsters using radiolabeled anticarcinoembryonic antigen immunoglobulin G. *Cancer Research* 34: 1–9.

178. Conti, P. S. et al. 2005. The role of imaging with (111)In-ibritumomab tiuxetan in the ibritumomab tiuxetan (zevalin) regimen: Results from a Zevalin Imaging Registry. *Journal of Nuclear Medicine: Official Publication, Society of Nuclear Medicine* 46: 1812–1818.

179. Terreno, E., F. Uggeri, and S. Aime. 2012. Image guided therapy: The advent of theranostic agents. *Journal of Controlled Release: Official Journal of the Controlled Release Society* 161: 328–337. doi:10.1016/j.jconrel.2012.05.028.

180. Bagalkot, V. et al. 2007. Quantum dot-aptamer conjugates for synchronous cancer imaging, therapy, and sensing of drug delivery based on bi-fluorescence resonance energy transfer. *Nano Letters* 7: 3065–3070. doi:10.1021/nl071546n.
181. Grange, C. et al. 2010. Combined delivery and magnetic resonance imaging of neural cell adhesion molecule-targeted doxorubicin-containing liposomes in experimentally induced Kaposi's sarcoma. *Cancer Research* 70: 2180–2190. doi:10.1158/0008-5472.CAN-09-2821.
182. Nasongkla, N. et al. 2006. Multifunctional polymeric micelles as cancer-targeted, MRI-ultrasensitive drug delivery systems. *Nano Letters* 6: 2427–2430. doi:10.1021/nl061412u.

2

Aptamer–Nanomaterial Conjugates for Medical Applications

Christine Reinemann and Beate Strehlitz

CONTENTS

2.1 Introduction

Due to their multiple attributes, aptamers are the most qualified nanosystems for the development of new biomedical devices for analytical, imaging, drug delivery, and many other medical applications. For instance, the common use of nanoparticles for drug delivering and bioimaging can be substantially improved by modification with aptamers to enhance the specific binding of the nanoparticles via the specific aptamer binding to their target molecule. Aptamers can serve as "drive propulsion system" for the transport of the nanoparticles to their site of action caused by the target affinity of the aptamers.

The following aptamer binding to the target "anchors" the nanoparticle–aptamer conjugate at its site of action. Furthermore, aptamer–nanoparticle conjugates can improve the measurement of medicinal relevant parameters and, hence, contribute to the development of intelligent medical devices. Despite their advantages and unique properties as molecular probes, however, aptamers have been used sparingly for medical applications because of the limited number of aptamers available for medical relevant target molecules.

This chapter gives a brief overview about recent relevant research on aptamers and their trends in different medical fields. The first part of the chapter provides a concise description of aptamers and their development and functionalities regarding nanoparticle modification. The main part is dedicated to the actual developments of aptamer-modified nanoparticles and their use in medical applications.

2.2 Aptamers

Aptamers are artificial nucleic acid ligands. They are short single-stranded DNA (ssDNA) or RNA oligonucleotides binding to their targets with high selectivity and sensitivity because of their 3D shape. The word "aptamer" was invented by Andrew Ellington and coworkers, derived from the Latin "aptus," to fit and to be a polymer that fits to its target [1]. Target molecules for aptamers can be virtually any class of substrate ranging from complex molecules, like whole cells, to large molecules, like proteins, over peptides to drugs and organic small molecules or even metal ions. Aptamers have the potential for application as specific binding molecules with high affinities in medical and pharmaceutical basic research, drug development, diagnosis, and therapy. Analytical and separation tools bearing aptamers as molecular recognition and binding elements are other big fields of application. Moreover, aptamers can be used for the investigation of binding phenomena in proteomics. The technology to evolve aptamers was discovered independently by three laboratories in 1990 [2–4] and was named "SELEX" (Systematic Evolution of Ligands by EXponential enrichment) by Larry Gold, who is the owner of the process patent. SELEX™ is a trademark of Gilead Sciences, Inc.

2.2.1 Structure and Functionality

The functionality of aptamers is based on their stable 3D structure, which depends on three different factors: (1) the primary sequence, (2) the length of the nucleic acid molecule (smaller than 100 nt), and (3) the environmental conditions. The specific and complex 3D structures of aptamers are characterized by stems, internal loops, bulges, hairpins, tetra loops, pseudoknots, triplicates, kissing complexes, or G-quadruplex structures. Binding of the aptamer to its target (Figure 2.1) results from a combination of complementarity in the geometrical shape, stacking interactions of aromatic rings and the nucleobases of the aptamers, electrostatic interactions between charged groups, van der Waals interactions, and hydrogen bondings [5]. The first aptamers developed consisted of unmodified RNA [2–4]. Later on, ssDNA aptamers for different targets were described [6], followed by aptamers containing chemically modified nucleotides [7]. Chemical modifications can introduce new features into the aptamers, that is, to improve their binding capabilities or enhance their stability [8]. An attempt to generate nuclease-resistant derivatives through a standard SELEX procedure was carried out using DNA enantiomers, so-called

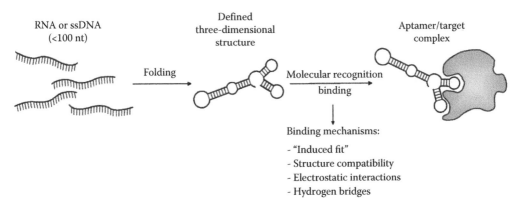

FIGURE 2.1
Schematic view of the aptamer molecular recognition principle. (Courtesy of Regina Stoltenburg.)

Spiegelmers [9]. In this case, the mirror image of the target is used for the aptamer selection. The natural D-DNA or D-RNA sequences for the mirror target are selected. The L-versions of the aptamers are the ligands of the natural targets. They are nuclease resistant and can be chemically synthesized. This strategy was successfully used first by Klussmann et al. [10] for RNA aptamer development for D-adenosine. Spiegelmer is a trademark of Noxxon Pharma AG, Germany [11]. With the Spiegelmer Technology, Noxxon has developed a diversified portfolio of clinical-stage Spiegelmer therapeutics. Some of them have reached phase III clinical trials now [11].

Aptamers are considered as "nucleic acid antibodies" with additional superior properties [12] sometimes. Many of the selected aptamers show affinities comparable to those observed for monoclonal antibodies. Aptamers can distinguish between chiral molecules and are capable of recognizing a distinct epitope of a target molecule [13–15]. Thus, the differentiation involving closely related targets (e.g., theophylline and caffeine [13]) is possible.

2.3 SELEX Process

Since the early phase of the SELEX technology, it became an important and widely used tool in molecular biological, pharmaceutical, and medical research. The technology was often modified to make the selection process more efficient and less time consuming or to select aptamers with particular binding features for different target molecules and for different applications [16]. Numerous variants of the original SELEX process were described to select aptamers with high affinities and specificities for their targets. In principle, the SELEX process is characterized by iterative cycles of *in vitro* selection and enzymatic amplification. Hence, the operation mimics a Darwinian-type process driving the selection toward relatively few (but optimized) structural motifs, which show the highest specificities and affinities to the selection target. The starting point of a typical SELEX process is a chemically synthesized random DNA oligonucleotide library that is consisting of about 10^{13}–10^{15} different sequence motifs [17]. Each DNA strand in this library contains a central random region flanked by specific sequences, which facilitate function as primer binding sites in the polymerase chain reaction (PCR). The central random regions of such libraries typically consist of 20–80 nt and the primer binding sites of 18–21 nt each. For the selection of RNA aptamers, the DNA library has to be converted into an RNA library prior to start the RNA SELEX process. A special sense primer with an extension at the 5′-end containing the T7 promoter sequence and an antisense primer are necessary to convert the ssDNA library into a dsDNA library by PCR. The dsDNA then is transcribed *in vitro* by the T7 RNA polymerase, resulting in a randomized RNA library, which is used to start an RNA SELEX. For the selection of DNA aptamers, the DNA library can be used directly in the first round of a SELEX process for the binding step to the target. Sense and antisense primers derived from the specific sequences at the 5′-end and 3′-end enable the amplification of the selected oligonucleotides in the SELEX rounds by PCR. In the first selection round, the complex RNA or DNA pool is incubated with the target. Subsequent stringent washing steps separate unbound and weakly bound oligonucleotides from the resulting binding complexes. The efficient removal of nonbinding oligonucleotides is one of the most crucial steps of an aptamer selection process and strongly affects the binding features of the aptamers to be selected. There are a variety of methods for effective sorting of unbound oligonucleotides out of the process. Depending on the possible target immobilization, an affinity matrix is generated, and affinity chromatography, magnetic separation, filtration, or centrifugation is performed. In case the target is not immobilized, selective capturing of the binding complexes or size separation is possible. Target-bound oligonucleotides are eluted and amplified by PCR (DNA SELEX) or RT-PCR (RNA SELEX). The resulting dsDNA has to be transformed into a new oligonucleotide pool by purifying the relevant ssDNA or by *in vitro* transcription and purifying the synthesized RNA. This new enriched pool of selected oligonucleotides is used in the following SELEX round for a binding reaction with the target. The change of concentration of bound oligonucleotides should be monitored during the consecutive rounds. Enrichment up to a saturation concentration of target-specific oligonucleotides indicates that the SELEX process is finished. The last SELEX round is stopped after the amplification step, and the PCR products are cloned to get individual aptamers from the selected pool. To avoid this labor and time-intensive step, sequencing of the aptamer pool by next-generation sequencing will become increasingly accepted [18]. After cloning, the individual aptamers are sequenced. Sequence analyzing is necessary thereafter as well as after next-generation sequencing. Representative aptamers are chosen and used in binding assays to characterize their binding features in more detail including the affinities and specificities.

2.3.1 SELEX for Nano-Medical Applications

In biomedical fields, the binding of the aptamer to a cell or to a specific epitope on the cell surface is desirable in many applications. Cell-SELEX is a variant of the SELEX procedure for the generation of cell-specific aptamers [19]. Whole cells are used as targets in this procedure. Moreover, it is possible to generate aptamers that can differentiate between two types of cells, for example, cancer cells from normal cells [20]. Therefore, counterselection steps are included for removing oligonucleotides binding to other domains on the cell surface than that which is responsible for the recognition of the desired cells. Partition of unbound oligonucleotides is possible by centrifugation. The oligonucleotides that have not bound to the unwanted cells are shifted to the selection step with the target cells. Afterward, bound oligonucleotides will be separated from the target cells and amplified in a PCR reaction.

Tissue SELEX is a further method for generating aptamers capable of binding to complex tissue targets such as collections of cells in diseased tissues [19,21]. Another variant is TECS-SELEX [22,23], which uses recombinant instead of the purified proteins displayed on the cell surface.

2.4 Advantages and Properties of Aptamers Regarding Their Use in Nanotechnologies for Medical Applications

Aptamers interact directly with their targets by binding reaction. The bound aptamer has an impact on the function of the target molecule. For example, it can inhibit the binding of this molecule to the cognate receptor. This effect extends the opportunity of using aptamers in pharmaceutical industries. The possible inhibition of the function of intracellular protein is relevant in cases of genetic diseases in which abnormal proteins are causing the problem for gene therapy [24].

In medical applications, aptamers are often compared and have to compete with antibodies. Taking the current state of commercialization of aptamer-based pharmaceuticals and clinical diagnostic tests, it seems that they are underdeveloped. But as Larry Gold, one of the inventors of aptamers, said in [25]: "It took 17 years after the first aptamer description for the first aptamer based drug to hit the market. That's still less time than it took for the first mAb drug to hit the market after César Milstein discovered how to make antibodies."

This first aptamer-based drug is predicated on an antivascular endothelial growth factor aptamer [26]. The PEGylated form of this aptamer (called pegaptanib) is used as the medicinal active component in a drug for treatment of age-related wet macular degeneration. Eyetech Pharmaceuticals was the company that developed and launched this first aptamer-based drug. The pharmaceutical product Macugen® (pegaptanib sodium injection) was approved by the U.S. Food and Drug Administration in December 2004. OSI Pharmaceuticals acquired Eyetech in 2005. Pfizer Limited obtained a license. The European Commission granted a marketing authorization valid throughout the European Union for Macugen to Pfizer Limited on January 31, 2006 [27–30].

Some more aptamers are in several stages of development, from preclinical studies to clinical trials [31].

Aptamers will not replace antibodies, particularly with regard to well-established applications. However, they are going to find their own niche of applications, which could be very substantial in many fields [25]. In comparison to antibodies, they show some advantages because of their molecular nature. Aptamers, especially DNA aptamers, are stable molecules that can be stored at room temperature. They can be heated up to 80°C or exposed to various solvents or harsh environments and will return to their original confirmation to providing a long shelf life [24]. They have very low immunogenicity and toxicity. Aptamers can be chemically modified to extend their lifetime in biological fluids, to immobilize them on surfaces, or to endow them with markers [13,27]. Chemical stabilization is possible so that their half-lives can be "tuned" to match their indication [25]. The naturally short shelf life of aptamers in the human body can be of advantage, for instance, for a therapeutic that acts as long as it is being infused and rapidly self-eliminates once the infusion is turned off (such as coagulants). In the same way, aptamer-based molecular imaging agents display short circulating half-lives in body. In contrast, antibody-based imaging agents may circulate for days until weeks [25]. A further advantage of using aptamers as drugs is the ability to design antidotes rationally. This was published for an anticoagulant/antidote combination with the aptamer REG1 binding to the coagulation factor IX and its antidote that can rapidly shut off the drug's effect [32].

Antibody therapies have mostly high prices. It was stated in [25] that aptamer therapy can go down to 10% of the antibody therapy because of the lower manufacturing costs of aptamers.

For clinical diagnostics, aptamers can complete antibody-based tests. They can be used in the same assay formats as antibodies. An example is the so-called enzyme-linked oligonucleotide assay (ELONA). This technique is basically an enzyme-linked immuno-sorbent assay (ELISA) in which aptamers are used instead of antibodies to detect the target molecules. The ELONA technique was first developed by the NeXstar company [24]. Fluorescent-labeled aptamers can be used in a variety of detection and diagnostic systems. For instance, by use of fluorescent-labeled aptamers that bind surface proteins of pathogens, it would be possible to detect any infection by viruses, bacteria, and even prions. Aptamers have been synthesized that are capable of clear distinction between the pathogenic and nonpathogenic forms of these proteins [33].

Using aptamers, a new generation of DNA microarrays called aptamer microarrays (aptamer chips) is under development, which uses both RNA and DNA oligoprobes [34,35]. Aptamers seem to be more advantageous than antibodies in microarrays. First, aptamers are regenerable multiple times without loss of sensitivity, while antibody microarrays suffer irreversible damage [36]. Second, the small size of aptamers provides a greater surface density of receptors at the chip surface, which can increase the signal efficiency of the microarray [36]. The functionality of an aptamer microarray is comparable to that of a DNA microarray. Different aptamer probes with specificity to the different target molecules in a cell extract are coated onto an array site of a glass slide. Each of the immobilized aptamer probes binds its target molecule, provided it is available in the cell extract. The bond target molecules will then be easily quantified at the specific aptamer site, comparable to the scanner-based readout of DNA microarrays [24]. These microarrays could potentially quantify 10^2–10^5 different proteins of a cell [37]. To realize such a powerful approach, the availability of specific aptamers for each of the cell proteins is an essential precondition. Furthermore, it should be noted that all aptamers immobilized on one chip have to undergo their target binding reaction under the same conditions (pH, buffer composition, temperature) and are applied for the microarray analysis.

Since aptamers can be selected for specific binding to small molecules, aptamer microarrays could also be constructed for detecting small molecule ligands, such as cellular metabolites. By combining different aptamers, it may suggest possibility to build an instrument that could analyze metabolites, proteins, and nucleic acids simultaneously [24].

Because of their outstanding properties, aptamers are used in combination with nanoparticles for biomedical sensing and detection and as aptamer–nanoparticle conjugates for smart drug delivery. The aptamer–nanoparticle conjugates enable an active controlled delivery of drugs, incorporated in the nanoparticles, once bound to a disease site because of the affinity of the aptamer to this site [12]. The following sections give an overview of actual developments concerning the use of aptamers and nanoparticles for medical applications.

2.5 Biomedical Sensing and Imaging Based on Aptamer–Nanomaterials

2.5.1 Detection of Medical Relevant Analytes by Use of Aptamer–Nanoparticle-Based Assays

2.5.1.1 Gold Nanoparticles

The unique physical properties of gold nanoparticles made them interesting in large vast of nano- and biotechnology applications. They are water soluble, nontoxic, simply synthesizable in various sizes and shapes, and easy to modify [38–40]. Furthermore, they exhibit unique optical properties, such as resistance to photobleaching, high surface plasmon resonance (SPR), and enhanced absorbance and scattering with quantum efficiency, which make them ideal as reporters for both *in vitro* and *in vivo* imaging [41–43].

Gold nanoparticles are ideal materials that allow colorimetric detection. Their very high extinction coefficient makes their colors distinguishable without any instrumentation at only a few nanomolar concentrations. Dispersed gold nanoparticles smaller than 100 nM in solution originally have a reddish color. If they aggregate, their color changes from red to purple or blue due to a shift in their plasmon resonance to a higher wavelength [44].

Based on this principle, there are two different strategies for designing target-induced colorimetric assays by use of oligonucleotide-modified gold nanoparticles: assembly and disassembly.

Huang et al. [45] used the assembly principle as highly specific sensing system to detect platelet-derived growth factors and receptors (PDGFs and PDGFRs) with aptamer-conjugated gold nanoparticles to reach detection limits of 3.2 nM. The applied PDGF DNA aptamers (K_D = 0.1 nM) featured a consensus motif that forms a three-way junction with a conserved single-stranded loop at the branch point [46]. The aptamers were coupled on the gold nanoparticle surface by thiol chemistry. In the presence of <400 nM PDGF, a color change occurred from red to purple. The authors assumed that PDGF molecules acted as bridging linkers for the aggregation of aptamer–gold nanoparticles.

The second strategy is the reverse disassembly strategy. It was used by Liu and Lu [47] for cocaine and adenosine detection assays. They constructed nanoparticle aggregates containing gold nanoparticles; thiol-modified aptamers for cocaine [48,49] and adenosine [50], respectively; and special aptamer linker sequences to form aggregates. These nanoparticle aggregates displayed a faint purple color in solution. Due to the structure-switching process of the aptamer upon target binding, one part of aptamer–nanoparticle that was constructed to release resulted in disassembly of the nanoparticles. A color change to red emerged. The detection range for cocaine was determined with 50–500 µM, for adenosine 0.3–2 mM. The authors explained the low sensitivity, due to the fact that first all of the linkages have to be broken, before the nanoparticle disassembly. This resulted in high background absorbance and allowed only qualitative or semiquantitative analysis by the naked eye. To overcome these limitations of the aptamer-based sensors in the solution phase, the sensor principle was additionally used for a swift and easy "dipstick" assay [51]. This test kit works according the lateral-flow technology like the well-known commercially available pregnancy test kit and its handling is notably user-friendly. The authors slightly modified the adenosine-responsive nanoparticle aggregates design mentioned previously by using two kinds of DNA-functionalized gold nanoparticles (particles 1 and 2 in Figure 2.2a) and two kinds of thiol-modified DNA to functionalize particle 2: biotinylated and nonbiotinylated. The aptamer-linked nanoparticle aggregates were spotted on the conjugation pad, and streptavidin was applied on the membrane as a thin line (Figure 2.2c). If the wicking pad of the device was dipped into a solution, the solution moved up along the device and rehydrated the aggregates. In the absence of adenosine, the rehydrated aggregates migrated to the bottom of the membrane where they stopped because of their large size (Figure 2.2d). In the presence of adenosine, the nanoparticles were disassembled owing to binding of adenosine by the aptamer. The dispersed nanoparticles can then migrate along the membrane and be captured by streptavidin to form a red line (Figure 2.2e).

Another example for the adaption of aptamer-coupled gold nanoparticles in heterogeneous systems is the dot-blot assay, which was developed by Wang et al. [52] for the detection of thrombin. The researchers first immobilized thrombin on a membrane. A color

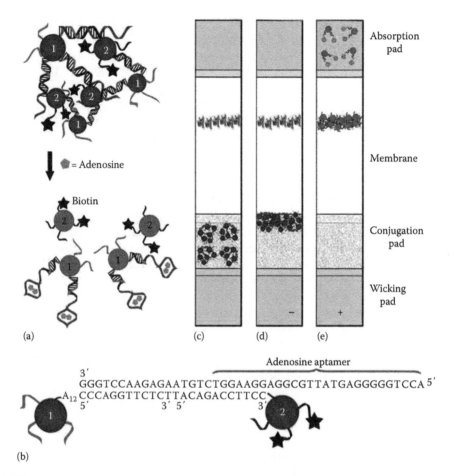

FIGURE 2.2

(See color insert.) Aptamer/nanoparticle-based lateral-flow device. (a) Adenosine-induced disassembly of nanoparticle aggregates into red-colored dispersed nanoparticles. Biotin is denoted as black stars. (b) DNA sequences and linkages in nanoparticle aggregates. Lateral flow devices loaded with the aggregates (on the conjugation pad) and streptavidin (on the membrane in cyan color) before use (c) and in a negative (d) or a positive (e) test. (From Liu, J., Mazumdar, D., and Lu, Y. A simple and sensitive "dipstick" test in serum based on lateral flow separation of aptamer-linked nanostructures. *Angew. Chem. Int. Eng. Edn.* 2006. 45. 7955–7959. Copyright Wiley-VCH Verlag GmbH & Co. KGaA. Reprinted with permission.)

change from colorless to red was produced when the aptamer–nanoparticles bound on the active side of the protein [53,54]. The aptasensor could be observed by the naked eye and had a detection limit of 14 pM after additional treatment with silver enhancement solution. Importantly, the suitability of the assay for thrombin measurements in complex biological samples, like human plasma, could be shown successfully.

Li et al. [55] developed an ultrasensitive densitometry cytokine detection method that uses gold nanoparticle–modified aptamers. The assay relies on aptamer pairs that bind to two distinct sites on the target protein (PDGF-BB, platelet-derived growth factor B-chain homodimer). Binding of the aptamers [46] to the target formed a sandwich complex that was immobilized onto a microplate via avidin. After silver enhancement, the signal was amplified and the absorbance was measured. The assay was linear from 1 fM to 100 pM. The analytical detection limit could be determined with 83 aM.

He et al. [56] developed a bioelectrochemical method for the detection of thrombin through directly detecting the redox activity of adenine nucleobases of aptamer using a pyrolytic graphite electrode. Thrombin captured by immobilized antithrombin antibody on microtiter plates was detected by aptamer–gold nanoparticles bio-barcodes. The adenine nucleobases were released by acid or nuclease from gold nanoparticles bound on microtiter plates. Differential pulse voltammetry was employed to investigate the electrochemical behaviors of the purine nucleobases. Because the nanoparticles carried a large number of aptamers per thrombin binding event, substantial signal amplification occurred. In this way, thrombin could be detected at a very low level (0.1 ng/mL). The method was validated by the detection of thrombin in fetal calf serum with minimum background interference.

A novel sandwich immunoassay was reported by Wang et al. [57]. This was designed to demonstrate the amplification effect of aptamer–gold nanoparticles conjugates for the detecting of human immunoglobulin E (IgE). Human IgE, captured by immobilized goat antihuman IgE on a SPR gold film, was sensitively detected by SPR spectroscopy with a detection limit of 1 ng/mL. Gold nanoparticles conjugated with antihuman IgE aptamer [36,58] were used as amplification reagent.

2.5.1.2 Silver Nanoparticles

Although silver exhibits many advantages over gold, such as higher extinction coefficients, sharper extinction bands, higher ratio of scattering to extinction, and extremely high field enhancements, it has been employed far less in the development of sensors, with the exception of sensors based on surface-enhanced spectroscopy. The reason for this is the lower chemical stability of silver nanoparticles when compared to gold. Nevertheless, recent developments include means of protecting efficiently silver nanoparticles that offer far improved chemical stabilities. As a consequence, silver nanoparticles are rapidly gaining in popularity, and several research groups have begun to explore alternative strategies for the development of optical sensors and imaging labels based on the extraordinary optical properties of the metal Ag nanoparticles [59–63].

Pagba et al. [64] showed aptamer–thrombin binding via surface-enhanced Raman spectroscopy (SERS) by utilizing silver nanoparticles functionalized with thiolated thrombin-specific aptamers [53]. They explicate that this technique is convenient for the detection of medical markers, of environment pollutants, and for biological and chemical threat agents due to the exceeding sensitivity of SERS and since aptamers are available for a lot of biomolecules with high specificity and affinity.

2.5.1.3 Platinum Nanoparticles

Platinum nanoparticles have excellent catalytic activity toward H_2O_2; they have been found to reduce the H_2O_2 oxidation/reduction overvoltages [65]. Consequently, electrodes modified with platinum nanoparticles have been frequently used to design amperometric enzyme biosensors since H_2O_2 was generated during the reaction between substrate (e.g., glucose) and enzyme (e.g., glucose oxidase) in the presence of oxygen [66–68].

Polsky et al. [69] used thrombin-aptamer-functionalized platinum nanoparticles for electroanalytic detection with a sensitivity limit corresponding to 1 nM. The functionalized nanoparticles acted as catalytic labels for the amplified electrochemical detection of the aptamer/target complex associated with an electrode. The thrombin aptamers [53] were used in thiol-modified manner to immobilize them on the platinum surface.

Higuchi et al. [70] chose another approach for sensing of thrombin and antithrombin antibodies by use of DNA aptamers that carry platinum nanoparticles. The aptamer–nanoparticle conjugates were prepared by reaction of thrombin-specific DNA aptamers [53] with $K_2[PtCl_4]$ in solution at 60°C–90°C. For this reaction, no aptamer functionalization with biotin, thiol, or other reactive groups was needed. The DNA–platinum complexes possessed peroxidase enzymatic activity while retaining the specific binding ability of the aptamers. Two types of DNAzyme-linked aptamer assays (DLAAs) were developed using these complexes, which successfully detected the targeted proteins. A sandwich type of DLAA was used for measurements of thrombin (detection of concentration limit of 0.4 mM) and a competitive DLAA type for detection of antithrombin antibodies in serum (with detection limits similar of the corresponding commercial ELISA kit). The DNA–platinum complexes retained their peroxidase enzymatic activity even after heat treatment.

2.5.1.4 Dye-Doped Silica Nanoparticles

Dye-doped silica nanoparticles feature several characteristics and advantages that make them suitable as a platform in bioanalysis and for biomedical applications. The high density of silica nanoparticles allows an easy separation during synthesis, modification, and detection steps that are conducted through centrifugation. Compared to, for example, polymer-based nanoparticles, silica nanoparticles show less aggregation and dye leakage. It is possible to encapsulate a large number of fluorophores inside single-silica nanoparticles, leading to intense fluorescence when excited. Such an enhanced fluorescence signal enables very sensitive measurements even of lower analyte concentrations. Indeed, by incorporation of organic dye inside a silica core, photobleaching and photodegradation are reduced. Silica nanoparticles are easy to synthesize in different shapes and degrees of porosity by well-known methods [71–73]. The silica surface provides a universally biocompatible and versatile substrate for biomolecule immobilization. Thus, different types of functional groups can be introduced into the nanoparticles for their conjugation with biomolecules such as aptamers [74,75].

Wang et al. [76] designed a simple and efficient silica nanoparticle–based fluorescent assay for ATP detection involving the principle of structure-switching-induced fluorescence change of the supernatant. Cy5-labeled aptamer [50] was released from the nanoparticle surface to the solution in presence of ATP, which was easy to be collected and recorded by fluorescence measurements. The linear response for ATP detection was observed from 0 to 6 mM with a detection limit of 34 µM.

The same group also reported on the development of sandwich assays for optical detection of thrombin and lysozyme in biological media (blood and protein mixtures). The assay worked with aptamer-functionalized silica nanoparticles as the sensory platform for thrombin and lysozyme, respectively. The presence of the targets induced the primary aptamer-functionalized nanoparticles to form a sandwich structure with fluorescein-labeled secondary aptamers. Thus, the specific protein–aptamer interaction resulted in fluorescent nanoparticles. A conjugated polyelectrolyte was further added to the nanoparticle solution to amplify the fluorescent signal. The assay showed high specificity for thrombin detection in serum with a detection limit of 1.06 nM. In a similar assay by using a lysozyme-specific aptamer [77,78], a detection limit for lysozyme was determined with approximately 0.36 µg/mL (linear range 0–22.5 µg/mL) and could be observed by the naked eye [79]. In these applications, the aptamer-functionalized silica nanoparticles held two functions—the first is to monitor the aptamer/protein-binding signal by fluorescence and the second function is separation of the analytes from the protein mixtures of the measured biological media.

Lin et al. [80] used the conjunction of split aptamers to develop a sensing method based on fluorescent silica nanoparticles. Therefore, $Ru(bpy)_3^{2+}$-doped silica nanoparticles (Ru-SNPs) were applied for measurements of thrombin in an aptamer-based electrochemiluminescence sensor. The sequence of the 15-base aptamer [53] had been split into two different fragments: capture probe and detection probe. The capture probe was modified with thiol and immobilized on the gold electrode, whereas the detection probe was functionalized with Ru-SNPs. In the presence of thrombin, both of these probes were combined to form a G-quadruplex structure for thrombin binding, allowing the electrochemiluminescence reagent to attach on the electrode surface. This resulted in an enhancement of the signals that could be readily detected. The electrochemiluminescence intensity had a linear relationship with the logarithm of thrombin concentration in the range of 330 fM–33 pM and a detection limit of 200 fM.

2.5.1.5 Magnetic Nanoparticles

Magnetic nanoparticles are a class of nanoparticles that can be manipulated using magnetic field. Such particles commonly consist of magnetic elements such as iron, nickel, and cobalt, and alloys of metal oxides. Iron oxide (ferrite) nanoparticles are the most explored magnetic nanoparticles up to date. If ferrite nanoparticles are smaller than 20 nM, they become superparamagnetic. This prevents self-agglomeration since they exhibit their magnetic behavior only when an external magnetic field is applied [81]. The surface of these nanoparticles can be modified through creation of a monolayer of organic polymer or inorganic metallic (e.g., gold) or oxide surfaces (e.g., silica or alumina). Such monolayers enable further functionalization by attachment of various bioactive molecules [82–85]. Functionalized in this way, magnetic nanoparticles exhibit two features: specificity and magnetism. Taking advantage of these features, an interaction with an external magnetic field and a positioning in a specific area becomes possible. Magnetic nanoparticles have been used for biomedical separations, magnetic resonance imaging (MRI), drug delivery, and magnetic fluid hyperthermia therapy, and many other applications [86].

Cross-linking of dextran-coated superparamagnetic iron oxide (CLIO) nanoparticles modified with different biomolecules had been used for detection of different targets. It has been shown that these functional CLIO nanoparticles assemblies create a distinctive magnetic phenomenon called magnetic relaxation switching and resulted in enhancement of spin–spin relaxation time of adjacent water protons [87].

Yigit et al. [88] prepared aptamer-conjugated CLIO nanoparticles to demonstrate a general method for aptamer-based biosensing using MRI. An adenosine aptamer-based design [50] served as model system for smart and functional CLIO contrast agents. A subsequent work [89] used the concept for selective detection of thrombin [53] down to 25 nM thrombin in 0.5-fold diluted human serum. The sensing strategy based on a thrombin-induced aggregated structure of aptamer CLIO nanoparticles. This nanoparticle aggregation led to decreases in the spin–spin relaxation time T_2, which resulted in decreased brightness of MRI. The advantage of the aptamer CLIO nanoparticles probe over other optical sensor for thrombin was that MRI signal is much less vulnerable to changes in background colors or fluorescence from biological media, such as serum.

Magnetic relaxation switch detection was used by Bamrungsap et al. [90] for lysozyme sensing. The detection method is based on the formation or dissociation of aggregates when magnetic nanoparticles bind to target molecules. The nanosensor for lysozyme detection was achieved using iron oxide nanoparticles conjugated with either lysozyme aptamer [77,78] or linker DNA. In the absence of lysozyme, the linker could hybridize with one part of the aptamer (seven bases of the aptamer plus a five-base extension) to form clusters. Upon the addition of lysozyme, the aptamer undergoes a structural switch for binding the target, leading to disassembly of clusters and an increase in the spin–spin relaxation time T_2 (a time constant characterizing the signal decay of transverse relaxation in magnetic resonance). In buffer, a linear relationship between lysozyme concentration and ΔT_2 was observed in the concentration range of 500 pM to 80 nM. The detection limit was determined with 500 pM. Detection in complex biological media was also demonstrated by spiking lysozyme into 100% human serum. The results showed that this nanosensor was able to detect lysozyme in serum in the range of nanomolar concentrations. A linear relationship between ΔT_2 and lysozyme concentration of 1–80 nM was achieved. To demonstrate the potential of the method for real clinical samples, the determination of lysozyme levels in different types of cancer cell lysates was also performed.

Wang et al. [91] presented an SPR-based bioassay using an amplification technique with iron oxide magnetic nanoparticles for an enhanced SPR signal. The functionality of the method was evaluated by an indirect competitive inhibition assay for detecting adenosine by employing Fe_3O_4 magnetic nanoparticles–antiadenosine aptamer [50] conjugates as the amplification agents. As a result, the authors could provide a sensitive approach to detect adenosine within the range of 10 nM to 10 mM.

2.5.1.6 Carbon Nanotubes

Carbon nanotubes have a high aspect ratio, high surface area, and excellent material properties including electrical and thermal conductivity and mechanical strength. The unique electronic and optical properties of carbon nanotubes, in conjunction with their size and mechanically robust nature, make these nanomaterials crucial to the development of the next generation of biosensing devices. The structure of carbon nanotubes can be visualized as the cylindrical roll-up of one or more flat graphene sheets containing carbon atoms in a honeycomb arrangement. Accordingly, nanotubes are categorized as single- or multiwalled carbon nanotubes (SWNTs or MWNTs, respectively). The incorporation of carbon nanotubes has the potential to address a variety of long-standing issues with respect to biosensing. In particular, the high surface area of single-walled carbon nanotubes while maintaining electrical conductivity is of particular interest to achieve high biomolecule densities suitable for device miniaturization. Carbon nanotubes can be readily dispersed as individual and lightly bundled nanotubes and assembled, screen-printed, and inkjet-printed to produce device configurations with controlled transparency. This is further amplified by an assortment of covalent and noncovalent functionalization strategies to link a variety of biological entities onto carbon nanotubes [92].

ssDNA interacts noncovalently with individual carbon nanotubes, through π-stacking interactions between the nucleotide bases and the SWNTs sidewalls. The ssDNA wraps around SWNTs with nearly complete DNA coverage to form stable complexes [74]. This phenomenon can be used, that is, for aptamer immobilization, for separation and assembly of nanotubes, and for biosensing. For example, Yang et al. [93] reported the development of a sensing platform that uses the property of SWNTs to act as fluorescence quenchers while interacting with dye-labeled DNA oligonucleotides. Upon hybridization of the probe to the target DNA, the dye was lifted from the nanotube surface, and the fluorescence signal was restored.

Based on the same phenomenon, Zhu et al. [94] constructed a photosensitizer–aptamer–SWNT complex, using protein-binding aptamers, photosensitizer, and SWNTs, to achieve singlet oxygen generation regulation. The photosensitizer was covalently linked with an aptamer and then wrapped onto the surface of SWNTs such the photosensitizer can only be activated by light upon target binding. To validate the principle, the researchers chose the human alpha-thrombin aptamer [53] and covalently linked it with chlorin e6, which is a second-generation photosensitizer. The results showed that SWNTs are better quenchers to singlet oxygen generation. In the presence of its target, the binding of target thrombin disturbed the DNA interaction with the SWNTs and caused the DNA aptamer to fall off the SWNT surface, resulting in the restoration of singlet oxygen generation.

Chen et al. [95] described a thrombin biosensing system based on the dissolution and aggregation of SWNTs directed by aptamer/protein interactions. Therefore, they used near-infrared absorption and photoluminescence properties of semiconductive SWNTs. Photoluminescence is observable only when SWNTs are individually dispersed. Aggregations of individual nanotubes into bundles lead to quenching of the fluorescence. The constructed optical thrombin assay based on reversible dissolution and aggregation of SWNTs via specific interactions of DNA aptamers [53] wrapped on the sidewall of SWNTs with the target protein. The assay showed a linear relationship between thrombin concentration from 0.2 to 6.3 nM and a detection limit of 0.1 nM.

The approach reported recently by Ouyang et al. [96] is working basically on the same principle. They engineered an anti-lysozyme aptamer [77,78] and luminescent europium(III) (Eu^{3+}) complex. The sensing scheme of the approach is based on the noncovalent assembly of SWNTs and ssDNA. The ability of the SWNT complex thus formed to both effectively quench the fluorophore and restore the luminescence signal in the presence of a target. Due to the highly specific target recognition ability of the aptamer and the powerful quenching property of SWNTs for luminescence reagents, the approach showed a good selectivity and high sensitivity for lysozyme. Thus, a linear measurement range from 10 nM to 2.0 μM lysozyme with detection limit of 0.9 nM was obtained. Additionally, the group could also show that the assay was able to detect lysozyme with low background signals in complex biological media like human urine.

Carbon nanotubes have been extensively studied as transducer elements of biosensors too. For example, Düzgün et al. [97] developed a solid-contact potentiometric aptasensor exploiting SWNTs as transducers for direct determination of proteins. As a model system to demonstrate the generic application of the approach, the SWNT surface had been modified by a thrombin aptamer [53]. The aptamer interacted with thrombin, and the affinity interaction gave rise to a direct potentiometric signal that could be easily recorded within 15 s. The dynamic linear range corresponded to the 100 nM–1 μM range of thrombin concentrations. The limit of detection was determined with 80 nM. The aptasensor displayed selectivity against elastase and bovine serum albumin and could easily regenerate by immersion in 2 M NaCl.

Maeshi et al. [98] fabricated another label-free protein biosensor based on aptamer-modified carbon nanotube field-effect transistors (CNT-FETs) for the detection of immunoglobulin E (IgE). By applying a voltage to a gate electrode, the semiconducting SWNTs can be switched from a conducting to an insulating state in a field-effect transistor. The effect is measurable in real time. The sensor system was designed by using IgE aptamers [36,58] covalently immobilized on the carbon nanotube channels via a 5′-amino-modification of the aptamers. The linear dynamic range of the sensor was determined to be from 250 pM to 20 nM with a detection limit of 250 pM. The authors compared the performances of the aptamer-conjugated CNT-FET with an IgE-mAb-modified CNT-FET and demonstrated that the performance of aptamer-based system provided better results than the antibody-combined CNT-FET under similar conditions.

Kara et al. [99] reported label-free bioelectronic detection of aptamer–thrombin interaction based on electrochemical impedance spectroscopy (EIS) technique by application of multiwalled carbon nanotubes (MWNTs). The MWNTs were used as modifiers of screen-printed carbon electrotransducers (SPCEs), showing improved characteristics compared to the bare SPCEs. First, 5′-amino linked aptamer sequences [53] were immobilized onto the modified SPCEs. Then, the binding of thrombin to the aptamer sequences was monitored by EIS transduction of the resistance to charge transfer in the presence of 5 mM $[Fe(CN)_6]^{3-/4-}$. A detection limit of 105 pM was obtained.

2.5.1.7 Quantum Dots

Quantum dots offer a number of advantages over standard fluorescent dyes for monitoring biological systems in real time, including better photostability, larger effective Stokes shifts, longer fluorescent lifetimes, and sharper emission bands than traditional organic fluorophores. In addition, quantum dots all respond to the same excitation wavelength, but emit light at different wavelengths; this should allow for the multiplex detection of different analytes in parallel. Thus, semiconductor quantum dots were implemented as optical labels for the analysis of biorecognition events or biocatalytic transformations. Particularly, quantum dot-stimulated fluorescence resonance energy transfer (FRET) reactions were used for the detection of biomolecules. Levy et al. [100] constructed a thrombin sensor based on quantum dot aptamer beacons. Therefore, in order to ensure that the conformational change would occur upon analyte binding, a quantum dot beacon was synthesized based on a two-piece aptamer beacon [101]. The final construct consisted of the biotinylated antithrombin aptamer [53] conjugated to a streptavidin-modified quantum dot and an oligonucleotide quencher conjugate that hybridized a disrupted aptamer structure (Figure 2.3). In the presence of the target thrombin, the quadruplex conformation of the aptamer was preferentially stabilized, resulting in a displacement of the antisense oligonucleotide quencher conjugate and a concomitant increase in fluorescence that could be easily measured in real time.

Freeman et al. [102] used a similar sensing principle for measurements of cocaine by aptamer-functionalized quantum dots that followed the approach, introduced by Stojanovic et al. [48]. Therefore, optimized sequences of cocaine aptamer fragments were engineered, which are able to self-assemble into supramolecular analyte–nucleic acid structures. This resulted in fluorescence quenching of the dye or quantum dot-labeled aptamer fragments. One of these aptamer subunits was linked to CdSe–ZnS quantum dots. The second dye-functionalized aptamer subunit self-assembled with the modified quantum dots in the presence of cocaine and stimulated FRET. This enabled the detection of cocaine with a detection limit corresponding to 1 μM.

Xiao et al. [103] developed a dual-aptamer strategy by taking the advantages of aptamers, the excellent separation ability of magnetic microparticles, and the high fluorescence emission features of quantum dots. Two aptamers (Apt1, Apt2) [104,105] were used, which recognize two distinct epitopes of prion proteins (PrP) and featured different affinities to the cellular prion protein (PrP^C) and the disease-associated isoform of prion protein (PrP^{Res}). These aptamers were coupled to the surfaces of magnetic microparticles and quantum dots, respectively, to form two different aptamer–particle complexes (Apt1–magnetic microbead complex and Apt2–quantum dot complex). These complexes could coassociate together through the specific recognitions of the two aptamers with their corresponding

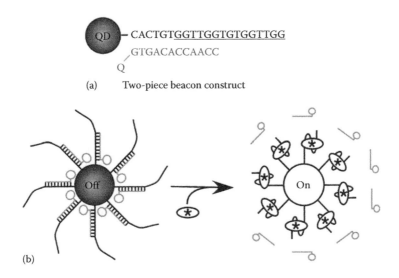

(a) Two-piece beacon construct

(b)

FIGURE 2.3
Design of a quantum dot aptamer beacon for the detection of thrombin. (a) Sequence of antithrombin aptamer (black) and quench oligonucleotide (gray). The portion of the antithrombin aptamer known to form a quadruplex has been underlined. (b) Schematic for detection. Binding of target (*) stabilizes the displacement of the quench oligonucleotide (gray) and enhances the fluorescence of the quantum dot. Since the quantum dots have multiple streptavidin molecules on their surfaces, multiple aptamers are associated with a single quantum dot. (From Levy, M., Cater, S. F., and Ellingtonm, A. D. Quantum dot aptamer beacons for the detection of proteins. *Chembiochem.* 2005. 6. 2163–2166. Copyright Wiley-VCH Verlag GmbH & Co. KGaA. Reprinted with permission.)

epitopes of PrP. In this way, it formed a sandwich structure that displays the strong fluorescence of quantum dots. Owing to the different binding affinities of the two aptamers with PrP^{Res} and PrP^C, the dual-aptamer strategy could be applied to discriminate PrP^{Res} and PrP^C successfully in human serum and brain homogenates.

2.5.1.8 Hybrid Aptamer–Dendrimers

Dendrimers are a new class of synthetic macromolecules. They are molecules with regularly branched treelike spherical morphologies and with monodisperse sizes. By regulating dendrimer synthesis, it is possible to precisely manipulate both their molecular weight and chemical composition, thereby allowing predictable tuning of their biocompatibility and pharmacokinetics. The unique structural properties of dendrimers, such as structural homogeneity, integrity, the controlled composition, the accretion of functional groups, and internal cavities, extend their use in biosensing applications [106].

Zhang et al. [107] developed a label-free and highly sensitive impedimetric aptasensor for thrombin determination based on a polyamidoamine dendrimer–modified gold electrode. Therefore, an amino-terminated polyamidoamine dendrimer was firstly covalently attached to the cysteine functionalized gold electrode through glutaraldehyde coupling. Subsequently, the dendrimer was activated with glutaraldehyde. An amino-modified thrombin-aptamer probe was immobilized onto the activated dendrimer monolayer film. After electrode preparation, the detection of thrombin was investigated using impedance technique. The results showed a linear relationship of the measured signals with the concentrations of thrombin in the range of 1–50 nM, and a detection limit as low as 0.01 nM was obtained.

2.6 *In Vitro* Detection of Cancer Cells

For a realistic prognosis of patients with cancer symptoms and an exact estimation of the stage of the cancer, shorter detection and diagnosis time lines are essential. This demands the development of quick and easy detection methods to determine the different cancer types. Much more than for other diseases, the consistent multitude different cancer cells need the implementation of personalized treatment. This suggests the discovery of new biomarkers that are specific to particular cancers, as well as the development of molecular probes that bind to these biomarkers, which are two of the most critical issues in the field of cancer biology today. Molecular recognition of disease-specific biomarkers, especially the recognition of proteins, or other biological molecules that differentiate between normal and abnormal cells, is a fundamental challenge in cancer cell biology [108]. Aptamers are very convenient for this purpose. It is possible to develop them *in vitro* and systematically for differentiation of cancer cells so that they are able to discriminate even small differences between molecules or cells. Aptamers have a low molecular weight, fast tissue penetration, and low toxicity. They can be specifically labeled with various reporters for molecular recognition. Despite their advantages and unique properties as molecular probes in cancer detection and treatment, aptamers have been developed sparingly for this application. There are only few aptamers generated directly from diseased cancerous cells. So far, most of the published aptamer–nanoparticle assays and sensors for cancer detection are constructed with only a handful of aptamers specific for T-cell leukemia and B-cell lymphoma. This panel of DNA aptamers was selected directly from cancer cells by cell-SELEX [109] for the recognition of molecular differences among leukemia patient samples [110]. With the generated aptamers, it is possible to differentiate between acute lymphoblastic T-cell leukemia (CCRF-CEM cells [23]) and B-cell lymphoma (Ramos cells, human Burkitt's lymphoma [111,112], and Toledo cells [110]). All of the aptamers show high specificity for their respective target cells and equilibrium dissociation constants in concentrations of nanomolar to submolar range.

Gold nanoparticles have been modified with these aptamers for their use as targeted probes for imaging and studying of lymphoma cancer cells. For example, Medley et al. [113] constructed a colorimetric assay for the direct detection of cancer cells. Therefore, gold nanoparticles were conjugated with thiol-modified DNA aptamers specific for CCRF-CEM cells and Ramos cells. The aptamer–gold nanoparticles assembled around the cell surface and caused a shift in extinction spectra. This led to a measurable red shift regulated by the particle size. A linear relationship between the amount of cells and the absorbance intensity at 650 nm for both target cells could be observed with a detection limit of 90 cells by spectroscopic detection and 1000 cells by the naked eye.

Another work used the described aptamers for a user-friendly application that enables cancer detection very easily and in shortest time with a strip-based assay [114] very similar to the "dipstick" assay for adenosine [51] described previously. In this case, a thiolated aptamer, specific for Ramos cells, was immobilized on gold nanoparticles and an additional Ramos aptamer (biotinylated) on the strip's test zone. When the sample solution containing Ramos cells was applied to the sample pad, the solution migrated by capillary action past the conjugate pad and then rehydrated the aptamer-conjugated gold nanoparticles. The Ramos cells interacted first with the modified nanoparticles and then continued migrating along the strip. By reaching the test zone, they got captured by a second reaction between the Ramos cells and the immobilized biotinylated aptamers. The visualization occurs by a characteristic red band due to the accumulation of gold nanoparticles in the test zone.

Under optimal conditions, the test strip was capable of detecting a minimum of 4000 Ramos cells only by visual judgment, without instrumentation and 800 cells, if there was a portable strip reader used. The measurements in buffer solution could be made within 15 min. The feasibility of the test system for the detection of cancer cells in biological fluids was evaluated, and a successful determination of Ramos cells in human blood could be shown.

Huang et al. [115] demonstrated the use of bimetallic gold–silver nanorods as an efficient and robust multivalent nanoplatform for molecular assembly aptamers for target cell recognition with the mentioned leukemia aptamers. As determined by flow cytometric measurements, simultaneous attachment of up to 80 single fluorophore-labeled aptamers onto the functionalized nanorod surface resulted in a much stronger fluorescence signal (excess of 300-fold) than that of an individual dye-labeled aptamer probe. Moreover, the molecular assembly of aptamers on the nanorod surfaces also significantly improved the binding affinity with cancer cells through simultaneous multivalent interactions with the cell membrane receptors. This leads to an affinity at least 26-fold higher than the intrinsic affinity of the original aptamer probes.

An electrochemical determination of cancer cells (Ramos) was developed by Ding et al. [116]. Gold nanoparticles bifunctionalized with aptamers and CdS nanoparticle were used for electrochemical signal amplification. The anodic stripping voltammetry technology employed for the analysis of cadmium ions dissolved from CdS nanoparticles on the aggregates provided a means to quantify the amount of the target cells. This electrochemical method could respond down to 67 cancer cells per mL with a linear calibration range from 1.0×10^2 to 1.0×10^5 cells per mL (measurements in PBS buffer), which shows very high sensitivity. In addition, the assay was able to differentiate between target and control cells based on the aptamer used in the assay, indicating the wide applicability of the assay for diseased cell detection.

Chen et al. [117] produced an aptamer-conjugated FRET nanoparticle assay that performs simultaneous multiplexed monitoring of cancer cells, in this case T-cell leukemia and B-cell lymphoma. Therefore, the authors firstly tuned the FRET-mediated emission signature by changing the doping ratio of three different dyes such that the nanoparticles would exhibit multiple colors upon excitation with a single wavelength. These FRET nanoparticles were then modified by a few aptamers specific for different cancer cell lines (Toledo, CCRF-CEM, and Ramos cells). Fluorescent imaging (Figure 2.4) and flow cytometry were used for cellular detection and demonstrated the selectivity and sensitivity of the method. Previously, Herr et al. [118] used similar dye-doped silica nanoparticles combined with aptamer-conjugated magnetic nanoparticles for a selective collection and detection of acute leukemia cells and were able to separate and monitor these CCRF-CEM cells from mixed cell and whole blood samples in this way.

The DNA aptamers specific for CCRF-CEM and Ramos cells have been used also for the development of an electrochemiluminescence assays for cancer cells based on dendrimer/CdSe–ZnS–quantum dot nanoclusters [119]. Due to the multitude of functional amine groups within the polyamidoamine (PAMAM) dendrimer, a large number of CdSe–ZnS–quantum dots could be incorporated, which significantly amplified the electrochemiluminescence signal. Moreover, magnetic beads for aptamer immobilization were combined with the dendrimer/quantum dot nanocluster probe, which simplified the separation procedures and favored for the sensitivity improvement of the sensor. In particular, a novel cycle-amplifying technique using a DNA device on magnetic microbeads was further employed in the assay, which greatly improved the sensitivity for detection of the cancer cells. Excellent discrimination between target and control cells could be demonstrated with the approach.

Toledo   CEM   Ramos

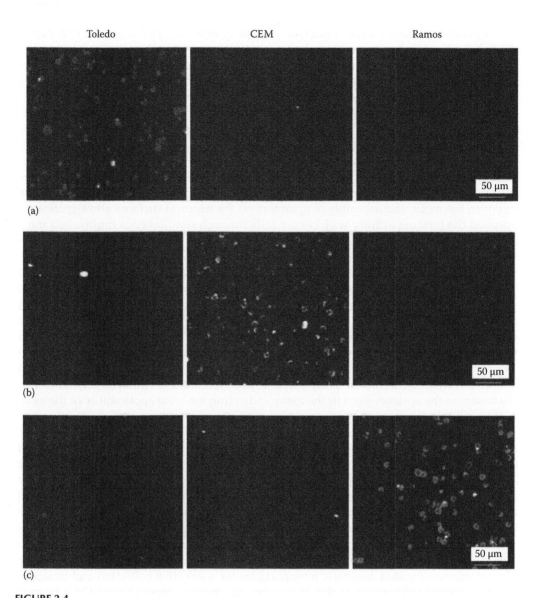

(a)

(b)

(c)

FIGURE 2.4
(See color insert.) Confocal microscopy images of individual nanoparticle–aptamer conjugates with the three different cells (Toledo, CCRF-CEM, and Ramos): (a) nanoparticle-(FAM)-aptamer T1, specific for Toledo; (b) nanoparticle-(FAM-R6G)-aptamer sgc8, specific for CCRF-CEM; and (c) nanoparticle-(FAM-R6GROX)-aptamer TDO5, specific for Toledo. (Reprinted with permission from Chen, X. et al., *Anal. Chem.*, 81, 7009. Copyright 2009 American Chemical Society.)

Besides the use of the mentioned DNA aptamers for leukemia cells, there were few nanoparticle-based cancer detection assays published. For example, Lu et al. [120] employed oval-shaped gold nanoparticles and were able to increase the sensitivity of an assay for breast cancer in that way. The group utilized multifunctional gold nanoparticles that were conjugated with monoclonal anti-HER2/c-erb-2-antibody and a S6 RNA aptamer. Applying this conjugate in two-photon scattering technique (TPS), it was possible to detect breast cancer SK-BR-3 cell lines very selectively and in a highly sensitive level (<100 cells/mL).

RNA aptamers directed against prostate-specific membrane antigen (PSMA) were developed by Lupold et al. [121]. These aptamers were the first reported RNA aptamers selected to bind a tumor-associated membrane antigen and the first application of RNA aptamers to a prostate specific cell marker. The set of aptamers is very specific and even able to discriminate different human prostate cancer cell lines. The dissociation constants of these PSMA aptamers were determined in a low nanomolar range. Walter et al. [122] used these aptamers to demonstrate the applicability of a laser-ablation-based *in situ* bioconjugation method to produce gold nanoparticles functionalized with aptamers. This *in situ* conjugation is a fast and simple one-step approach with the possibility of high throughput production of large amounts of different conjugated nanoparticles. The group demonstrated the suitability of the method firstly by using a DNA aptamer for streptavidin as a model. The applicability was confirmed by different analytical methods (agglomeration-based colorimetric assay, dot-blot assay). Afterward, the method was transferred to the RNA aptamer directed against PSMA. Successful detection of PSMA in human prostate cancer tissue was achieved by utilizing tissue microarrays.

A DNA aptamer for epithelial tumor marker mucin 1 (MUC1) was selected by Ferreira et al. [123]. MUC1 is a glycoprotein expressed on most epithelial cell surfaces and present in a variety of malignant tumors. The selected aptamers were demonstrated to detect MCF-7 breast cancer cells. Cheng et al. [124] reported an aptamer-based, quantitative detection protocol for MUC1 using a three-component DNA hybridization system with quantum dot labeling. The sensor is based on a construct of three specially designed DNA strands (quencher, quantum dot-labeled reporter, and the MUC1 aptamer stem), which allowed a strong fluorescence in the absence of the analyte. In the presence of MUC1 peptides, the fluorescence intensity decreased successively since the structure switch of the aptamer strand while binding MUC1. In that way, the quencher and fluorescence reporter were brought into close proximity, which leads to the occurrence of FRET, between the quencher and quantum dot. The detection limit for MUC1 with this approach was in the nanomolar (nM) level, and a linear response could be established for the approximate range found in blood serum. The method offers the possibility of improving it for the early diagnosis of different types of epithelian cancers.

The most advanced aptamer in the cancer setting is AS1411 (formerly known as AGRO100)—an experimental anticancer drug that was not generated using the SELEX process [125,126]. AS1411 is being administered systemically in clinical trials. This aptamer is a member of a novel class of antiproliferative agents known as G-rich oligonucleotides (GRO). These are non-antisense, guanosine-rich phosphodiester oligodeoxynucleotides that form stable G-quadruplex structures. The biological activity of GROs results from their binding to specific cellular proteins. One important target protein of GROs has been previously identified as nucleolin, a multifunctional protein expressed at high levels by cancer cells [127]. Ko et al. [128] used AS1411 for development of a two-color visualization system for specific biomarker targeting by different quantum dot nanoparticle probes labeled with an aptamer and a targeting peptide. The so-called fluorescence derby imaging used dual-color quantum dots conjugated by the AS1411 aptamer (targeting nucleolin) on one hand and the arginine-glycine-aspartic acid (targeting the integrin alpha(v)beta(3)) on the other. The simultaneous fluorescence imaging of cellular distribution of nucleolin and integrin using quantum dots enabled an easy monitoring of separate targets in cancer cells and in normal healthy cells. These results suggest the feasibility of a concurrent visualization of quantum dot-based multiple cancer biomarkers using small molecules such as aptamer or peptide ligands.

Aptamers combined with nanoparticles hold great potential for an effective detection of cancer even in an early stadium of the disease. However, there are only few aptamers available for different types of cancer cells currently. By development of further aptamers specific for cancer cell lines and markers, it will be possible to expand this field enormously.

2.7 Biomedical *In Vivo* Imaging

Aptamers were first introduced as imaging probes for *in vivo* studies in 1997, when Charlton et al. [129] used an aptamer selected against human neutrophil elastase labeled with a metastable isotope (^{99}Tc) for imaging inflammation processes in a rat model. This aptamer showed a higher signal-to-background ratio than its antibody counterpart, demonstrating the potential applications of aptamers as imaging probes for *in vivo* studies. Taking the advantages of rapidly expanding nanobiotechnology-based developments, aptamer–nanoparticle conjugation forms the basis of a new chemical and biological strategy for *in vivo* imaging. Because of their small size, nanoparticles can interact readily with biomolecules both on surface and within the cells. When conjugated with biomolecular affinity ligands, such as aptamers, they are considered as a revolutionary approach for detection of various diseases that can be combined with directed therapy strategies.

Targeted metallic nanoparticles modified with aptamers as targeting agents have shown potential as a platform for development of molecular-specific contrast agents. The study of Javier et al. [41] investigated the development of aptamer-based gold nanoparticles as contrast agents. The group devised a novel conjugation approach using an extended aptamer design where the extension was complementary to an oligonucleotide sequence attached to the surface of the gold nanoparticles. This conjugation approach was applied to create a contrast agent designed for the detection of PSMA, obtaining reflectance images of PSMA-positive and PSMA-negative cell lines treated with the anti-PSMA aptamer–gold conjugates. Recently, Kim et al. [130] presented a multifunctional drug-loaded aptamer–gold nanoparticle bioconjugate for combined computed tomography (CT) imaging and therapy of prostate cancer. By functionalizing the surface of gold nanoparticles with a PSMA-specific RNA aptamer [121], a targeted molecular CT imaging system was established which was capable of specific imaging of PSMA-positive prostate cancer cells with great specificity.

Zheng et al. [43] created a novel composite nanomaterial, termed aptamer nanoflares, that was used for a direct quantification of an intracellular analyte in a living cell. Aptamer nanoflares are gold nanoparticles functionalized with thiol-terminated aptamer sequences hybridized to a short complementary Cy5-labeled reporter strand. In the absence of the target, the gold nanoparticles are assembled and the Cy5 dye is quenched by the gold core. Upon binding the aptamer target, the Cy5-labeled reporter strand (flare) is released by a conformation change of the aptamer structure, resulting in an increase of fluorescence intensity. By using an aptamer for adenosine triphosphate (ATP) [50], the nanoflares readily entered cells, and were able to detect cytoplasmic ATP changes at physiologically relevant concentrations (0.1–3 mM), as observed by confocal microscopy and flow cytometry.

Aptamer-adapted silver nanoparticles were developed as a novel optical probe for simultaneous intracellular protein imaging and single-nanoparticle spectral analysis, wherein silver nanoparticles act as an illuminophore and the aptamer as a biomolecule specific recognition unit, respectively [131]. It was found that streptavidin-conjugated and aptamer-functionalized silver nanoparticles showed satisfactory biocompatibility and stability in cell culture medium and thus not only can act as a high-contrast imaging agent for both dark-field light scattering microscope and transmission electron microscopy (TEM) imaging but also can inspire supersensitive single-nanoparticle spectra for potential intercellular microenvironment analysis. The authors employed the nanomaterial for investigations of prion disease, a progressive neurodegenerative disorder. An imaging of the intracellular endocytic pathways of PrP in living cells became possible, and it has been shown that caveolae-related endocytosis is likely to convey a pathway for aptamer silver nanoparticles labeled PrP internalization in human bone marrow neuroblastoma cells.

To overcome limitations in targeted cell labeling regarding molecule size and instability of the detection molecules, Zhou et al. [132] generated a new type of sub-10 nm multifunctional nanomaterial by using the CCRF-CEM aptamer [23] and a dendrimer that additionally had the capacity to carry multiple fluorophore molecules. This hybrid DNA aptamer–dendrimer nanomaterial was applied for labeling acute leukemia cells. The results of binding studies by flow cytometry and fluorescence imaging microscopy showed high binding affinity and specificity of the constructed nanomaterial. Because of the very small size of the created aptamer–dendrimers, the authors assume a principal applicability as contrast agents for specific *in vivo* cancer imaging.

Most of these promising studies have been performed on the basis of *in vitro* cell assays. The *in vivo* functionality of these novel multifunctional nanoparticles has to be proven by further studies.

Nevertheless, there is an actual work that describes *in vivo* effects of aptamer–nanoparticles applied for cancer visualization in mouse models [133]. The AS1411 aptamer specific to nucleolin protein was conjugated to cobalt-ferrite nanoparticles surrounded by fluorescent rhodamine within a silica shell and to gallium-67 (^{67}Ga). These multimodal nanoparticles were administrated by intravenous injection into tumor-bearing nude mice, and their biodistribution was analyzed (Figure 2.5). The conjugates showed rapid blood clearance and accumulation in the tumor site as observed by scintigraphic images and MRI. Furthermore, accumulation of the conjugate was corroborated by fluorescence imaging of the tumor after organ extraction. However, the conjugate was also shown to accumulate nonspecific in the liver and intestine, which may be due to the size of the particles. But, one can foresee the promise of this technology once particle size can be optimizing to minimize nonspecific uptake.

2.8 Aptamer–Nanomaterial Conjugates as Smart Drug Delivery Agents

Apart from their use as sensing platform for bioanalysis discussed previously, aptamer–nanomaterial conjugates have also been applied in targeted drug delivery. When used as carriers bound with cargo molecules, such as drugs or functional proteins, either covalently or noncovalently. Covalent modifications of cargoes are generally achieved through standard gold–thiol chemistry, peptide bond formation, or similar methods. Electrostatic adsorption and hydrophobic interaction schemes are commonly employed to introduce cargo molecules noncovalently. Regardless of the methods for cargo molecule introduction, the biocompatibility of the conjugated nanomaterials themselves is considered as delivery vehicle for *in vivo* applications [74].

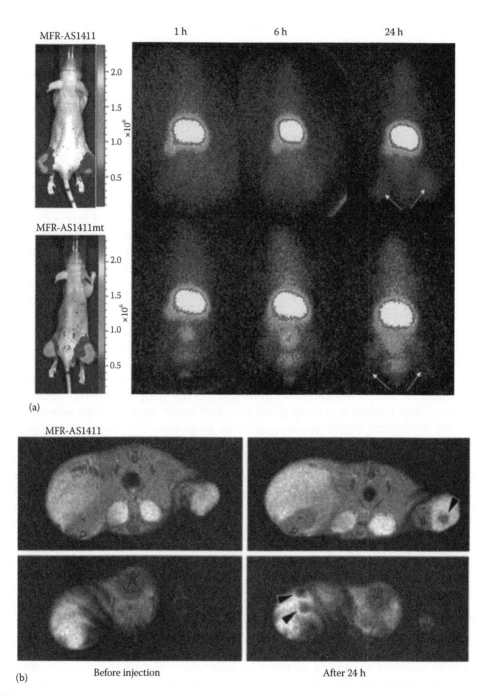

FIGURE 2.5
(See color insert.) *In vivo* multimodal cancer targeting and imaging using MFR-AS1411 particles. (a) MFR-AS1411 particles were intravenously injected into tumor-bearing mice, and radionuclide images were acquired at 1, 6, and 24 h after injection. Scintigraphic images of C6 tumors in mice that received MFR-AS1411 showed that C6 tumors had accumulated MFR-AS1411 at 24 h after injection but did not accumulate MFR-AS1411mt. Tumor growth patterns were followed using bioluminescence signals acquired from luciferase-expressing C6 cells. (b) MR images of tumor-bearing mice before and after injection of MFR-AS1411 were acquired. Dark signal intensities at tumor sites were detected in MFR-AS1411-injected mice (arrowhead).

MFR-AS1411mt   MFR-AS1411

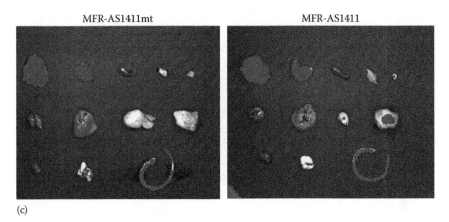

(c)

FIGURE 2.5 (continued)
(See color insert.) *In vivo* multimodal cancer targeting and imaging using MFR-AS1411 particles. (c) Tumors were isolated and their fluorescence verified using IVIS200 system. Fluorescence signal at tumor site injected with MFR-AS1411 was detected, compared with tumors injected with MFR-AS1411mt. Isolated organs in order from upper left to lower right were intestine, liver, spleen, muscle, fat, kidney, stomach, right tumor, left tumor, heart, lung, and tail. (Reprinted from Hwang do, W. et al., *J. Nucl. Med.*, 51, 98, 2010. With permission of the Society of Nuclear Medicine.)

Farokhzad and coworkers have extensively studied the use of the anti-PSMA A10 RNA aptamer, and its truncated version A10-3, to target nanoparticles. In a pioneering nanoparticle targeting study, they demonstrated that the A10-3 aptamer can be used to target poly(lactic acid)-block-polyethylene glycol (PEG) copolymer nanoparticles to PSMA positive prostate cancer cells. The resulting nanomaterial showed a 77-fold increase in binding to PSMA expressing prostate cancer cells in comparison to untargeted nanoparticles [134]. The A10-3 aptamer was again used to target modified poly(D,L-lactic-co-glycolic acid) (PLGA) nanoparticles to deliver docetaxel to prostate tumors *in vivo* where complete tumor regression was found in five out of seven mice after a single intratumoral injection. Moreover, all of the treated animals survived a 109 day study [135]. In the next step, the group optimized their protocol for production of PEGylated (PEG) PLGA nanoparticles and conjugated the resulting nanoparticles to the A10-3 aptamer and to docetaxel (and the related ^{14}C-paclitaxel). After systemic administration, the delivery of these nanoparticle conjugates to tumors was enhanced by 3.7-fold when compared to the nontargeted particles. Biodistribution patterns to the heart, lungs, and kidneys of the treated mice did not show substantial accumulation of nanoparticles. However, the presence of high aptamer surface density lead to increase in nanoparticle accumulation in liver and spleen. This was likely due to aptamer masking the PEG layers on the surfaces of the nanoparticles and compromising the nanoconjugates antibiofouling properties *in vivo*. Thus, in engineering targeted nanoparticles, it is to balance the tumor-targeting ligand surface density and the antibiofouling surface properties [136,137]. In a recent work, the Farokhzad group used a similar polymeric formulation to deliver cisplatin to PSMA expressing tumors by aptamer-functionalized Pt(IV) prodrug-PLGA-PEG nanoparticles. A dosage of 0.3 mg/kg of aptamer-targeted cisplatin nanoparticles was found to be more efficacious than a 1 mg/kg dosage of free cisplatin [138]. Furthermore, the same group demonstrated the feasibility of their systems for multiple drug therapy. Targeted dual-drug combination based on nanoparticles with hydrophobic docetaxel and hydrophilic Pt(IV) drug showed superior efficiency over single drug nanoparticle analogues or nontargeted nanoparticles [139].

In an additional study, they conjugated the anti-PSMA A10 aptamer with superparamagnetic iron oxide nanoparticles and with a doxorubicin cargo aiming dual functioning as combined prostate cancer imaging (with magnetic resonance) and therapy. The *in vitro* cytotoxicity assay showed that the nanoparticle-mediated doxorubicin delivery and the delivery of free doxorubicin are equally potent against PSMA-positive cancer cells. More importantly, the treatment with the aptamer-functionalized nanoparticles killed 47.5% of the PSMA-positive cells versus 22.8% of the PSMA-negative cells [140].

The results of *in vitro* imaging of cancer cells [119,124,128] indicate that aptamer-conjugated quantum dots have potential to be used for imaging and protein expression profiling of living cells and fixed tissue, as well as *in vivo* studies and drug delivery, although their use for the latter application may be limited by the relatively high cytotoxicity of quantum dots [75,141]. However, there are intentions for using this material for *in vivo*. For example, Savla et al. [142] attempted to use the MUC1 aptamers [123] to target quantum dots with a doxorubicin cargo for imaging and for the chemotherapy of ovarian cancer. Doxorubicin was attached to the quantum dots via a pH-sensitive hydrazone bond in order to provide the stability of the complex in systemic circulation and drug release in acidic environment inside cancer cells. In a mouse model bearing human ovarian cancer xenografts, more MUC1–aptamer quantum dots accumulated in the tumors when compared to nonmodified quantum dots. Ex vivo analysis of organs confirmed higher uptake in tumor and lower uptake in other organs. Data obtained demonstrate a high potential of targeted quantum dot conjugates in the treatment of cancer.

Besides other carriers, such as micelles and nanogels, aptamer-based drug delivery via nanoparticles hold the promise to represent new trends in the large vast of specific therapeutic applications and may contribute to the development of the next generation of nanoscale diagnostic and therapeutic modalities.

2.9 Summary and Outlook

Nanoparticles coupled with aptamers have been used for the detection of medical analytes that are mostly metal and magnetic nanoparticles, silica- or polymer-based nanoparticles, carbon nanotubes, and quantum dots. Gold nanoparticles offer unique properties that make them ideal detection materials. Particularly the color change caused by their aggregation or disaggregation and connected with the aptamer/target binding reaction has been shown to be used in the development of assays and tests. These changes in colors can be processed even by the naked eye without any device [51]. This application is especially user-friendly and in principle comparable to the well-known pregnancy test. Point-of-care diagnostics based on this principle will easily find the way soon to commercialization.

Besides gold nanoparticles, silver and platinum nanoparticles are used for optical, electrochemical, or SERS. For combined diagnostic and medical treatment, the nanoparticles have to be loaded with the drug. For this application, silica- or polymer-based nanoparticles are used. They can include additionally organic dyes for optical detection just as functional groups for their conjugation with aptamers [74,75].

Aptamer-modified magnetic nanoparticles offer the possibility of MRI as well as application of magnetic fluid hypothermia therapy [86]. Carbon nanotubes feature especially high mechanical stability, electrical conductivity, quenching property, and high surface area for loading with biomolecules. Screen- or inkjet-printed surfaces, also with transparency for optical detection, can be realized with carbon nanotubes.

Quantum dots broaden optical detection methods in various directions. They are particularly suitable for quantum dot-stimulated FRET-based (beacon) aptamer assays [75].

The huge potential of aptamer–nanoparticle-based detection of cancer cells and biomedical *in vivo* imaging has been shown with some examples. The development of more aptamers specific for cancer cells or marker proteins is urgently needed especially for these applications.

Aptamer-modified nanoparticles for medical applications are developed mostly for the detection of analytes, for biomedical *in vivo* imaging and for *in vivo* drug delivering.

Merging nanoparticles with aptamers permits additional or enhanced medical tools for many biomedical applications by combination of both complementary outstanding properties. Nanoparticles that are qualified for detection and loading with drugs can be directed to the site of their action by use of the affinity and specificity of aptamers that are conjugated to the nanoparticles or fixed on their surface. This is of high relevance particularly in cancer therapy. Most anticancer pharmaceuticals have destructive effects not only to the cancer cells but also to healthy cells. Aptamers can facilitate the cell-specific drug delivering in concise selective manner to the sick cancer cells because of their specificity binding. This can enhance therapeutic effects and diminish side effects. Moreover, simultaneous *in vivo* detection and therapy for cancer cells lowers burden of cancer patients.

Methods for aptamer-based nanoparticle application were developed but mostly by use of some few aptamers. The development of further specific aptamers is absolutely necessary. This is the precondition to provide these new detection, imaging, and therapy methods.

References

1. Ellington, A. On aptamers. http://ellingtonlab.org/blog/2011/03/06/on-aptamers/ (accessed on January 29, 2013).
2. Tuerk, C. and L. Gold. 1990. Systematic evolution of ligands by exponential enrichment: RNA ligands to bacteriophage T4 DNA polymerase. *Science* 249: 505–510.
3. Robertson, D. L. and G. F. Joyce. 1990. Selection in vitro of an RNA enzyme that specifically cleaves single-stranded DNA. *Nature* 344: 467–468, doi:10.1038/344467a0.
4. Ellington, A. D. and J. W. Szostak. 1990. In vitro selection of RNA molecules that bind specific ligands. *Nature* 346: 818–822.
5. Hermann, T. and D. J. Patel. 2000. Adaptive recognition by nucleic acid aptamers. *Science* 287: 820–825.
6. Ellington, A. D. and J. W. Szostak. 1992. Selection in vitro of single-stranded DNA molecules that fold into specific ligand-binding structures. *Nature* 355: 850–852.
7. Green, L. S. et al. 1995. Nuclease-resistant nucleic-acid ligands to vascular-permeability factor vascular endothelial growth-factor. *Chem. Biol.* 2: 683–695.
8. Gold, L., B. Polisky, O. Uhlenbeck, and M. Yarus. 1995. Diversity of oligonucleotide functions. *Annu. Rev. Biochem.* 64: 763–797.
9. Klussmann, S. 2006. *The Aptamer Handbook. Functional Oligonucleotides and Their Applications.* Wiley-VCH Verlag GmbH & Co. KGaA, Weinheim, Germany.
10. Klussmann, S., A. Nolte, R. Bald, V. A. Erdmann, and J. P. Furste. 1996. Mirror-image RNA that binds D-adenosine. *Nat. Biotechnol.* 14: 1112–1115.

11. NOXXON Pharma AG. http://www.noxxon.com (accessed on January 29, 2013).
12. Ozalp, V. C., Eyidogan, F., and H.A. Oktem. 2011. Aptamer-gated nanoparticles for smart drug delivery. *Pharmaceuticals* 4: 1137–1157.
13. Jenison, R. D., S. C. Gill, A. Pardi, and B. Polisky. 1994. High-resolution molecular discrimination by RNA. *Science* 263: 1425–1429.
14. Michaud, M. et al. 2003. A DNA aptamer as a new target-specific chiral selector for HPLC. *J. Am. Chem. Soc.* 125: 8672–8679.
15. Reinemann, C., R. Stoltenburg, and B. Strehlitz. 2009. Investigations on the specificity of DNA aptamers binding to ethanolamine. *Anal. Chem.* 81: 3973–3978, doi:10.1021/ac900305y.
16. Stoltenburg, R., C. Reinemann, and B. Strehlitz. 2007. SELEX-A (r)evolutionary method to generate high-affinity nucleic acid ligands. *Biomol. Eng.* 24: 381–403.
17. James, W. and R. A. Meyers. 2000. Aptamers. In *Encyclopedia of Analytical Chemistry*, pp. 4848–4871. John Wiley & Sons, Ltd., New York.
18. Schütze, T. et al. 2011. Probing the SELEX process with next-generation sequencing. *PLoS One* 6: e29604. doi:10.1371/journal.pone.0029604.
19. Daniels, D. A., H. Chen, B. J. Hicke, K. M. Swiderek, and L. Gold. 2003. A tenascin-C aptamer identified by tumor cell SELEX: Systematic evolution of ligands by exponential enrichment. *Proc. Natl. Acad. Sci. USA* 100: 15416–15421.
20. Sefah, K., D. Shangguan, X. L. Xiong, M. B. O'Donoghue, and W. H. Tan. 2010. Development of DNA aptamers using Cell-SELEX. *Nat. Protoc.* 5: 1169–1185, doi:DOI 10.1038/nprot.2010.66.
21. Morris, K. N., K. B. Jensen, C. M. Julin, M. Weil, and L. Gold. 1998. High affinity ligands from in vitro selection: Complex targets. *Proc. Natl. Acad. Sci. USA* 95: 2902–2907.
22. Ohuchi, S. P., T. Ohtsu, and Y. Nakamura. 2006. Selection of RNA aptamers against recombinant transforming growth factor-beta type III receptor displayed on cell surface. *Biochimie* 88: 897–904.
23. Shangguan, D. et al. 2006. Aptamers evolved from live cells as effective molecular probes for cancer study. *Proc. Natl. Acad. Sci. USA* 103: 11838–11843.
24. Vallian, S. and M. R. Khazaei. 2007. Medical applications of aptamers. *Res. Pharm. Sci.* 2: 59–66.
25. Carlson, B. 2007. Aptamers: The new frontier in drug development? *Biotechnol. Healthcare*: 31–35.
26. Tucker, C. E. et al. 1999. Detection and plasma pharmacokinetics of an anti-vascular endothelial growth factor oligonucleotide-aptamer (NX1838) in rhesus monkeys. *J. Chromatogr. B* 732: 203–212.
27. Lee, J. F., G. M. Stovall, and A. D. Ellington. 2006. Aptamer therapeutics advance. *Curr. Opin. Chem. Biol.* 10: 282–289.
28. Chapman, J. A. and C. Beckey. 2006. Pegaptanib: A novel approach to ocular neovascularization. *Ann. Pharmacother.* 40: 1322–1326.
29. Maberley, D. 2005. Pegaptanib for neovascular age-related macular degeneration. *Issues Emerg. Health Technol.* 76: 1–4.
30. Siddiqui, M. A. A. and G. M. Keating. 2005. Pegaptanib—In exudative age-related macular degeneration. *Drugs* 65: 1571–1577.
31. Ni, X., M. Castanares, A. Mukherjee, and S. E. Lupold. 2011. Nucleic acid aptamers: Clinical applications and promising new horizons. *Curr. Med. Chem.* 18: 4206–4214.
32. Dyke, C. K. et al. 2006. First-in-human experience of an antidote-controlled anticoagulant using RNA aptamer technology—A phase 1a pharmacodynamic evaluation of a drug-antidote pair for the controlled regulation of factor IXa activity. *Circulation* 114: 2490–2497.
33. Gilch, S., C. Kehler, and H. M. Schatzl. 2007. Peptide aptamers expressed in the secretory pathway interfere with cellular PrPSc formation. *J. Mol. Biol.* 371: 362–373, doi:DOI 10.1016/j.jmb.2007.05.052.
34. Brody, E. N. et al. 1999. The use of aptamers in large arrays for molecular diagnostics. *Mol. Diagn.* 4: 381–388.
35. Lee, M. and D. R. Walt. 2000. A fiber-optic microarray biosensor using aptamers as receptors. *Anal. Biochem.* 282: 142–146.

36. Liss, M., B. Petersen, H. Wolf, and E. Prohaska. 2002. An aptamer-based quartz crystal protein biosensor. *Anal. Chem.* 74: 4488–4495.
37. Cho, E. J., J. R. Collett, A. E. Szafranska, and A. D. Ellington. 2006. Optimization of aptamer microarray technology for multiple protein targets. *Anal. Chim. Acta* 564: 82–90.
38. Daniel, M. C. and D. Astruc. 2004. Gold nanoparticles: Assembly, supramolecular chemistry, quantum-size-related properties, and applications toward biology, catalysis, and nanotechnology. *Chem. Rev.* 104: 293–346, doi:10.1021/cr030698.
39. Rosi, N. L. et al. 2006. Oligonucleotide-modified gold nanoparticles for intracellular gene regulation. *Science* 312: 1027–1030, doi:10.1126/science.1125559.
40. Sperling, R. A., P. Rivera Gil, F. Zhang, M. Zanella, and W. J. Parak. 2008. Biological applications of gold nanoparticles. *Chem. Soc. Rev.* 37: 1896–1908, doi:10.1039/b712170a.
41. Javier, D. J., N. Nitin, M. Levy, A. Ellington, and R. Richards-Kortum. 2008. Aptamer-targeted gold nanoparticles as molecular-specific contrast agents for reflectance imaging. *Bioconjug. Chem.* 19: 1309–1312, doi:10.1021/bc8001248.
42. Li, N., T. Larson, H. H. Nguyen, K. V. Sokolov, and A. D. Ellington. 2010. Directed evolution of gold nanoparticle delivery to cells. *Chem. Commun. (Camb)* 46: 392–394, doi:10.1039/b920865h.
43. Zheng, D., D. S. Seferos, D. A. Giljohann, P. C. Patel, and C. A. Mirkin. 2009. Aptamer nano-flares for molecular detection in living cells. *Nano. Lett.* 9: 3258–3261, doi:10.1021/nl901517b.
44. Mirkin, C. A., R. L. Letsinger, R. C. Mucic, and J. J. Storhoff. 1996. A DNA-based method for rationally assembling nanoparticles into macroscopic materials. *Nature* 382: 607–609, doi:10.1038/382607a0.
45. Huang, C. C., Y. F. Huang, Z. Cao, W. Tan, and H. T. Chang. 2005. Aptamer-modified gold nanoparticles for colorimetric determination of platelet-derived growth factors and their receptors. *Anal. Chem.* 77: 5735–5741, doi:10.1021/ac050957q.
46. Green, L. S. et al. 1996. Inhibitory DNA ligands to platelet-derived growth factor B-chain. *Biochemistry* 35: 14413–14424.
47. Liu, J. and Y. Lu. 2005. Fast colorimetric sensing of adenosine and cocaine based on a general sensor design involving aptamers and nanoparticles. *Angew. Chem. Int. Ed. Engl.* 45: 90–94, doi:10.1002/anie.200502589.
48. Stojanovic, M. N., P. de Prada, and D. W. Landry. 2000. Fluorescent sensors based on aptamer self-assembly. *J. Am. Chem. Soc.* 122: 11547–11548.
49. Stojanovic, M. N., P. de Prada, and D. W. Landry. 2001. Aptamer-based folding fluorescent sensor for cocaine. *J. Am. Chem. Soc.* 123: 4928–4931.
50. Huizenga, D. E. and J. W. Szostak. 1995. A DNA aptamer that binds adenosine and ATP. *Biochemistry* 34: 656–665.
51. Liu, J., D. Mazumdar, and Y. Lu. 2006. A simple and sensitive "dipstick" test in serum based on lateral flow separation of aptamer-linked nanostructures. *Angew. Chem. Int. Ed. Engl.* 45: 7955–7959, doi:10.1002/anie.200603106.
52. Wang, Y. et al. 2008. Ultrasensitive colorimetric detection of protein by aptamer-Au nanoparticles conjugates based on a dot-blot assay. *Chem. Commun. (Camb)* 22: 2520–2522, doi:10.1039/b801055b.
53. Bock, L. C., L. C. Griffin, J. A. Latham, E. H. Vermaas, and J. J. Toole. 1992. Selection of single-stranded DNA molecules that bind and inhibit human thrombin. *Nature* 355: 564–566.
54. Wei, H., B. Li, J. Li, E. Wang, and S. Dong. 2007. Simple and sensitive aptamer-based colorimetric sensing of protein using unmodified gold nanoparticle probes. *Chem. Commun. (Camb)* 36: 3735–3737, doi:10.1039/b707642h.
55. Li, Y. Y. et al. 2007. Ultrasensitive densitometry detection of cytokines with nanoparticle-modified aptamers. *Clin. Chem.* 53: 1061–1066.
56. He, P., L. Shen, Y. Cao, and D. Li. 2007. Ultrasensitive electrochemical detection of proteins by amplification of aptamer-nanoparticle bio bar codes. *Anal. Chem.* 79: 8024–8029, doi:10.1021/ac070772e.
57. Wang, J., A. Munir, Z. Li, and H. S. Zhou. 2009. Aptamer-Au NPs conjugates-enhanced SPR sensing for the ultrasensitive sandwich immunoassay. *Biosens. Bioelectron.* 25: 124–129, doi:S0956-5663(09)00329-7.

58. Wiegand, T. W. et al. 1996. High-affinity oligonucleotide ligands to human IgE inhibit binding to Fc epsilon receptor I. *J. Immunol.* 157: 221–230.
59. Bae, D. R. et al. 2010. Lysine-functionalized silver nanoparticles for visual detection and separation of histidine and histidine-tagged proteins. *Langmuir* 26: 2181–2185, doi:10.1021/la9026865.
60. Guerrini, L., J. V. Garcia-Ramos, C. Domingo, and S. Sanchez-Cortes. 2009. Nanosensors based on viologen functionalized silver nanoparticles: Few molecules surface-enhanced Raman spectroscopy detection of polycyclic aromatic hydrocarbons in interparticle hot spots. *Anal. Chem.* 81: 1418–1425, doi:10.1021/ac8021746.
61. Lee, K. S. and M. A. El-Sayed. 2006. Gold and silver nanoparticles in sensing and imaging: sensitivity of plasmon response to size, shape, and metal composition. *J Phys. Chem. B* 110: 19220–19225, doi:10.1021/jp062536y.
62. Serra, A. et al. 2009. Non-functionalized silver nanoparticles for a localized surface plasmon resonance-based glucose sensor. *Nanotechnology* 20: 165501, doi:10.1088/0957–4484/20/16/165501.
63. Zhang, Y., K. Zhang, and H. Ma. 2009. Electrochemical DNA biosensor based on silver nanoparticles/poly(3-(3-pyridyl) acrylic acid)/carbon nanotubes modified electrode. *Anal. Biochem.* 387: 13–19, doi:10.1016/j.ab.2008.10.043.
64. Pagba, C. V., S. M. Lane, H. Cho, and S. Wachsmann-Hogiu. 2010. Direct detection of aptamer-thrombin binding via surface-enhanced Raman spectroscopy. *J. Biomed. Opt.* 15: 047006, doi:10.1117/1.3465594.
65. Chu, X., D. Duan, G. Shen, and R. Yu. 2007. Amperometric glucose biosensor based on electro-deposition of platinum nanoparticles onto covalently immobilized carbon nanotube electrode. *Talanta* 71: 2040–2047, doi:10.1016/j.talanta.2006.09.013.
66. Evans, S. A. et al. 2002. Detection of hydrogen peroxide at mesoporous platinum microelectrodes. *Anal. Chem.* 74: 1322–1326.
67. Hrapovic, S., Y. Liu, and J. H. Luong. 2007. Reusable platinum nanoparticle modified boron doped diamond microelectrodes for oxidative determination of arsenite. *Anal. Chem.* 79: 500–507, doi:10.1021/ac061528a.
68. Hrapovic, S., Y. Liu, K. B. Male, and J. H. Luong. 2004. Electrochemical biosensing platforms using platinum nanoparticles and carbon nanotubes. *Anal. Chem.* 76: 1083–1088, doi:10.1021/ac035143t.
69. Polsky, R., R. Gill, L. Kaganovsky, and I. Willner. 2006. Nucleic acid-functionalized Pt nanoparticles: Catalytic labels for the amplified electrochemical detection of biomolecules. *Anal. Chem.* 78: 2268–2271, doi:10.1021/ac0519864.
70. Higuchi, A. et al. 2008. Preparation of a DNA aptamer-Pt complex and its use in the colorimetric sensing of thrombin and anti-thrombin antibodies. *Anal. Chem.* 80: 6580–6586, doi:10.1021/ac8006957.
71. Arriagada, F. J. and K. Osseo-Asare. 1999. Synthesis of nanosize silica in a nonionic water-in-oil microemulsion: Effects of the water/surfactant molar ratio and ammonia concentration. *J. Colloid Interface Sci.* 211: 210–220, doi:10.1006/jcis.1998.5985.
72. DiRenzo, F., H. Cambon, and R. Dutartre. 1997. A 28-year-old synthesis of micelle-templated mesoporous silica. *Microporous Mater.* 10: 283–286.
73. Tapec, R., X. J. Zhao, and W. Tan. 2002. Development of organic dye-doped silica nanoparticles for bioanalysis and biosensors. *J. Nanosci. Nanotechnol.* 2: 405–409.
74. Chen, T. et al. 2011. Aptamer-conjugated nanomaterials for bioanalysis and biotechnology applications. *Nanoscale* 3: 546–556, doi:10.1039/c0nr00646g.
75. Lopez-Colon, D., E. Jimenez, M. You, B. Gulbakan, and W. Tan. 2011. Aptamers: Turning the spotlight on cells. *Wiley Interdiscip. Rev. Nanomed. Nanobiotechnol.* 3: 328–340, doi:10.1002/wnan.133.
76. Wang, Y. and B. Liu. 2008. Fluorescent detection of ATP based on signaling DNA aptamer attached silica nanoparticles. *Nanotechnology* 19: 415605.
77. Kirby, R. et al. 2004. Aptamer-based sensor arrays for the detection and quantitation of proteins. *Anal. Chem.* 76: 4066–4075.

78. Cox, J. C. and A. D. Ellington. 2001. Automated selection of anti-Protein aptamers. *Bioorg. Med. Chem.* 9: 2525–2531.
79. Wang, Y., K. Y. Pu, and B. Liu. 2010. Anionic conjugated polymer with aptamer-functionalized silica nanoparticle for label-free naked-eye detection of lysozyme in protein mixtures. *Langmuir* 26: 10025–10030, doi:10.1021/la100139p.
80. Lin, Z., L. Chen, X. Zhu, B. Qiu, and G. Chen. 2010. Signal-on electrochemiluminescence biosensor for thrombin based on target-induced conjunction of split aptamer fragments. *Chem. Commun. (Camb)* 46: 5563–5565, doi:10.1039/c0cc00932f.
81. Teja, A. S. and P. Y. Koh. 2009. Synthesis, properties, and applications of magnetic iron oxide nanoparticles. *Prog. Cryst. Growth Char.* 55: 22–45, doi:DOI 10.1016/j.pcrysgrow.2008.08.003.
82. Berry, C. C. 2009. Progress in functionalization of magnetic nanoparticles for applications in biomedicine. *J. Phys. D Appl. Phys.* 42.
83. Lacroix, L. M., D. Ho, and S. H. Sun. 2010. Magnetic nanoparticles as both imaging probes and therapeutic agents. *Curr. Top. Med. Chem.* 10: 1184–1197.
84. Nune, S. K. et al. 2009. Nanoparticles for biomedical imaging. *Expert Opin. Drug Deliv.* 6: 1175–1194.
85. Tanaka, K., N. Kitamura, and Y. Chujo. 2008. Properties of superparamagnetic iron oxide nanoparticles assembled on nucleic acids. *Nucleic Acids Symp. Ser. (Oxf)* 2008(52): 693–694.
86. McCarthy, J. R. and R. Weissleder. 2008. Multifunctional magnetic nanoparticles for targeted imaging and therapy. *Adv. Drug Deliv. Rev.* 60: 1241–1251, doi:10.1016/j.addr.2008.03.014.
87. Josephson, L., J. Lewis, P. Jacobs, P. F. Hahn, and D. D. Stark. 1988. The effects of iron-oxides on proton relaxivity. *Magn. Reson. Imaging* 6: 647–653.
88. Yigit, M. V. et al. 2007. Smart "turn-on" magnetic resonance contrast agents based on aptamer-functionalized superparamagnetic iron oxide nanoparticles. *ChemBioChem* 8: 1675–1678, doi:10.1002/cbic.200700323.
89. Yigit, M. V., D. Mazumdar, and Y. Lu. 2008. MRI detection of thrombin with aptamer functionalized superparamagnetic iron oxide nanoparticles. *Bioconjug. Chem.* 19: 412–417, doi:10.1021/bc7003928.
90. Bamrungsap, S., M. I. Shukoor, T. Chen, K. Sefah, and W. Tan. 2011. Detection of lysozyme magnetic relaxation switches based on aptamer-functionalized superparamagnetic nanoparticles. *Anal. Chem.* 83: 7795–7799. doi:10.1021/ac201442a.
91. Wang, J., A. Munir, Z. Zhu, and H. S. Zhou. 2010. Magnetic nanoparticle enhanced surface plasmon resonance sensing and its application for the ultrasensitive detection of magnetic nanoparticle-enriched small molecules. *Anal. Chem.* 82: 6782–6789, doi:10.1021/ac100812c.
92. Kim, S. N., J. F. Rusling, and F. Papadimitrakopoulos. 2007. Carbon nanotubes for electronic and electrochemical detection of biomolecules. *Adv. Mater.* 19: 3214–3228.
93. Yang, R. et al. 2008. Noncovalent assembly of carbon nanotubes and single-stranded DNA: An effective sensing platform for probing biomolecular interactions. *Anal. Chem.* 80: 7408–7413, doi:10.1021/ac801118p.
94. Zhu, Z. et al. 2008. Regulation of singlet oxygen generation using single-walled carbon nanotubes. *J. Am. Chem. Soc.* 130: 10856–10857, doi:10.1021/ja802913f.
95. Chen, H. et al. 2009. A novel near-infrared protein assay based on the dissolution and aggregation of aptamer-wrapped single-walled carbon nanotubes. *Chem. Commun. (Camb)* 33: 5006–5008, doi:10.1039/b910457g.
96. Ouyang, X. et al. 2011. New strategy for label-free and time-resolved luminescent assay of protein: Conjugate Eu3+ complex and aptamer-wrapped carbon nanotubes. *Anal. Chem.* 83: 782–789, doi:10.1021/ac103087z.
97. Duzgun, A., A. Maroto, T. Mairal, C. O'Sullivan, and F. X. Rius. 2010. Solid-contact potentiometric aptasensor based on aptamer functionalized carbon nanotubes for the direct determination of proteins. *Analyst* 135: 1037–1041, doi:10.1039/b926958d.
98. Maehashi, K. et al. 2007. Label-free protein biosensor based on aptamer-modified carbon nanotube field-effect transistors. *Anal. Chem.* 79: 782–787, doi:10.1021/ac060830g.

99. Kara, P. et al. 2010. Aptamers based electrochemical biosensor for protein detection using carbon nanotubes platforms. *Biosens. Bioelectron.* 26: 1715–1718.
100. Levy, M., S. F. Cater, and A. D. Ellington. 2005. Quantum-dot aptamer beacons for the detection of proteins. *Chembiochem* 6: 2163–2166.
101. Li, J. J., X. Fang, and W. Tan. 2002. Molecular aptamer beacons for real-time protein recognition. *Biochem. Biophys. Res. Commun.* 292: 31–40.
102. Freeman, R. et al. 2009. Self-assembly of supramolecular aptamer structures for optical or electrochemical sensing. *Analyst* 134: 653–656, doi:10.1039/b822836c.
103. Xiao, S. J. et al. 2010. Sensitive discrimination and detection of prion disease-associated isoform with a dual-aptamer strategy by developing a sandwich structure of magnetic microparticles and quantum dots. *Anal. Chem.* 82: 9736–9742, doi:10.1021/ac101865s.
104. Bibby, D. F. et al. 2008. Application of a novel in vitro selection technique to isolate and characterise high affinity DNA aptamers binding mammalian prion proteins. *J. Virol. Methods* 151: 107–115, doi:S0166-0934(08)00101-8.
105. Takemura, K. et al. 2006. DNA aptamers that bind to PrPC and not PrPSc show sequence and structure specificity. *Exp. Biol. Med.* 231: 204–214.
106. Lee, C. C., J. A. MacKay, J. M. J. Frechet, and F. C. Szoka. 2005. Designing dendrimers for biological applications. *Nat. Biotechnol.* 23: 1517–1526.
107. Zhang, Z. et al. 2009. A sensitive impedimetric thrombin aptasensor based on polyamidoamine dendrimer. *Talanta* 78: 1240–1245.
108. Phillips, J. A., D. Lopez-Colon, Z. Zhu, Y. Xu, and W. Tan. 2008. Applications of aptamers in cancer cell biology. *Anal. Chim. Acta* 621: 101–108.
109. Fang, X. and W. Tan. 2010. Aptamers generated from cell-SELEX for molecular medicine: A chemical biology approach. *Acc. Chem. Res.* 43: 48–57, doi:10.1021/ar900101s.
110. Shangguan, D. H., Z. H. C. Cao, Y. Li, and W. H. Tan. 2007. Aptamers evolved from cultured cancer cells reveal molecular differences of cancer cells in patient samples. *Clin. Chem.* 53: 1153–1155.
111. Mallikaratchy, P. et al. 2007. Aptamer directly evolved from live cells recognizes membrane bound immunoglobin heavy mu chain in Burkitt's lymphoma cells. *Mol. Cell Proteomics* 6: 2230–2238.
112. Tang, Z. W. et al. 2007. Selection of aptamers for molecular recognition and characterization of cancer cells. *Anal. Chem.* 79: 4900–4907.
113. Medley, C. D. et al. 2008. Gold nanoparticle-based colorimetric assay for the direct detection of cancerous cells. *Anal. Chem.* 80: 1067–1072, doi:10.1021/ac702037y.
114. Liu, G. et al. 2009. Aptamer-nanoparticle strip biosensor for sensitive detection of cancer cells. *Anal. Chem.* 81: 10013–10018, doi:10.1021/ac901889s.
115. Huang, Y. F., H. T. Chang, and W. Tan. 2008. Cancer cell targeting using multiple aptamers conjugated on nanorods. *Anal. Chem.* 80: 567–572, doi:10.1021/ac702322j.
116. Ding, C., Y. Ge, and S. Zhang. 2010. Electrochemical and electrochemiluminescence determination of cancer cells based on aptamers and magnetic beads. *Chemistry* 16: 10707–10714, doi:10.1002/chem.201001173.
117. Chen, X. et al. 2009. Using aptamer-conjugated fluorescence resonance energy transfer nanoparticles for multiplexed cancer cell monitoring. *Anal. Chem.* 81: 7009–7014, doi:10.1021/ac9011073.
118. Herr, J. K., J. E. Smith, C. D. Medley, D. Shangguan, and W. Tan. 2006. Aptamer-conjugated nanoparticles for selective collection and detection of cancer cells. *Anal. Chem.* 78: 2918–2924, doi:10.1021/ac052015r.
119. Jie, G., L. Wang, J. Yuan, and S. Zhang. 2011. Versatile electrochemiluminescence assays for cancer cells based on dendrimer/CdSe-ZnS-quantum dot nanoclusters. *Anal. Chem.* 83: 3873–3880, doi:10.1021/ac200383z.
120. Lu, W. et al. 2010. Multifunctional oval-shaped gold-nanoparticle-based selective detection of breast cancer cells using simple colorimetric and highly sensitive two-photon scattering assay. *ACS Nano.* 4: 1739–1749, doi:10.1021/nn901742q.

121. Lupold, S. E., B. J. Hicke, Y. Lin, and D. S. Coffey. 2002. Identification and characterization of nuclease-stabilized RNA molecules that bind human prostate cancer cells via the prostate-specific membrane antigen. *Cancer Res.* 62: 4029–4033.
122. Walter, J. G., S. Petersen, F. Stahl, T. Scheper, and S. Barcikowski. 2010. Laser ablation-based one-step generation and bio-functionalization of gold nanoparticles conjugated with aptamers. *J. Nanobiotechnol.* 8: 21, doi:1477–3155–8–21.
123. Ferreira, C. S., C. S. Matthews, and S. Missailidis. 2006. DNA aptamers that bind to MUC1 tumour marker: Design and characterization of MUC1-binding single-stranded DNA aptamers. *Tumour Biology:J. Int. Soc. Oncodev. Biology Med.* 27: 289–301, doi:10.1159/000096085.
124. Cheng, A. K., H. Su, Y. A. Wang, and H. Z. Yu. 2009. Aptamer-based detection of epithelial tumor marker mucin 1 with quantum dot-based fluorescence readout. *Anal. Chem.* 81: 6130–6139, doi:10.1021/ac901223q.
125. Girvan, A. C. et al. 2006. AGRO100 inhibits activation of nuclear factor-kappaB (NF-kappaB) by forming a complex with NF-kappaB essential modulator (NEMO) and nucleolin. *Mol. Cancer Ther.* 5: 1790–1799, doi:10.1158/1535–7163.MCT-05–0361.
126. Ireson, C. R. and L. R. Kelland. 2006. Discovery and development of anticancer aptamers. *Mol. Cancer Ther.* 5: 2957–2962.
127. Watanabe, T. et al. 2010. Nucleolin on the cell surface as a new molecular target for gastric cancer treatment. *Biol. Pharm. Bull.* 33: 796–803.
128. Ko, M. H. et al. 2009. In vitro derby imaging of cancer biomarkers using quantum dots. *Small* 5: 1207–1212, doi:10.1002/smll.200801580.
129. Charlton, J., J. Sennello, and D. Smith. 1997. In vivo imaging of inflammation using an aptamer inhibitor of human neutrophil elastase. *Chem. Biol.* 4: 809–816.
130. Kim, D., Y. Y. Jeong, and S. Jon. 2010. A drug-loaded aptamer-gold nanoparticle bioconjugate for combined CT imaging and therapy of prostate cancer. *ACS Nano* 4: 3689–3696, doi:10.1021/nn901877h.
131. Chen, L. Q. et al. 2010. Aptamer-based silver nanoparticles used for intracellular protein imaging and single nanoparticle spectral analysis. *J. Phys. Chem. B* 114: 3655–3659, doi:10.1021/jp9104618.
132. Zhou, J. et al. 2009. A hybrid DNA aptamer-dendrimer nanomaterial for targeted cell labeling. *Macromol. Biosci.* 9: 831–835, doi:10.1002/mabi.200900046.
133. Hwang do, W. et al. 2010. A nucleolin-targeted multimodal nanoparticle imaging probe for tracking cancer cells using an aptamer. *J. Nucl. Med.* 51: 98–105.
134. Farokhzad, O. C. et al. 2004. Nanoparticle-aptamer bioconjugates: A new approach for targeting prostate cancer cells. *Cancer Res.* 64: 7668–7672.
135. Farokhzad, O. C. et al. 2006. Targeted nanoparticle-aptamer bioconjugates for cancer chemotherapy in vivo. *Proc. Natl. Acad. Sci. USA* 103: 6315–6320.
136. Cheng, J. et al. 2007. Formulation of functionalized PLGA-PEG nanoparticles for in vivo targeted drug delivery. *Biomaterials* 28: 869–876.
137. Gu, F. et al. 2008. Precise engineering of targeted nanoparticles by using self-assembled biointegrated block copolymers. *Proc. Natl. Acad. Sci. USA* 105: 2586–2591.
138. Dhar, S., F. X. Gu, R. Langer, O. C. Farokhzad, and S. J. Lippard. 2008. Targeted delivery of cisplatin to prostate cancer cells by aptamer functionalized Pt(IV) prodrug-PLGA-PEG nanoparticles. *Proc. Natl. Acad. Sci. USA* 105: 17356–17361.
139. Kolishetti, N. et al. 2010. Engineering of self-assembled nanoparticle platform for precisely controlled combination drug therapy. *Proc. Natl. Acad. Sci. USA* 107: 17939–17944.
140. Wang, A. Z. et al. 2008. Superparamagnetic iron oxide nanoparticle-aptamer bioconjugates for combined prostate cancer imaging and therapy. *ChemMedChem* 3: 1311–1315, doi:10.1002/cmdc.200800091.
141. Michalet, X. et al. 2005. Quantum dots for live cells, in vivo imaging, and diagnostics. *Science* 307: 538–544.
142. Savla, R., O. Taratula, O. Garbuzenko, and T. Minko. 2011. Tumor targeted quantum dot-mucin 1 aptamer-doxorubicin conjugate for imaging and treatment of cancer. *J. Control. Release* 153: 16–22, doi:10.1016/j.jconrel.2011.02.015.

3

Recent Advances in Immobilization Strategies in Biomaterial Nanotechnology for Biosensors

Mambo Moyo, Jonathan O. Okonkwo, and Nana M. Agyei

CONTENTS

3.1 Introduction

A biosensor is an analytical device that converts the modification of the physical or chemical properties of a biomatrix into an electric or other kinds of signal whose amplitude depends on the concentration of defined analytes in the solution [1,2]. In other words, biosensors exhibit two elementary parts: the sensitive part where specific biomaterial events take place and a transducing system that mediates these biomaterial events into quantifiable signal. A schematic diagram of a biosensor is shown in Figure 3.1.

Biomaterials such as enzymes, antibodies, deoxyribonucleic acid (DNA), receptors, organelles and microorganisms, as well as animal and plant cells or tissues have been used as biological sensing elements recognizing different analytes after being carefully immobilized on the transducer. The specificity, storage, operational, and environmental factors are essential for an analyte to be detected and the stability of the biosensor depends on the selection of the typical biomaterial. The applications of biosensors in agricultural, horticultural, and veterinary analysis; pollution, water, and microbial contamination

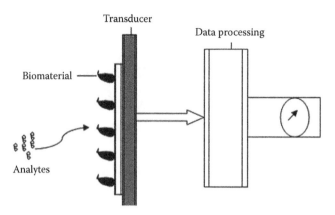

FIGURE 3.1
Scheme of a biosensor.

analysis; clinical diagnosis and biomedical diagnosis; fermentation analysis and control; industrial gases and liquids; mining and toxic gases; explosives and military arena; and flavors, essences, and pheromones have increased over the years [3–7]. The following characteristics have contributed to its widespread applications:

- "On-spot" analysis
- Inherent simplicity
- Rapid response
- Relative low cost
- Miniaturization

These applications allow for continuous monitoring and have been found to complement classical analytical methods such as atomic absorption spectrometry (AAS) and others.

3.2 Immobilization of Biomaterials

Biomaterials promote the rate of reactions but are not themselves consumed in the reactions. They may be used repeatedly for as long as they remain active. However, in most of the processes, biomaterials are mixed in a solution with substrates and cannot be economically recovered after the reaction and are generally wasted. Thus, immobilization of biomaterials in biosensing is a versatile tool that serves to increase the stability of a transducer system, allowing its application under extreme environmental conditions, its reuse, and the development of continuous monitoring. Boyukbayram et al. [8] reported that immobilized biomaterials have many operational advantages, some of which will be discussed later.

   Immobilization is a process in which the biological component/biomaterial (Figure 3.2) has to be properly attached to the transducer with maintained activity so that a viable biosensor is developed [9]. In other words, immobilization is "the imprisonment of a

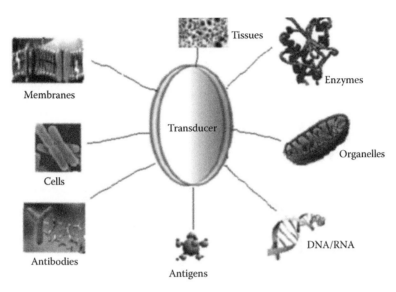

FIGURE 3.2
Biomaterials eligible for immobilization to the transducer surface.

biomaterial in a distinct phase that allows exchange with, but is separated from the bulk phase in which the substrate, inhibitor molecules are dispersed and monitored." The immobilization of these biomaterials onto the transducers represents a crucial step in the preparation and development of a biosensor.

Various immobilization groups for biomaterials can be grouped under physical and chemical methods. Physical methods include adsorption, entrapment, and encapsulation, while chemical methods include covalent linking and cross-linking [10,11]. Examples of the physical and chemical methods are shown in Figure 3.3. The main advantages, disadvantages, and major applications of each of these immobilization methods are presented in Table 3.1.

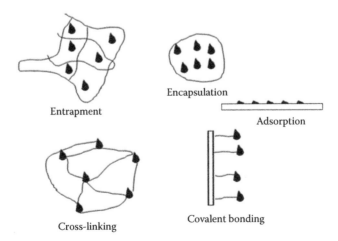

FIGURE 3.3
Schematic representations of different methods of immobilization of biomaterials (⬥ biomaterial).

TABLE 3.1

Advantages and Disadvantages of Different Immobilization Methods

Physical Methods	Advantages	Disadvantages	Applications
Adsorption	• Generally nondestructive • Minimal loss of biomaterial • Simple and easy • Little or no conformation change of the biomaterial	• Nonspecific adsorption • Desorption resulting from changes in temperature, pH, and ionic strength • Slow method	• Applicable to making environmental biosensors, for water pollution
Entrapment	• Several types of biomaterial can be immobilized within the same polymer • Easy to perform • No modification of the biological element • Increased operational and storage stability • Loss of biomaterial activity upon immobilization is minimized	• Diffusion barrier • Biomaterial leaching	• Water and microbial contamination analysis
Encapsulation	• Simple and rapid	• Requires a significant amount of biomaterial • As leaking of biocomponent • Diffusion barriers	• Clinical diagnosis and biomedical applications
Covalent coupling	• Short response time • Simple, mild, and wide applicability • Increased stability, less leakage of biomaterial • Low diffusional resistance	• Activity of biomaterial decreased • Matrix not regenerated • Coupling with toxic product	• Environmental biosensors, for water pollution and agriculture biosensors
Cross-linking	• Simple • Best used in conjunction with other methods • Very little desorption	• High biomaterial activity loss	• Agricultural, horticultural, and veterinary analysis biosensors

3.3 The Significance of Biomaterial Immobilization

The advantages of using some immobilized biomaterials are

- Reuse
- Continuous use
- Less labor-intensive
- Saving in capital cost
- Minimum reaction time

- Less chance of contamination in products
- More stability
- Improved process control
- High biomaterial to substrate ratio

However, some of the disadvantages of biomaterial immobilization include

- Effects on the stability and activity of biomaterials
- Possible solubility of substrate

3.4 Ideal Biosensor Characteristics after Immobilization

In recent years, several problems have been reported on biomaterials on electrodes. Some of the problems include, but not limited to, loss of biomaterial, need to reduce the time of response and offer disposable devices, maintenance of biomaterial stability, and shelf life of the biosensors [12,13]. Given these disadvantages, efforts have been geared toward developing several immobilization procedures.

The following desirable analytical characteristics are to be achieved after careful immobilization of the biomaterial using one or a combination of the following methods:

- Good operational ability
- Short response time to variation in analyte concentration
- High sensitivity to changes in signal and analyte concentrations (high $\Delta S/\Delta c_{analyte}$ (S = signal))
- Storage and prolonged stability to resist fouling, poisoning, or oxide formation that interferes with signal
- No hysteresis such that the signal is independent of prior history of measurements
- High selectivity to changes in target analyte concentration
- High reproducibility
- Simple calibration with standards

Sassolas et al. [14] reported that immobilized biomolecules should maintain their structure and function, retain their biological activity after immobilization, and remain tightly bound to the surface and not to be desorbed during the use of the biosensor. The intent of this chapter is to highlight the advances in the rapidly developing area of biomaterial immobilization methods used to develop optical, electrochemical, or gravimetric biosensors so that the ideal analytical characteristics highlighted earlier are captured.

3.5 Methods of Biomaterial Immobilization

The selection of immobilization techniques is mainly influenced by a variety of factors such as the nature of the biomaterial, the type of transducer used, the physicochemical properties of the analyte, and the operating conditions in which the biosensor is to function [2]. Singh et al. [15] reported that all these considerations should be taken into account for the biomaterial to exhibit maximum activity in its immobilized microenvironment. This section focuses on the theory of different classes of immobilization techniques that have been used for different biomaterials, while retaining the biological recognition properties of the biomaterial. The applicability of each of these techniques is discussed in each section.

3.5.1 Physical Methods

3.5.1.1 Adsorption

It is the oldest, simplest, and fastest way to prepare immobilized biomaterials on electrodes. It can be divided into two classes, namely, physical adsorption and chemical adsorption. Physical adsorption generally consists of simple deposition of biomaterials onto the surface of working electrode. The mechanism of attachment of some biomaterials is through weak bonds such as van der Waals forces and electrostatic interactions between the biomaterials and the transducer. Chemical adsorption is a stronger technique and the process involves the formation of covalent bonds, hydrogen, coordinated covalent, or even combination of any of these between the functional groups on the biomaterials and the immobilization surfaces.

Immobilization can be brought about by coupling a biomaterial either to external or internal surface of the carrier. The external surface binding method is advantageous as it does not involve pore diffusion. The disadvantages, however, include exposure of biomaterial to microbial attack and physical abrasion of biomaterial due to turbulence associated with the bulk solution. With the internal immobilization method, pore diffusion becomes a major problem. Examples of suitable surfaces are ion-exchange matrices, porous carbon, clay, hydrous metal oxides, glasses, and polymeric aromatic resins. Table 3.2 presents performances of some biosensors based on biomaterial-adsorption matrices and their characteristics.

3.5.1.1.1 Methods of Immobilization by Adsorption

There are four procedures for immobilization by adsorption, namely (a) *static process*, where the biomaterial is immobilized and in contact with the carrier without agitation (this is the most efficient technique but requires maximum time), (b) *the dynamic batch process* that involves the admixing of biomaterial with the carrier under constant agitation using mechanical shaker, (c) *the reactor loading process* whereby the carrier is placed into the reactor and biomaterial solution is transferred to the reactor with agitation of the whole content in the reactor, and (d) *the electrodeposition process* that involves the carrier being placed in the vicinity of one of the electrode in a biomaterial bath and electric current is applied leading to migration of biomaterial toward the carrier. This results in deposition of biomaterial on the surface of the carrier.

3.5.1.1.2 Applications

Recently, various configurations of adsorption have been utilized to immobilize different biomaterial groups. For example, glucose oxidase (GOx) was physically adsorbed on a polypyrrole (PPy) nanotube–modified electrode in order to develop an amperometric glucose biosensor [24]. A linear range for glucose detection between 5×10^{-4} M and 1.3×10^{-2} M was reported. A response time of about 3 s was observed. Reduced amounts of enzyme were used due to enhanced adsorption property; hence, production cost of the biosensor

TABLE 3.2

Biosensors Based on Biomaterials Adsorbed within a Polymer

Electrode and Biomaterial	Matrix for Immobilization	Substrate	Detection Mode	Detection Limit	Linearity	Remark	Reference
GCE; MDH	SWNT	NADH	Amperometry	0.5 μM	0.2–1 mM	A very efficient electrocatalytic behavior toward the oxidation of NADH at a low potential, a good operational stability, and high sensitivity	[16]
GCE; GOx	SWCNTs	β-D-Glucose	Amperometry	—		Physically adsorbed enzymes exhibit significant kinetic dispersion	[17]
GCE; HRP	AgNL	H_2O_2	Amperometry	0.13 μM	0.03–0.5 mM	Easy to fabricate and was stable over prolonged use	[18]
GCE/antibodies	Chitosan–carbon nanotubes–gold nanoparticles (CS–CNTs–GNPs)	CEA	Amperometry	0.04 ng mL^{-1}	0.1–2.0 ng mL^{-1} and from 2.0 to 200.0 ng mL^{-1}	Antibody immobilization matrix could exhibit good sensitivity, stability, and reproducibility	[19]
GCE/peroxidase labeled secondary anti-hCG antibody	AuNPs-dotted carbon nanotubes (MWCNTs)–graphene composite	Human chorionic gonadotropin (hCG)	DPV	0.0026 mIU mL^{-1}	0.005–500 mIU mL^{-1}	The immunosensor showed good precision, acceptable stability, and reproducibility	[20]

(continued)

TABLE 3.2 (continued)

Biosensors Based on Biomaterials Adsorbed within a Polymer

Electrode and Biomaterial	Matrix for Immobilization	Substrate	Detection Mode	Detection Limit	Linearity	Remark	Reference
Gold electrode; olfactory receptor	Octadecylmercaptan $(SH(CH_2)_{17}CH_3)$	—	Atomic force microscopy	—	—	Electrode hydrophobicity does affect the surface coverage	[21]
Graphite; yeast cells, *Hansenula anomala* microorganisms	Polyaniline (PANI)	—	Cyclic voltammetry	—	—	Demonstrates that microorganisms with suitable redox potential useful for wastewater treatment	[22]
Carbon fiber electrode (CFE) DNA	Polypyrrole (PPy) film	DA and EP	DPV	8.0×10^{-8} M; 6.0×10^{-8} M	3.0×10^{-7} to 1.0×10^{-5} M; 5.0×10^{-7} to 2.0×10^{-5} M	The template immobilization of DNA can also provide a novel route for fabrication of DNA sensors for hybridization detection	[23]

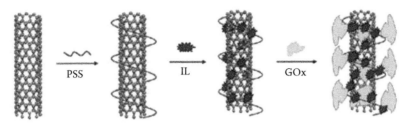

FIGURE 3.4
Schematic illustration of the immobilization of glucose oxidase (GOx) on the surface of single-walled nanotubes (SWNTs) using LBL deposition. PSS, polysodium 4-styrenesulfonate; IL, ionic liquid. (From Wu, X. et al., *J. Phys. Chem. B*, 113, 40, 13365, 2009. With permission.)

was reduced. Arvinte et al. [16] immobilized dehydrogenase enzymes on the surface of a single-walled nanotube (SWNT)–modified electrode to form an approximate monolayer film with this technique. Using this biosensor, a linear calibration curve was obtained for L-malic acid concentrations between 0.2 and 1 mM. In another study, Lyons et al. [17] demonstrated that adsorbed enzymes such as GOx exhibit significant kinetic dispersion and that the dispersion is best subjected to analysis via potential step chronoamperometry. A disposable H_2O_2 biosensor based on HRP adsorbed on AuNPs electrodeposited onto ITO electrode surface was developed [25]. The gold nanoparticles were very efficient to retain the enzyme activity and to promote electron transfer. A linear relationship in the concentration range extending from 8×10^{-6} to 3×10^{-3} M, with a detection limit of 2×10^{-6} M, was reported. This biosensor presented an excellent reproducibility, a high selectivity, and long-term stability (about 83% of the initial response remained after 12 weeks). Electrostatic immobilization of enzymes by the technique of layer-by-layer deposition was reported [26]. The surface of SWNTs positively charged by coating polysodium 4-styrenesulfonate combined with ionic liquids (ILs) and then immobilizing glucose onto the SWNTs is shown in Figure 3.4. The electrocatalytic activity of these electrodes based on the SWNT–enzyme complex was found to be affected by the nature of ILs.

Zheng et al. [27] modified a glassy carbon electrode (GCE) using multiwalled carbon nanotube (MWCNT). The modified electrode showed a pair of redox peaks that resulted from the oxygen-containing functional groups on the nanotube surface as evidenced by FTIR spectrum during their study. A recombinant thermostable dye-linked L-proline dehydrogenase (L-proDH) from hyperthermophilic archaeon (*Thermococcus profundus*) was further immobilized by physical adsorption. The GC/MWCNTs/L-proDH electrode exhibited an electrocatalytic signal for L-proline compared to bare GC, GC/L-proDH, and GC/MWCNTs electrodes, which suggested that the presence of MWCNTs efficiently enhances electron transfer between the active site of enzyme and electrode surface. A typical Michaelis–Menten catalytic response with lower apparent constant was reported. Firdoz et al. [28] reported a novel amperometric biosensor based on single-walled carbon nanotubes (SWCNTs) with acetylcholine esterase at pH 7.4 for the detection of carbaryl pesticides in water. The developed SWCNTs were dispersed in positively charged polyelectrolyte, polydiallyldimethylammonium chloride (PDDA), using layer-by-layer (LbL) self-assembly technique. The suggested dispersion mechanism was that of weak supramolecular interaction between electrostatically charged SWCNTs to the negatively charged acetylcholinesterase (AChE). This immobilization technique involving electrostatic interactions allowed the biosensor to have good sensitivity and stability toward the monitoring of carbaryl pesticides in water with the detection limit of 10^{-12} g L^{-1} and recovery of 99.8% ± 2.7% to 10^{-10} g L^{-1}. A new biosensor based on horseradish peroxidase (HRP) immobilized through adsorption on a silver nano-layer–modified glassy electrode denoted as HRP/AgNL/GCE was recently reported [18]. The resulting biosensor exhibited enhanced electrocatalytic activity toward reduction of H_2O_2. Obtained electrode biosensor exhibited a detection limit of 0.13 μM, and a linear range of 3.85×10^{-5}–5.2×10^{-4} M at a signal-to-noise ratio of three. The use of a sodium cholate suspension–dialysis method to adsorb the HRP onto SWCNTs was described [29]. The immobilized biomaterial on the modified electrode showed high sensitivity toward H_2O_2. A linear response to hydrogen peroxide measurement was obtained over the range from 1.0×10^{-12} to 1.0×10^{-11} M and an amperometric detection limit of 2.1×10^{-13} M due to its bioelectrocatalytic reduction based on direct electron transfer between gold electrode and the active site of the HRP.

The proper immobilization of DNA onto the transducer plays an important role in the overall performance of the DNA electrochemical detection. Double-stranded calf thymus DNA was physisorbed onto PPy–polyvinyl sulfonate (PPy–PVS) films electrochemically deposited onto indium tin oxide (ITO)-coated glass plates [30]. The DNA/PPy–PVS electrodes were found to have improved sensitivity to o-chlorophenol (0.1–25 ppm) and 2-aminoantharacene (0.01–15 ppm) solutions prepared in phosphate buffer 0.05 M (pH 7). The response time of the DNA/PPy–PVS electrodes has been found to be about 30 s. The ultrathin carbon nanotube (CNT)–DNA hybrid membrane formation by simple physical adsorption onto a thin alumina substrate was described [31]. In this approach, DNA was found to impart a hydrophilic nature to the CNTs, which enhanced the interaction between the nanotubes and hydrophilic porous alumina.

The creation of highly selective immunosensors is an important line of activity in biosensor engineering, because antibodies are capable of high-specific recognition of biologically active compounds after being carefully immobilized. The antibodies were adsorbed physically onto thin polymer films on gold electrodes [32]. These results indicated that the adsorbed antibodies kept their biological activity. Sensitivity down to 7 μg mL^{-1} and a resolution of 1.4 μg mL^{-1} were achieved.

Yang et al. [33] developed a label-free electrochemical impedance immunosensor by physical adsorption method to immobilize anti-*Escherichia coli* antibodies onto an ITO interdigitated array microelectrode (IDAM) for detection of E. coli O157. The detection range of the biosensor was from 4.3×105 to 4.36×108 cfu mL^{-1} with the detection limit of 106 cfu mL^{-1}. Phage as a molecular recognition element in biosensors immobilized by physical adsorption was reported [34]. This type of immobilization allowed the sensor to have a detection limit of a few nanomoles and a response time of a 100 s over the range of 0.003–210 nM. Cho et al. [35] physically adsorbed *E. coli* O157:H7 antibodies onto the carbon composite electrode. The porous carbon composite electrode was prepared by a solgel method with a mixture of graphite powder and tetraethyl orthosilicate/ethanol.

Gao et al. [19] fabricated a simple and controllable one-step electrodeposition method for the preparation of chitosan–CNTs–gold nanoparticles (CS–CNTs–GNPs). The nanocomposite film was used to fabricate an immunosensor for detection of carcinoembryonic antigen (CEA). The fabricated procedure of the immunosensor is shown in Figure 3.5. The immunosensor based on CS–CNTs–GNPs nanocomposite film as the antibody immobilization matrix exhibited good sensitivity, stability, and reproducibility for the determination of CEA.

Recently, Trnkova et al. [36] described the fabrication of a simple biosensor based on immobilization of MT to the surface of carbon paste electrode via chicken anti-MT antibodies. A novel *Shigella flexneri* immunosensor based on HRP-labeled antibodies to

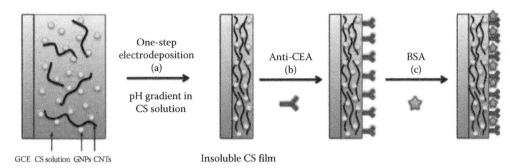

One-step
electrodeposition
(a)

pH gradient in
CS solution

Anti-CEA
(b)

BSA
(c)

GCE CS solution GNPs CNTs

Insoluble CS film

FIGURE 3.5
Schematic illustration of the fabrication procedure of the immunosensor. (a) One-step electrodeposition, (b) anti-carcinoembryonic antigen, and (c) bovine serum albumin. (From Gao, X. et al., *Talanta*, 85, 4, 1980, 2011. With permission.)

S. flexneri (HRP–anti-*S. flexneri*) immobilized by physical adsorption on the MWCNT/ sodium alginate (SA) composite–modified screen-printed electrode surface was successfully fabricated [37]. Under optimal conditions, the linear range of *S. flexneri* was from 104 to 1011 cfu mL^{-1} with a detection limit of 3.1×10^3 cfu mL^{-1} (S/N = 3). Lu et al. [20] reported on the Au nanoparticles (AuNPs) dotted CNTs–graphene composite that was immobilized on the working electrode, which can increase the surface area to capture a large amount of primary antibodies (Ab1) as well as improve the electronic transmission rate. The prepared bionanolabels, composed of mesoporous silica nanoparticles (MCM-41) coated with AuNPs through thionine linking, showed good adsorption of HRP-labeled secondary anti-hCG antibody (Figure 3.6). The immobilized enzyme showed good linear range and detection limit.

A crucial requirement in the development of receptors onto the transducing element of the biosensor in such a manner that their functionality is preserved is proper immobilization. Olfactory receptors immobilized by physical adsorption onto gold electrodes for electrical biosensor have been reported [21]. Microorganisms are also immobilized due to adsorptive interactions such as ionic, polar or hydrogen bonding, and hydrophobic interaction. Prasad et al. [36] reported the immobilization of the microorganisms by two different methods, namely, physical adsorption and covalent linkage.

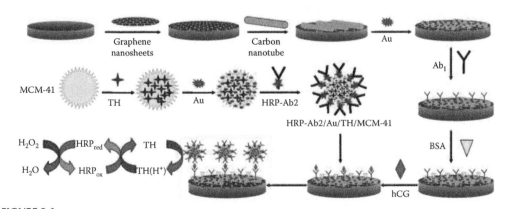

Graphene
nanosheets

Carbon
nanotube

Au

Ab$_1$

MCM-41

TH

Au

HRP-Ab2

HRP-Ab2/Au/TH/MCM-41

BSA

H_2O_2 HRP$_{red}$ TH

H_2O HRP$_{ox}$ TH(H$^+$)

hCG

FIGURE 3.6
(See color insert.) Fabrication processes of Au/TH/MCM-41 nanomaterials and measurement protocol of the electrochemical immunosensor. (From Lu, J. et al., *Biosens. Bioelectron.*, 33, 1, 29, 2012. With permission.)

3.5.1.2 Entrapping

The process involves the formation of a mixture of the biomaterial with monomer solution and then polymerized to a gel, trapping the biomaterial. The gels commonly used include polyacrylamide, starch gels, nylon, silastic gels, and conducting polymers [9]. The form and nature of matrices vary. Pore size of different immobilization matrix should be adjusted to prevent the loss of biomaterial from the matrix due to excessive diffusion. Occurrence of leakage of low-molecular-weight biomaterial such as enzymes from the gel is possible. Table 3.3 presents performances of some biosensors based on biomaterial-entrapping matrices.

3.5.1.2.1 Methods of Immobilization by Entrapping

3.5.1.2.1.1 *Occlusion within a Cross-Linked Gel*   The entrapment method involves the formation of a highly cross-linked gel as a result of the polymerization, which has a fine "wire mesh" structure and can hold smaller biomaterials in its cage more effectively. Some synthetic polymers such as polyarylamide, polyvinyl alcohol (PVA), and natural polymer are used to immobilize biomaterials using this technique.

3.5.1.3 Encapsulation

The process involves enclosing a droplet of solution of biomaterial in a semipermeable membrane capsule such as enzyme. The capsule is made up of cellulose nitrate and nylon. The method of encapsulation is cheap and simple but its effectiveness largely depends on the stability of biomaterial concerned although the biocatalyst is very effectively retained within the capsule. This technique is restricted to medical sciences only.

3.5.1.3.1 Applications

3D matrices such as an electropolymerized film, PPy [46–48], polyaniline (PANI) [49], a photopolymer [50], a silica gel [51], a polysaccharide [52], or a carbon paste [53,54] have been reported for the immobilization of enzymes.

Entrapping the protein on the polymeric membrane surface has been reported [55–57]. Recently, a selective and sensitive HRP-based biosensor was developed to detect a novel antiepileptic, levetiracetam (LEV) [58]. The response of the biosensor to LEV was based on chronoamperometric detection of reduced LEV formed upon HRP catalysis in the presence of oxidized LEV and H_2O_2. The enzyme was immobilized in a PPy film during electropolymerization by cyclic voltammetry. This biosensor showed a detection limit of 9.8×10^{-6} M. An amperometric biosensor was developed by entrapping cholesterol oxidase (CholOD) in poly 3,4-ethylenedioxypyrrole (PEDOP) for cholesterol detection [59]. The responses of the enzyme electrode were measured via monitoring the oxidation current of H_2O_2 at +0.7 V versus Ag/AgCl in the absence of any mediator. The detection limit was 4×10^{-4} M and the response time was 150 s.

Some research groups have reported the protocols using solgel microenvironment to effectively encapsulate the enzyme AChE [60]. Two acetylcholinesterases from different microorganisms (*Electrophorus electricus* and *Drosophila melanogaster*) were encapsulated, and in both cases, the enzyme retained high functionality while it could detect the model pesticide dichlorvos at very low concentrations. The biosensor had sensitivity of 2.5 mA mM^{-1}, a linear range of response between 1 and 3 mM, response time of about 30 s, and

TABLE 3.3

Biosensors Based on Biomaterials Entrapped and Encapsulated within a Polymer

Substrate	Immobilized Biomaterial	Detection Mode	Detection Limit	Linear Range	Remark	Reference
Glucose	Glucose oxidase	Amperometry	2.0×10^{-6} M	5.0×10^{-6} to 1.6×10^{-4} M	Studying the electron transfer of enzyme and the design of novel electrochemical biosensors	[38]
H_2O_2; O_2	Horseradish peroxidase (HRP)	Amperometry	4.0×10^{-7} mol L^{-1} for H_2O_2; 1.0×10^{-7} mol L^{-1} for O_2	1.0 μmol L^{-1} to 1.0 mmol L^{-1} for H_2O_2, 0.5–18.6 μmol L^{-1} for O_2	HRP retains its original structure and bioactivity in HRP–GO–Nafion solution, sensitivity, good reproducibility, and long-term stability	[39]
Dichlorvos	AChE	Amperometry	9.6×10^{-11} M	2×10^{-10} to 1×10^{-8}	Photopolymerization provides better detection of the enzyme	[40]
AChE	AChE/ChOD	Amperometry	6×10^{-7} M	1×10^{-6} to 1.5×10^{-3} M	Silica solgel membrane immobilization preserves the structure of the enzymes	[41]
Ethidium bromide	ssDNA	Amperometry	—	—	Detection of EtBr in spiked river water samples and daunomycin, an antitumor agent	[42]
Glucose	Glucose oxidase	Amperometry	50 μM	0.2–20 mM	Good reproducibility, rapid response, and stability achieved	[43]
Dopamine	HRP	SWV	9×10^{-6} M	9.9×10^{-5} to 1.6×10^{-3} M	Carbon paste provides a favorable environment for the working of HRP	[44]
SjAg	SjAg–HRP	Amperometry	0.06 mg mL^{-1}	0.11–22.4 mg mL^{-1}	The assay was competitive, nano-Au monolayer for consecutive assays were regenerated	[45]

sensor-to-sensor reproducibility (RSD) of 3%. In another study, glutamate dehydroge-nase was immobilized in a chitosan film in order to develop an optical biosensor for the determination of ammonium in water samples [52]. A linear response was obtained in the ammonium concentration range from 5×10^{-6} to 5×10^{-4} M. The biosensor was stable for at least one month when stored dry at 4°C.

Carbon paste, a mixture of carbon (graphite) powder and a binder (pasting liquid), is a popular electrode material used for the preparation of various electrodes, sensors, and detectors. An amperometric biosensor for continuous glucose monitoring based on MWCNTs/graphite/GOD packed needle-type electrode was also described [61]. The direct electrochemistry of GOx immobilized on the designed titanium carbide-Au nanoparti-cles–fullerene C_{60} composite film–modified GCE (TiC-AuNPs–C_{60}/GCE) [38].

Direct electron transfer of HRP encapsulated in graphene oxide (GO)–Nafion nanocom-posite film was reported [39]. As a new sensor with excellent electrocatalytic response to the reduction of H_2O_2 and O_2, calibrations with good linear relationships were obtained from 1.0 μmol L^{-1} to 1.0 mmol L^{-1} for H_2O_2 and from 0.5 μmol L^{-1} to 18.6 μmol L^{-1} for O_2 with the detection limits of 4.0×10^{-7} mol L^{-1} for H_2O_2 and 1.0×10^{-7} mol L^{-1} for O_2 at a signal-to-noise ratio of 3. Michaelis–Menten behavior with K_m^{app} values of 0.684 mmol L^{-1} for H_2O_2 and 0.0160 mmol L^{-1} for O_2 were obtained.

The composite of RTIL and GNPs-TNTs was immobilized on the electrode surface through the gelation of a small amount of HRP aqueous solution [62]. The HRP-immobilized electrode was investigated by cyclic voltammetry and chronoamperometry. The results from both techniques showed that the direct electron transfer between the nanocomposi-te-modified electrodes and heme in HRP could be realized. The biosensor responded to H_2O_2 in the linear range from 5×10^{-6} to 1×10^{-3} mol L^{-1} with a detection limit of 2.1×10^{-6} mol L^{-1} (based on the S/N = 3). Sun et al. [63] reported on a stable composite film composed of the ionomer Nafion, the ZnO nanoparticle, and the protein hemoglobin that was cast on the surface of an IL-modified carbon paste electrode (CILE) to establish a modified elec-trode denoted as Nafion/nano-ZnO/Hb/CILE. The electrochemical behaviors of hemoglo-bin (Hb) entrapped in the film were carefully investigated with cyclic voltammetry. A pair of well-defined and quasi-reversible redox voltammetric peaks for Hb Fe(III)/Fe(II) was obtained with the standard potential (E^0) located at −0.344 V (vs. SCE) in phosphate buffer solution (PBS, pH 7.0), which was attributed to the direct electron transfer of Hb with elec-trode in the microenvironments of ZnO nanoparticle and IL 1-butyl-3-methylimidazolium hexafluorophosphate (BMIMPF$_6$). The electrochemical parameters of Hb in the composite film were further carefully calculated with the results of the electron-transfer rate constant (k_s) as 0.139 s^{-1}, the charge transfer coefficient (α) as 0.413, and the number of electron trans-ferred (n) as 0.95. The Hb-modified electrode showed good electrocatalytic ability toward the reduction of trichloroacetic acid (TCA).

The immobilization of the DNA as a biomaterial using entrapment method has been reported [64]. A sensitive electrochemical detection of DNA hybridization using a paste electrode assembled by MWCNTs and immobilizing DNA probe within electropolymer-ized PPy was developed. The detection approach relied on entrapping of DNA probe within electropolymerized PPy film on the MWNT paste electrode and monitoring the current change generated from an electroactive intercalator of ethidium bromide (EB) after DNA hybridization. In another study, doping of nucleic-acid probes within electropolymerized PPy film onto carboxylic group-functionalized MWNT (MWNT–COOH)–modified electrode was used as basis for detection DNA hybridization [65].

Fu et al. [66] also fabricated a DNA biosensor by means of self-assembling colloidal Ag (Ag) to a thiol-containing solgel.

New label-free DNA recognition based on doping nucleic-acid probes within conducting polymer films was also reported [67]. The label-free approach relied on the doping of nucleic-acid probes within electropolymerized PPy films. The sensor gave a large increase or decrease in current for different oligonucleotides, coupled to the simplified operation and instantaneous response. The immobilization of microorganisms by entrapment can be achieved by either the retention of the cells in close proximity of the transducer surface using dialysis or filter membrane or in chemical/biological polymers/gels such as (alginate, carrageenan, agarose, chitosan, collagen, polyacrylamide, PVA, polyethylene glycol) polyurethane. The microbial immobilization method using PVA gel as an immobilizing material was improved and used for entrapment of activated sludge [68].

Methods for biomaterial immobilization should allow a stable bond between the sensorial surface and the biomaterial, without interfering with its biological activity. This is a key aspect in immunosensor assembly. Immobilizing the antibody or the antigen in the solid support frequently requires several wash steps and blocking in order to remove the excess reagent and to cover new immobilization sites. John et al. [69] first reported the incorporation of antibodies into conducting polymer films. Pyrrole was galvanostatically polymerized onto a platinum wire substrate from a solution that contained anti-human serum albumin (anti-HSA). The antibody was incorporated into the PPy film and the pyrrole anti-HSA electrode found to give a specific electrochemical response to HSA. A new strategy to construct amperometric immunosensor for *Schistosoma japonicum* antigen (SjAg) using antibodies loaded on an entrapped carbon paste electrode was reported [70]. The entrapped antibodies allowed the dynamic concentration range for SjAg assay from 0.11 to 22.4 μg mL^{-1} with a detection limit of 0.06 μg mL^{-1}. Stubbs et al. [71] have also reported that a hydrogel entrapment method of immobilization of antibodies on SAW devices is an efficient and reliable technique for incorporation into a biosensor. Susmel et al. [72] studied the performance of a piezoelectric immunosensor prototype in which the immobilization of the *Bacillus cereus* antibody on the crystalline surface was accomplished by simple entrapment within a thin Nafion film. In other words, an ion exchange is performed in a very porous polymer material that exhibits a good attachment to gold. O'Grady et al. [73] have demonstrated that antibody-doped PPy represents an electrically controllable sensing platform that can be exploited to collect rapid, repeated measurements of protein concentrations with molecular specificity. A novel immunosensor was recently proposed [74]. It is based on gold nanoparticles assembled onto the TMB/ Nafion film–modified electrode to provide active sites for the immobilization of antibody (anti-MIgG) molecules. The use of Nafion for antibody entrapment was explored in order to immobilize antibodies or antigens.

The development of a good biocompatible matrix for immobilization of cells is very crucial for improving the performance of functional biohybrids. Cell-based biosensors, bioelectronic portable devices containing plant living cells, have been used for monitoring some physiological changes induced by pathogen-derived signal molecules called flagellin. This new design employs the specific strain of tobacco cells [75]. The cells were immobilized on the surfaces of graphite electrodes by gel. The results have shown that plant cells might be immobilized on the electrode surfaces and can interact with the peptide virulent factors from bacteria. Figure 3.7 shows a schematic single cell-based biosensor with gel matrix.

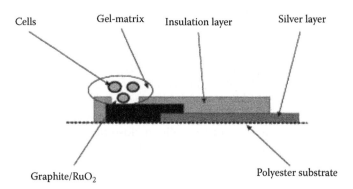

FIGURE 3.7
The scheme of the single cell–based biosensor with gel matrix. (From Oczkowski, T. et al., *Bioelectrochemistry*, 70, 1, 192, 2007. With permission.)

3.5.2 Chemical Methods

3.5.2.1 Covalent Linking

The basis of this method involves the formation of bond between a functional group in the biomaterial to the support matrix, resulting in a stable complex. Various functional groups from proteins (amino group, carboxylic group, phenol ring, indole group, imidazole group) and other polymeric support materials (amino and related groups of polysaccharides and silica gel, carboxylic acid and related groups of polyglutamic acid, carboxymethyl cellulose, aldehyde, and acetal groups of polymers) can be utilized. It should be stressed in this case that some functional groups, which are not essential for the catalytic activity of a biomaterial, could be covalently bonded to the support matrix, making it stronger. The binding mechanisms of the biomaterial to the solid support is generally carried out in two stages by (1) activation of the surface using polyfunctional reagents such as glutaraldehyde or carbodiimide, followed by biomaterial covalent coupling to the activated support. However, covalent immobilization can be performed directly onto the transducer surface or in carbon paste electrodes. For successful immobilization, mild conditions under which reactions are performed, such as low temperature, low ionic strength, and pH in the physiological range, are to be maintained. Table 3.4 presents performances of some biosensors based on covalently immobilized biomaterial.

3.5.2.1.1 Methods of Immobilization by Covalent Linking

Different methods of covalent bonding are (a) *peptide bond formation,* which involves bonding between the amino or carboxyl group of the support and amino or carboxyl group of the biomaterial, (b) *poly functional reagent,* where the use of a bifunctional or multifunctional reagent such as glutaraldehyde to form bonding between the amino group of the support and amino group of the biomaterial, (c) *group activation,* whereby cyanogen bromide is used as a support containing glycol group, that is, cellulose, Sephadex, Sepharose, and others, and (d) *diazoation,* which involves bonding between the amino group of the support such as aminobenzyl cellulose, aminosilanized porous glass, amino derivatives, and a tyrosyl or histidyl group of the biomaterial.

3.5.2.1.2 Applications

In recent years, covalent coupling of different biomaterials to polymeric supports has proved to be a popular chemical immobilization method used to develop sensors and biosensors. Covalent immobilizations of GOx [84], AChE [85,86], and uricase [87] have been reported.

TABLE 3.4
Biosensors Based on Covalently Immobilized Biomaterials

Substrate	Immobilized Biomaterial	Detection Method	Detection Limit	Linearity Range	Remark	Reference
Guaiacol	Laccase	Amperometry	0.04 μM	0.1–10 μM (lower concentration range) and 10–500 μM (higher concentration range)	A simple and novel approach with a fast response of 5 s, reusability (150 times), stability (5 months), interferrants, e.g., ascorbic showed a negligible interference	[76]
Trichoderma harzianum check	DNA	Cyclic voltammetry	1.0×10^{-19} mol L^{-1}	1.0×10^{-18} to 1.82×10^{-4} mol L^{-1}	The DNA biosensor detection approaches provided a quick, sensitive, and convenient method	[77]
Staphylococcus aureus	HRP–anti-RbIgG	Amperometry	1.4×10^4 cells mL^{-1}	1.3×10^3 to 7.6×10^4 cells mL^{-1}	The assay was competitive and with 30 min as time of detection, good recoveries after spiking to milk samples	[78]
H_2O_2	Cytochrome c	Amperometry	0.2 μM	3–700 μM	Enhanced sensitivity and selectivity for hydrogen peroxide in the presence of interference by dopamine, ascorbic acid, L-cysteine, and glucose, good reproducibility and long stability	[79]
α-Fetoprotein	HRP–anti-CEA	Cyclic voltammetry	—	—	The assay was competitive and with 35 min as time of detection, acceptable reproducibility, and the results obtained from clinical sera were in acceptable agreement with those from parallel single-analyte tests	[80]
NADH	Lactate dehydrogenase (LDH)	Differential pulse voltammetry	5 μM	50–500 μM	High sensitivity and reproducibility, and the minimal surface fouling combined with the attractive low potential of the NADH oxidation	[81]
17β-Estradiol	Estrogen receptor	Cyclic voltammetry	1.0×10^{-13} M	1.0×10^{-13} to 1.0×10^{-9} M	A good stability, in which the biosensor retained 88% of its initial activity after 3 weeks of storage	[82]
Antigens	Antibodies	Electrochemical immunosensor	10 ng mL^{-1}	25 ng mL^{-1} to 1 μg mL^{-1}	Cross-linking with glutaraldehyde method provided the highest signal	[83]

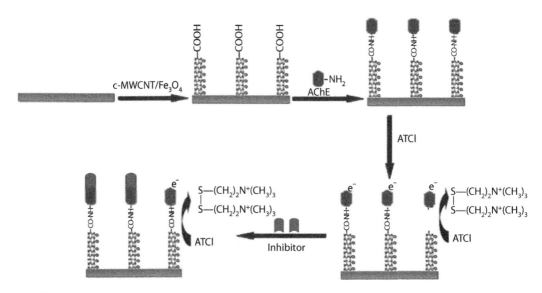

FIGURE 3.8
Scheme for the stepwise amperometric biosensor fabrication process and immobilized acetylcholinesterase (AChE) inhibition in pesticide solution. (From Chauhan, N. and Pundir, C.S., *Anal. Chim. Acta*, 701, 66, 2011. With permission.)

Tiwari et al. [88] have reported the use of prepared nanocomposite made up of silylated chitosan and MWCNTs (CHIT-SiO$_2$–MWCNTs) for immobilization of urease (Urs). Urs enzyme was covalently immobilized with the CHIT-SiO$_2$–MWCNTs matrix using glutar-aldehyde as a linker. The relatively low Michaelis–Menten constant of 0.15 mM indicated that the CHIT-SiO$_2$–MWCNTs matrix had a high affinity for the Urs enzyme and demonstrated the excellent electrocatalytic activity of MWCNTs. Chauhan et al. [89] fabricated an organophosphorus (OP) mediator–type biosensor based on covalently immobilizing the AChE purified from maize seedlings onto the composite of iron oxide nanoparticles and functionalized MWCNTs forming an AChE/Fe$_3$O$_4$/carboxylated MWCNTs (cMWCNT)/ Au electrode (Figure 3.8). The synergistic action of Fe$_3$O$_4$NP and cMWCNT showed excellent electrocatalytic activity at substantial low potential (+0.4 V). The biosensor exhibited good sensitivity (0.475 mA µM^{-1}), reusability (more than 50 times), and stability (2 months) as compared to the unmodified one.

Recently, Dhyani et al. [90] utilized a platform comprising electrophoretically prepared thioglycolic acid (TGA)–capped cadmium sulfide (CdS) quantum dots (QDs) deposited onto an ITO-coated glass plate for covalent immobilization of cholesterol oxidase for application to a cholesterol sensor. This type of immobilization allowed a low detection limit (1.87 mg dL^{-1}), high sensitivity (2.87 mA mg^{-1} dL^{-1} cm^{-2}), and a low value of Michaelis–Menten constant (K$_m$) (0.22 mM) revealing high enzyme affinity. Rawal et al. [76] described the construction of an amperometric biosensor for detection of phenolic compounds based on covalent immobilization of laccase (Lac) onto manganese dioxide nanoparticles (MnO$_2$NPs)–decorated cMWCNTs/PANI composite electrodeposited onto a gold (Au) electrode through N-ethyl-N'-3-dimethylaminopropyl carbodiimide (EDC) and N-hydroxysuccinimide (NHS) chemistry (Figure 3.9). The covalent immobilization of the biomaterial allowed the biosensor to exhibit an optimum response at pH 5.5 (0.1 M sodium acetate buffer) and 35°C, when operated at 0.3 V versus Ag/AgCl. Biosensor was successfully used to measure total phenolic content in tea leaves extract.

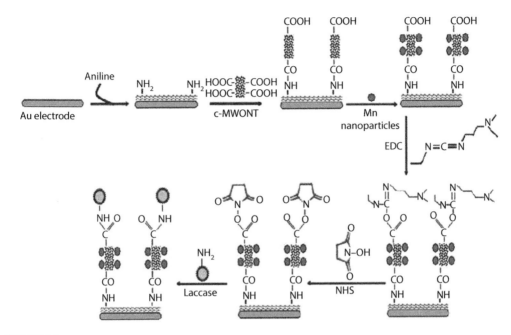

FIGURE 3.9
Schematic representation of laccase immobilized onto modified Au electrode (Lac/MnO₂NPs/cMWCNT/PANI/Au). (From Rawal, R. et al., *Int. J. Biol. Macromol.*, 51, 2,175, 2012. With permission.)

Teymourian et al. [81] immobilized lactate dehydrogenase (LDH) and NAD⁺ onto the Fe₃O₄/MWCNTs nanocomposite film by covalent bond formation between the amine groups of enzyme or NAD⁺ and the carboxylic acid groups of the Fe₃O₄/MWCNT film (Figure 3.10). The biosensor was found to have sensitivity of 7.67 μA mM⁻¹, and the applicability of the sensor for the analysis of lactate concentration in human serum samples was successfully demonstrated.

Covalent immobilization of AChE on iron oxide nanoparticles (Fe₃O₄NPs)–decorated cMWCNTs electrodeposited onto ITO-coated glass plate was reported [91].

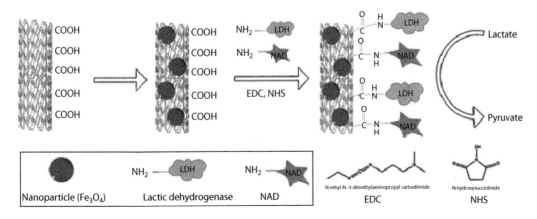

FIGURE 3.10
Schematic representation for the fabrication of Fe₃O₄/MWNTs/LDH/NAD⁺–modified GC electrode. (From Teymourian, H. et al., *Biosens. Bioelectron.*, 33, 1, 60, 2012. With permission.)

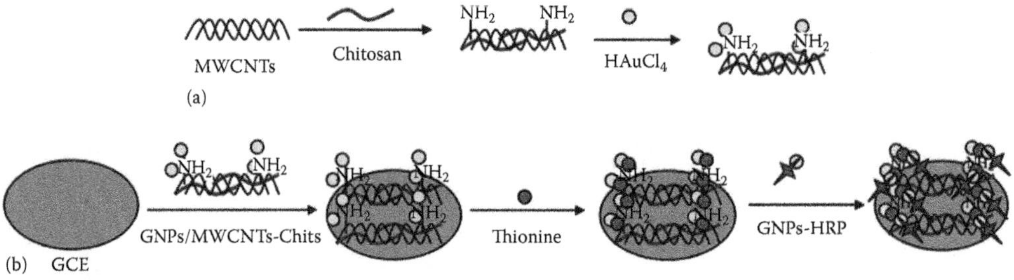

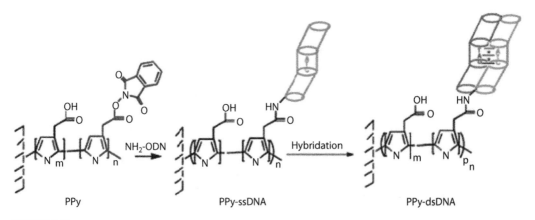

FIGURE 3.11
The fabricating procedures for (a) GNP/MWCNT–CHIT composite and (b) the biosensor. (From Li, S. et al., *Appl. Surf. Sci.*, 258, 7, 2802, 2012. With permission.)

The electrochemical biosensor was successfully used for the determination of pesticides in environmental and food samples. An electrochemical biosensor based on a mixture of gold nanoparticles (GNPs) and HRP was assembled onto the modified GCE by covalent bond (Figure 3.11) [92]. The linear response range of the biosensor to H_2O_2 concentration was from 5×10^{-7} to 1.5×10^{-3} mol L^{-1} with a detection limit of 3.75×10^{-8} mol L^{-1} (based on S/N = 3).

DNA probes bearing amine groups were covalently bonded on a supporting PPy matrix functionalized with activated ester groups (Figure 3.12) [93]. The functional electrodes fabricated by this method have been employed to monitor the DNA hybridization.

DNA covalent immobilization onto screen-printed electrode network for direct label-free hybridization detection was described [94]. Screen-printed electrode networks were functionalized electrochemically with 1-ethyl-3-3-dimethylaminopropyl carbodiimide. Single-stranded DNA with a C_6–NH_2 linker at the 5′ end was then covalently bound to the surface to act as probe for the direct, nonlabeled, and detection of complementary strands in a conductive liquid medium. An electrochemical DNA sensor based on silver nanoparticles/poly trans-3-3-pyridyl acrylic acid (PPAA)/MWCNTs with carboxyl group (MWCNT–COOH)–modified GCE was presented [95]. Thiol group end single-stranded DNA (HS–ssDNA) probe was easily covalently linked onto the surface of silver nanoparticles through a 5′ thiol linker. Differential pulse voltammetry (DPV) was used to monitor DNA hybridization events. Under the optimal conditions, the increase of reduction peak

FIGURE 3.12
Reaction of the grafted of ssDNA probe into polypyrrole (PPy) followed by hybridization of ssDNA target. (From Tlili, C. et al., *Talanta*, 68, 1, 131, 2005. With permission.)

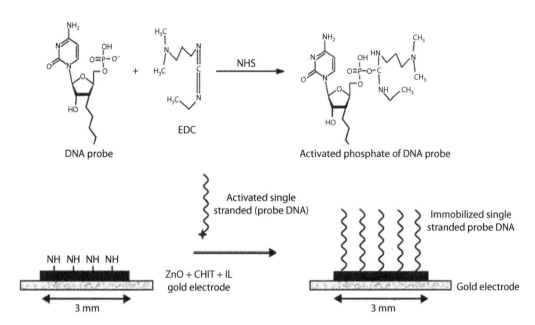

FIGURE 3.13
A schematic representation of the probe DNA-IL/ZnO/CHIT/Au electrode. (From Siddiquee, S., *J. Solid State Electrochem.*, 16, 1, 2012. With permission.)

current of Adriamycin was linear with the logarithm of the concentration of the complementary oligonucleotides from 9.0×10^{-12} to 9.0×10^{-9} M with a detection limit of 3.2×10^{-12} M. In addition, this immobilized DNA sensor exhibited an excellent reproducibility and stability during DNA hybridization assay.

Recently, a DNA electrochemical biosensor was developed based on the IL/ZnO nanoparticles/chitosan membrane–modified gold electrode [77]. The DNA was incorporated via the biocompatibility of nano-ZnO, the good film-forming ability of CHIT, and room-temperature IL (Figure 3.13). Under optimal conditions using cyclic voltammetry, the target DNA sequences could be detected in the concentration range of 1.0×10^{-18} to 1.82×10^{-4} mol L^{-1} and with the detection limit of 1.0×10^{-19} mol L^{-1}. The DNA biosensor was used in the identification of *Trichoderma harzianum*.

A novel protocol for development of DNA electrochemical biosensor based on thionine–graphene nanocomposite–modified gold electrode was presented [96]. An amino-substituted oligonucleotide probe was covalently grafted onto the surface of the thionine–graphene nanocomposite by the cross-linker glutaraldehyde (Figure 3.14). The complementary oligonucleotide could be quantified in a wide range of 1.0×10^{-12} to 1.0×10^{-7} M with a good linearity ($R^2 = 0.9976$) and a low detection limit of 1.26×10^{-13} M ($S/N = 3$).

Glucose–galactose receptor protein was immobilized onto an Au electrode through a genetically engineered cysteine residue [97]. An estrogen receptor was covalently attached to the carboxylic ends of 3-mercaptopropionic acid self-assembled monolayers by a coupling reaction based on 3-3-dimethylaminopropyl-1-ethylcarbodiimide hydrochloride/N-hydroxysuccinimide [82]. The biosensor gave a linear response ($r^2 = 0.992$) for 17β-estradiol concentration from 1.0×10^{-13} to 1.0×10^{-9} M with a detection limit of 1.0×10^{-13} M ($S/N = 3$). A good stability, in which the biosensor retained 88% of its initial activity after 3 weeks of storage in 50 mM phosphate buffer at pH 7.0, was demonstrated.

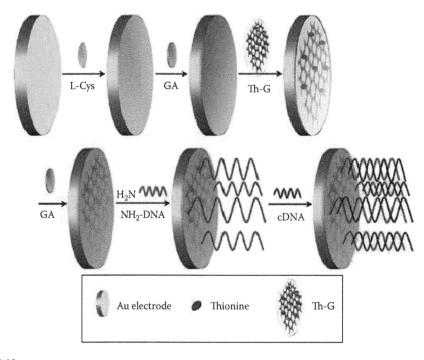

FIGURE 3.14
Schematic diagram of covalent immobilization of NH$_2$-substituted probe ssDNA on Th–G nanocomposite using glutaraldehyde as an arm linker. (From Zhu, L. et al., *Biosens. Bioelectron.*, 35, 1, 507, 2012. With permission.)

During past decades, electrochemical immunosensors based on the specificity of antigen–antibody interactions with electrochemical transduction have become an attractive subject. Human serum containing anti-Japanese encephalitis virus (JEV) antibodies was used to immobilize onto the surface of silanized interdigitated electrodes [83]. Covalent binding with a cross-linker of glutaraldehyde method provided the highest signal of the electrochemical immunosensor for detection of JEV antigens, with the linear range from 25 ng mL^{-1} to 1 µg mL^{-1}, and the limit of detection was about 10 ng mL^{-1}. Wilson [98] reported on capture antibodies that were immobilized on the porous iridium oxide electrodes by covalent attachment using 3-aminopropyl triethoxysilane and glutaraldehyde. Simultaneous detection of goat IgG and mouse IgG, and for the tumor markers CEA and AFP, was developed. The sensors had detection limits of 1, 2, 1.2, and 1 ng mL^{-1} for goat IgG, mouse IgG, CEA, and AFP, respectively. A polymeric material, poly-2-cyanoethylpyrrole, was used as for immobilizing antibody on gold electrode [99].

Recently, Amaro et al. [100] described a highly sensitive scano-magneto immunoassay using MWCNTs/AuNPs nanocomposites as a label in combination with magnetic beads (MBs). Conjugation of antibody to MWCNTs/AuNPs nanocomposite by covalent attachment was achieved. Based on this immobilization technique, scanometric detection, this platform could detect *S. enterica serovar Typhimurium* as low as 42 CFU mL^{-1} within 2.5 h assay time. A procedure in which the antibody molecules could be immobilized stably on the electrode surface attributed to the formation of covalent bond between the active –COOH of the carboxylated MWCNTs and the –NH$_2$ of antibody was described [101]. Immobilization of cytochrome c by formation of covalent linkage onto nickel oxide nanoparticles/cMWCNTs/PANI composite (NiO-NPs/cMWCNTs/PANI) electrodeposited on gold (Au) electrode was reported [79].

3.5.2.2 Cross-Linking

A device comprising an antigen or antibody species immobilized to a signal transducer is referred to as an immunosensor, which detects the binding of the complementary species. Selecting an efficient immobilization method that can yield high binding capacity with a specific target is the key element in achieving great diagnostic results. Cross-linking involves bridging between functional groups on the outer membrane of the microorganism's cells by multifunctional reagents such as glutaraldehyde and cyanuric chloride to form a network. The cells may be cross-linked directly onto the transducer surface or on a removable support membrane, which can then be placed on the transducer.

Cross-linking is characterized by covalent bonding between the various molecules of an enzyme via a polyfunctional reagent such as glutaraldehyde, diazonium salt, hexamethylene diisocyanate, and N, N'-ethylene bismaleimide. The demerit of using polyfunctional reagents is that they can denature the enzyme. This technique is cheap and simple but not often used with pure proteins because it produces very little of immobilized enzyme that has very high intrinsic activity. It is widely used in commercial preparation. Table 3.5 presents performances of some biosensors based on cross-linked biomaterials.

3.5.2.2.1 Applications

Qian et al. [108] designed a new amperometric biosensor for hydrogen peroxide based on cross-linking HRP by glutaraldehyde with MWCNTs/chitosan (MWNTs/CHIT) composite film coated on a GCE. The enzyme electrode exhibited excellent electrocatalytic activity and rapid response for H_2O_2. The linear range of detection was from 1.67×10^{-5} to 7.40×10^{-4} M with correlation coefficient of 0.998. The biosensor had good repeatability and stability for the determination of H_2O_2 and interference problems from other compounds were investigated. Gaikwad et al. [109] reported on GOD being immobilized on synthesized PANI-H_2SO_4 film by cross-linking via glutaraldehyde in phosphate and acetate buffer.

Du et al. [110] presented a simple and efficient method for immobilization of AChE on a MWNT cross-linked chitosan composite (CMC)–modified GCE. The constructed matrix used glutaraldehyde as cross-linker preventing leakage of enzyme from the film. Chitosan provided a biocompatible microenvironment around the enzyme and maintained its high biological activity. The immobilized enzyme had greater affinity for acetylthiocholine and excellent catalytic effect with a K_m^{app} value of 132 µmol L^{-1}, forming thiocholine, which was oxidized to produce detectable and rapid response.

A description of a novel glucose biosensor was proposed, based on the immobilization of GOx with cross-linking in the matrix of biopolymer chitosan (CS) on a GCE [111]. Tkac and coworkers developed a chitosan film containing SWNT (CHIT–SWNT) that was stabilized by chemical cross-linking with glutaraldehyde and free aldehyde groups producing a substrate used for covalent immobilization of galactose oxidase (GalOD) [112].

An AChE layer cross-linked by glutaraldehyde and another non-cross-linked were compared according to several parameters including background current, current before, and current after inhibition by paraoxon [113]. The results obtained confirmed the importance of glutaraldehyde cross-linking in the design of the acetylcholinesterase-based biosensor. A chitosan–glutaraldehyde-cross-linked uricase was immobilized onto Prussian blue nanoparticles (PBNPs) absorbed onto cMWCNT and PANI layer, electrochemically deposited on the surface of Au electrode [102]. The biosensor showed optimum response within 4 s at pH 7.5 and 40°C, when operated at 0.4 V versus Ag/AgCl. The linear working range for uric acid was 0.005–0.8 mM, with a detection limit of 5 µM.

TABLE 3.5

Biosensors Based on Cross-Linked Biomaterials

Substrate	Immobilized Enzyme	Detection Method	Detection Limit	Linear Range	Remark	Reference
Uric acid	Uricase	Amperometry	5 μM	0.005–0.8 mM	Sensor lost only 37% of its initial activity after its 400 uses over a period of 7 months, when stored at 4°C	[102]
Cd^{2+}	Alkaline phosphatase	Conductometry	0.5 ppm	0.5–50 ppm	The response time during 10 h was stable and good storage stability in buffer	[103]
Fusion gene	dsDNA	Cyclic voltammetry	6.7×10^{-13} M	1.0×10^{-12} to 1.0×10^{-11} M	Excellent specificity for single-base mismatch and complementary sequence (dsDNA) after hybridization	[104]
Fluoride	Tyrosinase	Amperometry	ND	1×10^{-6} to 2×10^{-5} M	Fast response (3 min) and good stability	[105]
Monosodium glutamate	L-glutamate OD and L-glutamate DH	Amperometry	0.02 mg L^{-1}	0.02–3 mg L^{-1}	The immobilization method provided good response time (2 min), neutral pH, and room-temperature conditions	[106]
Gliadin	Chicken anti-gliadin antibodies (IgY)	QCM	8 ppb	1×10^1 to 2×10^5 ppb gliadin	A potential alternative means of ensuring that people with wheat allergies and celiac patients have access to gliadin-free food	[107]

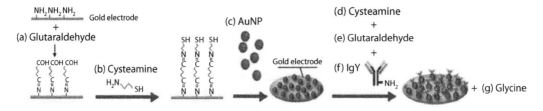

FIGURE 3.15
Production of the AuNP-modified electrode of QCM by adding (a) 2.5 M glutaraldehyde, (b) 0.5 M cysteamine, (c) AuNPs, (d) 0.5 M cysteamine, (e) 2.5 M glutaraldehyde, (f) anti-gliadin IgY, and (g) 1 M glycine. (From Chu, P. et al., *J. Agri. Food Chem.*, 60, 26, 6483, 2012. With permission.)

A label-free gliadin immunosensor that is based on changes in the frequency of a quartz crystal microbalance (QCM) chip was developed [107]. Chicken anti-gliadin antibodies (IgY) were immobilized directly on the AuNP-modified surface by cross-linking amine groups in IgY with glutaraldehyde (Figure 3.15). Sezgintürk et al. [114] described the advantages of the amino groups located on a dendrimers surface on which vascular endothelial growth factor receptor-1 (VEGF-R1) was immobilized via glutaraldehyde cross-linking. The immobilization technique allowed the fabricated biosensor to have good repeatability and reproducibility.

Covalent attachment of Abs on chemically activated sensor surfaces is the most common method for Ab immobilization. Numerous chemical coupling reagents (e.g., glutaraldehyde, carbodiimide, or succinimide ester) to cross-link mainly carboxylated functional groups have been tested to immobilize Abs onto various solid surfaces. Microorganisms have to be immobilized onto a transducer to develop a microbial biosensor. Mushroom (*Agaricus bisporus*) was homogenated and immobilized onto the GCE via gelatin and cross-linked by glutaraldehyde to construct an electrochemical tissue homogenate–based biosensor system [115]. The electrode response depends linearly on ethanol concentration between 1.0 and 10.0 mM.

3.6 Conclusion

The immobilization of biomaterials on electrode surfaces is an area of intense interest due to their wide practical applications, for instance, in the construction of biosensing devices. Numerous immobilization methods for biomaterials and different applications exist as discussed in this chapter. It can be concluded that the methods are generic. Physical adsorption being the simplest immobilization method since it consists of contacting the biomaterial with the surface for a definite period of time and consequently washing off non adsorbed biomaterials. However, this technique suffers from biomaterial desorption from the support due to weak binding and reversible nature of the binding. Cross-linking, on the other hand, is also a very popular immobilization technique for biomaterials but this method involves the use of a multifunctional reagent such as glutaraldehyde, which is toxic and causes loss of biomaterial functional properties. Covalent binding of biomaterials to a surface is another immobilization method that is carried out by initial activation of the support using a coupling agent, followed by biomaterial binding to the activated surface. Entrapment can also be used to immobilize biomaterials in 3D matrices. The technique offers the following advantages such as simplicity and no chemical reaction between a monomer and the biomaterial can affect the functional properties of the biomaterials. However, biomaterial leakage from the porous matrix can be encountered. There are interesting possibilities within the field of immobilization methods. Given the existing advances in immobilization methods, it is imminent that many conventional and modern analytical applications will be replaced by sensors and biosensors developed using these methods as witnessed by the advantages of the immobilization techniques highlighted in this chapter. A successful blending between these techniques in biomaterial immobilization will help to realize the full potential in future sensor and biosensor applications.

References

1. Xiang, C., Y. Zou, L. X. Sun, and F. Xu. 2009. Direct electrochemistry and enhanced electrocatalysis of horseradish peroxidase based on flowerlike ZnO-gold nanoparticle-Nafion nanocomposite. *Sensors and Actuators B: Chemical* 136: 158–162.
2. Moyo, M., J. O. Okonkwo, and N. M. Agyei. 2012. Recent advances in polymeric materials used as electron mediators and immobilizing matrices in developing enzyme electrodes. *Sensors* 12: 923–953.
3. Dewa, A. S. and W. H. Ko, Biosensors, in *Semiconductor Sensors*, ed. S.M. Sze (New York: Wiley Interscience, 1994), p. 415.
4. Elizabeth, A. H. H. (ed.) 1991. *Biosensors*, Prentice Hall, New York, p. 351.
5. Lowe, C. R. 1985. An introduction to the concepts and technology of biosensors. *Biosensors*, 1(1): 3–16.
6. Rogers, K. R. 1995. Biosensors for environmental applications. *Biosensors Bioelectronics* 10(6): 533–541.
7. Turner, A. P. F. 1989. Current trends in biosensor research and development. *Sensors Actuators* 17: 433–450.
8. Boyukbayram, A. E., S. Kiralp, L. Toppare, and Y. Yagci. 2006. Preparation of biosensors by immobilization of polyphenol oxidase in conducting co-polymers and their use indetermination of phenolic compounds in red wine. *Bioelectrochemistry* 69: 164–171.
9. Zhao, Z. and H. Jiang. 2010. Enzyme-based Electrochemical Biosensors. In: *Biosensors*, P. A. Serra (ed.), InTech, p.1.
10. Du, D., X. Huang, J. Cai, and A. Zhang. 2007. Amperometric detection of triazophos pesticide using acetylcholinesterase biosensor based on multiwall carbon nanotube–chitosan matrix. *Sensors and Actuators B: Chemical* 127: 531–535.
11. Andreescu, S., L. Barthelmebs, and J.-L. Marty. 2002. Immobilization of acetylcholinesterase on screen-printed electrodes: Comparative study between three immobilization methods and applications to the detection of organophosphorus insecticides. *Analytica Chimica Acta* 464: 171–180.
12. Amine, A., H. Mohammadi, I. Bourais, and G. Palleschi. 2006. Enzyme inhibition-based biosensors for food safety and environmental monitoring. *Biosensors and Bioelectronics* 21: 1405–1423.
13. Ahuja, T., I. A. Mir, D. Kumar, and R. Rajesh. 2007. Biomolecular immobilization on conducting polymers for biosensing applications. *Biomaterials* 28: 791–805.
14. Sassolas, A., L. J. Blum, and B. D. Leca-Bouvier. 2012. Immobilization strategies to develop enzymatic biosensors. *Biotechnology Advances* 30: 489–511.
15. Singh, M., N. Verma, G. A. K., and N. Redhu. 2008. Urea biosensors. *Sensors and Actuators B: Chemical* 134: 345–351.
16. Arvinte, A., L. Rotariu, and C. Bala. 2008. Amperometric low-potential detection of malic acid using single-wall carbon nanotubes based electrodes. *Sensors Actuators* 8: 1497–1507.

17. Lyons, M. E. G. and G. P. Keeley. 2008. Carbon nanotube based modified electrode biosensors. Part 1. Electrochemical Studies of the Flavin Group Redox Kinetics at SWCNT/Glucose oxidase composite modified electrodes. *International Journal of Electrochemical Science* 3: 819–853.

18. Rad, A. S., M. Jahanshahi, M. Ardjmand, and A. Safekordi. 2012. Hydrogen peroxide biosensor based on enzymatic modification of electrode using deposited silver nano layer. *International Journal of Electrochemical Science* 7: 2623–2632.

19. Gao, X., Y. Zhang, Q. Wu et al. 2011. One step electrochemically deposited nanocomposite film of chitosan–carbon nanotubes–gold nanoparticles for carcinoembryonic antigen immunosensor application. *Talanta* 85: 1980–1985.

20. Lu, J. et al. 2012. Ultrasensitive electrochemical immunosensor based on Au nanoparticles dotted carbon nanotube–graphene composite and functionalized mesoporous materials. *Biosensors and Bioelectronics* 33: 29–35.

21. Casuso, I., M. Pla-Roca, G. Gomila et al. 2008. Immobilization of olfactory receptors onto gold electrodes for electrical biosensor. *Materials Science and Engineering C* 28: 686–691.

22. Prasad, D., S. Arun, M. Murugesan et al. 2007. Direct electron transfer with yeast cells and construction of a mediatorless microbial fuel cell. *Biosensors and Bioelectronics* 22: 2604–2610.

23. Jiang, X. and X. Lin. 2005. Immobilization of DNA on carbon fiber microelectrodes by using overoxidized polypyrrole template for selective detection of dopamine and epinephrine in the presence of high concentrations of ascorbic acid and uric acid. *Analyst* 130: 391–396.

24. Ekanayake, E., D. M. G. Preethichandra, and K. Kaneto. 2007. Polypyrrole nanotube array sensor for enhanced adsorption of glucose oxidase in glucose biosensors. *Biosensors and Bioelectronics* 23: 107–113.

25. Wang, J., L. Wang, J. Di, and Y. Tu. 2009. Electrodeposition of gold nanoparticles on indium/tin oxide electrode for fabrication of a disposable hydrogen peroxide biosensor. *Talanta* 77: 1454–1459.

26. Wu, Y., J. Zheng, Z. Li, Y. Zhao, and Y. Zhang. 2009. A novel reagentless amperometric immunosensor based on gold nanoparticles/TMB/Nafion-modified electrode. *Biosensors and Bioelectronics* 24: 1389–1393.

27. Zheng, H., L. Lin, Y. Okezaki et al. 2010. Electrochemical behavior of dye-linked L-proline dehydrogenase on glassy carbon electrodes modified by multi-walled carbon nanotubes. *Beilstein Journal of Nanotechnology* 1: 135–141.

28. Firdoz, S., F. Ma, X. Yue et al. 2010. A novel amperometric biosensor based on single walled carbon nanotubes with acetylcholine esterase for the detection of carbaryl pesticides in water. *Talanta* 83: 269–273.

29. Wang, Y., J. Du, Y. Li et al. 2012. A amperometric biosensor for hydrogen peroxide by adsorption of horseradish peroxidase onto single-walled carbon nanotubes. *Colloids and Surfaces B: Biointerfaces* 90: 62–67.

30. Arora, K., A. Chaubey, R. Singhal et al. 2006. Application of electrochemically prepared polypyrrole–polyvinyl sulphonate films to DNA biosensor. *Biosensors and Bioelectronics* 21: 1777–1783.

31. Guo, M., W. Lv, S. Zhang et al. 2010. Ultrathin carbon nanotube–DNA hybrid membrane formation by simple physical adsorption onto a thin alumina substrate. *Nanotechnology* 21: 285601.

32. Cooreman, P., R. Thoelen, J. Manca et al. 2005. Impedimetric immunosensors based on the conjugated polymer PPV. *Biosensors and Bioelectronics* 20: 2151–2156.

33. Yang, L. J., Y. B. Li, and G. F. Erf. 2004. Interdigitated array microelectrode-based electrochemical impedance immunosensor for detection of *Escherichia coli* O157:H7. *Analytical Chemistry* 76: 1107–1113.

34. Nanduri, V., I. B. Sorokulova, A. M. Samoylov et al. 2007. Phage as a molecular recognition element in biosensors immobilized by physical adsorption. *Biosensors and Bioelectronics* 22: 986–992.

35. Cho, E. C., B. O. Jang, E. J. Kim, and K. Koo. 2008. Characterization of a carbon composite electrode for an electrochemical immunosensor. *Korean Journal of Chemical Engineering* 25: 548–552.

36. Trnkovaa, L., S. Krizkovab, V. Adamb, J. Hubalekc, and R. Kizeka. 2011. Immobilization of metallothionein to carbon paste electrode surface via anti-MT antibodies and its use for biosensing of silver. *Biosensors and Bioelectronics* 26: 2201–2207.

37. Zhao, G., X. Zhan, and W. Dou. 2011. A disposable immunosensor for Shigella flexneri based on multiwalled carbon nanotube/sodium alginate composite electrode. *Analytical Biochemistry* 408: 53–58.

38. Wang, M., J. Zheng, L. Wang, and S. Dong. 2012. Direct electrochemistry of glucose oxidase immobilized on titanium carbide-Au nanoparticles-fullerene C60 composite film and its biosensing for glucose. *Journal of the Chinese Chemical Society* 58: 1006–1014.

39. Zhang, L., H. Cheng, H. Zhang, and L. Qu. 2012. Direct electrochemistry and electrocatalysis of horseradish peroxidase immobilized in graphene oxide–Nafion nanocomposite film. *Electrochimica Acta* 65: 122–126.

40. Valdes-Ramirez, G., M. Cortina, M. Ramirez-Silva, and J.-L. Marty. 2008. Acetylcholinesterase-based biosensors for quantification of carbofuran, carbaryl, methylparaoxon, and dichlorvos in 5% acetonitrile. *Analytical and Bioanalytical Chemistry* 392: 699–707.

41. Yang, M., Y. Yang, Y. Yang, G. Shen, and R. Yu. 2005. Microbiosensor for acetylcholine and choline based on electropolymerization/sol–gel derived composite membrane. *Analytica Chimica Acta* 530: 205–211.

42. Matsumoto, Y., N. Terui, and S. Tanaka. 2006. Electrochemical detection and control of interactions between DNA and electroactive intercalator using a DNA–Alginate complex film modified electrode. *Environmental Science and Technology* 40: 4240–4244.

43. Salimi, A., G. R. Compton, and R. Hallaj. 2004. Glucose biosensor prepared by glucose oxidase encapsulated sol-gel and carbon-nanotube-modified basal plane pyrolytic graphite electrode. *Analytical Biochemistry* 333: 49–56.

44. Fritzen-Garcia, M. B. et al. 2009. Carbon paste electrode modified with pine kernel peroxidase immobilized on pegylated polyurethane nanoparticles. *Sensors and Actuators B: Chemical* 139: 570–579.

45. Lei, C. X., Y. Yang, H. Wang, G. L. Shen, and R. Q. Yu. 2004. Amperometric immunosensor for probing complement III (C3) based on immobilizing C3 antibody to a nano-Au monolayer supported by sol–gel-derived carbon ceramic electrode. *Analytica Chimica Acta* 513: 379–384.

46. Njagi, J. and S. Andreescu. 2007. Stable enzyme biosensors based on chemically synthesized Au-polypyrrole nanocomposites. *Biosensors and Bioelectronics* 23: 168–175.

47. Sohail, M. and S. Adeloju. 2008. Electroimmobilization of nitrate reductase and nicotinamide adenine dinucleotide into polypyrrole films for potentiometric detection of nitrate. *Sensors and Actuators B: Chemical* 133: 333–339.

48. Zhu, L., R. Yang, J. Zhai, and C. Tian. 2007. Bienzymatic glucose biosensor based on co-immobilization of peroxidase and glucose oxidase on a carbon nanotubes electrode. *Biosensors and Bioelectronics* 23: 528–535.

49. Shan, D., S. Wang, Y. He, and H. Xue. 2008. Amperometric glucose biosensor based on in situ electropolymerized polyaniline/poly(acrylonitrile-co-acrylic acid) composite film. *Material Science Engineering: C* 28: 213–217.

50. Corgier, B., C. A. Marquette, and L. J. Blum. 2005. Screen-printed electrode microarray for electrochemiluminescent measurements. *Analytica Chimica Acta* 538: 17.

51. Noorbakhsh, A., A. Salimi, and E. Sharifi. 2008. Fabrication of glucose biosensor based on encapsulation of glucose-oxidase on sol–gel composite at the surface of glassy carbon electrode modified with carbon nanotubes and Celestine blue. *Electroanalysis* 20: 1788–1797.

52. Azmi, N. E. et al. 2009. Biosensor based on glutamate dehydrogenase immobilized in chitosan for the determination of ammonium in water samples. *Analytical Biochemistry* 388: 28–32.

53. Yabuki, S. and F. Mizutani. 1995. Modifications to a carbon paste glucose-sensing enzyme electrode and a reduction in the electrochemical interference from L-ascorbate. *Biosensors and Bioelectronics* 10: 353–358.

54. Antiochia, R. and L. Gorton. 2007. Development of a carbon nanotube paste electrode osmium polymer-mediated for determination of glucose in alcoholic beverages. *Biosensors and Bioelectronics* 22: 2611–2617.

55. Jiang, X. et al. 2008. Immunosensors for detection of pesticide residues. *Biosensors and Bioelectronics* 23: 1577–1587.

56. Darain, F., S. U. Park, and Y. B. Shim. 2003. Disposable amperometric immunosensor system for rabbit IgG using a conducting polymer modified screen printed electrode. *Biosensors and Bioelectronics* 18: 773–780.

57. Liu, Y. C., C. M. Wang, K. P. Hsiung, and C. Huang. 2003. Evaluation and application of conducting polymer entrapment on quartz crystal microbalance in flow injection immunoassay. *Biosensors and Bioelectronics* 18: 937–942.

58. Alonso-Lomillo, M. A., O. Dominguez-Renedo, and M. J. Arcos-Martinez. 2009. Electrochemical determination of levetiracetam by screen-printed based biosensors. *Bioelectrochemistry* 74: 306–309.

59. Türkarslan, O., S. K. Kayahan, and L. Toppare. 2009. A new amperometric cholesterol biosensor based on poly(3,4-ethylenedioxypyrrole). *Sensors and Actuators B: Chemical* 484–488.

60. Sotiropoulou, S. and N. A. Chaniotakis. 2005. Tuning the sol–gel microenvironment for acetylcholinesterase encapsulation. *Biomaterials* 26: 6771–6779.

61. Jia, J., W. Guan, M. Sim, Y. Li, and H. Li. 2008. Carbon nanotubes based glucose needle-type biosensor. *Sensors* 8: 1712–1718.

62. Liu, X., H. Feng, R. Zhao, Y. Wang, and X. Liu. 2012. A novel approach to construct a horseradish peroxidase | hydrophilic ionic liquids | Au nanoparticles dotted titanate nanotubes biosensor for amperometric sensing of hydrogen peroxide. *Biosensors and Bioelectronics* 31: 101–104.

63. Sun, X., C. Zhai, and X. Wang. 2012. Recent advances in amperometric acetylcholinesterase biosensor. *Sensors and Transducers* 137: 199–214.

64. Qi, H., X. Li, P. Chen, and C. Zhang. 2007. Electrochemical detection of DNA hybridization based on polypyrrole/ss-DNA/multi-wall carbon nanotubes paste electrode. *Talanta* 72: 1030–1035.

65. Xu, Y., Y. Jiang, H. Cai, P. G. He, and Y. Z. Fang. 2004. Electrochemical impedance detection of DNA hybridization based on the formation of M-DNA on polypyrrole/carbon nanotube modified electrode. *Analytica Chimica Acta* 516: 19–27.

66. Fu, Y. et al. 2005. Electrochemical impedance behavior of DNA biosensor based on colloidal Ag and bilayer two-dimensional sol–gel as matrices. *Journal of Biochemical and Biophysical Methods* 62: 163–174.

67. Wang, J., M. Jiang, A. Fortes, and B. Mukherjee. 1999. New label-free DNA recognition based on doping nucleic-acid probes within conducting polymer films. *Analytica Chimica Acta* 402: 7–12.

68. Zhang, L., W. Wu, and J. Wang. 2007. Immobilization of activated sludge using improved polyvinyl alcohol (PVA) gel. *Journal of Environmental Sciences* 19: 1293–1297.

69. John, R., M. Spencer, G. G. Wallace, and M. R. Smyth. 1991. Development of a polypyrrole-based human serum-albumin sensor. *Analytica Chimica Acta* 249: 381–385.

70. Lei, C. X., F. C. Gong, G. L. Shen, and R. Q. Yu. 2003. Amperometric immunosensor for Schistosoma japonicum antigen using antibodies loaded on a nano-Au monolayer modified chitosan-entrapped carbon paste electrode. *Sensors and Actuators B: Chemical* 96: 582–588.

71. Stubbs, D. D., W. D. Hunt, S. H. Lee, and D. F. Doyle. 2002. Gas phase activity of anti-FITC antibodies immobilized on a surface acoustic wave resonator device. *Biosensors and Bioelectronics* 17: 471–477.

72. Susmel, S., R. Toniolo, A. Pizzariello, N. Dossi, and G. Bontempell. 2005. A piezoelectric immunosensor based on antibody entrapment within a non-totally rigid polymeric film. *Sensors and Actuators B: Chemical* 331: 111–338.

73. O'Grady, M. L. and K. K. Parker. 2008. Dynamic control of protein-protein interactions. *Journal of Surfaces and Colloids* 24: 316–322.

74. Wu, X., B. Zhao, P. Wu, H. Zhang, and C. Cai. 2009. Effects of ionic liquids on enzymatic cataly-sis of the glucose oxidase toward the oxidation of glucose. *The Journal of Physical Chemistry B* 113: 13365–13373.

75. Oczkowski, T., E. Zwierkowska, and S. Bartkowiak. 2007. Application of cell-based biosensors for the detection of bacterial elicitor flagellin. *Bioelectrochemistry* 70: 192–197.

76. Rawal, R., S. Chawla, P. Malik, and C. S. Pundir. 2012. An amperometric biosensor based on laccase immobilized onto MnO2NPs/cMWCNT/PANI modified Au electrode. *International Journal of Biological Macromolecules* 51: 175–181.

77. Siddiquee, S., N. A. Yusof, A. B. Salleh, S. G. Tan, and F. A. Bakar. 2012. Development of elec-trochemical DNA biosensor for Trichoderma harzianum based on ionic liquid/ZnO nanopar-ticles/chitosan/gold electrode. *Journal of Solid State Electrochemistry* 16: 273–282.

78. Escamilla-Gómez, V., S. Campuzano, M. Pedrero, and J. M. Pingarrón. 2008. Electrochemical immunosensor designs for the determination of *Staphylococcus aureus* using 3,3-dithiodipropi-onic acid di(N-succinimidyl ester)-modified gold electrodes. *Talanta* 77: 876–881.

79. Lata, S., B. Batra, N. Karwasra, and C. S. Pundir. 2012. An amperometric H2O2 biosensor based on cytochrome c immobilized onto nickel oxide nanoparticles/carboxylated multiwalled car-bon nanotubes/polyaniline modified gold electrode. *Process Biochemistry* 47: 992–998.

80. Wang, Z., Y. Tua, and S. Liu. 2008. Electrochemical immunoassay for α-fetoprotein through a phenylboronic acid monolayer on gold. *Talanta* 77: 815–821.

81. Teymourian, H., A. Salimia, and R. Hallaj. 2012. Low potential detection of NADH based on Fe3O4 nanoparticles/multiwalled carbon nanotubes composite: Fabrication of integrated dehydrogenase-based lactate biosensor. *Biosensors and Bioelectronics* 33: 60–68.

82. Byung, K. K., L. Jing, I. Ji-Eun et al. 2012. Impedometric estrogen biosensor based on estrogen receptor alpha-immobilized gold electrode. *Journal of Electroanalytical Chemistry* 671: 106–111.

83. Quang, H. T. et al. 2012. Development of electrochemical immunosensors based on different serum antibody immobilization methods for detection of Japanese encephalitis virus. *Advances in Natural Sciences: Nanoscience and Nanotechnology* 3: 1–7.

84. Li, J., Y. B. Wang, J. D. Qiu, S. D. C., and X. H. Xia. 2005. Biocomposites of covalently linked glu-cose oxidase on carbon nanotubes for glucose biosensor. *Analytical and Bioanalytical Chemistry* 383: 918–922.

85. Yin, H., S. Ai, W. Xu, J. Shi, and L. Zhu. 2009. Amperometric biosensor based on immobilized ace-tylcholinesterase on gold nanoparticles and silk fibroin modified platinum electrode for detec-tion of methyl paraoxon, carbofuran and phoxim. *Journal of Electroanalytical Chemistry* 637: 21–27.

86. Kandimalla, B. V. and H. Ju. 2006. Binding of acetylcholinesterase to multiwall carbon nano-tube-cross-linked chitosan composite for flow-injection amperometric detection of an organo-phosphorous insecticide. *Chemistry—A European Journal* 12: 1074–1080.

87. Ahuja, T., V. K. Tanwar, S. K. Mishra, D. Kumar, and A. M. Biradar. 2011. Immobilization of uri-case enzyme on self-assembled gold nanoparticles for application in uric acid biosensor. *Journal of Nanoscience and Nanotechnology* 11: 4692–4701.

88. Tiwari, A. A. 2009. A novel nanocomposite matrix based on silylated chitosan and multiwall carbon nanotubes for immobilization of urease. *Journal of Inorganic and Organometallic Polymers Materials* 19: 361–366.

89. Chauhan, N. and S. C. Pundir. 2011. An amperometric biosensor based on acetylcholinesterase immobilized onto iron oxide nanoparticles/multi-walled carbon nanotubes modified gold elec-trode for measurement of organophosphorus insecticides. *Analytica Chimica Acta* 701: 66–74.

90. Dhyani, H., M. A. Ali, M. K. Pandey, B. D. Malhotra, and P. Sen. 2012. Electrophoretically depos-ited CdS quantum dots based electrode for biosensor application. *Journal of Materials Chemistry* 22: 4970–4976.

91. Chauhan, N. and C. S. Pundir. 2012. An amperometric acetylcholinesterase sensor based on Fe3O4 nanoparticle/multi-walled carbon nanotube-modified ITO-coated glass plate for the detection of pesticides. *Electrochimica Acta* 67: 79–86.

92. Li, S., X. Zhu, W. Zhang, G. Xie, and W. Feng. 2012. Hydrogen peroxide biosensor based on gold nanoparticles/thionine/gold nanoparticles/multi-walled carbon nanotubes–chitosans composite film-modified electrode. *Applied Surface Science* 258: 2802–2807.

93. Tlili, C., H. Korri-Youssoufi, H. L. Ponsonnet, C. Martelet, and N. J. Jaffrezic-Renault. 2005. Electrochemical impedance probing of DNA hybridisation on oligonucleotide-functionalised polypyrrole. *Talanta* 68: 131–137.

94. Marquette, C. A., M. F. Lawrence, and L. J. Blum. 2006. DNA covalent immobilization onto screen-printed electrode networks for direct label-free hybridization detection of p53 sequences. *Analytical Chemistry* 78: 959–964.

95. Zhang, Y., K. Zhang, and H. Ma. 2009. Electrochemical DNA biosensor based on silver nanoparticles/poly(3-(3-pyridyl) acrylic acid)/carbon nanotubes modified electrode. *Analytical Biochemistry* 387: 13–19.

96. Zhu, L., L. Luo, and Z. Wang. 2012. DNA electrochemical biosensor based on thionine-graphene nanocomposite. *Biosensors and Bioelectronics* 35: 507–511.

97. Wang, J., L. A. Luck, and I. I. Suni. 2007. Immobilization of the glucose-galactose receptor protein onto a au electrode through a genetically engineered cysteine residue. *Electrochemical and Solid-State Letters* 10: R33-R36.

98. Wilson, M. S. 2005. Electrochemical immunosensors for the simultaneous detection of two tumor markers. *Analytical Chemistry* 77: 1496–1502.

99. Um, H. et al. 2011. Electrochemically oriented immobilization of antibody on poly-(2-cyanoethylpyrrole)-coated gold electrode using a cyclic voltammetry. *Talanta* 84: 330–334.

100. Amaro, M., S. Oaew, and W. Surareungchai. 2012. Scano-magneto immune assay based on carbon nanotubes/gold nanoparticles nanocomposite for *Salmonella enterica*. *Biosensors and Bioelectronics* 38: 157–162.

101. Cao, Y. et al. 2012. Ultrasensitive luminol electrochemiluminescence for protein detection based on in situ generated hydrogen peroxide as coreactant with glucose oxidase anchored AuNPs@MWCNTs labeling. *Biosensors and Bioelectronics* 31: 305–309.

102. Rawal, R., S. Chawla, N. Chauhan, T. Dahiya, and C. S. Pundir. 2012. Construction of amperometric uric acid biosensor based on uricase immobilized on PBNPs/cMWCNT/PANI/Au composite. *International Journal of Biological Macromolecules* 50: 112–118.

103. Berezhetskyy, A. L. et al. 2008. Alkaline phosphatase conductometric biosensor for heavy-metal ions determination. *ITBM-RBM* 29: 136–140.

104. Qian, P., S. Ai, H. Yin, and J. Li. 2010. Evaluation of DNA damage and antioxidant capacity of sericin by a DNA electrochemical biosensor based on dendrimer-encapsulated Au-Pd/chitosan composite. *Microchimica Acta* 168: 347–354.

105. Asav, E., E. Yorganci, and E. Akyilmaz. 2009. An inhibition type amperometric biosensor based on tyrosinase enzyme for fluoride determination. *Talanta* 78: 553–556.

106. Basu, A. K., P. Chattopadhyay, U. Roychudhuri, and R. Chakraborty. 2006. A biosensor based on co-immobilized L-glutamate oxidase and L-glutamate dehydrogenase for analysis of monosodium glutamate in food. *Biosensors and Bioelectronics* 21: 1968–1972.

107. Chu, P. T., C. S. Lin, W. J. Chen, C. F. Chen, and H. W. Wen. 2012. Detection of gliadin in foods using a quartz crystal microbalance biosensor that incorporates gold nanoparticles. *Journal of Agricultural and Food Chemistry* 60: 6483–6492.

108. Qian, L. and X. Yang. 2006. Composite film of carbon nanotubes and chitosan for preparation of amperometric hydrogen peroxide biosensor. *Talanta* 68: 721–727.

109. Gaikwad, P. D. et al. 2006. Immobilization of GOD on electrochemically synthesized PANI film by cross linking via glutaraldehyde for determination of glucose. *International Journal of Electrochemical Science* 1: 425–434.

110. Du, D. et al. 2007. An amperometric acetylthiocholine sensor based on immobilization of acetylcholinesterase on a multiwall carbon nanotube-cross-linked chitosan composite. *Analytical and Bioanalytical Chemistry* 387: 1059–1065.

111. Kang, X., Z. Mai, X. Zou, P. Cai, and J. Mo. 2007. Novel glucose biosensor based on immobilization of glucose oxidase in chitosan on a glassy carbon electrode modified with gold-platinum alloy nanoparticles/multiwall carbon nanotubes. *Analytical Biochemistry* 369: 71–79.
112. Tkac, J., J. W. Whittaker, and T. Ruzgas. 2007. The use of single walled carbon nanotubes dispersed in a chitosan matrix for preparation of a galactose biosensor. *Biosensors and Bioelectronics* 22: 1820–1824.
113. Pohanka, M., K. Kuča, and D. Jun. 2008. Optimization of acetylcholinesterase immobilization onto screen printed platinum electrode. *Journal of Applied. Biomedicine* 6: 27–30.
114. Sezgintürk, M. K. and Z. O. Uygun. 2012. An impedimetric vascular endothelial growth factor biosensor-based PAMAM/cysteamine-modified gold electrode for monitoring of tumor growth. *Analytical Biochemistry* 423: 277–285.
115. Akyılmaz, E. and O. İmer. 2011. Voltammetric determination of ethanol by using mushroom (*Agaricus bisporus*) Tissue homogenate-based biosensor. *Journal of Biology and Chemistry* 39: 65–70.

4

Nanofibers and Nanoparticles in Biomedical Applications

Ana Baptista, Paula Soares, Isabel Ferreira, and João Paulo Borges

CONTENTS

4.1 Introduction

Nanotechnology emerges from the physical, chemical, biological, and engineering sciences, where novel techniques are being developed to probe and manipulate single atoms and molecules. By interacting with biological molecules, nanotechnology opens up a promising and revolutionary field of application in medicine. Miniaturization provides cost-effective and more promptly operative mechanical, chemical, and biological components. In fact, the nanometer-sized objects also possess remarkable self-ordering and assembly behaviors under the control of forces different from macro objects [1]. The integration of nanomaterials with biology has led to the development of diagnostic devices, contrast agents, analytical tools, physical therapy applications, and drug delivery vehicles [2] useful for both *in vivo* and *in vitro* biomedical research and applications [3].

The considerable importance and progress in preparation and characterization of nano-materials and in its functionalization and application are the main reasons to review the research activity related to the processing of functional nanostructures focusing the nano-fiber and nanoparticle (NP) systems for biomedical applications.

4.1.1 Nanoparticles

In a general approach, nanostructures have at least one dimension in a range of 1–100 nm and can be classified according to their dimensional nanosize. Further, these nanostruc-tures exhibit new or enhanced size-dependent properties compared with larger particles of the same material. Although this definition is simple and not entirely false, a more thorough discussion is required to achieve an unambiguous and complete definition.

4.1.2 Definition

Some regulatory agencies such as International Organization of Standardization (ISO), American Society for Testing Materials (ASTM), and Scientific Committee on Emerging and Newly Identified Health Risks (SCENIHR) have provided their own definitions to clarify this issue. The ISO experts reported, for the first time, important definitions in ISO/TS 27687:2008 "Nanotechnologies—Terminology and definitions for nano-objects—Nanoparticle, nanofibre and nanoplate" [4]. According to this standard, a NP is defined as a "nano-object with all three dimensions in the nanoscale" being the nanoscale a "size range from approximately 1 nm to 100 nm." Two notes accompany this nanoscale definition: "Properties that are not extrapolations from a larger size will typically, but not exclusively, be exhibited in this size range. For such properties the size limits are considered approximate" and "The lower limit in this definition (approx-imately 1 nm) is introduced to avoid single and small groups of atoms from being designated as nano-objects or elements of nanostructures, which might be implied by the absence of a lower limit." This was the first standard published; currently, two other standards are being developed. Concerning NP definition, there are three impor-tant standards being developed: ISO/NP TS 12144 "Nanotechnologies—Core terms—Terminology and definitions" and ISO/NP TR 12802 "Nanotechnologies—Terminology and nomenclature—Framework" that provide a prioritized and systematic approach for developing further definitions.

Another regulatory agency is ASTM that has established the standard ASTM E2456-06 "Terminology for Nanotechnology" [5] where a NP is defined as "a sub-classification of ultrafine particle with lengths in two or three dimensions greater than 0.001 μm (1 nm) and smaller than about 0.1 μm (100 nm) and which may or may not exhibit a size-related intensive property." Accompanying this definition is a discussion relating the following: "This term is a subject of controversy regarding the size range and the presence of a size-related property. Current usage emphasizes size and not properties in the definition. The length scale may be a hydrodynamic diameter or a geometric length appropriate to the intended use of the nanoparticle." Thus, this definition is not clear about how prop-erties can change at the nanoscale and still does not include NPs within a size range from 100 to 999 nm.

So far, the previously cited standards define nanomaterials by their size, establishing an upper threshold of 100 nm. However, some authors [6–7] contested this limit and sug-gested an upper limit of 1000 nm for pharmaceuticals, arguing that the defined upper limit (100 nm) does not clearly distinguish nanomaterials from their bulk counterparts because it does not take into account other properties such as size, size distribution, and electrical, optical, and mechanical properties at nanoscale.

The European Commission also discussed the topic and issued a document entitled "Scientific Basis for the Definition of the Term 'nanomaterial,'" [8] stating a more complex approach. This document starts by considering the current purposes and defines the criteria that should be used in categorizing nanomaterials: size, size distribution, specific surface area (SSA), surface modifications, and other physical properties. Concerning the size factor, SCENIHR considers that it is a predominant factor, but that alone might not be sufficient as a measure as any size mentioned should be controllable and enforceable. In addition, not all techniques to measure the size are comparable, and the designation of one or more external dimensions does not capture aggregates, agglomerates, and complex assemblies. Regarding the size distribution, the authors consider that it should be expressed in number and not in mass, due to a possible misperception factor. The SSA is a complementary qualifier to distinguish nanostructured materials from non-nanostructured materials based on its integral material surface area per material volume. Finally, surface modification of the nanomaterials may change, remove, or include specific properties, for example, by coating a NP. Regarding the classification of the nanomaterials, as there is no scientific evidence for an upper cutoff of 100 nm, the term "nanomaterial" can be used for the entire nanoscale (1–999 nm). As such, the nanomaterials are divided in three categories:

- *Category 1*: Nanomaterials with a size range above 500 nm, in which a small part may present a size below 100 nm. In this case, the recommended approach is to confirm the nanomaterial properties by size distribution and specific properties.
- *Category 2*: Nanomaterials with a size range between 100 and 500 nm that are more likely nanomaterials but need a more detailed characterization and a nano-specific risk assessment.
- *Category 3*: Nanomaterials with a size range between 1 and 100 nm are considered nanomaterials and obligatorily require a nano-specific risk assessment. In this case, an SSA above 60 m^2/cm^3 may be used as an additional qualifier to indicate a size below 100 nm.

In conclusion, the SCENIHR states that the physicochemical properties may change with the size of the nanomaterials, but there is no scientific evidence to establish an upper and lower limit that reflects these differences for all them. In addition, there is no scientific evidence of a methodology (or group of tests) that may be applied for all the nanomaterials. Lastly, the size measurement is a universal tool to classify any nanomaterial; however, all the other properties should be taken into account, depending on the type of nanomaterial. Further, size distribution by number is the most relevant consideration.

4.1.3 Classification

Similar to NP definition, there is a disagreement in NP classification. First, it is important to define the parameters by which NPs are categorized. These parameters may include the type of material (metal, polymer, carbon, among others), morphology of the NPs, and their source. A classification of the NPs according to their origin, natural or accidental, with irregular shapes, and engineered NPs, with regular shapes, has been found [9]. Natural NPs are formed through natural processes and occur in the environment (e.g., volcanic dust, lunar dust, minerals, magnetotactic bacteria). Incidental NPs, also called waste or anthropogenic particles, occur as the result of man-made industrial processes (e.g., diesel exhaust, coal combustion, welding fumes).

Other authors have classified the NPs for toxicological purposes. An extensive study on health effects of the NPs classifies them according to four factors: dimensionality, morphology, composition, and uniformity/agglomeration [10]. The health effects were correlated with the source of the NP exposure: natural sources (dust storm, forest fires, volcanoes, ocean and water evaporation, organisms), anthropogenic sources (diesel and engine exhaust NPs, indoor pollution, cigarette smoke, building demolition, cosmetic and other consumer products, engineered NPs), environmental and occupational exposure (metals and other dusts, carcinogens, and poorly soluble [durable] particles), and aerosol pollution. The U.S. Environmental Protection Agency [11] also divides engineered NPs into four categories: carbon-based materials (nanotubes, fullerenes), metal-based materials (metal oxides and quantum dots), dendrimers, and composite NPs.

Relating to biomedical applications, the NPs were classified according to their therapeutic and diagnostic role into five categories [12] (Figure 4.1):

- *Type I*—NPs with endogenous contrast (e.g., iron oxide nanomaterials, quantum dots)
- *Type II*—NPs carrying an imaging agent (e.g., radio-labeled silica, fluorescent silica)
- *Type III*—NPs carrying a therapeutic agent (e.g., pH-responsive liposomes, RNAi NPs)
- *Type IV*—Labeled NPs carrying a therapeutic agent (fluorescent liposomes)
- *Type V*—NPs responsive to an external stimulus (e.g., gold nanorod, magnetic NPs)

Types I and II are self-containing NPs, allowing image-guided resection as their therapeutic role, but with a more specific role into diagnosis: Type I NPs may be useful for imaging with site specificity, border delineation, and phenotyping, while type II NPs may be used for border delineation and typing with site specificity. On the other hand, types III and IV involve the release of a therapeutic agent, being able to target the diseased area nonspecifically or specifically via homing ligands. Finally, type V NPs may be composed of any of the various elements of the other four types but have the specificity of only being activated in the presence of an external stimulus. These NPs may be active for photothermal therapy

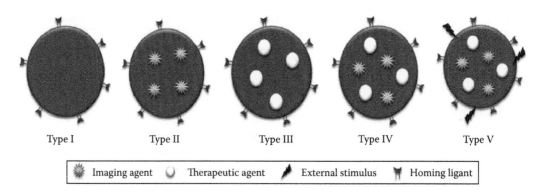

| Type I | Type II | Type III | Type IV | Type V |

| ✺ Imaging agent | ◗ Therapeutic agent | ➤ External stimulus | ♈ Homing ligand |

FIGURE 4.1
Classification of NPs according to their therapeutic and diagnostic roles. (From Jokerst, J.V. and Gambhir, S.S., *Acc. Chem. Res.*, 44, 1050, 2011.)

or selective drug release in the therapeutic domain or for thermal imaging, photoacoustic imaging, or computed tomography.

A different approach for categorizing NPs in the biomedical field was used [13], and 11 classes of NPs have been selected from the literature (up to 2010) with application in multimodality biomedical imaging and theranostics: metal oxide NPs, gold-containing nanostructures, quantum dots, silica NPs, molecular dots, carbon-based NPs, nanowires, biological NPs, dendrimers, and liposomes:

- *Metal oxide NPs*: In this category are included iron oxide NPs with polymeric coating—monocrystalline iron oxide nanocolloid (MION, 5–30 nm), cross-linked iron oxide (CLIO, 10–50 nm), and superparamagnetic iron oxide (SPIO, >50 nm). Although these NPs are conventional magnetic resonance imaging (MRI) agents, they might be affected by the enhanced permeability and retention (EPR) effect: this effect may be favorable or not, according to the NP destination. EPR effect is also called passive targeting and relies on the transport of nanosystems through the leaky tumor microvasculature into the tumor interstitium and cells [14]. Also, iron oxide NPs may suffer opsonization and have a potential toxicity.

- *Gold nanorods and nanoshells*: Depending on the NP definition considered, some of these structures may not be considered NPs. Nanorod is a nano-object with two similar external dimensions in the nanoscale, and the third dimension is significantly larger than the other two external dimensions. Nanoshells are hollow NPs and are nano-objects with all three dimensions in the nanoscale. Gold NPs have several advantages, depending on their structure: flexibility in tuning the wavelengths of optical absorption and scatter, from the visible to the near infrared (NIR), and ability to permit noninvasive photothermal ablation of tumors using ultrasounds and microwaves.

- *Quantum dots* are fluorescent semiconductor nanocrystals with a size range of 2–100 nm with unique tunable optical and targeting properties [15]. However, these NPs have several disadvantages such as UV decomposition, aggregation, reticuloendothelial system (RES) uptake, and a high potential for toxicity. *Molecular dots* are calcium phosphate nanocolloids with a size range of 20–100 nm. These biocompatible multilayer nanocomposites are more rigid than larger liposomes and have lower toxicity when compared to other NPs due to their composition. In addition, since these NPs are biocompatible and biodegradable, they could potentially be used as universal carriers.

- *Silica NPs* are very popular due to their biocompatibility, chemical and biological inertness, and thermal stability. Also, these systems are very easy to produce, and their synthesis may be performed in an aqueous media at room temperature [16]. However, these NPs are susceptible to aggregate, have low mechanical stability, and are potentially toxic.

- *Carbon nanotubes and nanowires* are nanostructures with only two dimensions in the nanoscale. Consequently, their characteristics are outranging our aim.

- *Biological NPs* may be composed of several types of biological materials: peptides, proteins, enzymes, lipoproteins, and natural polymers (e.g., chitosan, cellulose). The greatest advantages of these nanostructures are their biocompatibility and biodegradability. However, they often present weak mechanical properties, and it is difficult to control their size and shape. Some of these disadvantages may

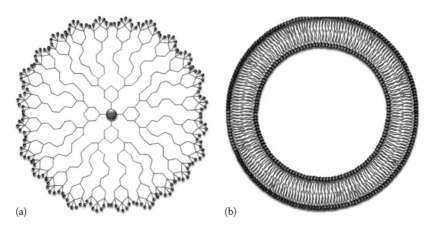

(a)   (b)

FIGURE 4.2
Schematic representation of a dendrimer (a) and a liposome (b).

be eliminated by using biodegradable and biocompatible synthetic polymers (e.g., poly(lactic acid) [PLA] and poly(vinyl alcohol) [PVA]). These polymers may be used to fabricate nanospheres for the drug delivery and controlled release of hydrophobic therapeutic drugs to their active site. In addition, these polymers may be used for coating inorganic-based NPs (gold, iron oxide NPs, quantum dots, silica NPs), increasing their stability and their blood half-life, and decreasing the RES uptake and their toxicity. Finally, coating these polymers with other functional polymers may lead to the production of "smart" polymers that become active after the application of an external stimulus such as pH, temperature, light, or ionic strength.

- *Dendrimers* (Figure 4.2a) are spherical macromolecular structures (3–30 nm) with highly branched (treelike) internal structures. Their terminal groups are responsible by their interaction with the external media. These structures are suitable to encapsulate therapeutic agents both in their interior and in their surface. Further, dendrimers are hydrophilic, flexible, biocompatible, and eliminated via renal filtration. As a disadvantage, dendrimers have a limited synthesis, making their production expensive.

- *Liposomes* (Figure 4.2b) are composed of phospholipid bilayers, giving them a large payload capacity and a straightforward synthesis. However, they have poor mechanical stability and may suffer the EPR effect and opsonization. Liposomes have the potential to encapsulate in their lipid membrane hydrophilic drugs, diagnostic agents, peptides, antibodies, hormones, and macromolecules.

Although a more general approach about NP categorization was provided, some types of NPs are still unclassified, like core–shell NPs. These examples show that it is difficult to arrange a classification method that includes all the NP types but separates them according to their specific characteristics.

4.1.4 Iron Oxide Nanoparticles

Among the different types of NPs, magnetic NPs are an important class due to their unique characteristics. Magnetic NPs are generally composed of magnetic elements such as iron, nickel, cobalt, and their oxides. These can be manipulated by an external magnetic field, which is a great advantage since it avoids invasive therapeutics and diagnostics. Iron oxide NPs are however superior to other metal oxide NPs for their biocompatibility and stability and are the most commonly employed magnetic NPs for biomedical applications [17]. From all the iron oxides, Fe_3O_4 (magnetite) possesses the most interesting properties because of the presence of iron cations in two valence states, Fe^{2+} and Fe^{3+}, in the inverse spinel structure [18]. In the bulk form, this material is ferromagnetic due to the antiferromagnetic coupling among Fe (III) in its inverse spinel structure. However, at nanoscale (<20 nm), iron oxide NPs are superparamagnetic at room temperature, that is, when the size of the particle is reduced enough, the magnetite NPs (mNPs) become a single magnetic dipole and therefore maintain one large magnetic moment. Consequently, these NPs are easily magnetized under the application of an external magnetic field and unmagnetized once the field is removed. Another interesting property of iron oxide NPs is their ability to generate heat when subjected to an alternating magnetic field (AMF). This thermal phenomenon may be due to energy losses during the demagnetization process (specific loss power). This property is crucial for hyperthermia applications [19–21].

Due to an extensive range of applications, several authors have focused on their production and characterization. Iron oxide NPs can be produced by a relatively simple synthesis. There are several methods available, although none of them is universal for all the types of magnetic NPs. When choosing a method for iron oxide NPs, it is important to consider the purpose of the NP. The coprecipitation method is probably the simplest method to produce iron oxide NPs. It is widely used for biomedical applications because of ease of implementation and need for less hazardous materials. By a careful control of the reaction parameters, it is possible to obtain a fine suspension of NPs with sizes as small as 5 nm. For magnetite preparation, a base is added to a mixture of Fe^{2+} and Fe^{3+} solution (molar ratio of 1:2) under an inert atmosphere:

$$Fe^{2+}_{(aq)} + Fe^{3+}_{(aq)} + 8OH^{-}_{(aq)} \rightarrow Fe_3O_{4(s)} + 4H_2O_{(l)} \qquad (4.1)$$

The main disadvantages of this method are the inherent difficulty of controlling particles size, distribution, and morphology. Also, an oxygen-free atmosphere is required to avoid oxidation of ferrous ion in the solution [22–23]. Iron oxide NPs may be produced also through water-in-oil (W/O) microemulsion systems that are composed of fine microdroplets of the aqueous phase trapped within assemblies of surfactant molecules dispersed in a continuous oil phase. The surfactant-stabilized microcavities (of approximately 10 nm range) provide the necessary confinement to limit particle nucleation, growth, and agglomeration [22]. Thermal decomposition is another approach for iron oxide NP synthesis that is based on the decomposition of metal oxysalts (e.g., nitrates, carbonates, and acetates) when heated to a certain temperature. Since it is a more controlled environment, experiments showed good size control of the samples, narrow size distribution, and crystalline individual and disperse magnetic iron oxide NPs [22]. Thermolysis of $Fe(acac)_3$ in diphenyl ether in the presence of small amounts of hexadecane-1,2-diol gives very fine Fe_3O_4 NPs (around 1 nm), which can be enlarged by adding excess $Fe(acac)_3$ into the reaction mixture [18].

The superior magnetic properties of iron oxide NPs, along with their inherent biocompatibility and inexpensiveness, have made iron oxide NPs the main choice for theranostic agents. Iron oxide NPs are superparamagnetic; if occurs when a particle is small enough for thermal fluctuations to cause random flipping of magnetic moments. Randomization of the orientation of these magnetic moments results in an average magnetization of zero in the absence of an applied magnetic field. The characteristic time from one moment-flip to the next is given by the Néel–Arrhenius relation:

$$\tau = \tau_0 \exp\left(\frac{KV}{k_B T}\right) \tag{4.2}$$

where
  τ_0 is the time between flip attempts (typically between 10^{-9} and 10^{-12} s)
  K is the magnetic anisotropy energy density
  V is the particle volume
  k_B is the Boltzmann constant
  T is temperature

Through the analysis of this relation, it is possible to conclude that the relaxation time decreases, or flipping frequency increases, as the particle size decreases. Considering the previously cited threshold of 20 nm, particles with an upper size are composed by multiple magnetic domains rather than a single large crystal [24].

Typically, once the iron oxide NPs are administrated within the body, they are internalized by the cells via endocytosis where, clustered in liposomes, the NPs are degraded into iron ions by an array of hydrolyzing enzymes at low pH according to endogenous iron metabolism pathways. However, NP physical characteristics (size, surface charge, etc.) determine their biodistribution in the body. Large NPs (>200 nm) are usually retained in the spleen via mechanical filtration followed by phagocytosis, while small NPs (<10 nm) are rapidly removed through extravasation and renal clearance. Therefore, particles within a range diameter of 10–100 nm are optimal for intravenous administration [17].

Considering the iron oxide core, bare iron oxide NPs have shown some toxic effects, while coated iron oxide NPs have been found to be relatively nontoxic. Poly(ethylene glycol) (PEG)-coated NPs have shown biocompatibility since exposed cells remained more than 99% viable relative to control at an upper concentration of 1 mg/mL [25]. On the other hand, bare iron oxide NPs induced 25%–50% loss in fibroblast viability at 250 μm/mL. In a more extensive study, the same group [26] found that iron oxide NP toxicity is dose-dependent. Iron oxide NPs caused a 20% reduction in cell viability at the lowest concentration tested (0.05 mg/mL). Further reductions were seen at higher concentrations, with the highest concentration tested (2.0 mg/mL), resulting in about 60% loss of cell viability.

Iron oxide NPs may cause toxicity by inducing redox cycling and catalytic chemistry via the Fenton reaction:

$$H_2O_2 + Fe^{2+} \rightarrow Fe^{3+} + HO^- + HO^* \tag{4.3}$$

This reaction is the most prevalent source of reactive oxygen species (ROS) in biological systems [17]. However, *in vivo* toxicity studies on iron oxide NPs show no long-term implications of their use when administrated at clinically relevant concentrations via relevant routes. For example, although iron oxide NP deposits were detected in the prostates of prostate cancer patients after 1 year of magnetic hyperthermia therapy in phase I clinical trials, there were no evidences of toxic effects due to this accumulation [27].

Considering all the previously described characteristics, iron oxide NPs are exciting candidates as ideal platform materials for targeted imaging of biological materials.

Additionally, their mode of action is noninvasive and remote for controlled drug release and cell signaling in the therapeutic field. Their greater magnetic properties allow for targeted imaging of specific receptors in disease processes and for monitoring treatment efficiency. The development of theranostic agents based on iron oxide NPs is a relatively recent approach but is an exciting application of these NPs. If by one hand it is still too early to predict the success of these technologies, the higher interest in this field has led to a higher investment in the area, and exciting results are expected soon.

The main biomedical applications of NPs will be further discussed in Section 4.3.

4.2 Nanofibers

A nanofiber is a 2D nanomaterial (according to ISO/TS 27687:2008), typically with diameters between tens and hundreds of nanometers. However, the third dimension (length), depending on the production process, can reach several kilometers. These long fibers have a very large SSA, and in particular polymeric nanofibers are suitable for many applications being used as individual fiber, multicomponent fibers, or fiber mats [28–29] (Figure 4.3).

Several methods such as *template, drawing, phase separation, self-assembly,* and *electrospinning* are employed to obtain nanofibers [29]. The *template* synthesis method uses a nanoporous membrane to make solid or hollow nanofibers whose diameter is controlled by the porous size of the template [30]. Various raw materials can be used with this technique, such as electronically conducting polymers, metals, semiconductors, and carbon-based materials. The *drawing* method is a process in which one fiber at a time is drawn. A sharp

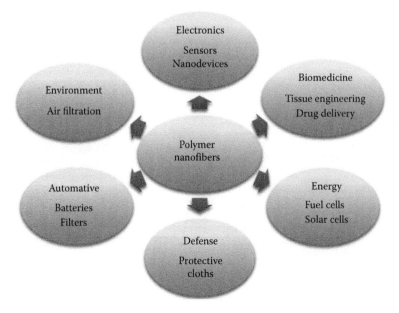

FIGURE 4.3
Fields of application of polymer nanofibers.

tip from atomic force microscope (AFM), to which is applied a voltage, is used to pull a fiber from a droplet of a viscoelastic solution writing one-by-one single nanofiber at a constant rate, being possible to develop 3D networks of suspended fibers [31]. An alternative process uses a glass micropipette filled with a polymeric solution. In contact with an adequate substrate, it leaves a drop, which is pulled by moving the pipette from one point to the other of the substrate forming the nanofiber. The liquid fiber is then solidified by rapid evaporation of the solvent [32]. These processes are possible only in viscoelastic materials that undergo strong deformations being cohesive enough to support the stress developed during the drawing. The *phase separation* process is obtained by mixing a polymer with the solvent before undergoing gelation. After phase separation, the solvent is removed by drying, and a porous nanofibrous structure is created [29]. However, the process takes relatively long time to transfer the solid polymer into the nanoporous structure. The *self-assembly* is a process in which individual, preexisting components organize themselves into desired patterns and functions [33]. The intermolecular forces are the main mechanism responsible for the assembling of molecule units, and it determines the macromolecular shape of nanofiber. Aiming the production of membranes or high-throughput fiber formation, the previously mentioned methods have relevant limitations concerning the materials, costs, and production rates. In comparison, *electrospinning* overcomes these limitations.

In electrospinning process, an electric field is applied between a tip of a capillary and a collector [28]. This electric field causes a jet of the solution to be drawn toward a grounded collector (Figure 4.4). During this process, a rapid evaporation of solvent of the jet occurs, and the fibers solidify into the collector with high deposition rate. A wide variety of polymers and precursors can be used to obtain electrospun nanofibers with several shapes and dimensions. Besides this, the process also enables the production of monocomponent or composite fiber mats [34]. Typically, nonwoven fiber mats composed by randomly oriented fibers connected together by physical entanglements or bonds between individual fibers are produced. The morphology of electrospun fibers depends upon process parameters such as the solution properties (viscosity, conductivity, and surface tension), the processing conditions (the flow rate of the solution through the syringe, the voltage applied to the

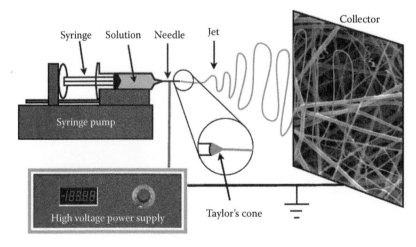

FIGURE 4.4
Schematic representation of a typical electrospinning setup, some of the processes involved, and the production of nanofibers.

needle, and the distance between the needle and the collector), and the environmental conditions (temperature and humidity) [34,35]. Each of the earlier parameters, separately or combined, can control the fiber diameter and its uniformity and shape, and consequently electrospun fibers with a wide variety of cross-sectional shapes such as cylindrical, ribbons, or helices can be obtained [36]. The versatility of electrospinning process enables the production of continuous nanofibers of natural and synthetic polymers [37], such as PLA [38], polyurethanes [39], silk fibroin [40], collagen [41], cellulose and its derivatives [36,42], composites [43], and ceramics [44].

Over the last century, a great progress has been made in the development of various designs and modifications of the electrospinning setup [45–47] in order to produce core/shell composite fibers, multichannel tubular structure or porous fibers. Thus, the field of application has also been enlarged, in particular for biomedical applications such as tissue engineering scaffolds, drug delivery systems, and biodevices. The next section will review specific biomedical applications.

4.3 Biomedical Applications

The great interest in the last decade in nanotechnology is due to their wide range of applications, especially in the biomedical field. The exceptional properties of the materials at the nanoscale make them perfect candidates for drug delivery systems, as scaffolds for tissue regeneration or for biodevice development or as theranostic agents. NP applications may be generally divided in three categories: drug delivery, tissue regeneration, and theranostics. In theory, all the NP types described in Section 4.1.3 have numerous applications in these three categories. Moreover, the conjugation between different types of NPs or between NPs and other nanostructures may lead to the development of nanoplatforms suitable to "do it all in one." In this behalf, theranostic is an emerging concept that refers to the ability of a nanomaterial to both diagnose and treat diseases [48]. On the other hand, electrospinning technique allows the production of structures composed by nanofibers with a large SSA, which can be easily functionalized. Moreover, a variety of biodegradable and biocompatible polymers can be electrospun by conventional or modified electrospinning methods. These specificities together with the possible combination of different materials and morphological structures can be a powerful tool for biomedical applications. The functionalization of electrospun fibers with biomolecules or bioactive molecules is possible by *physical adsorption, blend electrospinning, coaxial electrospinning*, or *covalent immobilization* [45]. The *physical adsorption* method consists in the attachment of biomolecules to the electrospun fibers via electrostatic forces when it is immersed into the solution, which contains the biomolecules (Figure 4.5a). On the other hand, when biomolecules are mixed and homogenized within a polymer solution and used in the *blend electrospinning*, a hybrid structure is obtained (Figure 4.5b). Another approach is the use of *coaxial electrospinning* in which the polymer solution and the biological one are coaxially and simultaneously electrospun, producing composite nanofibers with a core–shell structure (Figure 4.5c). The *immobilization* of the biomolecules onto the nanofiber surface via chemical bond (Figure 4.5d) is a different way to achieve functionalized nanofiber mats. In all mentioned approaches, the most frequently used as biomolecules are enzymes, proteins, drugs, viruses, and bacteria.

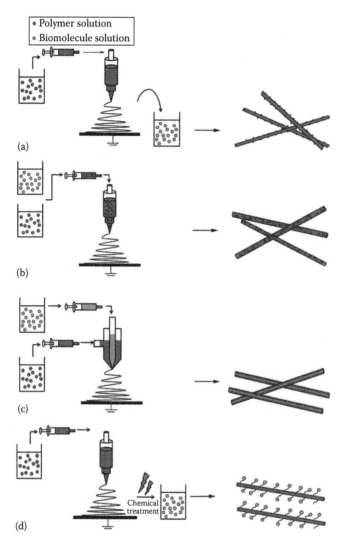

FIGURE 4.5
(See color insert.) Schematic approaches to produce nanofiber with bioactive molecules: (a) physical adsorption, (b) blend electrospinning, (c) coaxial electrospinning, and (d) covalent immobilization. (From Ji, W. et al., *Pharmaceutical Research*, 28, 1259, 2011.)

4.3.1 Tissue Regeneration

Tissue engineering has emerged as an interdisciplinary field that merges engineering and life sciences and aims the development of biological substitutes for restoring, maintaining, or improving the function of human tissue. One of the main challenges concerns the designing and engineering of scaffolds or polymeric matrices that mimic the structure and biological functions of the natural extracellular matrix (ECM).

A functional scaffold has to obey to basic requirements: A scaffold should possess a high degree of porosity, with an appropriate pore size distribution, and a large surface area; biodegradability is often a requirement; it must possess the required structural integrity to prevent the pores of the scaffold from collapsing during neo-tissue formation, with the appropriate mechanical properties; and finally the scaffold should be nontoxic to cells and

TABLE 4.1

Types of NPs and Their Applications in Tissue Regeneration

Type of NPs	Application in Tissue Regeneration
Titania NPs	Reinforcement phase in scaffolds for bone regeneration
Quantum dots	Used as amplifiers, markers, and biological sensors
Hydroxyapatite NPs	Reinforcement phase in scaffolds for bone regeneration, delivery system
Polymeric NPs	Targeted delivery systems
Stimuli-sensitive polymeric NPs	Delivery system

Source: Adapted from Engel, E. et al., *Trends Biotechnol.*, 26, 39, 2008.

biocompatible, interacting with the cells to promote cell adhesion, proliferation, migration, and differentiated cell function [49].

From NP point of view, this field has brought huge expectation for a great number of current worldwide human illnesses (e.g., Alzheimer's disease, Parkinson's disease, osteoporosis, or cancer). The importance of NPs in this field (Table 4.1) is addressed mainly toward the development of delivery systems for genetic material, biomolecules (e.g., growth and differentiation factors), and bone morphogenetic proteins. Another application of NPs in tissue regeneration is as reinforcing- or bioactivity-enhancement phase for polymeric matrices in 3D scaffolds [50–51].

Controlled delivery of biomolecules is crucial for support and enhancement of tissue growth. The use of nanoscale particles enables the intracellular delivery of molecules and the possibility of reaching targets normally inaccessible (e.g., blood–brain barrier, tight junctions, and capillaries), whereas the control over biomolecule dosage and period of delivery is increased [51]. As an example of this application, a simple and effective strategy for the treatment of osteoarticular tuberculosis, a frequent extrapulmonary complication of tuberculosis, was already demonstrated [52]. Due to the poor blood supply in the osseous lesion developed by tubercle bacillus, it is hard to achieve high efficacy for the patients when they are administrated intravenously with antitubercular drugs. In addition, the patients usually are submitted to debridement operation, leaving a residual cavity, where generally resides the tubercle bacillus. Consequently, it is important to fill and dose the cavity to promote tissue regeneration. For that reason, a tricalcium polyphosphate bioceramics (TPB)-based system for both tissue regeneration and osteoarticular tuberculosis therapy was proposed. The system was composed of TPB and gelatin hydrogel binary composite inside, coated with a biodegradable polymer. An appropriate amount of antitubercular drug (rifampicin) was loaded into gelatin hydrogel-containing composites to maintain a local drug level above a minimal inhibitory concentration. Therefore, while the gelatin hydrogel acts as a platform for controlled release of rifampicin, the TPB composite promotes tissue regeneration in the osseous lesion.

In fact, TPB are widely used in scaffolds for tissue engineering, acting as the reinforcing phase of a polymer matrix and improving scaffold bioactivity. The ECM of bone tissues consists of a continuous organic phase (mostly collagen type I) and a dispersed mineral phase (mainly hydroxyapatite (HA), $Ca_{10}(PO_4)_6(OH)_2$). As such, the use of HA NPs dispersed in a continuous phase (e.g., hydrogel or biocompatible polymer) leads to the construction of a scaffold that mimics the ECM of bone tissues. In addition, the presence of HA in the scaffold improves its osteoconductivity through the high affinity of the apatitic crystals for proteins (such as growth factors) that have an essential role in modeling attachment, proliferation, and differentiation of osteoblasts [51,53]. With this basis, several authors have focused on the development of scaffolds containing HA NPs. The synthesis of nanocomposites of gold NPs, HA NPs, and graphene (carbon-based 2D nanostructures) for bone regeneration [54] was reported. This method is based on the *in situ* synthesis of few-layer graphenes over a gold/HA catalyst through radio-frequency chemical vapor deposition. The referred authors evaluated the multilayer nanocomposite toxicity and found that these structures present very low toxicity and that toxicity is dose-dependent. Moreover, these structures were found to have good biocompatibility toward the bone cell proliferation. In the same year, a drug delivery system combining composite scaffolds of collagen and HA with bisphosphonate derivative liposomes was produced [55]. The authors investigated the drug release profile of three different drugs (doxorubicin [DOX], lysozyme, carboxyfluorescein) from the composite and concluded that this scaffold is promising for application in bone tissue engineering and regenerative medicine, mainly due to the HA biocompatibility and the versatility of liposomes to encapsulate a variety of pharmacological agents. More recently, HA NPs were either directly loaded into oligo(poly(ethylene glycol) fumarate) (OPF) hydrogel or loaded into gelatin microspheres, which were subsequently integrated in the OPF matrix. Both HA-containing composites showed a considerable capacity to mineralize upon soaking in simulated body fluid (SBF), meaning that these structures are suitable for applications in bone regeneration [53].

Although the development of NPs in tissue regeneration seems a very promising approach, effective application in human therapy has encountered small progress. This may be due to some questions addressed to the use of NPs in the human body. There is a need of a comprehensive understanding of their secondary effects and cytotoxicity once they enter the body.

Among nanostructured matrices, electrospun nanofibrous scaffolds have exhibited great performance in cell attachment, proliferation, and penetration, through both *in vitro* and *in vivo* trials. In literature, many studies can be found using scaffolds made of electrospun nanofibers mimicking tissues such as blood vessels [56–58], bones [59], and muscles [60–62].

4.3.3.1 Blood Vessels

Tissue engineering applied to blood vessels is focused on the development of small-diameter vascular grafts. Small-diameter arteries, arteries of less than 5 mm in diameter, include the coronary arteries and many peripheral arteries found in the arms and legs. Thus, electrospinning can provide the construction of vascular scaffolds due to the simplicity of shaping tubular constructs using rotation and translational motion. A sequential multilayering electrospinning with a rotating mandrel-type collector to produce scaffolds that mimic morphological and mechanic architecture of a blood vessel was used [56]. In the media layer of the native blood vessels, the smooth muscle cells (SMCs) and collagen fibrils have a marked circumferential orientation to provide the mechanical strength necessary to withstand the circulatory high pressures. Additionally, the intima layer of the native blood vessels consists in endothelial cells coating the vessel's internal surface. Between those two layers exists an internal elastic lamina mainly composed of elastin, which confers elastic properties to the blood vessels (Figure 4.6). Owing to this, a bilayered tubular scaffold composed by oriented and stiff PLA fibers in the outside and random and elastic PCL fibers in the inside was fabricated [56]. The desirable level of malleability

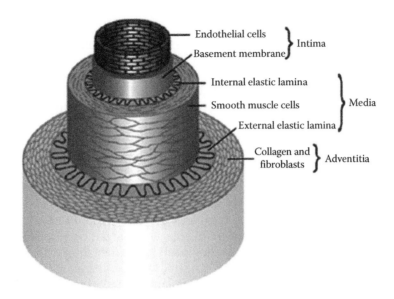

FIGURE 4.6
Representative drawing of the arterial wall illustrating the three primary layers of the tissue: intima, media, and adventitia. (From Sarkar, S. et al., *Med. Biol. Eng. Comput.*, 45, 327, 2007.)

(elastic up to 10% strain) was obtained and proved the capability of promoting cell growth and proliferation. Also the electrospinning of collagen and elastin had become a reality [63]. These two biopolymers are the main structural components of the vascular ECM. Collagen is responsible for the structural integrity and tensile strength of tissues, and elastin gives them elasticity. However, the electrospinning fibers of pure collagen or elastin have shown poor mechanical properties. The electrospun collagen or elastin dissolves in water instantaneously if uncross-linked. Hence, the incorporation of biodegradable synthetic polymers into the structure imparts immediate strength to the scaffolds and simultaneously maintains the bioactivity of the biopolymer (collagen or elastin). The fabrication of a composite vascular scaffolding system by electrospinning of polycaprolactone (PCL) and collagen (type I) blends has been proved [57]. These composite scaffolds were designed to provide biomechanical ability and accommodate vascular endothelial and SMCs for use in vascular tissue. PCL is aliphatic polyester that degrades slowly and possesses high tensile and elongation properties for vascular grafts. The resulting composite scaffold showed good biocompatibility and acts as a support for cell growth and proliferation *in vitro*. The fabrication of a fibrous tubular scaffold by electrospinning of polydioxanone (PDO) and elastin blends was reported [58]. This synthetic polymer is bioresorbable with a slow resorption rate and induces a low inflammatory response. *In vitro* experiments demonstrated that PDO–elastin composites have the mechanical integrity and strength of PDO combined with the bioactivity and minimal energy loss (due to elasticity) of elastin. The preparation of electrospun collagen and elastin meshes from aqueous solutions with addition of PEO has been described [41]. The addition of PEO was necessary to increase the viscosity of the solution and obtain continuous and homogeneous fibers. The produced nonwoven meshes were stabilized by cross-linking with *N*-(3-dimethylaminopropyl)-*N'*-ethylcarbodiimide hydrochloride (EDC) and *N*-hydroxysuccinimide (NHS). Upon cross-linking, PEO is completely removed. A successful scaffold was achieved, and *in vitro* assays showed promising results in SMC growth.

4.3.3.2 Bones

Electrospun nanofiber meshes do not possess, on their own, mechanical properties similar to that of bone tissue; however, they have been studied extensively in bone tissue engineering for their ability to support cell growth and differentiation into osteogenic phenotypes. A key cellular phenotype is the osteoblast, which is the cell type that lays down the ECM of bone tissue and thus is the target cell type for recruitment and differentiation in bone reconstruction.

Synthetic polymers and their copolymers, natural polymers, composites of natural and synthetic polymers, inorganic materials such as bioactive glass and HA, and composites of inorganic materials with synthetic and natural polymers are the major nanofibers used in osseous tissue engineering. To promote cell growth and proliferation, the scaffold needs to degrade gradually along with the construction of new tissue to be completely replaced by the new tissue. The use of poly-L-lactide (PLLA), PLLA/HA, and PLLA/collagen/HA as electrospun nanofiber scaffolds for bone tissue regeneration has been proved [59]. PLLA is one of the most utilized polymers in tissue engineering scaffold, drug delivery, and fixation in orthopedics and as bioabsorbable membrane due to its biodegradability and biocompatibility. HA is currently applied to biomedical implant due to its biodegradability, bioactivity, and osteoconductive properties. However, as a bioceramic, HA cannot be easily shaped in the bone defect sites due to its natural rigidity, but if it is used together with synthetic biodegradable polymers, the mechanical properties of HA can be significantly improved. The *in vitro* assays showed that comparing to PLLA scaffold, the introduction of HA in the polymer matrix (PLLA/HA scaffolds) has contributed to enhance the cell proliferation and in the presence of collagen the cell attachment and proliferation are accelerated. The enhanced adhesion and growth of osteoblasts on PLLA/collagen/HA nanofibers lead to a mineral deposition 57% higher than that observed on PLLA/HA nanofibers. Thus, the combination of structural, mechanical, and biological properties of the materials chosen for the scaffold plays an important role for bone tissue regeneration.

4.3.3.3 Muscles

Skeletal muscle tissue is composed of bundles of highly oriented and densely packed muscle fibers, each with multinucleated cells derived from myoblasts. The fibers are densely packed in ECM to form an organized muscle tissue that generates longitudinal contraction. The engineering of functional muscle tissue for reconstruction requires the design and fabrication of scaffolds that allow cellular organization mimicking native individual fiber formation with unidirectional orientation. Thus, the electrospun nanofibers have appropriate biocompatible characteristics for skeletal muscle, to allow cell adhesion and growth, degradability as muscle cells mature into tissue, and elasticity for contractile function.

The PCL-collagen-based nanofibers have been studied as a scaffold system for implantable-engineered muscle [60]. The influence of orientation of the electrospun PCL/collagen nanofibers on the morphology, adhesion, proliferation, differentiation, organization of human skeletal muscle cells, and the myotube formation was investigated. The unidirectional orientation nanofibers were compared to randomly oriented nanofibers, and the results have shown that aligned nanofiber scaffolds remarkably induce muscle cell alignment and myotube formation providing so implantable functional muscle tissues for patients with large muscle defects. An elastomer polymer, poly(lactide-*co*-glycolide) (PLGA), was also electrospun to obtain a scaffold for skeletal tissue regeneration [61]. The fibers of the scaffold stayed well aligned when produced at high-speed rotating mandrel collector. The so well-organized nanofiber scaffolds provided the topographical cues and contact guidance for the alignment of myoblasts. *In vitro* assays have demonstrated its ability to support the attachment, proliferation, and differentiation of mice myoblasts.

The engineering of complex composite tissues with coordinated functions, for instance, tissue interfaces [65], is a topic of increased interest. Recently, muscle–tendon junction (MTJ) tissues were proposed [62]. Indeed, the design of a continuous scaffold suitable for both MTJ tissue types is challenging for engineering. MTJs require a seamless interface to allow force transfer from muscle to tendon. Combining the fabrication of a dual scaffold that has differentiated regional mechanical property, it is possible to mimic the performance of native MTJ. The co-electrospinning was used to obtain such combinatory effect with three distinct scaffold regions: a PCL/collagen region (one end side), a PLLA/collagen region (opposite end side), and an overlap region (center). Both polymers were blended with collagen to improve cell attachment. The results of mechanical properties showed the PLLA/collagen side of the scaffold, the stiffest one with the lowest strain; the PCL/collagen side, the most compliant with the highest strain; and the middle region that possesses an intermediate stiffness and strain, which are analogous to the tendon, muscle, and junction, respectively. These distinct mechanical properties are attractive for MTJ tissue engineering improvement.

4.3.2 Drug Delivery

Drug delivery consists in a controlled release system, where the active substance is loaded into a carrier or device and then released at a predictable rate *in vivo* when administered by an injected or noninjected route. This is an important field of research, which has caught the attention of pharmaceutical researchers, medical doctors, and industry. The main reason for this interest is that a safe and targeted drug delivery could improve the performance of some classic drugs already in the market. To develop appropriate carriers, it is important to have knowledge on drug incorporation and release, formulation stability and shell life, biocompatibility, biodistribution and targeting, and functionality. In addition, it is important to consider the fate of the NPs after the drug is completely released; in this respect, biodegradable NPs with a limited half-life in blood would be optimal [66–68].

Drug delivery may be achieved by two types of targeting: passive targeting and active targeting. Passive targeting is achieved through the EPR effect or by localized delivery, in other words, by directly administrating the formulation in the target site. This last approach is very common in solid tumors (prostate, head, and neck cancer). Active targeting requires the conjugation of receptor-specific ligands (e.g., antibodies) that can promote site-specific targeting [68].

NPs due to their small size can efficiently penetrate across barriers into individual cells, thus allowing efficient drug accumulation in the target site. Therefore, the unwanted side effects and toxicity of the therapeutic agent are diminished, and its therapeutic efficacy is enhanced. Concerning drug delivery, NPs have the following advantages: increase drug solubility in water, protection from degradation (and, consequently, increase half-life in blood), and provide a prolonged and controlled drug release and decrease the drug side effects [68]. That said, NPs have a huge potentiality as drug delivery systems (Table 4.2), although more research on their mode of action and toxicity is required.

Gold NPs are one of the NP types used in drug delivery systems due to their attractive features that include surface plasmon resonance, controlled interaction with thiol groups, and nontoxic nature. Gold NPs may be conjugated with pharmaceutical compounds, like methotrexate. Methotrexate is an analogue of folic acid that has the ability to destroy folate

TABLE 4.2

Some Examples of NPs Approved for Clinical Usage

Product	Composition	Target Disease	Advantages
Doxil	PEGylated liposome	Ovarian cancer and multiple myeloma	Enhanced circulation time and is up to six times more effective than free DOX
Abraxane	Albumin NPs	Lung and breast cancer	Enhanced cytotoxicity, shorter infusion time, low dose required
Nanoxel	Nanoparticulate delivery system for paclitaxel	Breast and ovarian cancer	Cremophor-free water-soluble formulation, greater efficacy, and decreased cytotoxicity
Verigene platform	DNA-functionalized gold NPs	Diagnostics	Gold NPs enable efficient diagnosis for methicillin-resistant *Staphylococcus aureus*
AmBisome	Liposomal amphotericin	Fungal and parasite infection	Improve the tolerability profile of amphotericin B deoxycholate, decreased nephrotoxicity

Sources: Parveen, S. et al., *Nanomedicine: Nanotechnology, Biology and Medicine* 8, 147, 2012; Moen, M. et al., *Drugs* 69, 361, 2009.

metabolism of cells, and it is used as a cytotoxic anticancer drug and to treat autoimmune diseases. A new formulation of methotrexate bound to gold NPs to serve as drug carriers was developed [70] and their cytotoxic effect *in vitro* and antitumor effect *in vivo* evaluated. The results showed that the accumulation of methotrexate is faster and higher in tumor cells with methotrexate–gold NPs than with free methotrexate. In addition, the NP composites showed higher cytotoxicity effects on several tumor cell lines compared to an equal dose of free methotrexate. Surface modification of gold NPs is another approach commonly used. The main advantages of this approach are as follows: increase circulation lifetime of the NPs by preventing its uptake by RES, attach to the desired targeting and therapeutic molecules, improve stability, and decrease agglomeration and toxicity. Another interesting feature of gold NPs is their strong plasmon resonance with light, which may be used as an external stimulus to release the drug from the NPs. This might provide an innovative route to deliver material directly into the cytoplasm or nucleus of target cells [71].

An NP system capable of targeting the heart after myocardial infarction composed of a nanosized PEGylated liposome able to carry a therapeutic drug and a ligand specific to AT1 (angiotensin II type I) has been proposed [72]. There, the targeting is based on the overexpression of AT1 receptor in the infracted heart and has found that the NPs only target the injured heart and that AT1 targeting further enhanced delivery to the injured myocardium. Therefore, the proposed system is a favorable approach to decrease heart injury after myocardial infarction.

The dextran NPs composed of a thiolated dextran derivative and a PEGylated thiol derivative sequentially mixed with a lipid-modified dextran derivative have proved to enhance the binding efficiency of PEG chains to the dextran hydrogel [73]. With this study, the authors concluded that the dextran-loaded DOX NPs had a curative effect on multidrug-resistant (MDR) osteosarcoma cell lines by increasing the amount of drug accumulation in the nucleus via P-glycoprotein (Pgp)—its overexpression is one of the main responsible for MDR) independent pathway, thus increasing apoptosis when compared to DOX alone.

An injectable *in situ* gelling drug delivery system composed of an ammonium derivative of chitosan (*N*-(2-hydroxypropyl)-3-methylammonium chitosan chloride [HTCC]) and thermosensitive chitosan/gelatin blend hydrogels and HTCC NPs loaded with bovine serum albumin (BSA) showed that the release rate of BSA precursor gels was affected by the properties of the NPs [74], thus confirming that these NPs may be used as a vehicle for controlled drug release of proteins.

Magnetic drug targeting is based on the ability to provide selective attachment to a functional molecule and confer magnetic properties to the target area and allows manipulation and transportation to the target site through the control of a magnetic field produced by an electromagnet or permanent magnet. A very interesting review [75] about magnetically targeted drug delivery during cardiopulmonary resuscitation (CPR) pointed out a pharmaceutical treatment as common method for assigning the heart and circulating during CPR and often return of spontaneous circulation. This procedure, based on the application of external magnets near the heart and intravenous administration of magnetic NPs (e.g., iron oxide NPs) loaded with the pharmaceutical agents required during CPR, is extremely important due to the poor survival of cardiac arrest victims.

A different yet advantageous approach for drug delivery systems is achieved by using electrospun nanofibers. The high SSA of polymer nanofibers and the modulation of nanofiber morphology, porosity, composition, and simplicity of the electrospinning process are great advantages that provide the ability to conveniently incorporate therapeutic compounds into the electrospun fibers for preparing useful drug delivery systems [76]. Indeed, recent developments have been made in this field. A number of methods such as *coating*, *embedding*, and *encapsulating* by coaxial and emulsion electrospinning can be utilized for load drugs onto electrospun fibers:

- *Coating*: The coating method consists in the adsorption or cross-linkage of drug molecules to the surface of electrospun fibers via a physical or chemical method, thus obtaining drug-loaded fibers. The chemical conjugation of a recombinant human epidermal growth factor (rhEGF) with the surface of electrospun nanofibers for *in vivo* wound healing treatment of diabetic ulcers has been reported [77]. Current treatments for diabetic foot ulcers include the administration of disinfectants followed by application of epidermal growth factor (EGF)-containing gels around the lesions. A better solution was obtained through the electrospinning of biocompatible nanofibers with functional amine groups on the surface (PCL and PCL/PEG, block copolymer). By immersing it in an aqueous solution, the exposed functional amino groups on the surface of the nanofibers were chemically conjugated to rhEGF by activating the carboxylic groups of the protein. Human primary keratinocytes were cultivated on EGF-conjugated nanofibers in order to investigate the effect of EGF nanofibers on the differentiation of keratinocytes. Thus, wound healing effects of the EGF nanofibers were successfully confirmed in diabetic animals with dorsal wounds.

- *Embedding*: A polymer solution containing drugs can be directly electrospun to form drug-loaded fibers in one step. In this process, it is required that the drug solution and polymer solution are either miscible liquids or solid drug particles well dispersed into the polymer solution. The use of PEG/poly(D,L-lactide) copolymer electrospun fibers used as a drug delivery system is an example of this loading process [78]. That copolymer was combined with paracetamol (acetaminophen, N-(4-hydroxy-phenyl)acetamide), which was chosen as the model drug since it is widely used as analgesic and antipyretic drug. *In vitro* matrix degradation profiles of these fibers were characterized by measuring their weight loss, the molecular weight decrease, and their morphology change. From these studies, it was observed that scaffolds with smaller fiber diameter have a higher contact area between polymer and water, which, consequently, accelerates the matrix breakdown. Additionally, *in vitro* drug release assays confirmed that the release behavior mainly depends on polymer matrix degradation and drug diffusion. Owing this, the drug release rate can be controlled by polymer degradation, which can be done by adjusting the electrospun fiber diameter and its porosity.

- *Encapsulating*: The previously mentioned embedding method faces some difficulties related to the electrospinning of macromolecular dispersions. These difficulties were partially solved with the electrospun of core–shell- and hollow-structured fibers. Using the encapsulating method, it has been proposed the development of a core–shell-structured fiber as a carrier for drug delivery [79]. For that was utilized a coaxial electrospinning and selected PLLA and tetracycline hydrochloride (TCH) as the shell and core materials, respectively. The PLLA was used as the shell polymer due to its good biocompatibility and appropriate biodegradation rate, whereas TCH was employed as a model drug because it is a broad-spectrum antibiotic for inhibition of bacteria growth. The release profile of TCH from the electrospun fibers *in vitro* studies was as follows: in a first stage, the drug was released from the carrier at a low rate due to the poor degradation of PLLA without any burst release; later, a faster release was detected along PLLA degradation. Thus, these core–shell fibers are promising for drug delivery system, for suturing materials, and for wound dressing applications.

4.3.3 Biodevices

Biomedical technology usually requires various portable, wearable, easy-to-use, and implantable devices that can interface with biological systems. Currently, small batteries with limited lifetime power implantable medical microsystems. Although the scientific progress in this area has enabled a decrease in the electrical requirements of the miniaturized devices, the development of a suitable power source remains a major challenge for many devices in the bioengineering and medical fields [80,81].

Indeed, the use of electrospun fibers for the development of functional devices opens a new path for the creation of novel, lightweight, and flexible nanostructures. These functional nanostructures can be used in the development of sensing devices such as implantable biofuel cells, self-powered biosensors, and autonomously operated devices [82–84]. Owing their extremely high surface-to-volume ratio and tunable porosity, electrospun membranes can be successfully applied in the fabrication of nanostructures for energy conversion devices.

Our research team is currently working on the development of a bio-battery based on biocompatible electrospun membranes [85]. The bio-battery reported by us is composed of an ultrathin monolithic structure in which the separator and the electrodes are physically integrated into thin and flexible polymeric structure. A highly porous structure is produced by electrospinning to work as a bio-battery after the deposition of metallic layers (electrodes) in each faces (Figure 4.7). In order to power electronic medical implants, power-supply systems must be able to operate independently over a prolonged period, without the need of external recharging or refueling. This cellulose-based structure demonstrated the ability to generate electrical energy from physiological fluids showing a power density of 3 μW cm^{-2} [85]. This is a very promising achievement since a typical power required for a pacemaker operation is around 1 μW. Besides the supplying of low-power-consumption devices, biochemical monitoring systems and artificial human muscle stimulation mechanisms can also be foreseen as potential field of applications where it is desirable this kind of implantable micropower sources.

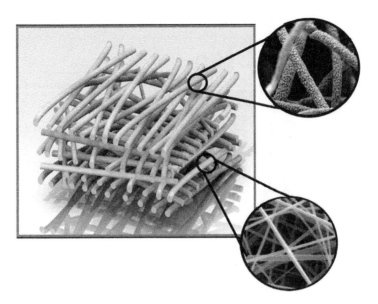

FIGURE 4.7
Schematic bio-battery structure.

4.3.4 Theranostics

NP development is highly focused on the development of both therapeutic and diagnostic agents. Ideally, these two approaches should be combined in a single nano-agent, because of its potential ability to simultaneously image and treat disease at the cellular level. The term theranostics is used to define ongoing efforts to develop more specific, individualized therapies for various diseases and to combine diagnostic and therapy capabilities into a single agent. These nanostructures may present differences in size, shape, functionality, targeting mechanism, and imaging modality; however, some components are common between them (Figure 4.8):

- *Detection component* for noninvasive imaging—can be metallic or magnetic compounds for MRI, fluorescent compounds for optical imaging, and radioisotopes for positron emission tomography (PET) and single-photon emission computed tomography (SPECT).
- *Polymer coating*—this component is essential for the colloidal stability and to provide functional groups for bioconjugation.
- *Drug loading capabilities*—this may be achieved by physical encapsulation in the polymeric matrix and electrostatic interaction with the other components of the nanostructure or by covalent biding.

The rationale for theranostics arose from the fact that diseases, such as cancer, are very heterogeneous and the existing treatments are effective only in a small part of the patient population and at selective stages of disease development. The theranostic approach is expected to provide therapeutic protocols that are more specific to individuals and, therefore, more likely to improve prognostics [86]. This integration of diagnostic imaging capabilities with therapeutic intervention is crucial to overcome cancer challenges. Recent advances on cancer research have shown a huge heterogeneity among the different

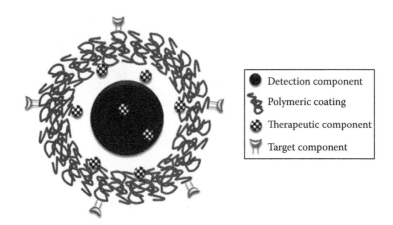

●	Detection component
🖌	Polymeric coating
⊞	Therapeutic component
⊽	Target component

FIGURE 4.8
Basic components of a theranostic nanostructure.

cancer types. Furthermore, significant differences have been found in the same cancer type, between the primary tumor and its metastasis foci, and even differences between cells that constitute individual tumors [87,88]. That said, cancer has become one of the major applications of theranostic nanostructures.

Due to the exciting potential of these theranostic nanostructures, many researchers have been focused on this issue. Theranostic nanostructures can be roughly divided by their detection and therapeutic component (Figure 4.8).

4.3.4.1 Detection Component

The detection component through molecular imaging allows the characterization of biological processes at cellular and subcellular level, without any damage to the living tissues. As so, by selecting a specific probe or a contrast agent, it is possible to create a powerful tool to both detect and characterize early stage disease and monitor the treatment progress. There are vast possibilities for the detection component, however two materials are the most used in theranostic nanostructures: (a) *iron oxide NPs* and (b) *quantum dots*.

4.3.4.1.1 Iron Oxide NPs

Iron oxide NPs have been used clinically as MRI contrast agents mainly in the liver (e.g., Resovist, Feridex), due to the high magnetic moments that make them efficient in reduction T2 relaxation time, leading to signal attenuation on a T2 or T2*-weighted map.

The idea of using magnetic micro- or nanoparticles as drug carriers for cancer treatment dates back to the late 1970s [89], when magnetic micro- and nanoparticles attached to cytotoxic drugs were developed. Since then, several authors tried to improve this technique. For example, some authors have used magnetic NPs to detect apoptosis, an active process of cellular self-destruction present in many disorders, such as neurodegenerative diseases (e.g., Alzheimer, Parkinson), by MRI [90–92]. More recently, another research group has proposed nanocomposites based on the inclusion of magnetite nanocrystals into NPs constructed by self-assembling molecules of squalenoyl gemcitabine (SQgem) bioconjugate. These nanocomposites, after subcutaneous injection on mice tumor model, were magnetically guided to the target site and showed high anticancer activity. The superior therapeutic activity and enhanced tumor accumulation were successfully visualized using T2-weighted imaging in MRI. This method was further actualized by the design of squalene-based NPs containing the T1 Gd3+ contrast agent instead of magnetite and by using different chemotherapeutic drugs such as DOX, cisplatin, and paclitaxel.

The use of MRI contrast agents has another advantage, that is, treatment monitoring. By using oleic acid–coated iron oxide and pluronic-stabilized mNPs loaded with DOX and paclitaxel, highly synergetic antiproliferative activity in breast cancer cells was demonstrated [93], the NP accumulation and time of circulation in blood being monitored by MRI. Thus, that work showed that the incorporation of cytotoxic agents in mNPs does not affect its physical (size and surface) properties or magnetization characteristics.

4.3.4.1.2 Quantum Dots

Quantum dots are nanostructures with favorable characteristics for theranostic applications as a detector component, to identify disease via targeting ligands. However, their innate toxicity limits the applications in biomedical field [12,86]. To overcome quantum-dot toxicity, hydrophobic quantum dots and iron oxide NPs were co-encapsulated along with DOX, into micelles formed with PEGylated phospholipid [94]. The conjugates were further coupled to a tumor-homing peptide F3 and injected into an MDA-MB-435 xenograft model. The results showed successful tumor targeting by both optical and MRI modalities. The use of quantum dots in biomedical applications is very promising owing to their unique physicochemical properties such as broad absorption spectra, size-dependent narrow and stable emission spectra, and photostability. However, the inherent toxicity is a large limitation of their application. Moreover, there is a need to understand the quantum-dot faith once they enter the body [86].

4.3.4.2 Coating of the Theranostic Agent

As seen by the given examples, the diagnostic component must be coated with a biocompatible material in order to be used as a theranostic agent. In the case of iron oxide NPs, these are generally coated with a biocompatible material (e.g., polysaccharide, synthetic polymer, and lipid protein), creating a composite morphology often referred as core–shell structure. An ideal coating biomaterial for iron oxide NPs is the one that satisfies the following requirements [95]:

- Prevent opsonization and, consequently, increase the blood half-life of the iron oxide NPs in circulation
- Avoid agglomeration of the NPs in physiological conditions
- Achieve the desired surface charge for NPs' function
- Preserve the functionalities of the NPs
- Ensure the biocompatibility of the iron oxide NPs
- Serve as a platform for NPs' surface functionality

Several organic and inorganic coatings that meet these requirements have been used on the iron oxide NP surface. One example of a polysaccharide-based coating is given by chitosan. Chitosan is the *N*-deacetylated product of chitin, the second most abundant polysaccharide in nature, and possesses favorable characteristics: nontoxicity, biocompatibility, biodegradability, cationic, and hydrophilic. In addition, chitosan possesses amine and hydroxyl groups, suitable for NP functionalization with targeting, imaging, and therapeutic agents.

On the other hand, if the polymeric component is composed of a "smart" polymer, the theranostic nanostructure will only become activated when an external stimulus (temperature, pH, redox potential, or specific biomolecules) is present. For example, a theranostic nanostructure for a specific solid tumor composed by a polymer sensitive to a pH variation is capable of only releasing the therapeutic drug in the tumor microenvironment, which is characteristically more acidic than physiological conditions. This pH variation will induce changes in physical conformation (swelling or deswelling) of the polymeric coating, leading to drug release [96].

Another type of coating with promising results as a part of a theranostic agent is liposomes. Liposomes are very versatile structures because of its capacity to incorporate pharmaceutical drugs, water-soluble macromolecules (hydrophilic), or lipidic macromolecules (hydrophobic). Consequently, liposomes are a promising approach as theranostic agents [97]. Liposome application as theranostic agents was achieved by combining delivery and MRI of DOX-loaded liposomes in a Kaposi's sarcoma *in vivo* [98]. MRI was used both to determine liposome tissue distribution and to monitor DOX delivery and release. The results showed that Cd3 (high-affinity neural cell adhesion molecule [NCAM] binding ligand)-coated liposomes loaded with both DOX and gadolinium-DOTAMA(C18) were internalized and induced apoptosis of Kaposi's cells and tumor endothelial cells significantly more efficiently than did with uncoated PEG liposomes. In addition, the incorporation of DOX in PEG-coated liposomes reduced its cardiotoxicity while increasing its therapeutic efficacy. Furthermore, the use of gadolinium-labeled liposomes allowed the concomitant MRI visualization of the drug delivery in the tumor region.

Similarly to liposomes, dendrimers are versatile structures capable of acting as drug and imaging agents with applications in tumor regression, gene delivery, and molecular imaging. However, their toxicity is a current barrier to its usage in this field [99]. To overcome this, several researchers have focused on chemical modification of dendrimers. For example, new dendrimer was designed and synthesized as based building block that is water soluble and can be functionalized in a highly controllable and orthogonal fashion [100]. This dendrimer was functionalized with an NIR-cyanine dye. The results showed fluorescence in the NIR region with a large Stokes shift and relatively high quantum yields. In addition, the dendrimers showed no toxicity toward T98G human cells and, therefore, are promising for theranostic applications.

4.3.4.3 Target Component

After the detection component and coating of the theranostic nanostructure, another important issue is the target component. Although passive targeting is possible due to the NPs' small dimensions, it is not suitable for all diseases. As so, specific targeting is generally required. The specific targeting may be achieved using numerous biomarkers, specifically overexpressed by the target cells. Monoclonal antibodies were the first targeting agents to exploit molecular recognition to deliver mNPs [101]. Magnetic nanocrystals conjugated to Herceptin were used as cancer-targeting monoclonal antibody for breast cancer treatment and successfully monitored *in vivo* selective targeting events of human cancer cells implanted in live mice [102]. Also, radioconjugate NPs have been developed using recombinant antibody fragments, di-scFv-c (111 In-DOTA-di-scFv-NP), for imaging and therapy of anti-MUC-1 10 expressing cancers [103]. Further conjugation with fluorescent dye-labeled antibodies enables both *in vitro* and *ex vivo* optical detection of cancer, as well as *in vivo* MRI, which are potentially applicable for an advanced multimodal detection system.

Depending on NP characteristic, all types of NPs previously described may be applied for theranostics as a component of the theranostic nanostructure. Concerning this matter, iron oxide NPs are among the most used NPs as theranostic agents. As such, iron oxide NPs may have two main functions: MRI contrast agents and hyperthermia agents. Hyperthermia is an old technique, which is recognized as a possible treatment option for cancer. Hippocrates, the father of medicine, treated surface tumors by cauterizing them with a hot iron. More recently, advanced hyperthermia methods have been employed in cancer treatment such as hot bath water, pyrogens (e.g., a mixture of bacterial toxins), perfusion heating, high-frequency radiation, and magnetic fluid hyperthermia [26]. The theory behind hyperthermia is to kill cancer cells without the use of drugs (chemotherapy) or radiation (radiotherapy) just by increasing temperature in the tumor site. This idea is based on the higher sensitivity of cancer cells to temperature oscillations when compared to normal cells, thereby decreasing the side effects [104].

However, more recently, a concept of intracellular hyperthermia emerged in which the particles are concentrated at the tumor site and remotely heated using an applied magnetic field to the required hyperthermic temperatures (42°C–45°C). This concept is based on the principle that alternating the magnetic field rapidly would cause particles to quickly flip their magnetic polarity. This flipping is accompanied with some hysteretic loss, which manifests as heat. The amount of generated heat is affected by the magnitude of the field, the size and characteristic of the iron oxide NPs, the depth of the tumor within the body, and the concentration of iron oxide NPs in the tumor site. The magnetic field has the advantages of not being absorbed by the living tissues and to penetrate deep into those tissues. The important properties of magnetic particles for inducing hyperthermia are nontoxicity, biocompatibility, injectability, high-level accumulation in the target tumor, and effective absorption of the energy of the AMF [104,105].

Clinical application of hyperthermia is established not as a single treatment modality but as a way to improve chemotherapy and radiotherapy results, when applied in repetitively short intervals. More so, local and regional radio-frequency hyperthermia can be regarded as a well-established, nontoxic treatment, which is used according standardized protocols worldwide. Currently, hyperthermia is approved to treat advanced cancer, such as recurrent rectal cancer, prostate cancer, uterine cancer, head and neck cancer, lung cancer, and breast cancer, in conjugation with chemotherapy and radiotherapy [106,107].

4.4 Conclusion

Recent developments of a wide spectrum of nanoscale technologies are starting to offer new options for the diagnostic and treatment of several diseases, such as cancer. In this context, nanomedicine is an emerging area that includes NPs and nanofibers as biomaterials acting as biological mimetics.

The definition of a nanomaterial is not entirely clear, although some regulatory agencies have already started working on this subject. However, these definitions are generally simple and widespread, not taking into account some important properties of the nanoscale materials. Therefore, a more precise and rigorous definition of the nanoscale materials is required.

Although there is a huge discussion about nanomaterial definition, it is established that NPs have all three dimensions in the nanoscale, while nanofibers only have two dimensions in the nanoscale. Owing to their very large SSA, nanofibers are optimal nanoscale materials for the development of nanostructures, such as scaffolds for biomedical applications. Nanofibers can be produced by several methods; however, electrospinning offers a versatile and simple method for the production of continuous nanofibers. The main applications of electrospun nanofibers are focused on the development of scaffolds for tissue regeneration, drug delivery, and production of biodevices. There are several types of NPs depending on their composition (metal, polymer, etc.), their morphological characteristics, their source, their application, etc. As such, NP application in the biomedical field is very broad; furthermore, through the conjugation of different materials (e.g., core–shell NPs that are composed of at least two different materials), these applications can be exponentially increased.

Although the development of nanoscale materials, such as NPs and nanofibers, seems to have great potential for several biomedical applications, the concern about their safety has not been addressed in a sufficient and satisfactory manner. Furthermore, the large surface area of nanoscale materials makes the very reactive in the biological environment, raising a need for a better understanding about the faith of these materials once they enter the body.

In conclusion, nanotechnology is an area with a broad range of applications in the biomedical field that are still starting to be discovered. Consequently, the toxic effects of these materials are still unexplored. Therefore, parallel to the research of the potential application of nanomaterials, toxicological research must be made.

Acknowledgments

The authors' work was partially supported by Portuguese Science and Technology Foundation (FCT-MCTES) through the Strategic Project PEst-C/CTM/LA0025/2011. Ana Baptista and Paula Soares also acknowledge FCT-MCTES for the doctoral grants SFRH/BD/69306/2010 and SFRH/BD/79302/2011, respectively.

References

1. Boisseau, P. and B. Loubaton. 2011. Nanomedicine, nanotechnology in medicine. *Comptes Rendus Physique* 12: 620–636, doi:10.1016/j.crhy.2011.06.001.
2. Jain, K. K. 2007. Applications of nanobiotechnology in clinical diagnostics. *Clinical Chemistry* 53: 2002–2009, doi:10.1373/clinchem.2007.090795.
3. Shi, J., A. R. Votruba, O. C. Farokhzad, and R. Langer. 2010. Nanotechnology in drug delivery and tissue engineering: From discovery to applications. *Nano Letters* 10: 3223–3230, doi:10.1021/nl102184c.
4. ISO/TS 27687:2008 Nanotechnologies-Terminology and definitions for nano-objects-Nanoparticles, nanofibre and nanoplate, *ISO Standards*, Geneva 20, Switzerland, 2008.
5. Materials, A. S. T. 2006. *Terminology*. Vol. E 2456 – 06, p. 4, ASTM International, West Conshohocken, PA.
6. U.S. Food and Drug Administration Nanotechnology Task Force. Nanotechnology Report, 2007.
7. Brouwer, N., M. Weda, D. van Riet-Nales, and D. de Kaste. 2010. Nanopharmaceuticals: Implications for the European pharmacopoeia. *Pharmeuropa* 22: 5–7.

8. SCENIHR-Scientific Committe on Emerging and Newly Identified Health Risks. Scientific Basis for the Definition of the Term "nanomaterial". European Union, 2010.

9. Kumar, A. P., D. Depan, N. S. Tomer, and R. P. Singh. 2009. Nanoscale particles for polymer degradation and stabilization—Trends and future perspectives. *Progress in Polymer Science* 34: 479–515.

10. Buzea, C., I. I. P. Blandino, and K. Robbie. 2007. Nanomaterials and nanoparticles: Sources and toxicity. *Biointerphases* 2: MR17–MR172.

11. Jackson, L. P. U.S. Environmental Protection Agency, Washington, DC, http://www.epa.gov/, 2012.

12. Jokerst, J. V. and S. S. Gambhir. 2011. Molecular imaging with theranostic nanoparticles. *Accounts of Chemical Research* 44: 1050–1060.

13. Choi, H. S. and J. V. Frangioni. 2010. Nanoparticles for biomedical imaging: Fundamentals of clinical translation. *Molecular Imaging* 9: 291–310.

14. Abeylath, S. C., S. Ganta, A. K. Iyer, and M. M. Amiji. 2011. Combinatorial-designed multifunctional polymeric nanosystems for tumor-targeted therapeutic delivery. *Accounts of Chemical research* 44: 1009–1017.

15. Juzenas, P. et al. 2008. Quantum dots and nanoparticles for photodynamic and radiation therapies of cancer. *Advanced Drug Delivery Reviews* 60: 1600–1614.

16. Ambrogio, M. W., C. R. Thomas, Y.-L. Zhao, J. I. Zink, and J. F. Stoddart. 2011. Mechanized silica nanoparticles: A new frontier in theranostic nanomedicine. *Accounts of Chemical Research* 44: 903–913.

17. Shubayev, V. I., T. R. Pisanic II, and S. Jin. 2009. Magnetic nanoparticles for theragnostics. *Advanced Drug Delivery Reviews* 61: 467–477.

18. Gubin, S. P. 2009. Introduction—Magnetic oxides. In *Magnetic Nanoparticles*, (ed. S. P. Gubin) pp. 14–15. Wiley-VCH Verlag GmbH & Co, Weinheim, Germany.

19. Sun, C., J. S. H. Lee, and M. Zhang. 2008. Magnetic nanoparticles in MR imaging and drug delivery. *Advanced Drug Delivery Reviews* 60: 1252–1265.

20. Ho, D., X. Sun, and S. Sun. 2011. Monodisperse magnetic nanoparticles for theranostic applications. *Accounts of Chemical Research* 44: 875–882.

21. Dias, A. M. G. C., A. Hussain, A. S. Marcos, and A. C. A. Roque. 2011. A biotechnological perspective on the application of iron oxide magnetic colloids modified with polysaccharides. *Biotechnological Advances* 29: 142–155.

22. Indira, T. K. and P. K. Lakshmi. 2010. Magnetic nanoparticles: A review. *International Journal of Pharmaceutical Sciences and Nanotechnology* 3(3): 1035–1042.

23. McBain, S. C., H. H. Yiu, and J. Dobson. 2008. Magnetic nanoparticles for gene and drug delivery. *International Journal of Nanomedicine* 3: 169–180.

24. Cole, A. J., V. C. Yang, and A. E. David. 2011. Cancer theranostics: The rise of targeted magnetic nanoparticles. *Trends in Biotechnology* 29: 323–332.

25. Gupta, A. K. and A. S. G. Curtis. 2004. Surface modified superparamagnetic nanoparticles for drug delivery: Interaction studies with human fibroblasts in culture. *Journal of Materials Science: Materials in Medicine* 15: 493–496, doi:10.1023/B:JMSM.0000021126.32934.20.

26. Gupta, A. K. and M. Gupta. 2005. Synthesis and surface engineering of iron oxide nanoparticles for biomedical applications. *Biomaterials* 26: 3995–4021, doi:10.1016/j.biomaterials.2004.10.012.

27. Johannsen, M. et al. 2007. Morbidity and quality of life during thermotherapy using magnetic nanoparticles in locally recurrent prostate cancer: Results of a prospective phase I trial. *International Journal of Hyperthermia* 23: 315–323.

28. Ramakrishna, S. et al. 2006. Electrospun nanofibers: Solving global issues. *Materials Today* 9: 40–50, doi:10.1016/s1369-7021(06)71389-x.

29. Zhang, Y., C. T. Lim, S. Ramakrishna, and Z.-M. Huang. 2005. Recent development of polymer nanofibers for biomedical and biotechnological applications. *Journal of Materials Science: Materials in Medicine* 16: 933–946.

30. Che, G., B. B. Lakshmi, C. R. Martin, and E. R. Fisher. 1998. Chemical vapor deposition based synthesis of carbon nanotubes and nanofibers using a template method. *Chemistry of Materials* 10: 260–267.

31. Harfenist, S. A. et al. 2004. Direct drawing of suspended filamentary micro- and nanostructures from liquid polymers. *Nano Letters* 4: 1931–1937.

32. Nain, A. S., J. C. Wong, C. Amon, and M. Sitti. 2006. Drawing suspended polymer micro-/ nanofibers using glass micropipettes. *Applied Physics Letters* 89: 183105, doi:10.1063/1.2372694.

33. Hwang, W. et al. 2009. Surface induced nanofiber growth by self-assembly of a silk-elastin-like protein polymer. *Langmuir: The ACS Journal of Surfaces and Colloids* 25: 12682–12686, doi:10.1021/la9015993.

34. Huang, Z.-M., Y. Z. Zhang, M. Kotaki, and S. Ramakrishna. 2003. A review on polymer nano-fibers by electrospinning and their applications in nanocomposites. *Composites Science and Technology* 63: 2223–2253, doi:10.1016/s0266-3538(03)00178-7.

35. Li, D. and Y. Xia. 2004. Electrospinning of nanofibers: Reinventing the wheel? *Advanced Materials* 16: 1151–1170.

36. Canejo, J. P. et al. 2008. Helical twisting of electrospun liquid crystalline cellulose micro- and nanofibers. *Advanced Materials* 20: 4821–4825, doi:10.1002/adma.200801008.

37. Liang, D., B. S. Hsiao, and B. Chu. 2007. Functional electrospun nanofibrous scaffolds for biomed-ical applications. *Advanced Drug Delivery Reviews* 59: 1392–1412, doi:10.1016/j.addr.2007.04.021.

38. Yang, F., R. Murugan, S. Wang, and S. Ramakrishna. 2005. Electrospinning of nano/micro scale poly(L-lactic acid) aligned fibers and their potential in neural tissue engineering. *Biomaterials* 26: 2603–2610, doi:10.1016/j.biomaterials.2004.06.051.

39. Zhuo, H., J. Hu, S. Chen, and L. Yeung. 2008. Preparation of polyurethane nanofibers by elec-trospinning. *Journal of Applied Polymer Science* 109: 406–411, doi:10.1002/app.28067.

40. Min, B.-M. et al. 2004. Electrospinning of silk fibroin nanofibers and its effect on the adhe-sion and spreading of normal human keratinocytes and fibroblasts in vitro. *Biomaterials* 25: 1289–1297, doi:10.1016/j.biomaterials.2003.08.045.

41. Buttafoco, L. et al. 2006. Electrospinning of collagen and elastin for tissue engineering applica-tions. *Biomaterials* 27: 724–734, doi:10.1016/j.biomaterials.2005.06.024.

42. Kim, C.-W., D.-S. Kim, S.-Y. Kang, M. Marquez, and Y. L. Joo. 2006. Structural studies of electro-spun cellulose nanofibers. *Polymer* 47: 5097–5107, doi:10.1016/j.polymer.2006.05.033.

43. Pimenta, A. F. R. et al. 2012. Electrospinning of ion Jelly fibers. *Materials Letters*, doi:10.1016/j.matlet.2012.04.146.

44. Franco, P. Q., C. F. C. João, J. C. Silva, and J. P. Borges. 2012. Electrospun hydroxyapatite fibers from a simple sol–gel system. *Materials Letters* 67: 233–236, doi:10.1016/j.matlet.2011.09.090.

45. Ji, W. et al. 2011. Bioactive electrospun scaffolds delivering growth factors and genes for tissue engineering applications. *Pharmaceutical Research* 28: 1259–1272, doi:10.1007/s11095-010-0320-6.

46. Sahay, R., V. Thavasi, and S. Ramakrishna. 2011. Design modifications in electrospinning setup for advanced applications. *Journal of Nanomaterials* 2011: 1–17, doi:10.1155/2011/317673.

47. Teo, W. E. and S. Ramakrishna. 2006. A review on electrospinning design and nanofibre assem-blies. *Nanotechnology* 17: R89–R106, doi:10.1088/0957-4484/17/14/R01.

48. McCarthy, J. R. and R. Weissleder. 2008. Multifunctional magnetic nanoparticles for targeted imaging and therapy. *Advanced Drug Delivery Reviews* 60: 1241–1251, doi:10.1016/j.addr.2008.03.014.

49. Agarwal, S., J. H. Wendorff, and A. Greiner. 2008. Use of electrospinning technique for biomedi-cal applications. *Polymer* 49: 5603–5621.

50. Khang, D., J. Carpenter, Y. W. Chun, R. Pareta, and T. J. Webster. 2010. Nanotechnology for regenerative medicine. *Biomedical Microdevices* 12: 575–587.

51. Engel, E., A. Michiardi, M. Navarro, D. Lacroix, and J. A. Planell. 2008. Nanotechnology in regenerative medicine: The materials side. *Trends in Biotechnology* 26: 39–47.

52. Xue, M. et al. 2012. Biodegradable polymer-coated, gelatin hydrogel/bioceramics ternary composites for antitubercular drug delivery and tissue regeneration. *Journal of Nanomaterials* 2012: 8.

53. Nejadnik, M. R., A. G. Mikos, J. A. Jansen, and S. C. G. Leeuwenburgh. 2012. Facilitating the mineralization of oligo(poly(ethylene glycol) fumarate) hydrogel by incorporation of hydroxy-apatite nanoparticles. *Journal of Biomedical Materials Research A* 100A: 1316–1323.

54. Biris, A. R. et al. 2011. Novel multicomponent and biocompatible nanocomposite materials based on few-layer graphenes synthesized on a gold/hydroxyapatite catalytic system with applications in bone regeneration. *The Journal of Physical Chemistry C* 115: 18967–18976.

55. Wang, G., M. E. Babadagli, and H. Uludag. 2011. Bisphosphonate-derivatized liposomes to control drug release from collagen/hydroxyapatite scaffolds. *Molecular Pharmaceutics* 8: 1025–1034.

56. Vaz, C. M., S. van Tuijl, C. V. Bouten, and F. P. Baaijens. 2005. Design of scaffolds for blood vessel tissue engineering using a multi-layering electrospinning technique. *Acta Biomaterialia* 1: 575–582, doi:10.1016/j.actbio.2005.06.006.

57. Lee, S. J. et al. 2008. Development of a composite vascular scaffolding system that withstands physiological vascular conditions. *Biomaterials* 29: 2891–2898, doi:10.1016/j.biomaterials.2008.03.032.

58. Smith, M. J. et al. 2008. Suture-reinforced electrospun polydioxanone-elastin small-diameter tubes for use in vascular tissue engineering: A feasibility study. *Acta Biomaterialia* 4: 58–66, doi:10.1016/j.actbio.2007.08.001.

59. Prabhakaran, M. P., J. Venugopal, and S. Ramakrishna. 2009. Electrospun nanostructured scaffolds for bone tissue engineering. *Acta Biomaterialia* 5: 2884–2893, doi:10.1016/j.actbio.2009.05.007.

60. Choi, J. S., S. J. Lee, G. J. Christ, A. Atala, and J. J. Yoo. 2008. The influence of electrospun aligned poly(epsilon-caprolactone)/collagen nanofiber meshes on the formation of self-aligned skeletal muscle myotubes. *Biomaterials* 29: 2899–2906, doi:10.1016/j.biomaterials.2008.03.031.

61. Avis, K. J., J. E. Gough, and S. Downes. 2010. Aligned electrospun polymer fibres for skeletal muscles regeneration. *European Cells and Materials* 19: 193–204.

62. Ladd, M. R., S. J. Lee, J. D. Stitzel, A. Atala, and J. J. Yoo. 2011. Co-electrospun dual scaffolding system with potential for muscle-tendon junction tissue engineering. *Biomaterials* 32: 1549–1559, doi:10.1016/j.biomaterials.2010.10.038.

63. Sell, S. A., M. J. McClure, K. Garg, P. S. Wolfe, and G. L. Bowlin. 2009. Electrospinning of collagen/biopolymers for regenerative medicine and cardiovascular tissue engineering. *Advanced Drug Delivery Reviews* 61: 1007–1019, doi:10.1016/j.addr.2009.07.012.

64. Sarkar, S., T. Schmitz-Rixen, G. Hamilton, and A. Seifalian. 2007. Achieving the ideal properties for vascular bypass grafts using a tissue engineered approach: A review. *Medical & Biological Engineering & Computing* 45: 327–336.

65. Yang, P. and J. Temenoff. 2009. Engineering orthopedic tissue interfaces. *Tissue Engineering: Part B* 15: 127–141.

66. Jong, W. H. D. and P. J. A. Borm. 2008. Drug delivery and nanoparticles: Applications and hazards. *International Journal of Nanomedicine* 3: 133–149.

67. Doane, T. L. and C. Burda. 2012. The unique role of nanoparticles in nanomedicine: Imaging, drug delivery and therapy. *Chemical Society Reviews* 41: 2885–2911.

68. Parveen, S., R. Misra, and S. K. Sahoo. 2012. Nanoparticles: A boon to drug delivery, therapeutics, diagnostics and imaging. *Nanomedicine: Nanotechnology, Biology and Medicine* 8: 147–166.

69. Moen, M., K. Lyseng-Williamson, and L. Scott. 2009. Liposomal amphotericin B: A review of its use as empirical therapy in febrile neutropenia and in the treatment of invasive fungal infections. *Drugs* 69: 361–392.

70. Chen, Y.-H. et al. 2007. Methotrexate conjugated to gold nanoparticles inhibits tumor growth in a syngeneic lung tumor model. *Molecular Pharmaceutics* 4: 713–722.

71. Pissuwan, D., T. Niidome, and M. B. Cortie. 2011. The forthcoming applications of gold nanoparticles in drug and gene delivery systems. *Journal of Controlled Release* 149: 65–71.

72. Dvir, T. et al. 2011. Nanoparticles targeting the infarcted heart. *Nano Letters* 11: 4411–4414.

73. Susa, M. et al. 2009. Doxorubicin loaded polymeric nanoparticulate delivery system to overcome drug resistance in osteosarcoma. *BMC Cancer* 9: 399, doi:10.1186/1471-2407-9-399.

74. Chang, Y. and L. Xiao. 2010. Preparation and characterization of a novel drug delivery system: Biodegradable nanoparticles in thermosensitive chitosan/gelatin blend hydrogels. *Journal of Macromolecular Science, Part A: Pure and Applied Chemistry* 47: 608–615.

75. Xanthos, T., M. Chatzigeorgiou, E. O. Johnson, and A. Chalkias. 2012. Magnetically targeted drug delivery during cardiopulmonary resuscitation and the post-resuscitation period. *Resuscitation* 83: 803–805.

76. Yu, D.-G. 2009. Electrospun nanofiber-based drug delivery systems. *Health* 01: 67–75, doi:10.4236/health.2009.12012.
77. Choi, J. S., K. W. Leong, and H. S. Yoo. 2008. In vivo wound healing of diabetic ulcers using electrospun nanofibers immobilized with human epidermal growth factor (EGF). *Biomaterials* 29: 587–596, doi:10.1016/j.biomaterials.2007.10.012.
78. Peng, H. et al. 2008. In vitro degradation and release profiles for electrospun polymeric fibers containing paracetamol. *Colloids and Surfaces. B, Biointerfaces* 66: 206–212, doi:10.1016/j.colsurfb.2008.06.021.
79. He, C.-L. et al. 2006. Coaxial electrospun poly(L-lactic acid) ultrafine fibers for sustained drug delivery. *Journal of Macromolecular Science, Part B: Physics* 45: 515–524, doi:10.1080/00222340600769832.
80. Lueke, J. and W. A. Moussa. 2011. MEMS-based power generation techniques for implantable biosensing applications. *Sensors (Basel)* 11: 1433–1460, doi:10.3390/s110201433.
81. Harb, A. 2011. Energy harvesting: State-of-the-art. *Renewable Energy* 36: 2641–2654, doi:10.1016/j.renene.2010.06.014.
82. Sawicka, K., P. Gouma, and S. Simon. 2005. Electrospun biocomposite nanofibers for urea biosensing. *Sensors and Actuators B: Chemical* 108: 585–588, doi:10.1016/j.snb.2004.12.013.
83. Hansen, B. J., Y. Liu, R. Yang, and Z. L. Wang. 2010. Hybrid nanogenerator for concurrently harvesting biomechanical and biochemical energy. *ACS Nano* 4: 3647–3652.
84. Kim, J., H. Jia, and P. Wang. 2006. Challenges in biocatalysis for enzyme-based biofuel cells. *Biotechnology Advances* 24: 296–308, doi:10.1016/j.biotechadv.2005.11.006.
85. Baptista, A. C. et al. 2011. Thin and flexible bio-batteries made of electrospun cellulose-based membranes. *Biosensors & Bioelectronics* 26: 2742–2745, doi:10.1016/j.bios.2010.09.055.
86. Xie, J., S. Lee, and X. Chen. 2010. Nanoparticle-based theranostic agents. *Advanced Drug Delivery Reviews* 62: 1064–1079.
87. Soares, P. I. P., I. M. M. Ferreira, C. M. M. Novo, S. Dias, and B. J. Paulo. 2012. Doxorubicin vs. ladirubicin: Methods for improving osteosarcoma treatment. *Mini-Reviews in Medical Chemistry* 12(12): 1239–1249.
88. Sumer, B. and J. Bao. 2008. Theranostic nanomedicine for cancer. *Nanomedicine* 3: 137–140.
89. Widder, K., A. Senyei, and D. Scarpelli. 1978. Magnetic microspheres: A model system for site specific drug delivery *in vivo*. *Proceedings of the Society for Experimental Biology and Medicine* 58: 141–146.
90. Zhao, S.-H. et al. 2010. N-(2-hydroxyl) propyl-3-trimethyl ammonium chitosan chloride nanoparticle as a novel delivery system for parathyroid hormone-related protein 1–34. *International Journal of Pharmaceutics* 393: 268–272, doi:10.1016/j.ijpharm.2010.04.034.
91. Poptani, H. et al. 1998. Monitoring thymidine kinase and ganciclovir induced changes in rat malignant glioma in vivo by nuclear magnetic resonance imaging. *Cancer Gene Therapy* 5: 101–109.
92. Blankenberg, F. et al. 1997. Quantitative analysis of apoptotic cell death using proton nuclear magnetic resonance spectroscopy. *Blood* 89: 3778–3786.
93. Jain, T. et al. 2008. Magnetic nanoparticles with dual functional properties: Drug delivery and magnetic resonance imaging. *Biomaterials* 29: 4012–4021.
94. Park, J.-H., G. v. Maltzahn, E. Ruoslahti, S. N. Bhatia, and M. J. Sailor. 2008. Micellar hybrid nanoparticles for simultaneous magnetofluorescent imaging and drug delivery. *Angewandte Chemie International Edition* 47: 7284–7288.
95. Laurent, S., S. Dutz, U. O. Häfeli, and M. Mahmoudi. 2011. Magnetic fluid hyperthermia: Focus on superparamagnetic iron oxide nanoparticles. *Advances in Colloid and Interface Science* 166: 8–23.
96. Caldorera-Moore, M. E., W. B. Liechty, and N. A. Peppas. 2011. Responsive theranostic systems: Integration of diagnostic imaging agents and responsive controlled release drug delivery carriers. *Accounts of Chemical Research* 44: 1061–1070.

97. Al-Jamal, W. T. and K. Kostarelos. 2011. Liposomes: From a clinically established drug delivery system to a nanoparticle platform for theranostic nanomedicine. *Accounts of Chemical Research* 44: 1094–1104.
98. Grange, C. et al. 2010. Combined delivery and magnetic resonance imaging of neural cell adhesion molecule–targeted doxorubicin-containing liposomes in experimentally induced Kaposi's sarcoma. *Cancer Research* 70: 2180–2190.
99. Janib, S. M., A. S. Moses, and A. MacKay. 2010. Imaging and drug delivery using theranostic nanoparticles. *Advanced Drug Delivery Reviews* 62: 1052–1063.
100. Ornelas, C., R. Pennell, L. F. Liebes, and M. Weck. 2011. Construction of a well-defined multifunctional dendrimer for theranostics. *Organic Letters* 13: 976–979.
101. Cerdan S, Lotscher H R, Kunnecke B, and S. J. 1989. Monoclonal antibody-coated magnetite particles as contrast agents in magnetic resonance imaging of tumors *Magnetic Resonance in Medicine* 12: 151–163.
102. Huh, Y.-M. et al. 2005. In vivo magnetic resonance detection of cancer by using multifunctional magnetic nanocrystals. *Journal of American Chemical Society* 127: 12387–12391.
103. Natarajan, A., S. J. Denardo, G. L. Denardo, and C. Gruettnre. Site-specific conjugation of ligands to nanoparticles. U.S. patent 2010.
104. Soares, P. I. P., I. M. M. Ferreira, R. a. G. B. N. Igreja, C. M. M. Novo, and J. P. M. R. Borges. 2012. Application of hyperthermia for cancer treatment: Recent patents review. *Recent Patents on Anti-Cancer Drug Discovery* 7: 64–73.
105. Berry, C. C. 2009. Progress in functionalization of magnetic nanoparticles for applications in biomedicine. *Journal of Physics D: Applied Physics* 42: 224003, doi:10.1088/0022–3727/42/22/224003.
106. Hildebrandt, B. et al. 2011. Induced hyperthermia in the treatment of cancer. In *Cancer Management in Man: Chemotherapy, biological therapy, hyperthermia and supporting measures,* Cancer Growth and Progression 13, (ed. B. Minev), Springer Science and Business Media B.V, New York.
107. Matsumine, A. et al. 2007. Novel hyperthermia for metastatic bone tumors with magnetic materials by generating an alternating electromagnetic field. *Clinical & Experimental Metastasis* 24: 191–200, doi:10.1007/s10585–007–9068–8.

5

Nanoemulsions as a Vaccine Adjuvant

Grace Ledet, Levon A. Bostanian, and Tarun K. Mandal

CONTENTS

5.1 Introduction

5.1.1 Nanoemulsions and the Immune System

Nanotechnology has touched nearly every aspect of pharmaceutical development, and vaccine research and development is no exception. This chapter is intended to introduce the reader to nanoemulsions and their use as vaccine adjuvants, particularly highlighting the adjuvants approved around the world in addition to those adjuvant candidates in clinical development. Of particular interest are the commonalities between those formulations approved for use in humans and insight into their modes of action as vaccine adjuvants. In order to understand the use of nanoemulsions in immunology, one must have an understanding of the immune system and how vaccines take advantage of the immune system in order to prevent infection. Thus, this section includes a brief overview of the immune system and its response to foreign pathogens, an introduction to the role of adjuvants in enhancing vaccine efficiency, and the science behind nanoemulsions and their formation.

5.1.2 Overview of the Immune System and Vaccination

The purpose of the immune system is to prevent disease and infection, which can be caused by a wide variety of extrinsic sources such as viruses, bacteria, fungi, and parasites. The human immune system is composed of two interrelated components called innate immunity and adaptive immunity [1]. The two types of immune responses are interrelated because they use the same effector cells. The innate immune response is nonspecific and is the body's first line of defense against infection from pathogens. The primary function of the innate immune response is to eliminate or control infection by foreign pathogens [1]. As the first response to a foreign body, the innate immune system is responsible for deciding whether or not to respond to a particular antigen. Antigens are typically proteins or polysaccharides, such as the protein coat of a virus, cell wall components, or the DNA or RNA of bacteria, viruses, protozoa, or fungi, but can be any foreign molecule or cell that the innate immune cells do not recognize as "self." The cells of the innate immune response recognize foreign pathogens and their antigens with pattern recognizing receptors (PRRs), which recognize molecules present in a wide range of microbes called pathogen-associated molecular patterns (PAMPs) [2]. Among the PRRs on the surface of cells in the innate immune system, Toll-like receptors (TLRs) play a major role in immune recognition. Ten TLRs have been identified in mice and humans [3]. TLRs, similar to all PRRs, are receptors specific to structurally conserved molecules derived from microbes [4].

Once PRRs recognize PAMPs, cells trigger signaling pathways, which lead to the activation of phagocytes (which engulf and destroy pathogens) and to the initiation of the inflammatory process. This is accomplished by the secretion of various cytokines, such as interleukins, interferons, and chemokines, which in turn trigger the release of other cytokines and stimulate cell actions [1]. The cells active in the innate immune response include macrophages, dendritic cells, mast cells, neutrophils, eosinophils, and natural killer cells [5]. If the cells of the innate immune system are unable to quell an infection, the adaptive immune system (also referred to as specific immune response) is activated to help combat the infection. The adaptive immune system has the ability to remember and adapt to infections, but the adaptive immune response is also referred to as the acquired immune response because this cellular memory is not transferred from one generation to another, except for limited transfer of antibodies from a mother to a fetus. Thus, every individual must *acquire* this immunologic memory over their lifetime.

Dendritic cells, which are found in almost all tissues of the body, are the main link between the innate and adaptive immune systems. Dendritic cells, one of the classes of antigen-presenting cells, are largely responsible for alerting the cells of the adaptive immune system by the expression of costimulatory molecules [6]. Dendritic cells recognize foreign pathogens with their surface receptors, such as TLRs, and undergo dendritic cell maturation, which involves the migration of the dendritic cells to the lymph nodes in order to alert antigen-specific T cells [7]. The main cell types of adaptive immunity can only be activated once antigen-presenting cells (such as dendritic cells) uptake, process, and present an antigen to naïve lymphocytes in the lymph nodes [1]. B lymphocytes differentiate into plasma cells, which secrete antibodies in the blood, which signal other cells to initiate the defense against the opposing antigen or cells bearing them. Some B lymphocytes do not differentiate into plasma cells but remain as memory B cells to elicit a rapid and large release of antibodies if the antigen reoccurs by subsequent infections following the initial recognition of the antigen. Cytotoxic T lymphocytes (also referred to as CD8$^+$ cells) directly attack and kill cells bearing the antigen [8]. Antibody-mediated responses are also called humoral responses and are the main defense against bacteria, viruses, microbes, and toxins. Cell-mediated responses are those that do not involve antibodies, such as the action of antigen-specific T lymphocytes, natural killer cells, and other cells that release various cytokines in response to an antigen. There are two different types of T-helper cells (also referred to as CD4+ cells), Th1 and Th2, which are characterized by the release of different and distinct cytokines [9,10]. The pattern of cytokine production determines the type of immune response, whether predominantly a humoral or cellular response. Understanding what kind of response is needed to confer immunity and what mechanisms lead to a Th1 or Th2 response are critical to the development of vaccines.

In summary, the main effects of activating the adaptive immune system are the release of antibodies and cell-mediated immune responses, which are triggered by the exposure to antigens. The response of the adaptive immune system increases in magnitude with subsequent exposures to the same antigen—immunologic memory [11]. This immunologic memory by the adaptive immune system is the foundation of prophylactic vaccination against infection. Traditional vaccines consist of small quantities of living or dead microbes or molecules derived from the foreign microbe that activates the adaptive immune response and the induction of memory B cells, which will produce a rapid, effective immune response to future infection by that microbe. For instance, influenza vaccines are either live attenuated or inactivated vaccines. The inactivated vaccines have three different forms of antigen preparation—inactivated whole virus, subvirions where purified virus particles are disrupted to solubilize the lipid envelope ("split" vaccine), or as a subunit vaccine (purified surface proteins, which are known antigens) [12]. The external variables that can impact the efficacy of a vaccine include the route of administration, the dose, the dosage schedule, and the immunologic status of the recipient.

5.1.3 Role of Adjuvants in Vaccine Development

Adjuvants are additives included in vaccine formulations to enhance, modulate, and/or direct the immunogenicity of the vaccine while causing minimal toxicity [13]. Adjuvants can be generally classified by their mode of action as either (1) immunostimulatory agents, which act on the immune system to increase a vaccine's efficacy, or (2) vehicles, which deliver the antigen to the immune system [14]. In some cases, an adjuvant can fit into both of these broad categories. PAMPs are the basis for many immunostimulatory adjuvants [15]. Adjuvants can modulate the immune response to a vaccine in many ways including creating a depot for the prolonged release of an antigen, engagement with PRRs to increase the activity of dendritic cells, and induction of cytotoxic T lymphocytes and/or T-helper lymphocytes [16]. Some of the benefits of adjuvants include (1) an increased immunologic response to a vaccine, especially for those individuals with reduced responsiveness to nonadjuvanted vaccines; (2) the chance to reduce the amount of antigen needed to elicit the desired immune response, which allows for more doses to be administered with the same amount of antigen (dose sparing); (3) the reduction or elimination of multiple doses while still achieving adequate protection; and (4) the ability to alter the type of immune response in order to provide the most appropriate response to the specific antigen in question [17–20]. Overall, adjuvants are necessary to enhance the immunogenicity and efficacy of many vaccines, especially vaccines for those populations that are more susceptible to a specific disease or less able to fight off a specific disease. For example, elderly adults are particularly susceptible to disease because the ability of the immune system to mount a strong and efficacious defense decreases with age. Also, infants and children are more susceptible to certain transmittable diseases because they lack the fully developed immune system of an adult. Finally, immunocompromised individuals need effective, reliable vaccines to prevent opportunistic infections and other transmittable diseases for which their immune system is ill-equipped to fight.

Adjuvants are becoming increasingly important as vaccine technology progresses beyond the use of live attenuated vaccines and toward the use of isolated antigens, such as the subunit influenza vaccine mentioned in the previous section. In general, live virus vaccines and inactivated viruses or bacteria do not require an adjuvant to increase their immunogenicity. Highly purified recombinant subunit antigens reduce the risk of toxicity associated with live viruses but are often less immunogenic than traditional live virus vaccines. For those diseases in which live or inactivated vaccines are unavailable, impractical, or unsafe, recombinant vaccines are the only option and thus necessitate the development of appropriate adjuvants.

Adjuvants must be tailored to their specific target antigen because of the complexity of the immune response, the diversity of pathogens, and the many routes of entry into the body available to pathogens [1]. The ultimate goal of optimizing an adjuvant for a specific antigen is to maximize the vaccine potency. Some criteria considered when selecting the appropriate type of adjuvant required for a specific vaccine include understanding the nature of the antigen, the type of immune response required to best provide seroprotection, and the delivery route of the vaccine. The optimal adjuvant should also be biocompatible, biodegradable, safe, stable for a long shelf life, and inexpensive to manufacture [14]. The benefits of including an adjuvant in a vaccine formulation must be balanced against the risk of adverse side effects of the adjuvant. Reactions to adjuvants can be either local (e.g., pain, inflammation, swelling, or granulomas at the site of injection) or systemic (e.g., nausea, fever, allergic reactions, or organ toxicity) [21]. The balance between potency and adverse reactogenicity is a constant battle in the fight to find appropriate adjuvants for vaccines currently under development.

5.1.4 Nanoemulsions

Emulsions are two-phase colloidal systems where oil droplets are dispersed in an aqueous medium to form an oil-in-water emulsion (o/w) or the reversal of that to form a water-in-oil emulsion (w/o). The constituent that makes up the "droplet" is alternatively referred to as the internal phase, dispersed phase, or discontinuous phase, while the immiscible material surrounding the droplets is referred to as the continuous phase or external phase [22]. Multiple emulsions are also possible, such as w/o/w (water-in-oil-in-water) or o/w/o (oil-in-water-in-oil) emulsions. Nanoemulsions are those emulsions with an internal droplet size in the nanometer range. Generally, nanoemulsions have a particle size of ≤100 nm, but the term "nanoemulsion" is also used to refer to emulsions up to the size of 1000 nm (i.e., 1 μm). Nanoemulsions are kinetically stable. Microemulsions are also emulsions with a droplet size around 100 nm but differ from nanoemulsions in that they are thermodynamically stable. While semantically the terms "nanoemulsion" and "microemulsion" may seem contradictory, these terms are the standard terminology in the literature, and the difference between the two systems is according to their stability and assembly, not their size, even though the names may imply a difference in particle size [23]. Even though nanoemulsions are less stable than thermodynamically stable microemulsions, nanoemulsions can still be stable for several years. Also, nanoemulsions can be infinitely dilutable, while microemulsions cannot maintain their thermodynamic stability upon dilution. Optically, nanoemulsions can range from milky white opacity to transparent with a blue hue (the slight blue color comes from Rayleigh scattering, which occurs when the droplets in the sample are smaller than the wavelength of the light passing through the sample) [24].

Nanoemulsions are formed by the addition of a stabilizer or surfactant that surrounds and disperses the internal phase within the continuous phase. The surfactant adsorbs to the surface of the emulsion droplets to prevent the droplets from aggregating and separating from the continuous phase. Additionally, the surfactant reduces the interfacial tension at the oil–water interface, allowing for the formation of small droplet sizes [25]. The interfacial tension at the phase boundaries in nanoemulsions is usually 1–10 mN m^{-1}, which is significantly higher than the 10^{-1} to 10^{-3} mN m^{-1} of microemulsions [26,27]. Nanoemulsions can require significantly less surfactant to form nano-sized droplets than microemulsions, but nanoemulsions require energy to be put into the system to reach a small particle size and stability, whereas microemulsions self-assemble by spontaneous emulsification when the concentration of their constituents reach a thermodynamic balance [24]. The type of emulsion, whether o/w or w/o, is dependent on the choice of excipients, the concentration of each excipient, the physical characteristics of the oil and surfactants, and the manufacturing procedure. The choice of surfactant is determined by the selected oil and by the type of emulsion desired. The hydrophilic–lipophilic balance (HLB) is used to describe the emulsifying characteristics of the surfactant. A high HLB value ($\geq$10) refers to hydrophilic surfactants, and a low HLB value ($\leq$10) designates a lipophilic nature. Oils may have a reported optimal HLB value for surfactants, but these are guidelines, not absolute requirements for emulsification. Generally, a hydrophilic surfactant is used to make an o/w nanoemulsion, and a lipophilic surfactant is required for a w/o nanoemulsion. Another distinction between microemulsions and nanoemulsions is that the surfactant in microemulsions is generally soluble in both phases. Most often, nanoemulsions are formulated with a combination of a lipophilic surfactant and a hydrophilic surfactant to further reduce the interfacial tension at the oil–water interface resulting in an even smaller particle size and increased stability.

The selection of oil, surfactant, and aqueous phase is dependent on the final purpose of the nanoemulsion, the physical and chemical properties of any additional components (e.g., the antigen for a nanoemulsion vaccine adjuvant), and the final desired physiochemical properties of the nanoemulsion. Thus, no specific information can be given about the most suitable oil or surfactant for a nanoemulsion because their properties and purposes can vary broadly and are all interdependent on the other component choices.

5.1.4.1 Manufacturing Techniques

In order to break larger droplets of the internal phase into smaller droplets to form a nanoemulsion, energy must be put into the system, the basic process of emulsification. To significantly deform a droplet into smaller droplets, the applied shear force must exceed the pressure within the droplet, which is called the Laplace pressure. Laplace pressure is the pressure exerted on the molecules within the droplet by the curved interface of the droplet. Large droplets have a lower Laplace pressure than smaller droplets, which means that as particles are broken down into smaller droplets, more shear force is needed to break them down further. Additionally, more surfactant is needed to cover the larger surface-to-volume ratio of increasingly smaller droplets [28]. Thus, when formulating a nanoemulsion, excess surfactant must be present to ensure that the desired droplet size can be achieved as the surface-to-volume ratio increases.

Several different types of homogenizers exist to reduce the particle size of an emulsion, such as high-pressure homogenizers, colloid mills, microfluidizers, and ultrasonic homogenizers. They essentially serve the same purpose—provide enough shear stress to reduce the particle size into the nanometer range. These are sometimes collectively referred to as high-energy emulsification methods. Microfluidizers use high pressures (500–20,000 psi) to force the emulsion through microchannels to an impingement area that breaks the emulsion particle size down further into the nanometer range. Often, the two phases of the emulsion are combined and processed though an in-line homogenizer, yielding a coarse initial emulsion, prior to passage through the microfluidizer to form the nano-emulsion [29]. Several manufacturing parameters can be adjusted to optimize the size and distribution of the nanoemulsion droplets. Some of these parameters include the following: the number of times the sample is passed through the homogenizer, the temperature at the time of homogenization, the valve pressure of the homogenizer, and the type of homogenizer.

Recently, low-energy methods have been proposed because of benefits such as being less destructive to sensitive compounds and energy saving for large-scale production [30]. Low-energy methods are based on the phase inversion temperature (PIT) of the emulsion formulation. With certain formulations, the curvature of the surfactant layer changes with temperature, and when a critical temperature is reached (i.e., the PIT), the surface curvature reaches zero and the interfacial tension decreases, causing the emulsion to invert to a reverse emulsion [31,32]. The PIT method takes advantage of such surfactant properties by preparing the emulsion at a high temperature and cooling the emulsion through the PIT to invert the emulsion and produce a nanoemulsion. Obviously, this method is highly dependent on surfactant selection and is therefore not applicable to all formulations. Additionally, while this method avoids the destructive shear forces experienced during high-energy methods, the formulation and any additional compounds included in the formulation are exposed to heat, which could still damage sensitive compounds.

5.1.4.2 Characterization of Nanoemulsions

One of the most important characteristics of a nanoemulsion is its particle size. This is principally due to the fact that the size of the droplets within the nanoemulsion impacts the stability of the nanoemulsion, its optical properties, and its rheology [33]. The goal with most nanoemulsions is to achieve an emulsion with a very uniform particle size, known as a monodispersed system. When the distribution of the particle size results in droplets with multiple sizes, the nanoemulsions are described as polydispersed or a multimodal distribution. The polydispersity index is a measure of the size distribution, is typically defined as the standard deviation divided by the mean, and is utilized as a measure of homogeneity. A polydispersity index of ≤0.200 is generally regarded as a monodispersed distribution [24]. The particle size and distribution are important because changes in particle size or distribution are indicative of instability of the formulation. In general, a small particle size and homogeneous distribution are more stable than an emulsion with significant percentage of large particles and/or a large size distribution [34]. A uniform particle size in the nanometer range is important for stability because the droplets will not be subject to creaming or sedimentation because the Brownian motion of the particles exceeds the gravitational forces on the particles [33].

Another important characteristic of nanoemulsions related to their kinetic stability is the surface charge of the internal phase. The charge on the surface of an emulsion droplet is important because it determines the interactions that will occur between individual droplets and between droplets and other components of the formulation. The zeta potential is a measure of the electrical potential between the internal phase and the interfacial layer of external phase associated around the droplet. The zeta potential is significant for colloidal systems, such as nanoemulsions, because a large zeta potential (> ±30 mV) leads to significant repulsive forces between droplets, making the nanoemulsions resistant to flocculation and coalescence [35].

5.1.4.3 Nanoemulsion Stability

Stability is a term often used to describe nanoemulsions and generally refers to the emulsion's ability to resist physical and chemical changes over time [22]. Once a nanoemulsion is formed and the applied energy is removed, the nanoemulsion is still subject to further internal and external forces during storage, such as gravitational, thermal, and interfacial stresses. Instability can manifest in many different forms—creaming, sedimentation, coalescence, flocculation, and Ostwald ripening (Figure 5.1).

Creaming and sedimentation are the result of gravitational forces. Creaming occurs when the emulsion droplets are less dense than the continuous phase and condense at the upper surface of the nanoemulsion. Sedimentation is the opposite, in that gravity pulls the more dense droplets to the bottom of the nanoemulsions, where subsequent coalescence or

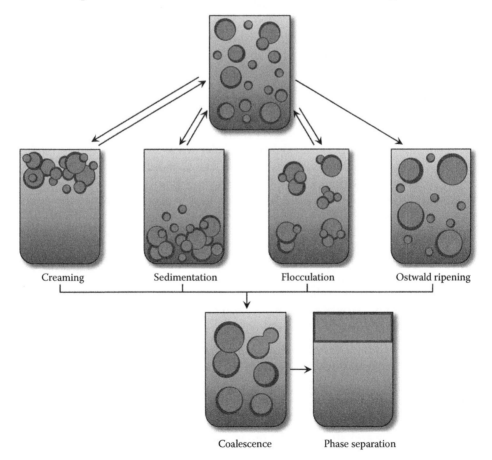

FIGURE 5.1
Nanoemulsions are subject to many different forms of instability. Some can be reversible such as creaming, sedimentation, or flocculation, while some forms of instability are not reversible and result in permanent changes to the physical characteristics of the nanoemulsion. Ostwald ripening is the main instability issues faced by nanoemulsions.

flocculation can occur [22]. Generally, nanoemulsions are not subject to sedimentation [36]. Coalescence is the fusion of smaller droplets into a larger droplet, which can eventually lead to complete phase separation. Coalescence can be prevented by choosing a surfactant or surfactants that provide a strong repulsion (high zeta potential) between droplets and can be detected by changes in the size distribution of the nanoemulsion. Flocculation is the reversible aggregation of droplets together but not necessarily the fusion of those droplets as with coalescence [24]. When manufacturing nanoemulsions, reducing the number of large particles as much as possible is paramount because larger droplets within an emulsion can act as a nucleation point for further coalescence [37]. Another form of instability occurs when the dispersed phase is partially soluble in the external phase. The smaller droplets, with their high internal pressure, disperse into the continuous phase, and larger droplets slowly form by mass transport of the dispersed material through the continuous phase. This process is known as Ostwald ripening [38]. For o/w emulsions, the rate of Ostwald ripening can be reduced by choosing a dispersed phase with very little solubility in the continuous phase or by choosing a binary internal phase, where the solubility of one of the components is far less than the solubility of the other. While there are specific actions that can be taken to alter a formulation in order to avoid a certain form of instability, one must always keep in mind that all of these forms of instability are interrelated and that one type of instability can accelerate or cause the occurrence of another.

5.2 Historical Perspective

Why are nanoemulsions appropriate as vaccine adjuvants? First, the preferred adjuvants for most adjuvanted vaccines, aluminum salts, are not appropriate for all antigens and do not induce a strong cell-mediated immune response. Secondly, emulsions have a proven history to be safe for human use for both cosmetic and medicinal applications. Thirdly, nanoemulsions can be stable for a long storage life, and their creation is scalable to large-scale manufacturing. Finally, emulsions have a history of success as vaccine adjuvants, although prior to the 1990s that success was confined to veterinary formulations [34]. The ideal nanoemulsion for application as a vaccine adjuvant should display all those characteristics listed in Section 5.1.3 as advantageous for vaccine adjuvanticity in addition to the ability to tailor its properties to elicit the optimal immune response for the chosen antigen. The following sections chronicle the progression of nanoemulsions as vaccine adjuvants and the current state of their development.

5.2.1 Approved Vaccine Adjuvant Emulsions

Prior to the 1990s, the only approved vaccine adjuvants were aluminum salts, first described as adjuvants in 1926 [39]. In 70 years since the institution of aluminum salts (also referred to as alum) as the preferred adjuvant for human vaccines, no other adjuvant has proved to be both safe and effective for human use—until the reemergence of emulsions in the 1990s with MF59® (Novartis, Basel, Switzerland) (Figure 5.2). Worth noting is the development of Freud's incomplete adjuvant, a w/o emulsion, which was an extension of the veterinary adjuvant Freud's complete adjuvant and which was a human vaccine candidate but was discontinued in the 1960s due to adverse side effects such as abscesses or cysts at the site of inoculation [40,41]. Prior to the stoppage of studies involving this

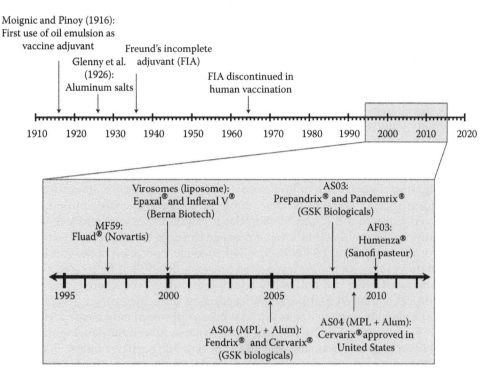

FIGURE 5.2
The history of approved adjuvants accelerated tremendously starting in the 1990s [39,46]. The majority of twentieth century vaccine adjuvant technology was dominated by the use of aluminum salts, but nanoemulsions account for four of the six approved adjuvants (excluding aluminum salts) licensed since 1997. Of the adjuvants listed in this figure, only AS04 is licensed for use in the United States. The rest of the adjuvants listed are approved in Europe and other parts of the world.

mineral oil-based emulsion, approximately 900,000 doses were administered to patients in Britain as an adjuvanted influenza vaccine [42–44]. With improvements in the quality of the mineral oil and surfactants for formulations like Freud's incomplete adjuvant, new mineral oil w/o emulsions are being tested yet again in humans (see Section 5.3 concerning Montanide ISA 51 and Montanide ISA 720). Additional adjuvants besides aluminum salts are necessary because alum is a poor stimulator of T lymphocyte responses, which is critical for those infections that require a strong cell-mediated immune response to confer long-lasting immunity [14]. For example, the three diseases that result in the most global mortality each year (i.e., malaria, tuberculosis, and HIV) are all fought within the body by cellular immune response, a Th1 immune response, which alum is not effective at generating [45].

Since the approval of MF59, other adjuvants have emerged, including immunomodulatory compounds and liposomes, but the majority of the adjuvants investigated are emulsion based or a combination of nanoemulsions and immunomodulatory agents. This section will cover those nanoemulsion adjuvants approved for human use. Perhaps the biggest leap forward for nanoemulsion vaccine adjuvant development came in 2009 with the H1N1 influenza pandemic (swine flu) when 180 million doses of pandemic influenza vaccine were produced from June to December 2009 [47]. To date, no adjuvanted influenza vaccines have been approved in the United States.

5.2.1.1 MF59

MF59 was the first nonalum adjuvant approved for human use in Europe in 1997. MF59 is in the influenza vaccine FLUAD® (Novartis, Basel, Switzerland) and is an o/w emulsion composed of 5% squalene, 0.5% Tween 80 (polysorbate 80), and 0.5% Span 85 (sorbitan trioleate) in sodium citrate–citric acid aqueous buffer (pH 6.5) [37,48]. MF59 was originally developed as a vehicle for the immunostimulant muramyl tripeptide phosphatidylethanolamine (MTP-PE), but MTP-PE was removed from the formulation due to toxicity, while MF59 was found to have its own adjuvant properties independent of this additive [49]. MF59 is prepared by first dissolving Span 85 in squalene and Tween 80 in the aqueous phase. A coarse emulsion is made by combining these two phases in an in-line homogenizer. A microfluidizer is then used to produce an emulsion with a small, uniform particle size. To further refine the particle size distribution, the emulsion is filtered through a 0.22 μm filter under nitrogen to remove any larger oil droplets remaining in the emulsion following passage through the microfluidizer, removing 99.5% of droplets larger than 1.2 μm in size. The final incorporation of the vaccine antigen depends on the stability of the antigen in the presence of MF59. If long-term stability has been demonstrated, then the antigen is combined with MF59, filtered, and packaged; otherwise, MF59 is filtered and packaged separately from the antigen [37,50]. Visually, MF59 is a milky white, homogeneous liquid with an approximately mean particle size of 150 nm. MF59 can be stored at 2°C–8°C for up to 3 years without the development of large particles above the manufacturing specification (1×10^7 particles larger than 1.2 μm). However, the nanoemulsion is sensitive to extremes in temperatures and thus must be kept at 2°C–8°C.

Since its introduction to the market in 1997 as an influenza vaccine for the elderly, more than 27 million patients received FLUAD in the first 9 years of its use. Analysis of these subjects demonstrated that FLUAD has a good safety profile in the elderly population in addition to infants, children, and adults [51]. Although FLUAD was associated with increased pain and swelling at the injection site after immunization, it also induced higher antibody titers than nonadjuvanted influenza vaccines in elderly subjects, which includes those who have shown decreased antibody responses to conventional inactivated influenza vaccines. MF59 has also been studied in clinical trials for vaccine candidates against herpes simplex virus 2 (HSV), human immunodeficiency virus (HIV), cytomegalovirus (CMV), hepatitis B virus (HBV), hepatitis C virus (HCV), *Escherichia coli*, parvovirus, and human papillomavirus (HPV) [49].

While the immunogenicity and safety of MF59 have made it a great vaccine adjuvant, the biggest disadvantage of MF59 is the stability of the nanoemulsion; although stable at 2°C–8°C for years, MF59 cannot be frozen or kept at room temperature. Since a cold chain must be maintained for most vaccine antigens, this is not a problem for most applications, but its temperature sensitivity does limit its use in remote and developing countries. As with other nanoemulsion vaccines, the mode of action of MF59 has not been completely elucidated. However, more is known about MF59 than other emulsion adjuvants. MF59 increases the humoral response to a vaccine adjuvant, producing 5–250 times higher antibody response compared to nonadjuvanted vaccines [50]. Mosca et al. (2008) concluded that MF59 has significant immunostimulatory activity at the site of injection and causes the activation of dendritic cells in the muscle tissue [52]. Also, the complete MF59 emulsion is required for adjuvanticity—only complete MF59 induces cell migration events in the muscle tissue, while its excipients do not demonstrate immunostimulatory activity comparable to that of complete MF59 [49,53].

5.2.1.2 AS03

GlaxoSmithKline (GSK) has been working on alternative adjuvants to alum for over two decades. Their adjuvant systems include a variety of adjuvant types, such as aluminum salts, emulsions, liposomes, and immunomodulatory compounds. They have had particular success with combinations of these adjuvants. For example, their adjuvant AS04 is a combination of the TLR agonist 3-O-desacyl-4'-monophosphoryl lipid A (MPL) and aluminum hydroxide or aluminum phosphate; AS04 has been approved for the hepatitis B vaccine FENDrix® (GlaxoSmithKline Biologicals S.A., Rixensart, Belgium). Their first nanoemulsion product to be registered is also a combination product—AS03 is a nanoemulsion with the addition of α-tocopherol, which has proven to have immunostimulatory properties in veterinary vaccines [54–56]. The emulsion component of AS03 consists of 5% v/v squalene, 5% v/v α-tocopherol, and 1.8% v/v Tween 80 in phosphate–saline buffer (PBS) (pH 6.8) and is a whitish homogeneous liquid with a particle size of 150–155 nm [12,57]. To make this precursor nanoemulsion referred to as SB62, the oil phase (squalene and α-tocopherol) is mixed with the aqueous phase (Tween 80 and PBS) at a ratio of 10% the total volume of the mixture and stirred at room temperature for 15 min. The pre-emulsion is then passed through a microfluidizer for 8 cycles at 15,000 psi to allow the shear, impact, and cavitation forces of the microfluidizer to reduce the particle size distribution to between 100 and 200 nm. Then, nanoemulsion SB62 is sterilized by passage through a 0.22 μm filter and stored at 2°C–8°C. To transform the bulk precursor nanoemulsion SB62 into the final adjuvant AS03, PBS is mixed with SB62, stirred for 15–20 min, and sterilized by filtration using a 0.22 μm filter. AS03 is stored at 2°C–8°C until it is loaded into sterile syringes for packaging. The vaccine antigen is packaged separately and combined with AS03 immediately prior to intramuscular injection [12]. Prepandrix™ (GlaxoSmithKline Biologicals S.A., Rixensart, Belgium) is the approved prepandemic influenza (H5N1) vaccine, which includes AS03, and Pandemrix™ (GlaxoSmithKline Biologicals S.A., Rixensart, Belgium) is approved as a pandemic influenza A (H1N1)v vaccine with AS03. Both are approved in Europe.

Since its approval in September 2009 for vaccination against H1N1, at least 30.8 million Europeans were immunized with Pandemrix by September 2010 with millions of additional doses given in other parts of the world. Its proven cross-reactive humoral responses and dose-sparing ability made AS03 ideal for the 2009 influenza A/H1N1 pandemic [58]. In a Phase I/II, observer-blinded, randomized study evaluating different formulations of the H1N1 2009 pandemic influenza vaccine with and without AS03, both the adjuvanted and nonadjuvanted formulations had a clinically acceptable safety profile for both young (18–64 years old) and elderly (>64 years old). The data from this study suggested that a single dose (3.75 μg) of AS03-adjuvanted vaccine was sufficiently immunogenic and safe, and to reach these same endpoints with the nonadjuvanted vaccine, a higher antigen content (15 μg) was required. The humoral immune response elicited after the first immunization dose with the AS03-adjuvanted vaccine persisted for 6 months post immunization [59]. As seen with other nanoemulsion adjuvant systems, AS03-adjuvanted vaccines result in more frequent patient reporting of local symptoms, such as pain, redness, and swelling, compared to the nonadjuvanted H1N1 vaccine, but these were transient and mostly mild to moderate in intensity [60]. However, the Finnish National Institute for Health and Welfare recommended the discontinuation of Pandemrix vaccinations in 2010 because of reported cases of narcolepsy, a rare sleep disorder that causes a person to fall asleep suddenly and unexpectedly, in children and adolescents following vaccination with Pandemrix [61]. The vaccine was not identified as the cause of the cases but was reported to have a temporal association with the onset of the narcolepsy cases. In 2011, the European Medicines Agency recommended restricting the use of Pandemrix in people under 20 years of age; in this age group, Pandemrix is only to be administered in the absence of seasonal trivalent influenza vaccines. However, they also concluded that the overall benefit–risk balance of Pandemrix remains positive [62].

The mode of action of AS03 as an adjuvant has been investigated in mice and human cell lines by Morel et al. (2011). Interestingly, they demonstrated that AS03's functionality extends beyond that of an antigen delivery system because its adjuvanticity is not dependent on its physical association with the antigen (not dependent on mixing AS03 with the antigen prior to injection). Its adjuvant activity is local to the injection site (induction of local cytokine production) because injection at a completely different location than the antigen did not improve the immune response to the antigen. These results suggest that AS03 stimulates a localized innate immune response. AS03 promotes dendritic cell and monocyte migration to draining lymph nodes. Furthermore, AS03 promotes higher antigen-specific antibody responses than alum, and α-tocopherol is essential to the AS03 formulation to achieve an optimal antigen-specific adaptive response [58].

5.2.1.3 AF03

Sanofi Pasteur's adjuvant AF03 is a component of the vaccine Humenza® (Sanofi Pasteur S.A., Lyon, France), which had been approved in Europe for influenza prophylaxis in the event of a pandemic (Sanofi Pasteur voluntarily withdrew Humenza in 2011 for commercial reasons) [63]. AF03 is composed of approximately 2.8% w/v squalene, 0.53% w/v polyoxyethylene cetostearyl ether, 0.42% w/v Span 80 (sorbitan oleate), and 0.51% w/v mannitol with a phosphate- or citrate-buffered aqueous phase [64]. This emulsion is thermoreversible, which means that the emulsion transitions from a w/o to an o/w emulsion when heated to its PIT. Thermoreversible nanoemulsions are very stable because of the homogeneous particle size of the resulting emulsion, with at least 50% of the droplets with a size ≤110 nm. The PIT for AF03 and similar emulsions is preferably between 50°C and 65°C to allow stability at "high"-temperature storage conditions (e.g., 37°C) and to prevent damage to any heat-sensitive components with excessively high temperatures (e.g., 80°C) [65]. Additionally, AF03 could contain a TLR4 agonist molecule. As discussed in Section 5.1.4, this method of nanoemulsion production has several favorable qualities, such as ease of large-scale production, reduced physical agitation for sensitive molecules, and resultant small droplet size. Similar to MF59, AF03 cannot be frozen and must be stored at 2°C–8°C.

The adjuvanticity of AF03 is best demonstrated in children younger than 3 years of age. When comparing the administration of the nonadjuvanted Panenza® (Sanofi Pasteur S.A., Lyon, France) in the United States and Humenza in Europe, both induced high antibody responses after one dose, and with two injections, both could achieve close to 100% seroprotection in children. The difference in the adjuvanted and nonadjuvanted vaccines was evidenced in the 3 years or younger demographic, where one dose of the adjuvanted Humenza resulted in 96%–98% seroprotection as opposed to only 33%–34% with Panenza [60,66]. Also, AF03 manifested no safety concerns—no dose–response effect observed, similar incidence of reactions in vaccine and placebo groups for children, and only a slightly higher incidence of mild reactions with AF03.

5.2.2 Ubiquitous Use of Squalene in Nanoemulsion Adjuvants

One common link among all of the nanoemulsion vaccine adjuvants described in Section 5.2.1 is the use of squalene as the oil phase of the formulations. Why the focus on one oil excipient when so many possible candidates are available? Squalene is naturally occurring and found most notably in shark livers and in olives, and its hydrogenated form squalane is also found in human sebum (a skin surface lipid). Squalene is a liner hydrocarbon and is a precursor of cholesterol (Figure 5.3). Squalene and squalane are extensively used in the cosmetic industry as an emollient, and perhaps the most compelling reason to use squalene or squalane as the lipid phase of nanoemulsions for vaccine adjuvants is the impressive safety record for their use in cosmetics and drug delivery. In 1976, 18 cosmetic products included squalene, and 400 products contained squalane; by 2001, the FDA had 29 products containing squalene and 595 products with squalane [67]. The first documentation of squalene or squalane as an emulsion for the delivery of antigens occurred in the 1980s [68,69]. Squalene has proven its safety and tolerability in MF59 with millions of doses administered to every demographic (from infants to the elderly) without adverse effects. The only notable possible adverse effect related to squalene was brought to light in 2000—Asa et al. (2000) showed a correlation between Gulf War syndrome, a chronic illness manifested in Gulf War veterans, and high-squalene antibodies, possibly the result of being immunized with vaccines containing squalene [70]. This link was later completely disproven because squalene was not used in the vaccines given to those service men in the study and because vaccines that do contain squalene (MF59) did not elicit antibodies against squalene [49].

The excellent safety profile of squalene is especially important in light of the fact that the limiting factor for every vaccine adjuvant candidate since alum has been adverse safety profiles (e.g., Freud's incomplete adjuvant). Squalene is used for its proven safety and biocompatibility—it is nonirritating, nonallergenic, and poorly absorbed through the GI tract and has no toxicity by all routes of administration [71]. Squalene has very low water solubility, reducing the likelihood of Ostwald ripening, thus improving the stability of a nanoemulsion [36]. Additionally, squalene tends to lead to smaller droplet sizes in emulsions because of its high surface tension [72,73]. While other oils may be suitable for vaccine adjuvant nanoemulsions, based on the current body of research, squalene and squalane will continue to be the preferred choice for vaccine adjuvant candidate products.

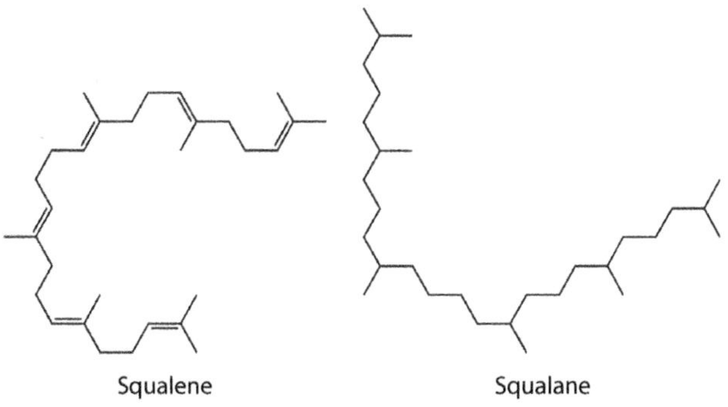

Squalene   Squalane

FIGURE 5.3
Squalene and squalane have been extensively used in cosmetics and medicine. Squalane is the hydrogenated form of squalene.

5.3 Current State of Development

NanoBio Corporation (Ann Arbor, Michigan), a biopharmaceutical company established in 2000 based on research from the University of Michigan's Center for Biologic Nanotechnology, has extensively researched their nanoemulsion vaccine adjuvant for a variety of indications. Initially investigating their nanoemulsion as an antimicrobial [74,75], they have since studied their nanoemulsion formulation as a vaccine adjuvant in animal models for anthrax [76], hepatitis B [77], HIV [78], smallpox [79], and respiratory syncytial virus [80].

As with the other nanoemulsions for vaccine adjuvancy, the thrust of their research and their first clinical trial for their formulation is for influenza vaccination [81–83]. $W_{80}5EC$ differs from other nanoemulsions currently on the market or under investigation because its oil phase is soybean oil rather than the traditional choice of squalene. The composition of $W_{80}5EC$ is 64% soybean oil, 1% cetylpyridinium chloride (CPC), 5% Tween 80, and 8% ethanol in water [84]. The nanoemulsion is manufactured with a high-speed emulsifier, which achieves an average droplet size of 200–600 nm. The final composition of $W_{80}5EC$ was arrived at after balancing the need for FDA-approved excipients and the desired characteristics of the formulation. The prototype formulations X8P and 20N10 both functioned as vaccine adjuvants in studies with mice but did not exclusively contain ingredients that were "generally recognized as safe" (GRAS) by the FDA [82]. The final, optimized formulation $W_{80}5EC$ was the formulation chosen to advance to clinical trial [83]. Figure 5.4 illustrates the evolution of $W_{80}5EC$ through the published literature and gives some insight into the incremental optimizations that occur during nanoemulsion development.

Makidon et al. [84] showed that $W_{80}5EC$ is stable following dilution with either buffered or nonbuffered saline and is stable for 10 months at room temperature. In the Phase I clinical trial with 140 adults, $W_{80}5EC$ when administered with Fluzone® (approved inactivated seasonal influenza antigen; Sanofi Pasteur S.A., Lyon, France) was safe and produced no

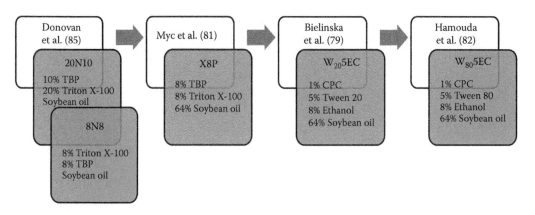

FIGURE 5.4
The figure illustrating the progression of nanoemulsion formulations through published literature over time from the University of Michigan's Center for Biologic Nanotechnology and NanoBio Corp. Note the transition to GRAS excipients such as Tween 80, which is also found in other nanoemulsion vaccine adjuvants. The optimization of a nanoemulsion is continual during the preclinical stages of development as more is learned about the mode of action of the nanoemulsion on the immune system and the interaction between different antigens and the nanoemulsion formulation [79,81,82,85]. CPC, cetylpyridinium chloride; TBP, tributyl phosphate.

dose-limiting toxicity up to the highest concentration evaluated (20% $W_{80}5EC$). Additionally, $W_{80}5EC$ with the Fluzone antigen induced both systemic and mucosal immune responses. This clinical trial, as well as most of the preclinical studies of this nanoemulsion and its predecessors, has studied intranasal delivery of the adjuvant and antigen [76–81,83,84,86]. The mucosal immune response is attributed to the internalization of the emulsion droplets by the epithelial cells within the nasopharyngeal mucosa and subsequent activation of TLR-2 and TLR-4 [87].

The Infectious Disease Research Institute (IDRI, Seattle, WA) has also published extensively about nanoemulsion formulations for vaccines. Fox et al. (2008) studied different source materials and structures for emulsion components and determined that synthetic and plant-derived components (squalene derived from olive oil) can be substituted with animal-derived components (squalene derived from sharks) and maintain physicochemical and biological properties [88]. Stable emulsion (SE) is composed of 10% (v/v) squalene, 1.8% (v/v) glycerol, 1.9% egg phosphatidylcholine, and 0.09% (w/v) Pluronic F68 in 25 mM ammonium phosphate buffer (pH 5.5) [89]. Once the aqueous phase and oil phase are combined, they are mixed with a laboratory emulsifier at 5,000–10,000 rpm for several minutes. Then, a high-pressure homogenizer is used to produce the final nanoemulsion with 12 passes at 30,000 psi. The formulation is stored at 2°C–8°C [88]. The particle size of the emulsion is around 95 nm with a polydispersity index of 0.04 [89]. Many preclinical studies have examined SE (alternatively referred to as EM001 with slight formulation differences) as a vaccine adjuvant, including studies with vaccines for polio, influenza, tuberculosis, and Leishmania [89–92]. Additionally, Baldwin et al. (2011) were the first to attempt at adjuvanting inactivated poliovirus vaccine with the nanoemulsion SE [89]. EM001 combined with synthetic MPL (the TLR4 agonist found in GSK's AS04), referred to as EM005, has proven to enhance the humoral immune response, generate higher-quality antibodies, and generate more long-lived plasma cells as an adjuvant for Fluzone in mice. Overall, EM005 improved antibody responses and enhanced the Th1 cellular immune response for Fluzone [91]. Baldwin et al. studied the inclusion of six different TRL agonists with EM005 (alternatively called GLA-SE) for a tuberculosis vaccine candidate [92]. For tuberculosis, GLA-SE has been tested in challenge models and demonstrated that boosting bacillus Calmette–Guérin (BCG), the current TB vaccine on the market, with a novel recombinant subunit vaccine mixed with GLA-SE reduced pathology and produced long-term protection against tuberculosis in a guinea pig model [93]. Finally, comparing GLA-SE (EM005) to SE as adjuvants for TB, GLA-SE induced a Th1 response, whereas SE induced a Th2-biased response in mice; GLA-SE improved survival in guinea pig challenge model, whereas SE worsened survival; and, finally, GLA-SE reduced lung pathology, while SE did not [94]. These results emphasize the importance of identifying the active elements within a formulation, especially when both immunomodulators, such as MPL, and delivery vehicles, such as SE, are included in the same adjuvant formulation.

As with other GSK products, their vaccine candidate AS02 is a combination product with different known immunostimulatory agents. AS02 is an o/w emulsion with MPL and QS21. MPL is the TLR4 agonist included in AS04, and QS21 (derived from the bark of a South American tree, *Quillaja saponaria*) has been identified as an immunomodulator, which improves the antigen presentation to antigen-presenting cells [95]. Compositionally, the emulsion portion of AS02 is identical to that of AS03. The emulsion precursor to both AS02 and AS03 is SB62 and consists of 5% v/v squalene, 5% v/v α-tocopherol, and 1.8% v/v Tween 80 in PBS (pH 6.8). The particle size of the nanoemulsion (pre dilution) is 150–155 nm [12]. For more information about the manufacturing procedure for AS02, see Section 5.2.1.

AS02 is designed primarily for malaria but has also been evaluated for tuberculosis, hepatitis B, HIV, and cancer vaccines [96]. For the malaria vaccine, AS02 is combined with a multimeric antigen called RTS,S. AS02 was determined to be the best GSK Adjuvant System in a Phase I/IIa clinical trial, comparing two other adjuvant systems to AS02 by their efficacy to prevent infection by *Plasmodium falciparum* delivered by mosquito bites [97]. This initial trial, in which six out of seven volunteers treated with AS02 and RTS,S did not contract malaria, was the first report of RTS,S/AS02 providing significant protection against *P. falciparum* malaria in humans in a laboratory setting. Further Phase II studies in areas of Africa where malaria is endemic have demonstrated that RTS,S/AS02 can elicit high specific antibody titers, reduce the incidence of severe disease in children (1–4 years of age) by 57.7% over 6 months, and induce $CD4^+$ T cell activity [98–103]. Also, studying the safety of RTS,S/AS02 in children aged 1–4 years in Mozambique, the vaccine has been shown to be safe and well tolerated when given in three doses (0, 1, 2 months) to semi-immune children in endemic areas, though local reactogenicity (mild to moderate pain and swelling) increased with increasing the number of doses and was reported more frequently with the RTS,S/AS02 group than the controls. Over a 21 month surveillance period, fewer cases of serious adverse reactions were reported in the children who received RTS,S/AS02 [104]. Also, RTS,S/AS02 confers at least 18 months of partial protection to children in rural endemic areas against a range of clinical diseases caused by *P. falciparum* [105]. For some of studies, the preservative thimerosal was added to the vaccine (denoted RTS,S/AS02$_A$), but the adjuvant AS02 was identical to that used in the thimerosal-free formulations.

A group from the University of Georgia aims to use a nanoemulsion to deliver plasmid DNA as a vaccine. Their initial approach was to exploit the higher density of low-density lipoprotein (LDL) receptors present in cancer cells and tumors. Due to the increased number of receptors present, LDLs could be used as a targeting pathway to deliver active ingredients to tumors. Shawer et al. (2001) formulated an artificial lipoprotein system, coined "VLDL-resembling phospholipid submicron emulsions," to deliver active ingredients to rat brain glioma cells utilizing the LDL pathway [106]. Using the same system but adapted for vaccine delivery, Alanazi et al. (2004) attempted to deliver plasmid DNA vaccine containing a gene for a rabies virus glycoprotein. Specifically, their artificial lipoprotein system is a nanoemulsion composed of 70% w/w triolein, 22.7% w/w egg phosphatidylcholine, 2.3% w/w lysophosphatidylcholine, 3% w/w cholesterol oleate, and 2% w/w cholesterol. After this lipid mixture is combined by solvent evaporation, the lipid mixture is mixed with the aqueous phase (2.4 M NaCl) at a ratio of 1:100 w/v. The pre-emulsion is sonicated and passed through a high-pressure homogenizer [107]. The resulting nanoemulsion has a particle size of 54.3 ± 24.8 nm and zeta potential of −43.67 ± 1.38 mV. To associate the negatively charged DNA plasmids with the nanoemulsion droplets, positively charged palmitoyl poly-L-lysine (p-PLL) is complexed with the nanoemulsion to neutralize the negatively charged nanoemulsion and make it suitable as a DNA carrier. Testing the transfection efficiency of the nanoemulsion–DNA complex in cell culture, the artificial lipoprotein carrier resulted in 4.2-fold increase in transfection efficiency compared to lipofectamine, a commercially available gene delivery system [106].

Montanide ISA 720 (Seppic, S.A., Paris, France) is a w/o emulsion containing squalene as the oil phase and mannide monooleate as the emulsifier. The emulsion is made at 30:70 aqueous to oil mixture based on weight, and the aqueous phase is generally PBS containing the candidate antigen. This mixture can be emulsified by different methods—vortexed for 30 min, homogenized at 6000 rpm with a steel saw tooth rotor stator generator probe attached to a homogenizer, or manually pushed through a syringe coupling piece for 20 strokes. This produces an emulsion that is stable for at least 1 year at 4°C but with a particle size around 1 μm, which is generally not classified as a nanoemulsion [108]. Montanide ISA 720 produces mild to moderate local reactions and some mild general adverse events, which are transient. Some severe reactions, such as granulomas, can also occur [109]. Montanide ISA 720 has shown promise as adjuvant for malaria vaccines, for which aluminum salts have generally elicited poor responses [110]. Montanide ISA 720's success might be due to a depot effect at the injection site, which results in slower, longer release of the antigen [111]. Generally targeted as a malaria adjuvant, Montanide ISA 720 has been shown to stimulate both humoral and cellular immune responses [112–115]. In a study comparing an aluminum salt, Montanide ISA 720, and AS02 by assessing the safety and immunogenicity of each adjuvant, the type and magnitude of immune response elicited by the three adjuvants were different. Alhydrogel™ (Brenntag, Mülheim an der Ruhr, Germany), the aluminum salt, is known to produce a Th2 immune response, while Montanide ISA 720 and AS02 both skewed toward a Th1 response. AS02 produced the most systemic adverse events, but all were transient and primarily mild to moderate in nature. Montanide resulted in abscesses in 2 of the 10 volunteers after three immunizations [110]. Montanide ISA 51 is another w/o emulsion from SEPPIC, which is formulated with mineral oil and is mainly targeted to cancer vaccine applications [116]. Montanide ISA 51 has been tested in 25 clinical trials in over 4,000 patients and 40,000 doses. From these studies, Montanide ISA 51 generally elicits mild to moderate local reactions such as granuloma, local pain, tenderness, and erythema. General adverse reactions include nausea, vomiting, and pain [109]. The biggest hindrance for advancement beyond clinical trials for both Montanide ISA 720 and Montanide ISA 51 is their reactogenicity. For example, a trial conducted with a malaria vaccine candidate formulated with Montanide ISA 720 was suspended due to severe adverse events [115]. Montanide ISA 51 is a derivative of Freund's incomplete adjuvant, which was discontinued because of adverse reactogenicity (although this is attributed to the impurity of the mineral oil in the formulation).

5.4 Conclusion and Future Perspective

Nanoemulsions have enormously impacted vaccine adjuvant research. They are inexpensively and easily produced, are stable for long storage periods, have an extensive safety record (>140 million people vaccinated including children), allow for dose sparing of vaccine antigens, and are flexible in design to allow for additional immunomodulatory compounds. Some of their disadvantages include an increased incidence of mild adverse reactions at the site of injection and limited understanding of the role they play in immunity. While nanoemulsions seem to have proven their versatility and effectiveness, there is still much more research and growth needed before they are established as the *preferred* vaccine adjuvant—for example, the United States is yet to license an adjuvanted influenza vaccine even though several have been approved abroad.

Yet, as the global population and international travel increase, the global threat of pandemics, emerging diseases, and biological warfare intensifies. In the past 30 years, more than 30 emerging (e.g., AIDS, severe acute respiratory syndrome [SARS], and swine influenza) or reemerging infections (e.g., anthrax, malaria, and tuberculosis) have been identified [117,118]. As new vaccine technologies emerge to counter such infections, adjuvants must improve proportionally. The versatility of nanoemulsions is their best asset to aid researchers in creating effective vaccines for emerging diseases. Not only do new vaccines need to address the needs of a population with a different immune response from previous generations (i.e., a larger elderly population), researchers are also tasked with finding possible vaccines for age-related diseases for this growing demographic in developed countries. Will nanoemulsions such as MF59, approved specifically for the elderly, be the answer to the question of vaccines for an aging and immunocompromised population? Based on the rapid growth of nanoemulsions precipitated by the 2009 influenza pandemic, further research into this question is a certainty.

References

1. Sztein, M. B. 2004. Recent advances in immunology that impact vaccine development. In *New Generation Vaccines*, 3rd edn, Revised and Expanded, eds. M. M. Levine, J. B. Kaper, R. Rappuoli, M. A. Liu, and M. F. Good, 159–178. New York: Marcel Dekker.
2. Medzhitov, R. and C. A. Janeway, Jr. 2002. Decoding the patterns of self and nonself by the innate immune system. *Science* 296: 298.
3. Underhill, D. M. and A. Ozinsky. 2002. Toll-like receptors: Key mediators of microbe detection. *Current Opinion in Immunology* 14: 103–110.
4. Kawai, T. and S. Akira. 2007. TLR signaling. *Seminars in Immunology* 19: 24–32.
5. Janeway, C. A., Jr. and R. Medzhitov. 2002. Innate immune recognition. *Annual Review of Immunology* 20: 197–216, doi:10.1146/annurev.immunol.20.083001.084359.
6. Guermonprez, P., J. Valladeau, L. Zitvogel, C. Thery, and S. Amigorena. 2002. Antigen presentation and T cell stimulation by dendritic cells. *Annual Review of Immunology* 20: 621.
7. Banchereau, J. and R. M. Steinman. 1998. Dendritic cells and the control of immunity. *Nature* 392: 245.
8. Harty, J. T., A. R. Tvinnereim, and D. W. White. 2000. CD8+ T cell effector mechanisms in resistance to infection. *Annual Review of Immunology* 18: 275.
9. Jankovic, D., Z. Liu, and W. C. Gause. 2001. Th1- and Th2-cell commitment during infectious disease: Asymmetry in divergent pathways. *Trends in Immunology* 22: 450–457.
10. McGuirk, P. and K. H. G. Mills. 2002. Pathogen-specific regulatory T cells provoke a shift in the Th1/Th2 paradigm in immunity to infectious diseases. *Trends in Immunology* 23: 450–455.
11. Masopust, D. and L. J. Picker. 2012. Hidden memories: Frontline memory T cells and early pathogen interception. *The Journal of Immunology* 188: 5811–5817, doi:10.4049/jimmunol.1102695.
12. D'Hondt, E., N. Hehme, E. J. Hanon, and J. Stephenne. Influenza Vaccine. U.S. patent US 2007/0141078 A1, June 21, 2007.
13. Wack, A. and R. Rappuoli. 2005. Vaccinology at the beginning of the 21st century. *Current Opinion in Immunology* 17: 411–418.
14. Reed, S. G., S. Bertholet, R. N. Coler, and M. Friede. 2009. New horizons in adjuvants for vaccine development. *Trends in Immunology* 30: 23–32, doi:10.1016/j.it.2008.09.006.
15. Ishii, K. J. and S. Akira. 2007. Toll or toll-free adjuvant path toward the optimal vaccine development. *Journal of Clinical Immunology* 27: 363–371, doi:10.1007/s10875–007–9087-x.
16. Byars, N. E. and A. C. Allison. 1987. Adjuvant formulation for use in vaccines to elicit both cell-mediated and humoral immunity. *Vaccine* 5: 223–228.
17. Banzhoff, A. et al. 2009. MF59-adjuvanted H5N1 vaccine induces immunologic memory and heterotypic antibody responses in non-elderly and elderly adults. *PLoS One* 4: e4384, doi:10.1371/journal.pone.0004384.
18. Boyle, J. et al. 2007. The utility of ISCOMATRIX™ adjuvant for dose reduction of antigen for vaccines requiring antibody responses. *Vaccine* 25: 2541–2544.

19. Mastelic, B. et al. 2010. Mode of action of adjuvants: Implications for vaccine safety and design. *Biologicals: Journal of the International Association of Biological Standardization* 38: 594–601.

20. Schwarz, T. F. et al. 2009. Single dose vaccination with AS03-adjuvanted H5N1 vaccines in a randomized trial induces strong and broad immune responsiveness to booster vaccination in adults. *Vaccine* 27: 6284–6290.

21. Petrovsky, N. and J. C. Aguilar. 2004. Vaccine adjuvants: Current state and future trends. *Immunology and Cell Biology* 82: 488–496, doi:10.1111/j.0818–9641.2004.01272.x.

22. McClements, D. J. 2007. Critical review of techniques and methodologies for characterization of emulsion stability. *Critical Reviews in Food Science and Nutrition* 47: 611–649, doi:10.1080/10408390701289292.

23. Anton, N. and T. F. Vandamme. 2011. Nano-emulsions and micro-emulsions: Clarifications of the critical differences. *Pharmaceutical Research* 28: 978–985, doi:10.1007/s11095–010–0309–1.

24. Mason, T. G., J. N. Wilking, K. Meleson, C. B. Chang, and S. M. Graves. 2006. Nanoemulsions: Formation, structure, and physical properties. *Journal of Physics: Condensed Matter* 18: R635.

25. Goloub, T. and R. J. Pugh. 2003. The role of the surfactant head group in the emulsification process: Single surfactant systems. *Journal of Colloid and Interface Science* 257: 337–343.

26. Afshar, S. and A. Yeung. 2011. Considerations when determining low interfacial tensions. *Journal of Colloid and Interface Science* 364: 276–278.

27. Kumar, P. and K. L. Mittal. 1999. *Handbook of Microemulsion Science and Technology.* Boca Raton, FL: CRC Press. LLC, Taylor & Francis Group.

28. Mason, T. G. 1999. New fundamental concepts in emulsion rheology. *Current Opinion in Colloid and Interface Science* 4: 231–238.

29. Shah, P., D. Bhalodia, and P. Shelat. 2010. Nanoemulsion: A pharmaceutical review. *Systematic Reviews in Pharmacy* 1: 24–32, doi:10.4103/0975–8453.59509.

30. Anton, N. and T. F. Vandamme. 2009. The universality of low-energy nano-emulsification. *International Journal of Pharmaceutics* 377: 142–147.

31. Shinoda, K. and H. Saito. 1968. The effect of temperature on the phase equilibria and the types of dispersions of the ternary system composed of water, cyclohexane, and nonionic surfactant. *Journal of Colloid and Interface Science* 26: 70–74.

32. Shinoda, K. and H. Saito. 1969. The stability of O/W type emulsions as functions of temperature and the HLB of emulsifiers: The emulsification by PIT-method. *Journal of Colloid and Interface Science* 30: 258–263.

33. Tadros, T. F. 2009. Emulsion science and technology: A general introduction. In *Emulsion Science and Technology,* ed. T. F. Tadros, pp. 1–56. Weinhein, Germany: Wiley-VCH Verlag GmbH & Co. KGaA.

34. Aucouturier, J., L. Dupuis, and V. Ganne. 2001. Adjuvants designed for veterinary and human vaccines. *Vaccine* 19: 2666–2672, doi:10.1016/S0264–410X(00)00498–9.

35. Kim, D. M., S. S. Hyun, P. Yun, C. H. Lee, and S. Y. Byun. 2012. Identification of an emulsifier and conditions for preparing stable nanoemulsions containing the antioxidant astaxanthin. *International Journal of Cosmetic Science* 34: 64–73, doi:10.1111/j.1468–2494.2011.00682.x.

36. Tadros, T., P. Izquierdo, J. Esquena, and C. Solans. 2004. Formation and stability of nano-emulsions. *Advances in Colloid and Interface Science* 108–109: 303–318, doi:10.1016/j.cis.2003.10.023.

37. Ott, G., R. Radhakrishnan, J.-H. Fang, and M. Hora. 2000.The adjuvant MF59: A 10-year perspective. In *Vaccine Adjuvants: Preparation Methods and Research Protocols,* ed. D. T. O'Hagan, pp. 211–228. Totowa, NJ: Humana Press, Inc.

38. Wooster, T. J., M. Golding, and P. Sanguansri. 2008. Impact of oil type on nanoemulsion formation and Ostwald ripening stability. *Langmuir* 24: 12758–12765, doi:10.1021/la801685v.

39. Glenny, A. T., G. A. H. Buttle, and M. F. Stevens. 1931. Rate of disappearance of diphtheria toxoid injected into rabbits and guinea-pigs: Toxoid precipitated with alum. *The Journal of Pathology and Bacteriology* 34: 267–275, doi:10.1002/path.1700340214.

40. Dekker, C. L., L. Gordon, and J. Klein. Dose optimization strategies for vaccines: The role of adjuvants and new technologies. February 5–6, National Vaccine Advisory Committee Meeting, U.S. Department of Health and Human Services.

41. Henle, W. and G. Henle. 1945. Effect of adjuvants on vaccination of human beings against influenza. *Proceedings of the Society for Experimental Biology and Medicine.* 59: 179–181.
42. Salk, J. E., M. L. Bailey, and A. M. Laurent. 1952. The use of adjuvants in studies on influenza immunization: II. Increased antibody formation in human subjects inoculated with influenza virus vaccine in a water-in-oil emulsion. *American Journal of Epidemiology* 55: 439–456.
43. Lindblad, E. B. 2000. Freund's adjuvants. In *Vaccine Adjuvants: Preparation Methods and Research Protocols*, ed. D. T. O'Hagan, pp. 49–63. Totowa, NJ: Humana Press, Inc.
44. Beebe, G. W., A. H. Simon, and S. Vivona. 1972. Long-term mortality follow-up of army recruits who received adjuvant influenza virus vaccine in 1951–1953. *American Journal of Epidemiology* 95: 337–346.
45. Brewer, J. M. 2006. (How) do aluminium adjuvants work? *Immunology Letters* 102: 10–15.
46. Moignic, L. and Pinoy. 1916. Les vaccins en emulsion dans les corps gras ou 'lipo-vaccins'. *Comptes Rendus de la Societe de Biologie* 79: 201–203.
47. Stephenson, I. et al. 2010. Report of the fourth meeting on "Influenza vaccines that induce broad spectrum and long-lasting immune responses", World Health Organization and Wellcome Trust, London, U.K., November 9–10, 2009. *Vaccine* 28: 3875–3882.
48. Podda, A. 2001. The adjuvanted influenza vaccines with novel adjuvants: Experience with the MF59-adjuvanted vaccine. *Vaccine* 19: 2673–2680.
49. Schultze, V. et al. 2008. Safety of MF59 adjuvant. *Vaccine* 26: 3209–3222, doi:10.1016/j.vaccine.2008.03.093.
50. Ott, G., G. L. Barchfeld, and G. Van Nest. 1995. Enhancement of humoral response against human influenza vaccine with the simple submicron oil/water emulsion adjuvant MF59. *Vaccine* 13: 1557–1562.
51. Pellegrini, M., U. Nicolay, K. Lindert, N. Groth, and G. Della Cioppa. 2009. MF59-adjuvanted versus non-adjuvanted influenza vaccines: Integrated analysis from a large safety database. *Vaccine* 27: 6959–6965.
52. Mosca, F. et al. 2008. Molecular and cellular signatures of human vaccine adjuvants. *Proceedings of the National Academy of Sciences of the United States of America* 105: 10501–10506, doi:10.1073/pnas.0804699105.
53. Seubert, A. et al. 2011. Adjuvanticity of the oil-in-water emulsion MF59 is independent of Nlrp3 inflammasome but requires the adaptor protein MyD88. *Proceedings of the National Academy of Sciences of the United States of America* 108: 11169–11174, doi:10.1073/pnas.1107941108.
54. Tengerdy, R. P. and N. G. Lacetera. 1991. Vitamin E adjuvant formulations in mice. *Vaccine* 9: 204–206.
55. Tengerdy, R. P., E. Ameghino, and H. Riemann. 1991. Serological responses of rams to a Brucella ovis-vitamin E adjuvant vaccine. *Vaccine* 9: 273–276.
56. Hogan, J. S. et al. 1993. Vitamin E as an adjuvant in an *Escherichia coli* J5 vaccine. *Journal of Dairy Science* 76: 401–407, doi:10.3168/jds.S0022–0302(93)77359–2.
57. European Medical Agency. European public assessment report: Prepandrix. Last updated: December 22, 2011.
58. Morel, S. et al. 2011. Adjuvant System AS03 containing alpha-tocopherol modulates innate immune response and leads to improved adaptive immunity. *Vaccine* 29: 2461–2473, doi:10.1016/j.vaccine.2011.01.011.
59. Ferguson, M. et al. 2012. Safety and long-term humoral immune response in adults after vaccination with an H1N1 2009 pandemic influenza vaccine with or without AS03 adjuvant. *The Journal of Infectious Diseases* 205: 733–744, doi:10.1093/infdis/jir641.
60. Girard, M. P., J. M. Katz, Y. Pervikov, J. Hombach, and J. S. Tam. 2011. *Report of the 7th Meeting on Evaluation of Pandemic Influenza Vaccines in Clinical Trials*, World Health Organization, Geneva, Switzerland, February 17–18, 2011. *Vaccine* 29: 7579–7586.
61. Finnish National Institute for Health and Welfare (THL). Press Release: National Institute for Health and Welfare Recommends Discontinuation of Pandemrix Vaccinations. August 25, 2010.
62. European Medicines Agency. Press Release: European Medicines Agency Recommends Restricting Use of Pandemrix. July 27, 2011.

63. European Medicines Agency. Public statement on: Humenza (Pandemic Influenza Vaccine (H1N1) Split Virion, Inactivated, Adjuvanted) Withdrawal of the Marketing Authorisation in the European Union. June 20, 2011.
64. European Medicines Agency. European Public Assessment Report: Humenza. Last updated: June 30, 2011.
65. Klucker, M.-F., J. Haensler, P. Probeck-Quelleec, and P. Chaux. Thermoreversible oil-in-water emulsion. U.S. patent US 2007/0191314 A1, August 16, 2007.
66. Plennevaux, E. et al. 2011. Influenza A (H1N1) 2009 two-dose immunization of US children: An observer-blinded, randomized, placebo-controlled trial. *Vaccine* 29: 1569–1575, doi:10.1016/j.vaccine.2010.12.116.
67. Cosmetic Ingredient Review Expert Panel. 2003. Annual Review of Cosmetic Ingredient Safety Assessments–2001/2002. *International Journal of Toxicology* 22 (Suppl 1): 1–35.
68. Allison, A. C. and N. E. Byars. 1986. An adjuvant formulation that selectively elicits the formation of antibodies of protective isotypes and of cell-mediated immunity. *Journal of Immunological Methods* 95: 157–168.
69. Carelli, C., F. Audibert, and L. Chedid. 1981. Persistent enhancement of cell-mediated and antibody immune responses after administration of muramyl dipeptide derivatives with antigen in metabolizable oil. *Infection and Immunity* 33: 312–314.
70. Asa, P. B., Y. Cao, and R. F. Garry. 2000. Antibodies to squalene in Gulf War syndrome. *Experimental and Molecular Pathology* 68: 55–64, doi:10.1006/exmp.1999.2295.
71. Fox, C. B. 2009. Squalene emulsions for parenteral vaccine and drug delivery. *Molecules* 14: 3286–3312, doi:10.3390/molecules14093286.
72. Chung, H., T. W. Kim, M. Kwon, I. C. Kwon, and S. Y. Jeong. 2001. Oil components modulate physical characteristics and function of the natural oil emulsions as drug or gene delivery system. *Journal of Controlled Release* 71: 339–350, doi:10.1016/S0168-3659(00)00363-1.
73. Wang, J.-J., K. C. Sung, O. Y.-P. Hu, C.-H. Yeh, and J.-Y. Fang. 2006. Submicron lipid emulsion as a drug delivery system for nalbuphine and its prodrugs. *Journal of Controlled Release* 115: 140–149.
74. Hamouda, T. et al. 1999. A novel surfactant nanoemulsion with broad-spectrum sporicidal activity against *Bacillus* species. *The Journal of Infectious Diseases* 180: 1939–1949, doi:10.1086/315124.
75. Hamouda, T. et al. 2001. A novel surfactant nanoemulsion with a unique non-irritant topical antimicrobial activity against bacteria, enveloped viruses and fungi. *Microbiological Research* 156: 1–7.
76. Bielinska, A. U. et al. 2007. Mucosal immunization with a novel nanoemulsion-based recombinant anthrax protective antigen vaccine protects against *Bacillus anthracis* spore challenge. *Infection and Immunity* 75: 4020–4029, doi:10.1128/IAI.00070-07.
77. Makidon, P. E. et al. 2008. Pre-clinical evaluation of a novel nanoemulsion-based hepatitis B mucosal vaccine. *PLoS One* 3: e2954, doi:10.1371/journal.pone.0002954.
78. Bielinska, A. U. et al. 2008. Nasal immunization with a recombinant HIV gp120 and nanoemulsion adjuvant produces Th1 polarized responses and neutralizing antibodies to primary HIV type 1 isolates. *AIDS Research and Human Retroviruses* 24: 271–281, doi:10.1089/aid.2007.0148.
79. Bielinska, A. U. et al. 2008. A novel, killed-virus nasal vaccinia virus vaccine. *Clinical and Vaccine Immunology* 15: 348–358, doi:10.1128/CVI.00440-07.
80. Lindell, D. M. et al. 2011. A novel inactivated intranasal respiratory syncytial virus vaccine promotes viral clearance without Th2 associated vaccine-enhanced disease. *PLoS One* 6: e21823, doi:10.1371/journal.pone.0021823.
81. Myc, A. et al. 2003. Development of immune response that protects mice from viral pneumonitis after a single intranasal immunization with influenza A virus and nanoemulsion. *Vaccine* 21: 3801–3814.
82. Hamouda, T. et al. 2010. Efficacy, immunogenicity and stability of a novel intranasal nanoemulsion-adjuvanted influenza vaccine in a murine model. *Human Vaccines* 6: 585–594.
83. Stanberry, L. R. et al. 2012. Safety and immunogenicity of a novel nanoemulsion mucosal adjuvant W805EC combined with approved seasonal influenza antigens. *Vaccine* 30: 307–316, doi:10.1016/j.vaccine.2011.10.094.

84. Makidon, P. E. et al. 2010. Characterization of stability and nasal delivery systems for immunization with nanoemulsion-based vaccines. *Journal of Aerosol Medicine and Pulmonary Drug Delivery* 23: 77–89, doi:10.1089/jamp.2009.0766.

85. Donovan, B. W. et al. 2000. Prevention of murine influenza A virus pneumonitis by surfactant nano-emulsions. *Antiviral Chemistry and Chemotherapy* 11: 41–49.

86. Hamouda, T., J. A. Sutcliffe, S. Ciotti, and J. R. Baker, Jr. 2011. Intranasal immunization of ferrets with commercial trivalent influenza vaccines formulated in a nanoemulsion-based adjuvant. *Clinical and Vaccine Immunology* 18: 1167–1175, doi:10.1128/CVI.00035–11.

87. Bielinska, A. U. et al. 2010. Induction of Th17 cellular immunity with a novel nanoemulsion adjuvant. *Critical Reviews in Immunology* 30: 189–199.

88. Fox, C. B. et al. 2008. Monitoring the effects of component structure and source on formulation stability and adjuvant activity of oil-in-water emulsions. *Colloids and Surfaces. B, Biointerfaces* 65: 98–105, doi:10.1016/j.colsurfb.2008.03.003.

89. Baldwin, S. L. et al. 2011. Increased potency of an inactivated trivalent polio vaccine with oil-in-water emulsions. *Vaccine* 29: 644–649, doi:10.1016/j.vaccine.2010.11.043.

90. Bertholet, S. et al. 2009. Optimized subunit vaccine protects against experimental leishmaniasis. *Vaccine* 27: 7036–7045.

91. Baldwin, S. L. et al. 2009. Enhanced humoral and Type 1 cellular immune responses with Fluzone adjuvanted with a synthetic TLR4 agonist formulated in an emulsion. *Vaccine* 27: 5956–5963, doi:10.1016/j.vaccine.2009.07.081.

92. Baldwin, S. L. et al. 2009. Intradermal immunization improves protective efficacy of a novel TB vaccine candidate. *Vaccine* 27: 3063–3071, doi:10.1016/j.vaccine.2009.03.018.

93. Bertholet, S. et al. 2010. A defined tuberculosis vaccine candidate boosts BCG and protects against multidrug-resistant *Mycobacterium tuberculosis*. *Science Translational Medicine* 2: 53ra74, doi:10.1126/scitranslmed.3001094.

94. Baldwin, S. L. et al. 2012. The importance of adjuvant formulation in the development of a tuberculosis vaccine. *Journal of Immunology (Baltimore, Md.: 1950)* 188: 2189–2197, doi:10.4049/jimmunol.1102696.

95. Kensil, C. R. and R. Kammer. 1998. QS-21: A water-soluble triterpene glycoside adjuvant. *Expert Opinion on Investigational Drugs* 7: 1475–1482, doi:10.1517/13543784.7.9.1475.

96. Garcon, N., P. Chomez, and M. Van Mechelen. 2007. GlaxoSmithKline Adjuvant Systems in vaccines: concepts, achievements and perspectives. *Expert Review of Vaccines* 6: 723–739, doi:10.1586/14760584.6.5.723.

97. Stoute, J. A. et al. 1997. A preliminary evaluation of a recombinant circumsporozoite protein vaccine against *Plasmodium falciparum* malaria. *New England Journal of Medicine* 336: 86–91, doi:10.1056/NEJM199701093360202.

98. Alonso, P. L. et al. 2004. Efficacy of the RTS,S/AS02A vaccine against *Plasmodium falciparum* infection and disease in young African children: Randomised controlled trial. *Lancet* 364: 1411–1420, doi:10.1016/S0140–6736(04)17223–1.

99. Pinder, M. et al. 2004. Cellular immunity induced by the recombinant *Plasmodium falciparum* malaria vaccine, RTS,S/AS02, in semi-immune adults in The Gambia. *Clinical and Experimental Immunology* 135: 286–293.

100. Reece, W. H. et al. 2004. A CD4+ T-cell immune response to a conserved epitope in the circumsporozoite protein correlates with protection from natural *Plasmodium falciparum* infection and disease. *Nature Medicine* 10: 406–410, doi:10.1038/nm1009.

101. Epstein, J. E. et al. 2004. Safety, tolerability, and antibody responses in humans after sequential immunization with a PfCSP DNA vaccine followed by the recombinant protein vaccine RTS,S/AS02A. *Vaccine* 22: 1592–1603, doi:10.1016/j.vaccine.2004.01.031.

102. Lalvani, A. et al. 1999. Potent induction of focused Th1-type cellular and humoral immune responses by RTS,S/SBAS2, a recombinant *Plasmodium falciparum* malaria vaccine. *The Journal of Infectious Diseases* 180: 1656–1664, doi:10.1086/315074.

103. Sun, P. et al. 2003. Protective immunity induced with malaria vaccine, RTS,S, is linked to *Plasmodium falciparum* circumsporozoite protein-specific CD4+ and CD8+ T cells producing IFN-gamma. *Journal of Immunology (Baltimore, Md.: 1950)* 171: 6961–6967.
104. Sacarlal, J. et al. 2008. Safety of the RTS,S/AS02A malaria vaccine in Mozambican children during a Phase IIb trial. *Vaccine* 26: 174–184, doi:10.1016/j.vaccine.2007.11.003.
105. Alonso, P. L. et al. 2005. Duration of protection with RTS,S/AS02A malaria vaccine in prevention of *Plasmodium falciparum* disease in Mozambican children: Single-blind extended follow-up of a randomised controlled trial. *Lancet* 366: 2012–2018, doi:10.1016/S0140-6736(05)67669-6.
106. Shawer, M., P. Greenspan, S. Oie, and D. R. Lu. 2002. VLDL-resembling phospholipid-submicron emulsion for cholesterol-based drug targeting. *Journal of Pharmaceutical Sciences* 91: 1405–1413, doi:10.1002/jps.10117.
107. Alanazi, F., Z. F. Fu, and D. R. Lu. 2004. Effective transfection of rabies DNA vaccine in cell culture using an artificial lipoprotein carrier system. *Pharmaceutical Research* 21: 675–682.
108. Miles, A. P. et al. 2005. Montanide ISA 720 vaccines: Quality control of emulsions, stability of formulated antigens, and comparative immunogenicity of vaccine formulations. *Vaccine* 23: 2530–2539, doi:10.1016/j.vaccine.2004.08.049.
109. Aucouturier, J., S. Ascarateil, and L. Dupuis. 2006. The use of oil adjuvants in therapeutic vaccines. *Vaccine* 24 (Suppl 2): S2–44–45.
110. Roestenberg, M. et al. 2008. Safety and immunogenicity of a recombinant *Plasmodium falciparum* AMA1 malaria vaccine adjuvanted with Alhydrogel, Montanide ISA 720 or AS02. *PLoS One* 3: e3960, doi:10.1371/journal.pone.0003960.
111. Aucouturier, J., L. Dupuis, S. Deville, S. Ascarateil, and V. Ganne. 2002. Montanide ISA 720 and 51: A new generation of water in oil emulsions as adjuvants for human vaccines. *Expert Review of Vaccines* 1: 111–118, doi:10.1586/14760584.1.1.111.
112. Malkin, E. et al. 2008. A Phase 1 trial of PfCP2.9: An AMA1/MSP1 chimeric recombinant protein vaccine for *Plasmodium falciparum* malaria. *Vaccine* 26: 6864–6873.
113. Wu, Y. et al. 2008. Phase 1 trial of malaria transmission blocking vaccine candidates Pfs25 and Pvs25 formulated with Montanide ISA 51. *PLoS One* 3: e2636.
114. Hu, J. et al. 2008. Safety and immunogenicity of a malaria vaccine, *Plasmodium falciparum* AMA-1/MSP-1 chimeric protein formulated in montanide ISA 720 in healthy adults. *PLoS One* 3: e1952.
115. Herrera, S. et al. 2011. Phase I safety and immunogenicity trial of *Plasmodium vivax* CS derived long synthetic peptides adjuvanted with montanide ISA 720 or montanide ISA 51. *The American Journal of Tropical Medicine and Hygiene* 84: 12–20, doi:10.4269/ajtmh.2011.09–0516.
116. Seppic S. A. Vaccine Adjuvants, Your solutions for therapeutic vaccines, Montanide ISA 51 VG, Montanide ISA 720 VG. Report No. VA/4572/GB/01/March 2010. March 2010.
117. Morens, D. M., G. K. Folkers, and A. S. Fauci. 2004. The challenge of emerging and re-emerging infectious diseases. *Nature* 430: 242–249.
118. Rappuoli, R., C. W. Mandl, S. Black, and E. De Gregorio. 2011. Vaccines for the twenty-first century society. *Nature Reviews. Immunology* 11: 865–872, doi:10.1038/nri3085.

6

Inorganic Nanomaterials of Carbonate Apatite as Intracellular Nucleic Acid/Drug Transporters

Sharif Hossain, Ezharul Hoque Chowdhury, and Toshihiro Akaike

CONTENTS

6.1 Introduction

Introducing nucleic acids (genes or gene-silencing elements) to mammalian cells is a powerful way to modulate the cellular functions and signal transduction mediated by proteins. Following delivery to the cytoplasm, a foreign gene enters the nucleus and is transcribed to the corresponding mRNA, which is subsequently transported to the cytoplasm for translation into a specific protein. However, a gene-silencing element, such as an antisense oligonucleotide or a small interfering RNA, blocks the transcription of a target mRNA. Nucleic acid delivery has been an essential tool to turn on and off the expression of a particular gene in basic research laboratories and is highly promising for the development of new therapeutic concepts, such as gene therapy and DNA vaccination, which are likely to have an impact on clinical medicine and biotechnology in this century [1,2].

Traditionally, gene delivery systems have been divided into viral and nonviral vectors and can be generally classified into four major groups: viral carriers, organic cationic compounds, recombinant proteins, and inorganic nanoparticles. In viral carriers, part of the original DNA or RNA sequence is replaced with the gene of interest to be inserted and delivered. Carriers consisting of cationic compounds mainly include cationic lipids, cationic polysaccharides, and polycationic polymers, where the positive charges are the result of protonation of various amino/imino groups. A particular type of DNA vectors receiving attention recently is recombinant proteins [3]. These proteins mimic the various viral properties by combining diverse peptide segments that are required for efficient gene delivery into a single molecule through protein engineering. For example, the multifunctional proteins may include polylysine segments, protamine, histones, and amphipathic cationic peptides to bind DNA to form thermodynamically stable complexes. They may also contain antibodies or antibody segments for targeting cell-surface receptors and some short peptide sequences acting as nuclear localization signals. Although the formation of vector–DNA complexes follows the similar mechanism among the cationic compounds, protein vectors are much more biocompatible. Inorganic nanoparticles as new nonviral carriers have attracted much attention only recently. Many inorganic materials, such as carbonate apatite, calcium phosphate, gold, carbon materials, silicon oxide, iron oxide, and layered double hydroxide (LDH), have been studied. Viral carriers are to date the most effective; however, severe side effects (e.g., immune response and insertional mutagenesis) limit the successful application in cellular delivery. Cationic carriers (lipids and polymers) may avoid such problems but are often toxic to the cells. In contrast, inorganic nanoparticles show low toxicity and promise for controlled delivery properties, thus presenting a new alternative to viral and cationic carriers. Here, negatively charged DNA molecules are usually condensed with cationic reagents to allow formation of complexes carrying net positive charges. The resulting complexes can interact electrostatically with anionic heparan sulfate proteoglycans (syndecans) on the cell surface and reach the cytoplasm in the form of endosomes, through endocytosis [4]. The extremely low pH, as well as the enzymes within the late endosomes, usually brings about the degradation of entrapped DNA and the associated complexes. Finally, DNA that survives both endocytic processing and cytoplasmic nucleases must dissociate from the condensed complexes either before or after nuclear translocation through a nuclear pore or during cell division [2]. The precise mechanism of DNA release from macromolecular complexes is yet to be fully elucidated. However, the competitive binding of anionic lipids, heparin, RNA, and proteins in the endosomal membrane or in the cytosol with the cationic vectors can help DNA unpackaging [5,6]. Many therapeutic applications require a vehicle with the capability of delivering a transgene(s) to only a selected cell type, in order to increase the expression efficacy and prevent any side effects. A common strategy in nonviral cases involves the attachment of a targeting moiety to a polycation (lipid or polymer) backbone, which finally condenses the DNA through ionic interactions. Polylysine, the first backbone used for gene delivery, has been conjugated to a diverse set of cell-specific ligands, such as asialoorosomucoid [7], transferrin [8], EGF [9], mannose [10], fibroblast growth factor (FGF) [11], and antibodies [12] in order to target, respectively, hepatocytes via asialoglycoprotein receptors, transferrin receptor–positive cells, EGF receptor–carrying cells, macrophages through membrane lectins, and FGF receptor–bearing cells and lymphocytes via surface-bound antigens. In a similar way, polymers such as polyethylenimine (PEI) and liposomes have been coupled to other cell-surface receptor–specific ligands in addition to those described earlier, such as integrin-binding peptide conjugated to PEI to target integrins on cell surfaces [13] and vitamin folate conjugated to liposomes through a polyethylene spacer to target folate receptor–bearing cells [14]. However, it is noted that the cellular transfer efficiency with existing inorganic nanoparticles is relatively low at this stage.

6.2 Nanoparticles in Drug and Gene Delivery

Drug delivery is one of the most advanced application areas for nanoparticles. The primary research goals of nanoparticles in therapeutics delivery include highly specific drug targeting and delivery, reduced toxicity while maintaining therapeutic effects, greater safety and biocompatibility, and faster development of new safe medicines. The basic prerequisites for the design of new materials for drug delivery include drug incorporation, cellular interaction and intracellular release, formulation stability and shelf life, biocompatibility, biodistribution and targeting, and functionality. Moreover, the possible adverse effects of residual material after drug delivery should be considered as well when used solely as a carrier. Therefore, biodegradable nanoparticles with a life span as long as therapeutically needed would be optimal. The basic objective of drug incorporation within the nanoparticles is to retain the enhanced delivery to, or uptake by, target cells and/or to reduce the toxicity of the free drug to nontarget organs. Nanoparticulate delivery will moderate the margin between the doses, resulting in therapeutic efficacy; for example, tumor cell death and toxicity to other organ systems, which will result in an increase of the therapeutic index. Therefore, design of target-specific and long-circulation nanoparticles is needed. Most of the compounds recently used for drug delivery are biodegradable polymers resulting in drug release after degradation. One of the problems associated with the use of particulate drug carriers, including nanomaterials, is the entrapment in the mononuclear phagocytic system as present in the liver and spleen [15–18]. However, targeting strategies of nanoparticles may be favorable during treating liver diseases such as tumor metastasis or hepatitis. It was found that surface modification of nanoparticles with polyethylene glycol (PEG) resulted in long circulation in the bloodstream by inhibiting recognition and phagocytosis by the mononuclear phagocytic system [19–21]. In addition to altering the distribution in the body, the PEG modification also reduced *in vitro* toxicity when gold nanorods were modified using PEG. Coating of nanoparticles may also prevent agglomeration, which was reviewed recently [22]. Sometimes, nanoparticle size can influence the distribution of nanoparticles as demonstrated for lipid vesicles for which a lower liver uptake was found for the smaller vesicles (200–300 vs. 25–50 nm). Even the actual distribution may be influenced by the small size differences and thus affect the overall bioavailability of nanoparticles [23].

6.3 Essential Properties and Biointeractions of Inorganic Nanoparticles

In order to aid the understanding of the fundamental surface properties and functionalization of nanoparticles and their biointeractions, it is useful to conceptualize some generic scenarios of nanoparticles and biomolecular interactions. This will give a proper understating about the overall processes of biomolecule uptake and the cellular transfer pathways. The following section will address the critical factors affecting the transfection efficiency and effectiveness of nanohybrids.

6.3.1 Surface Functionalities of Inorganic Nanoparticles

A significant number of inorganic nanoparticles have been reported, which can potentially be used as carriers for the cellular delivery of various drugs including genes and proteins.

However, most inorganic nanoparticles have to be subjected to chemical and/or biological modification to meet the stringent requirements for cellular delivery, such as good biocompatibility, strong affinity between the carriers and biomolecules, high charge density of the nanohybrids, site specificity, etc. The functionalization or modification depends, to a large degree, on the types of nanoparticles that provide specific functional groups on the surface.

6.3.2 Physical Interactions between Biomolecules and Nanoparticles

For successful loading of biomolecules in nanoparticles, the interaction between nanoparticles and biomolecules is one of the important factors that directly influence the overall cellular transfection of biomolecules (drugs, genes or proteins, etc.) as well as the controllable release in the targeted cells. Surface modification of inorganic nanoparticles can improve the interactions between nanoparticles and biomolecules. Generally, the interactions between modified nanoparticles and biomolecules are primarily electrostatic, while hydrophilic (polar) and hydrophobic interactions do exist among the organic chains. The negatively charged DNA chains interact with positive amino groups, resulting in adsorption (loading) of DNA on the modified inorganic nanoparticle. The biomolecule loading is hence dependent on the charge density, the modifier structure, and the chain length. It is worth noting that most modifiers provide a very good protection for biomolecules, DNA, for example, from being degraded by enzymes in the plasma, which is largely due to the steric effect and electrostatic repulsion from the modifiers. In gene delivery, this is especially important since nucleic acids are prone to be attacked by enzymes in the plasma.

6.3.3 Driving Forces Involved in Cellular Transport of Nanodevices

The driving forces involved in cellular transport of organic/inorganic nanohybrids into targeted cells matter a lot. In a normal drug administration, drug molecules will distribute to any targeted and untargeted cells via blood circulation, which means there is no preference for drug molecules to recognize targeted cells. Modified nanoparticles as carriers of drugs and biomolecules have several distinguished advantages. First of all, we can use external forces to concentrate organic/inorganic nanohybrids to a targeted region and thus increase their possibility to interact with the targeted cells. Secondly, modified nanoparticles with positive charges (positive zeta potential) can drive the particles to approach a normally negatively charged cell membrane via electrostatic interactions. Thirdly, inorganic nanoparticles are modified so as to adjust the adhesion of organic/inorganic nanohybrids to the cell membrane. On the other hand, nanoparticles coated with PEG can also improve their nonspecific cellular uptake. Finally, grafting specific biomolecules onto nanoparticles can provide site-specific delivery into cells.

6.3.4 Cellular Internalization of Organic/Inorganic Nanohybrids

Once nanoparticles adhere to the surface of a cell, they may be internalized into the cell. Although ion channels exist in the cell membrane, the nanohybrids are too large to pass through. One possible pathway for such nanohybrids to transfect into the cells is by either phagocytosis or endocytosis [24]. Phagocytosis generally involves the uptake of particles larger than 500 nm. Therefore, modified inorganic nanoparticles with biomolecules are thought to be mainly endocytosed since their particle size is normally less than 500 nm. Although endocytosis has not been well understood, microscopic observations indicate the changes in cell morphology and cytoskeleton structure during the process. The nano-hybrid is gradually embedded by deforming the membrane and endocytosed into the cell. The endocytosis may be a receptor-mediated or non-receptor-mediated process depending on ligand presentation [24]. The endocytosis process can be promoted by certain functionalities on the nanoparticle surface. However, it can also be prohibited to some degree by similar modifications that cause a strong adhesion to the cell receptor and stop it from entering the cell [25].

6.3.5 Intracellular Targeting and Release of Carried Biomolecules

Apart from the cellular targeting, the fate of the nanoparticles along with the incorporated therapeutics at the cellular level is of great importance. Efficient delivery of small interfering RNA (siRNA), for example, to the cytosol of target cells depends on both the translocation of the nonviral vectors through the plasma membrane and their subsequent escape from endosomal/lysosomal compartments. Endocytosis, the vesicular uptake of extracellular macromolecules, has been established as the main mechanism for the internalization of nonviral vectors into the cells. However, after endocytosis, the internalized molecules tend to be trapped in intracellular vesicles and eventually fuse with lysosomes where they are degraded. It is, therefore, important for a delivery system to avoid the fate of lysosomal degradation by facilitating release of the internalized siRNA into the cytoplasm. The release of therapeutics from the carrier is essential for binding with specific mRNA (in the case of siRNA) in the cytoplasm for effective silencing.

Once the biomolecule-loaded nanoparticles are internalized, an endosomal compartment is formed. The nanohybrids containing DNA or biomolecules may be released from this endosome into the cytoplasm via disrupting the endosomal membrane. The rupture could be conducted by the disrupting peptides, but more possibly, via a so-called proton sponge process [26,27]. Since the endosome is generally accompanied by acidification through proton pump or fusion with lysosomes that leads to a low pH to activate the enzymes, thus the loaded DNA is prone to be degraded in this acidic environment. Fortunately, the modifier on the inorganic nanoparticle surface normally contains some secondary/tertiary amines, which can buffer the falling pH by protonation of secondary/tertiary amines and thus increase the time before the degradation of DNA and the likelihood of its spontaneous translocation into cytoplasm. As the buffering continues, the reflux of counter ions may induce the osmotic swelling and rupture the endosomal membrane to release the nanohybrids. The quick rupture may enhance escape of the nanohybrids from the endosomal degradation and thus the gene transfection efficiency.

6.4 Key Issues to Be Considered during Cellular Delivery Using Nanoparticle

Since selective delivery of drugs for avoiding the adverse side effects to healthy organs still remains the major hurdle, especially for cancers and other diseases, the use of nanoparticles has become popular to combat such challenges. Moreover, some therapeutics such as DNA, RNA, and proteins that cannot enter the cells by normal diffusion require incorporation into nanoparticles for effective delivery. Therefore, after considering the earlier mentioned properties of inorganic nanoparticles, the following parameters should be taken into account during cellular delivery using nanoparticle:

- Effective particle size for endocytosis
- Zeta potential for cellular interaction with nanoparticles and surface functionalization for targetability
- Biocompatibility and cytotoxicity of nanoparticles [28]
- Protection capability from enzyme and controlled release of therapeutics for consequent efficacy

Although nanoparticle-based drug delivery ensures fascinating advancement in cancer treatment, rigorous *in vivo* assessment is required to establish the therapeutic efficacy and potential toxicity of these inorganic nanoparticles. The nanoparticles can theoretically be targeted to many tissues or cells through passive or active targeting. Passive targeting to the tumor can be made possible by means of precisely controlling the size of particles. Active targeting can be made by coupling one or more cell-recognizable protein(s) on the nanoparticle surface. Moreover, drug delivery to the central nervous system (CNS) is made difficult by the existence of the blood–brain barrier (BBB), which is formed by the complex

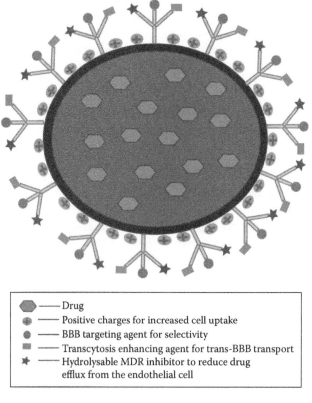

- ⬡ —— Drug
- ⊕ —— Positive charges for increased cell uptake
- ● —— BBB targeting agent for selectivity
- ▬ —— Transcytosis enhancing agent for trans-BBB transport
- ✦ —— Hydrolysable MDR inhibitor to reduce drug efflux from the endothelial cell

FIGURE 6.1
(See color insert.) A model drug nanoparticle concept for overcoming blood–brain barrier (BBB).

tight junctions between the endothelial cells of the brain capillaries and the low endocytic activity of the vertebrate brain. Several approaches for delivering drugs to the CNS have been developed to enhance the capacity of therapeutic molecules to cross the BBB by modifying the drug itself or by coupling it to a vector for receptor-mediated, carrier-mediated, or adsorption-mediated transcytosis. The current challenge is to develop drug delivery systems that ensure the safe and effective passage of drugs across the BBB. Therefore, nanoparticles able to ferry drugs across the BBB to CNS tumors have been reported and some of the necessary characteristics of such tools have been defined [29]. Chemical derivatization or encapsulation into polymeric nanoparticles has been evaluated as a possibility for enhancing drug selectivity. Depending on the key issues to be considered, the proposed structure of a model BBB drug nanoparticle delivery system has been illustrated in Figure 6.1.

6.5 Biodegradable Functional Inorganic Nanoparticles of Carbonate Apatite

Carbonate apatite is a natural component of the body and is usually found in the hard tissue, such as bone and teeth. The newly developed carbonate apatite, as with hydroxyapatite, adsorb DNA, but, unlike the latter, it can prevent the growth of its crystals to a significant extent, enabling the synthesis of nanosized crystals to effectively carry the associated DNA, siRNA, or drug across the cell membrane [30–33]. It also possesses a high dissolution rate in endosomal acidic pH, leading to the rapid release of the bound DNA for a subsequent high level of protein expression. Carbonate apatite facilitates many therapeutic applications as illustrated in Figure 6.2.

6.5.1 Molecular Aspects of DNA Delivery with Nanocrystals of Carbonate Apatite

Being inspired by the need to develop a superior nonviral approach in terms of both efficiency and biocompatibility, the development of an inorganic DNA nanocarrier based on *in vitro*–synthesized carbonate apatite, a major component of hard tissues in the body, has recently been reported [33]. Particles are generated in the presence of DNA in a bicarbonate-buffered solution or medium containing appropriate concentrations of phosphate and calcium salts through incubation at a high temperature. The chemical reaction required for particle formation takes place with Ca^{2+}, PO_4^{3-}, and HCO_3^-, and DNA that is negatively charged can be electrostatically associated with the cationic (Ca^{2+} rich) domains of the particles [34]. The number of particles as well as their size, two crucial factors for determining transfection potency, are dramatically influenced by the concentrations of calcium, phosphate, and bicarbonate and, additionally, by the pH of the medium (or solution) and the incubation temperature and time. Increasing the concentration of any of the reactants (calcium, phosphate, and bicarbonate), in general, accelerates particle formation by providing a stronger driving force for the chemical reaction while the other parameters (pH, incubation temperature, and time) are constant. On the other hand, an increase in pH and temperature (or incubation time) shifts the ionization equilibrium of phosphate and bicarbonate toward the forward direction

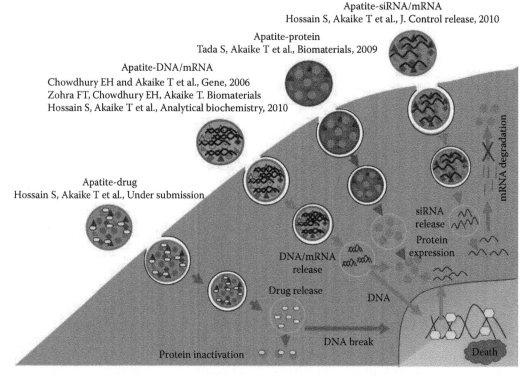

FIGURE 6.2
(See color insert.) Super efficient carbonate apatite as nucleic acids/drug transporter: mode of action.

and consequently favors particle generation by increasing the rate of the reaction. The induction of particle formation is generally accompanied by growth and aggregation of the particles, leading to larger-sized crystals [34]. However, bicarbonate, which is a minor component of the final apatite structure, $Ca_{10}(PO_4)_{6-x}(CO_3)_x(OH)_2$, prevents particle aggregation and generates small-sized crystals in a dose-dependent manner. Due to the small size of the crystal, which has an average diameter of 50–300 nm, and strong binding with DNA (almost 100% efficiency), carbonate apatite can be rapidly internalized with the associated DNA into the cell by endocytosis, following close contact with the cell surface through electrostatic forces. The efficiency of cellular uptake of DNA has been estimated to be at least 10 times more rapid than that achieved by classical CaP method [33,34].

6.5.2 Transfection Efficiency and the Role of Crystal Dissolution

Transgene expression results only after multiple barriers are overcome, such as endocytosis, DNA release from the particles and the endosomes, and translocation to the nucleus. Once inside the endosomes, exposure to an increasingly acidic environment results in the consumption of the excess H^+ by phosphate and carbonate ions in the particle, leading to particle dissolution and consequential DNA release. Therefore, inorganic crystals with high acid solubility or low crystallinity would enable quicker DNA release in endosomes, compared with crystals with low solubility or high crystallinity. In addition, the high dissolution rate may contribute to the destabilization of endosomes, to achieve DNA release into the cytoplasm, as vacuolar proton-pumping ATPase-driven massive proton accumulation that causes crystal dissolution could also lead to passive chloride influx to endosomes and subsequent endosome swelling and rupture. Finally, released DNA can enter the nucleus either through the nuclear pore or during cell division and be heavily expressed both in primary and cancer cell lines with an efficiency 5–100 times higher than with existing techniques.

6.5.3 Bio-Recognition Devices Associated with the Inorganic Crystals for Cell-Targeted Delivery

As mentioned before, a gene delivery system should have a precise cell-targeting capability for many therapeutic applications. To achieve this with the inorganic nano-carrier now under discussion, which delivers transgenes in a nonspecific manner depending on ionic interactions between the particles and the cell surface, an organic–inorganic hybrid nanocarrier was synthesized by coating the DNA-embedded particle surface. The surface was coated sequentially with a cell-recognizable protein, such as asialofetuin for asialoglycoprotein receptors on hepatocytes or transferrin for transferrin receptors on several cancer cell lines, and a highly hydrophilic protein, such as serum albumin for preventing nonspecific interactions of the particle surface with cell-surface proteoglycans, other receptors, or serum proteins [35]. The composite particles, with dual surface properties, have been shown to accelerate transgene delivery and expression only in the corresponding receptor-bearing cells. Moreover, mimicking the natural mineralization process, extracellular matrix proteins, such as collagen or fibronectin, have been successfully attached to the nanocrystals for integrin-specific efficient delivery of DNA [36]. A remarkable success has been achieved for transfecting embryonic stem cells, where particles initially did not show any significant transfection activity owing to the lack of particle interactions with the cell surface. However, when the particles were complexed with a naturally occurring fibronectin and genetically engineered E-cadherin–Fc together with DNA, a synergistic effect resulted in a dramatic enhancement both in transgene delivery and expression in mouse embryonic stem cells that possessed both transmembrane fibronectin-specific integrin and E-cadherin [37].

6.5.4 Carbonate Apatite as Novel Nanocarrier for Therapeutics Delivery

Since gene therapy through intracellular delivery of a functional gene or a gene-silencing element is a promising approach to treat critical diseases, therefore, RNA interference is one of the most powerful and specific mechanisms for gene silencing by degradation of mRNA. Although RNA interference is highly promising, the success of RNAi is highly dependent on the delivery of exogenous siRNA, and clinical trials of therapeutic siRNA are prohibited, because the efficiency of siRNA or DNA delivery into cytosol of target cells using a nonviral delivery system is still inadequate. Efficient delivery of siRNA to the cytosol of target cells depends on both the endosomal entry and their subsequent escape from endosomal/lysosomal compartments ensuring high gene knockdown efficiency. Thus, intracellular siRNA or DNA delivery approach offering high efficiency is highly expected. Besides gene delivery, contemporary cancer therapy, particularly with respect to drug delivery, has begun an evolution from traditional methodology. Delivery of drugs into the cancer cells to achieve high cytotoxicity with minimal side effect is a desirable objective for offering a potential approach for cancer therapy. To achieve these goals, the focus is the development of novel carriers for both existing and new drugs and defining better therapeutic targets relative to the molecular changes in the cancer cells

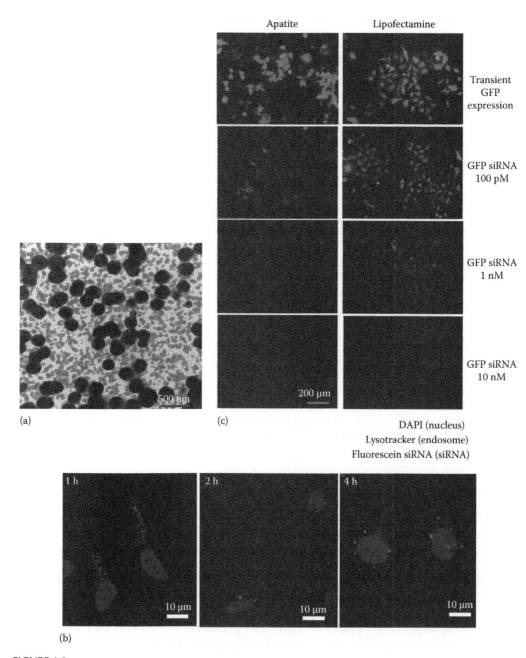

FIGURE 6.3
(See color insert.) (a) Desirable nanosize, (b) endosomal entry, time-dependent exit of siRNA, and (c) finally effective knockdown of reporter gene (GFP) mediated by carbonate apatite nanoparticles. Scale bar 500 nm.

through intracellular drug delivery to get high cytotoxic efficiency for better killing of cancer cells in the tumor level.

Therefore, recently, we have developed pH-sensitive carbonate apatite to efficiently deliver DNA [31,33], siRNA [28,30], mRNA [38], protein [39], and anticancer drugs (under publication) into the mammalian cells by the virtue of its high-affinity interactions with the siRNA or other genetic materials and the desirable size of the resulting siRNA/apatite complexes. The size of siRNA/apatite complex is important as it affects its cellular uptake through endocytosis. Endocytosis of a large complex is usually less efficient than that of a small one. Therefore, the size distribution of siRNA/apatite complexes was measured using TEM. Moreover, following internalization into cells, siRNA was found to escape from the endosomes in a time-dependent manner. This quick dissolution in endosomal pH is a very interesting phenomenon for drug delivery system so that siRNA can be released easily from the carrier and escaped from the endosome. It is, therefore, important for a delivery system to avoid the fate of lysosomal degradation by facilitating release of the internalized siRNA into the cytoplasm. A confocal microscopic analysis was performed on HeLa cells after the delivery of fluorescein siRNA/apatite complexes and observed the cellular localization of siRNA for 1–4 h. Endosomes/lysosomes were stained with LysoTracker™ Red. Most of the fluorescein siRNAs (Green) were co-localized in stained endosomes and/or lysosomes after 1 h incubation of HeLa cells with siRNA/apatite complexes. However, interestingly after 2 and 4 h, most of the siRNAs escaped the endocytic vesicles as shown in Figure 6.3. Therefore, finally, carbonate apatite more efficiently silenced reporter genes at a low dose than commercially available lipofectamine [30]. Besides, nanosized carbonate apatite was successfully fabricated for the first time for intracellular delivery of anticancer drug doxorubicin (DXR) to colon and ovarian cancer cell lines achieving significant tumor growth inhibition as revealed by an *in vivo* study (unpublished data). We found that the positively charged anticancer drugs like cisplatin or doxorubicin can electrostatically bind with the negatively charged domains of carbonate apatite, and after effective cellular interaction, the loaded drugs can effectively be released in the acidic compartment of endosome for subsequent anticancer action. The proposed mechanism of drug binding to carbonate apatite and intracellular delivery of anticancer drug like cisplatin has been illustrated in Figure 6.4.

6.6 Conclusion

Despite intensive efforts for the last three decades, there has been a lack of proper understanding of the molecular and cellular barriers to gene delivery, which could assist in developing a superior nonviral technique. Therefore, an increasing number of inorganic nanoparticles have been studied as drug or gene delivery carriers, because of their versatile physicochemical properties. Inorganic nanoparticles are readily available, can be easily functionalized, and possess good biocompatibility and low cytotoxicity. It is evident that the versatility of inorganic nanoparticles makes them very suitable as potential delivery carriers. However, the transfection efficiency of these inorganic nanohybrids is relatively low at this stage. Therefore, an advanced technology for gene delivery to mammalian cells based on the properties of carbonate apatite has been developed that prevent crystal growth, for the generation of nanoscale particles.

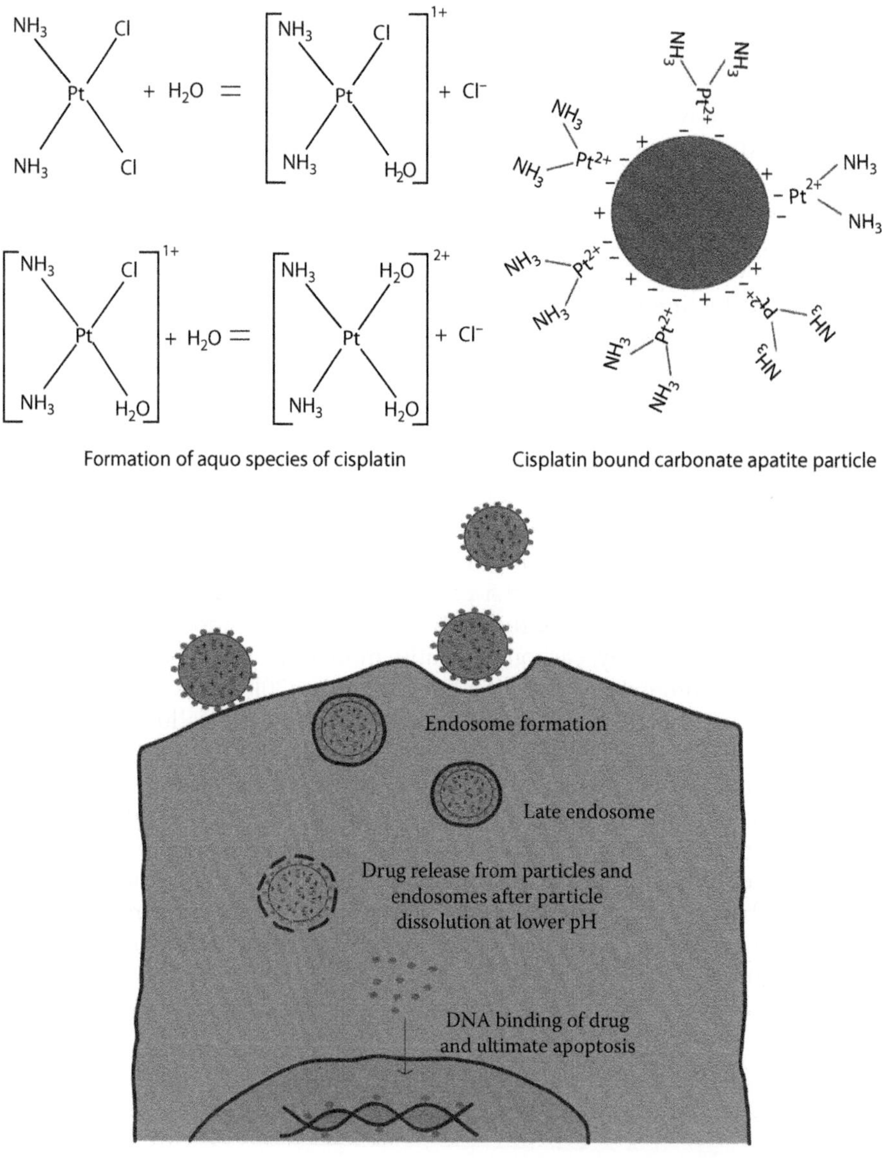

FIGURE 6.4
The proposed mechanism of anticancer drug binding to carbonate apatite and intracellular delivery.

These properties are beneficial for efficient endocytosis and the high-affinity binding of DNA and protein molecules for delivering DNA to the cell in either a specific or non-specific manner and allow a fast dissolution rate in endosomal acidic compartments to facilitate DNA release from the particles and endosomes. These novel approaches could pave the way to the wide potential applications of carbonate apatite, from laboratories to clinical medicine, and provide insights to create a new era for inorganic crystal-based gene and drug delivery.

References

1. Luo, D. and W.M. Saltzman. 2000. Synthetic DNA delivery systems. *Nature Biotechnology* 18: 33–37.
2. Chowdhury, E.H. and T. Akaike. 2005. Bio-functional inorganic materials: An attractive branch of gene-based nano-medicine delivery for 21st century. *Current Gene Therapy* 5: 669–676.
3. Aris, A. and A. Villlaverde. 2004. Modular protein engineering for non-viral gene therapy. *Trends in Biotechnology* 22: 371–377.
4. Kopatz, I., J.S. Remy, and J.P. Behr. 2004. A model for non-viral gene delivery: Through syndecan adhesion molecules and powered by actin. *Journal of Gene Medicine* 6: 769–776.
5. Xu, Y. and F.C. Szoka. 1996. Mechanism of DNA release from cationic liposome/DNA complexes used in cell transfection. *Biochemistry* 35: 5616–5623.
6. Bertschinger, M., G. Backliwal, A. Schertenleib et al. 2006. Disassembly of polyethylenimine-DNA particles in vitro: Implications for polyethylenimine-mediated DNA delivery. *Journal of Controlled Release* 116: 96–104.
7. Wu, G.Y., and C.H. Wu. 1987. Receptor-mediated in vitro gene transformation by a soluble DNA carrier system. *Journal of Biological Chemistry* 262: 4432–4439.
8. Cotten, M., T. F. Langle-Rouault, H. Kirlappos et al. 1990. Transferrin-polycation-mediated introduction of DNA into human leukemic cells: Stimulation by agents that affect the survival of transfected DNA or modulate transferrin receptor levels. *Proceedings of the National Academy of Sciences of the United States of America* 87: 4033–4037.
9. Schaffer, D.V. and D.A. Lauffenburger. 1998. Optimization of cell surface binding enhances efficiency and specificity of molecular conjugate gene delivery. *Journal of Biological Chemistry* 273: 28004–28009.
10. Erbacher, P., M.T. Bousser, J. Raimond et al. 1996. Gene transfer by DNA/glycosylated polylysine complexes into human blood monocyte-derived macrophages. *Human Gene Therapy* 7: 721–729.
11. Hoganson, D.K., L.A. Chandler, G.A. Fleurbaaij et al. 1998. Targeted delivery of DNA encoding cytotoxic proteins through high-affinity fibroblast growth factor receptors. *Human Gene Therapy* 9: 2565–2575.
12. Buschle, M., M. Cotten, H. Kirlappos et al. 1995. Receptor-mediated gene transfer into human T lymphocytes via binding of DNA/CD3 antibody particles to the CD3 T cell receptor complex. *Human Gene Therapy* 6: 753–761.
13. Erbacher, P., J.S. Remy, and J.P. Behr. 1999. Gene transfer with synthetic virus-like particles via the integrin-mediated endocytosis pathway. *Gene Therapy* 6: 138–145.
14. Wang, S., R.J. Lee, G. Cauchon et al. 1995. Delivery of antisense oligodeoxyribonucleotides against the human epidermal growth factor receptor into cultured KB cells with liposomes conjugated to folate via polyethylene glycol. *Proceedings of the National Academy of Sciences of the United States of America* 92: 3318–3322.
15. Demoy, M., S. Gibaud, J.P. Andreux et al. 1997. Splenic trapping of nanoparticles: Complementary approaches for in situ studies. *Pharmaceutical Research* 14: 463–468.
16. Gibaud, S., M. Demoy, J.P. Andreux et al. 1996. Cells involved in the capture of nanoparticles in hematopoietic organs. *Journal of Pharmaceutical Sciences* 85: 944–950.
17. Moghimi, S.M., A.C. Hunter, and J.C. Murray. 2001. Long-circulating and target-specific nanoparticles: Theory to practice. *Pharmacological Reviews* 53: 283–318.
18. Lenaerts, V., J.F. Nagelkerke, T.J. Van Berkel et al. 1984. In vivo uptake of polyisobutyl cyanoacrylate nanoparticles by rat liver Kupffer, endothelial, and parenchymal cells. *Journal of Pharmaceutical Sciences* 73: 980–982.
19. Bazile, D., C. Prud'homme, M.T. Bassoullet et al. 1995. Stealth Me.PEG-PLA nanoparticles avoid uptake by the mononuclear phagocytes system. *Journal of Pharmaceutical Sciences* 84: 493–498.

20. Niidome, T., M. Yamagata, Y. Okamoto et al. 2006. PEG-modified gold nano rods with a stealth character for in vivo applications. *Journal of Controlled Release* 114: 343–347.
21. Peracchia, M.T., E. Fattal, D. Desmaele et al. 1999. Stealth PEGylated polycyanoacrylate nanoparticles for intravenous administration and splenic targeting. *Journal of Controlled Release* 60: 121–128.
22. Gupta, M. and A.K. Gupta. 2004. In vitro cytotoxicity studies of hydrogel pullulan nanoparticles prepared by AOT/N-hexane micellar system. *Journal of Pharmacy and Pharmaceutical Sciences* 7: 38–46.
23. Seki, J., S. Sonoke, A. Saheki et al. 2004. A nanometer lipid emulsion, lipid nano-sphere(LNS), as a parenteral drug carrier for passive drug targeting. *International Journal of Pharmaceutics* 273: 75–83.
24. Rolland, A. 1999. *Advanced Gene Delivery: From Concepts to Pharmaceutical Products*, Harwood, Amsterdam.
25. Gupta, A.K. and A.S. Curtis. 2004. Lactoferrin and ceruloplasmin derivatized superparamagnetic iron oxide nanoparticles for targeting cell surface receptors. *Biomaterials* 25: 3029–3040.
26. Azzam, T. and A.J. Domb. 2004. Current developments in gene transfection agents. *Current Drug Delivery* 1: 165–193.
27. Pouton, C.W. and L.W. Seymour. 1998. Key issues in non-viral gene delivery. *Advanced Drug Delivery Review* 34: 3–19.
28. Hossain, S., E.H. Chowdhury, and T. Akaike. 2011. Nanoparticles and toxicity in therapeutic delivery: The ongoing debate. *Therapeutic Delivery* 2: 125–132.
29. Hossain, S., T. Akaike, and E.H. Chowdhury. 2010. Current approaches for drug delivery to central nervous system. *Current Drug Delivery* 7: 389–397.
30. Hossain, S., A. Stanislaus, M.J. Chua et al. 2010. Carbonate apatite-facilitated intracellularly delivered siRNA for efficient knockdown of functional genes. *Journal of Controlled Release* 147: 101–108.
31. Hossain, S., S. Tada, T. Akaike et al. 2010. Influences of electrolytes and glucose on formulation of carbonate apatite nanocrystals for efficient gene delivery to mammalian cells. *Analytical Biochemistry* 397: 156–161.
32. Haque, A., B. Hexig, S. Hossain et al. 2011. The effect of recombinant E-cadherin substratum on the differentiation of endoderm-derived hepatocyte-like cells from embryonic stem cells. *Biomaterials* 32: 2032–2042.
33. Chowdhury, E.H., A. Maruyama, A. Kano et al. 2006. pH-sensing nano-crystals of carbonate apatite: Effects on intracellular delivery and release of DNA for efficient expression into mammalian cells. *Gene* 376: 87–94.
34. Chowdhury, E.H. and T. Akaike. 2007. High performance DNA nano-carriers of carbonate apatite: Multiple factors in regulation of particle synthesis and transfection efficiency. *International Journal of Nanomedicine* 2: 101–106.
35. Chowdhury, E.H. and T. Akaike. 2005. A bio-recognition device developed onto nano-crystals of carbonate apatite for cell-targeted gene delivery. *Biotechnology and Bioengineering* 90: 414–421.
36. Chowdhury, E.H., M. Nagaoka, K. Ogiwara et al. 2005. Integrin-supported fast rate intracellular delivery of plasmid DNA by extracellular matrix protein embedded calcium phosphate complexes. *Biochemistry* 44: 12273–12278.
37. Kutsuzawa, K., E.H. Chowdhury, M. Nagaoka et al. 2006. Surface functionalization of inorganic nano-crystals with fibronectin and E-cadherin chimera synergistically accelerates trans-gene delivery into embryonic stem cells. *Biochemical and Biophysical Research Communications* 350: 514–520.
38. Zohra, F.T., E.H. Chowdhury, and T. Akaike. 2009. High performance mRNA transfection through carbonate apatite-cationic liposome conjugates. *Biomaterials* 30: 4006–4013.
39. Tada, S., E.H. Chowdhury, C.S. Cho et al. 2010. pH-sensitive carbonate apatite as an intracellular protein transporter. *Biomaterials* 31: 1453–1459.

7

Bioceramic Nanomaterials in Medical Applications

Jagat R. Kanwar, Ganesh Mahidhara, Kislay Roy, and Rupinder K. Kanwar

CONTENTS

7.1 Introduction

Nanotechnology is an emerging interdisciplinary area, which has many applications including drug delivery. Nanocarrier drug delivery involves target drugs enclosed in a particular ceramic, polymer, and/or amphiphilic matrix. Controlled-release, nano-platform availability for combinatorial therapy and specific targeting are the promises of nanotechnology. Many bio-inspired ceramics, capable of bioactive responses and biodegradation, have been innovated in past three to four decades. Ceramic nanomaterials with mesoporous and microporous surface morphology have proved to possess desired characteristics that are used in nanomedicine for targeted delivery of bio-macromolecules. Biomaterial is a substance that has been engineered to take a form that, alone or as part of a complex system, is used to direct, by control of interactions with components of living systems, the course of any therapeutic or diagnostic procedure, in human or veterinary medicine [1]. In broad sense, biomaterials are designed to treat disorders of a biological system, which perform their function by interfering with the system without disturbing its anatomical and/or physiological integrity. The major difference of biomaterials from other classes of materials is their ability to remain in a biological environment without damaging the surroundings and without being damaged in that process. Thus, biomaterials are solely associated with the health-care domain and must have an interface with tissues or tissue components [2,3]. The last three to four decades represent the scientific advancement in innovative use of ceramics in biomedical applications such as skeletal repair and construction. Ceramics can be defined as solid compounds that are formed by chemical reaction between a metal and a nonmetallic elemental solid or between a nonmetal and nonmetallic elemental solids [4–11].

The materials within this class of medical implant are often referred to as "bio-ceramics" [4]. With the improved bioactivity and biocompatibility, bio-ceramics have been exploited in various medicinal applications such as tissue regeneration and drug delivery.

7.2 Bio-Ceramics Used in Medical Applications

Most ceramics used in medical applications such as dentistry and orthopedics are based on calcium phosphates. Various forms of calcium phosphate ceramics are being used for the last four decades, which includes hydroxyapatite, beta-tricalcium phosphate (β-TCP), biphasic calcium phosphate (BCP), amorphous calcium phosphate, carbonated apatite, and calcium-deficient hydroxyapatite [5]. Bio-ceramics with microporous and mesoporous surface architecture have been developed and the list includes crystalline ceramics, bio-glasses, and ceramic hybrids.

7.2.1 Crystalline Ceramics

Crystalline ceramics have been widely employed in the areas of orthopedics and dentistry. These include synthetic apatites like hydroxyl apatite, substituted apatites, biphasic mixtures of calcium phosphates and calcium phosphate cements. Structural and chemical analogue of bone mineral component, which has hexagonal symmetry making it suitable to be used as a biomaterial, is hydroxyapatite (OHyAp), $Ca_{10} (PO_4)_6(OH)_2$ [6]. Applications of liquid mix technique, precipitation of aqueous solutions, and aerosol synthetic techniques are widely used for production of sub micrometric particles of OHyAp [7,8]. However, these synthetic apatites do not exactly mimic bone mineral in all aspects especially in the carbonate content in sublattice arrangement. Synthetic procedures cannot incorporate carbonate ions in OHyAp in lower amounts than seen in bone mineral [9]. Ions such as Na^+, K^+, Mg^{2+}, Sr^{2+}, Cl^-, F^-, and $[HPO4]^{2-}$ can be substituted in sublattices of OHyAp [10]. This improves tissue growth as younger tissues can develop and grow faster while storing other minerals that the body requires during growth. These ionic substitutions can vary in the surface integrity of OHyAp and with positive outcome on the material in the biological milieu. In addition, the presence of ions such as carbonate and strontium aids in dissolution of apatite for its biodegradation. Bioactive responses, mechanical strength, and porosity of the ceramic particles can be improved by doping silicates into the structure [11,12].

Biphasic mixtures of OHyAp and β-TCP have been used to synthesize mineral components of bones. These two different phases (stable OHyAp and resorbable β-TCP), under physiological conditions, form carbonate hydroxyl apatite. The released carbonate and phosphate ions from this biphasic mixtures facilitates to the bone regrowth. Solid state, treatment of natural bone, combustion precipitation, and blending of different calcium phosphates are the various techniques employed in the preparation of these mixtures [13,14]. In another development, the composites of OHyAp with 2.5 and 5 wt% of a double oxide (50 mol% CaO and 50 mol% P_2O_5) glass were prepared using the conventional powder mixing, sintering method in order to synthesize and analyze microstructural properties of the complexes prepared from fine mixed powders of OHyAp and bio-glasses to be able to develop a versatile tissue engineering material [15].

Calcium, phosphate, carbonate, and sulfate salt-based cements attracted much consideration in bone tissue engineering and dentistry owing to their excellent biocompatibility and bone repair properties [16]. They consist of powder phase of calcium and/or phosphate salts together with an aqueous phase, react at room or body temperature, and form a calcium phosphate crystal that sets by entanglement of crystals. These cements are easy to handle and can be manipulated by varying different constituents as per the choice of the required physicochemical characteristics [17]. BCPs composed of OHyAp and β-TCP have been intensively investigated in the past decades. In a recent study, novel BCP powders composed of alpha-tricalcium phosphate (alpha-TCP) and OHyAP were prepared by thermal decomposition of carbonated amorphous calcium phosphates (CACP), in order to produce bio-mimicking minerals [18].

7.2.2 Bio-Glasses

Usually bio-glasses are used for assisting bone matrix regeneration and bone growth. Bioactive glasses are typically composed of SiO_2, Na_2O, CaO, and P_2O_5 and prepared by solgel process, because of which key properties such as composition and texture could be modified as per the necessities. Also, due to their bioactivity, bio-glasses are eligible as bioactivity accelerators of mineral apatites or as bioactivity inductors of magnetic materials for hyperthermia treatments of osseous tumors. Growth factors and polypeptides can be included in bio-glasses to be able to achieve controlled-release profiles.

Bioactive glasses could be used as classic melt glasses, used in fillings or cavities in stomatology. *In vitro* tests in simulated body fluid (SBF) showed the formation of carbonate hydroxyapatite (CHA); nanocrystals show bioactivity of melt glasses [19]. Solgel glasses are formed by ternary $CaO–P_2O_5–SiO_2$ or binary $CaO–SiO_2$ system, which is more time consuming than that of traditional melt glasses [20]. Network modifiers such as Mg or Zn can be used to dope into this system to enhance mechanical and/or resorption capacities [21]. It has been observed that in ternary system, formation of Si–OH groups hinders formation of amorphous calcium phosphate, which means CHA crystallization requires more time in ternary systems compared to binary systems [6]. These glasses were shown to have excellent biocompatibility *in vitro* as well as *in vivo* [22]. Because of this property, mixed materials have been produced, in which glass component provides bioactivity. Solgel glass–polymer–drug hybrids [23] can be used for obtaining sustained therapeutic action; solgel glass magnetic component–OHyAp-based materials can be used for hyperthermia treatments in bone tumors [24,25].

7.2.3 Mixed Ceramic Hybrids

Polymeric organic components could be inserted in ceramic/solgel glass component to form organic–inorganic hybrids, which have improved mechanical strength, elasticity, and biodegradation. Varying formulations of polymers and inorganic components have been tested with a wide range of performances in SBF with respect to bioactivity. The mixed ceramic hybrids can be broadly classified into two classes, out of which, type I hybrids have weak interactions between organic and inorganic components. Polyvinyl alcohol (PVA) or polyhydroxyethylmethacrylate (pHEMA) has been reported to form hybrids of type I [26]. Type II hybrids are classified as hybrids formed by strong chemical linkage such as covalent or ionic covalent linkage. Polymethyl methacrylate (PMMA), poly ε-caprolactone (PCL), gelatin, and acetyl cellulose along with silica were observed to form type II hybrids [27].

Mesoporous materials have pore size in the range of 2–50 nm with different geometries. Ordered mesoporous materials have been reported to have advanced applications in controlled delivery of bio-macromolecules [28]. Improved biomedical applications of these materials could be a result of increased surface silane groups, which in turn improves the formation of CHA for enhanced bioactive response [29]. Also, star gels, which are comprised of organic core and inorganic overhangs, were reported to have intermediate mechanical behavior between glasses and rubbers. Thus, these materials may well be used as good candidates in orthopedics, dentistry, or drug delivery with little modification to their bioactivity.

In order to produce bio-ceramics with improved bioactivity and porosity (hybrid character of solgel glass and mesoporous materials), template glasses have been synthesized. These template glasses showed high bioactivity, homogeneity, and biodegradability compared to either solgel glasses or mesoporous materials [30].

7.3 Uses of Bio-Ceramics as Nanoparticles

Improving therapeutic index while reducing the side effects of the bio-macromolecular formulations could be achieved by designing relatively inert and nontoxic carriers for *in vivo* delivery, which represents an optimistic approach. Astonishing efforts in the field of nanobiotechnology have been made towards improving slow/controlled release, target specificity, improved efficiency, lower side effects, ease of delivery of drugs and pharmaceuticals by developing novel nanotechnology platforms (Table 7.1). The chemistry of inorganic nanoparticles has been reported to be advanced for cellular uptake, compared to other formulations. In addition, improved bioavailability and biodegradation because of the biomaterial component further adds up to the reduced side effects as discussed in the earlier section. Thus, many classes of inorganic nanoparticles have been used as carriers [31–33]. The tested inorganic materials for macromolecular delivery include biodegradable

TABLE 7.1

Summary of Ceramic Nanoparticles Used in Bio-Macromolecular Delivery

S.No.	Ceramic Used	Size Range	Formulation	Comments
1.	Calcium phosphate	400 nm	Alginate, chitosan, and calcium phosphate	Controlled release of serratiopeptidase (STP) [49]
2.	Hydroxyapatite	200 nm–1 μm	Hydroxyapatite and alginate	Controlled release of insulin was successfully tested in simulated intestinal fluid (SIF) [57].
3.	Calcium phosphate	350 nm	Alginate, chitosan, lactoferrin, and calcium phosphate	*In vitro* and *in vivo* efficacy, toxicity, and controlled release of anticancer proteins [61]
4.	Calcium phosphate	25–50 nm	Calcium and phosphate	Transfection of DNA in HeLa and MC3T3 cells was studied [42].
5.	Silica	225 nm	Silica nanoparticles	Enhanced transfection of DNA in COS-7 cells [35]
6.	Silica	100 nm	NaOH-modified silica	Enhanced DNA transfection in HT 1080 cells [36]
7.	Silica	250 nm	Silica and polyamidoamine	DNA transfection in CHO cells [37]

calcium phosphate, carbon nanotubes, silica, gold, magnetite, quantum dots, strontium phosphate, magnesium phosphate, manganese phosphate, and double hydroxides (anionic clays). Inorganic nanoparticles possess some outstanding characteristics such as resistance to microbial attack, ease of preparation, low toxicity, and good storage stability.

7.3.1 Ceramic Nanoparticles as Drug Delivery Systems

These are the nanoparticles made up of inorganic–ceramic compounds such as calcium phosphates, silica, titanium, and aluminum, that are known to have wide applicability in terms of macromolecular delivery. Ceramic nanoparticles have size less than 50 nm, which helps them invade the reticuloendothelial system (RES) of the body and exploit enhanced permeability and retention (EPR) effect posed by pathological angiogenic vasculatures [31]. As discussed earlier, changes in porosity and/or surface modifications in the ceramic nanoparticles are insignificant with variations in pH, temperature, and enzyme actions, thus making these particles ideal candidates for delivery of macromolecular drugs such as proteins and enzymes [32]. As summarized in Table 7.2, there have been a few patents with respect to the controlled release of drugs from ceramic materials, which are gaining pace in the recent past. Moreover, because of enhanced biocompatibility and surface modifications, these ceramic particles could have implications in effective targeting [33]. Silica nanoparticles have been prepared by well conventional solgel methods [34]. A three-component transfection system has been developed for efficient DNA transfection, a first successful step toward this modular synthetic DNA delivery system [35]. Cationic surface modifications have been

TABLE 7.2

List of Patents Developed for Controlled Release of Drugs Using Bio-Ceramics

S.No.	Ceramic Material	Significance	Reference
1.	Silica	Effective encapsulation of the drug HPPH	[67]
2.	Calcium phosphate with Ca/P ratio 1.45–1.55	Drug implanted into porous ceramic was reported to have controlled-release profile	[68]
3.	Iridium oxide and titanium nitrate	Gene transfer for smooth muscle cells using platinum iridium alloy as underlying layer	[69]
4.	Titanium oxide, zirconium oxide, scandium oxide, cerium oxide, and yttrium oxide	Oral, sustained-release dosage form, which includes a combination of a drug, a ceramic structure, and a polymer coating, has been worked out	[70]
5.	Silica fibers and colloidal silica	Fibrous and colloidal ceramic structures were used in implants for controlled delivery of drugs	[71]
6.	Hydroxyapatite fluorapatite, iodapatite, carbonate-apatite	Hybrid imaging particles for enhanced diagnostics using NMR, x-ray, or echogenic characteristics	[72]
7.	Oxides of titanium, zirconium, scandium, cerium, and yttrium	Sustained release of opioid agonists (e.g., oxycodone)	[73]
8.	Ca aluminate, Ca silicate	Chemically bonded or sintered ceramics produce microporous structures for drug encapsulation.	[74]

observed in efficient binding, condensation, and protection of the DNA [35]. Silica nanoparticles along with DNA having a size range of 10–100 nm modified with sodium chloride have been reported to show about 70% of transfection efficiency *in vitro* and *in vivo* [36]. In yet another interesting study demonstrated that mesoporous silica nanoparticles were covalently modified with polyamidoamine dendrimers. DNA combined with these nanoparticles showed maximum transfection efficiency in neuroglia, human cervical cancer and CHO cells compared to commercial reagents [37]. Increase in uptake of DNA with the aid of silica nanoparticles could be attributed to increase in sedimentation because of high density and also to the possible exertion of mechanical pressure on cell membrane [38]. In a recent study, cancer cell–targeted gene silencing was observed with a magnetic-nanoparticle platform (magnetism-engineered iron oxide [MEIO]) on which a fluorescent dye (Cy5), siRNA for green fluorescent protein (GFP) and an arginine, glycine and aspertic acid (RGD) peptide targeting moiety were attached. This platform has promising implications for simultaneous molecular imaging and gene-silencing properties [39]. Calcium phosphates (CaPs), being important constituent in bones and other hard tissues, has been used in nanoparticulate form for transfection of bio-macromolecules. It has been reported that calcium phosphate occurs in CHA form as nanoparticles, with the exception of enamel [40]. This lessens any concerns regarding biocompatibility, when calcium phosphate nanoparticles are used as external therapeutic aid. In a study, CaP nanoparticles with varying ratios of Ca/P have been developed and studied for transfection efficiencies in HeLa and MC3T3-E1 cell lines. It has been observed that nanoparticles with 25–50 nm size and Ca/P ratio between 100 and 300 showed maximum transfection efficiency [41]. DNA-loaded CaP nanoparticles have been successfully tested for transfection in CNE-2 cells [42]. In an interesting development, Kakizava et al. have used block copolymer, polyethylene glycol-*block*-poly(methacrylate)–modified CaP-siRNA nanoparticles for obtaining high transfection efficacy, tolerability in the biological medium and low toxicity [43]. Rapid precipitation and immediate surface fictionalization techniques have been employed in preparation of 80 nm CaP nanoparticles bearing either DNA or oligonucleotides [44]. Improved transfection efficiencies because of enhanced protection from intracellular degradation were observed for oligonucleotides protected in the core of CaP [44]. Encapsulation of yet another biologically fragile molecule, that is proteins, has been studied extensively by using ceramic calcium phosphates. Specific targeting ligands, usually monoclonal antibodies, may be attached to the carrier surface, thus promoting a diagnostic molecular recognition and the immune carrier character of the particles. Bovine serum albumin (BSA) protein controlled-release system with poly(lactic-glycolic acid) (PLGA) biodegradable polymer on the surface of OHyAp nanocarrier has been worked out. Controlled-release profiles and *in vitro* cytotoxicity studies in CCL163 cell lines represent the biocompatibility of the nano-formulation [45]. In addition, a method for injecting methotrexate, a cancer therapeutic chemical, into degradable CaP non-agglomerating particles prepared by reverse micelle technique having 250 nm in size has been reported. Cellular internalization studies revealed that these nanoparticles could act as intracellular delivery systems [46]. Folic acid with its high affinity for folate receptors that are often overexpressed in cancer cells presents another compound that may be added to CaP nanoparticle carriers [47]. Another study reported the preparation of CaP nanoparticles conjugated with *cis*-diamminedichloroplatinum (CDDP, cisplatin). These nanoparticles were found to exhibit twice higher IC50 values when compared to the free drug in an *in vitro* cell proliferation assay using a human ovarian cancer cell line [48]. Recently Saraf et al. [5,49] have developed ceramic nanoparticles, which can be used for delivery of peptides. Magnetic nanoparticles have been used commonly for various biomedical applications, especially in the diagnosis and treatment of cancer. In recent past, different magnetic composites were made out of ferrimagnetic bio-glass ceramics (FBCs) [24,50] that provide magnetic properties to be able to use as MRI contrasting agents as well as for hyperthermia treatment. Limitation of this particular technique is the inability to synthesize nanosized particles due to crystal growth, which happens because of the requirement of high temperature in this process. However, this limitation was overcome by the development of magnetic hydroxyapatite nanoparticles (m-OHyAp)—by replacement of Ca^{+2} ions in the core of apatite with Fe^{+2} ions. It has been observed that administration of hyperthermia in tumor-bearing mice treated with m-OHyAp nanoparticles significantly reduced tumor volume compared to other controls [51]. Multifunctional ceramic nanoparticles were investigated for simultaneous assessment of fluorescent imaging and hyperthermia treatment [52]. Theragnostic HAP nanocrystals doped with Eu+3 and Gd+3 were engineered to enable simultaneous contrast enhancement for different imaging techniques, including X-ray diffraction (XRD), magnetic resonance imaging (MRI), and near-infrared fluorescence imaging alongside acting as therapeutic agents by being conjugated with the cell-membrane receptor ligand folic acid and by having their surface ammonized with a dendigraft polymer, poly ethelene imine (PEI) [53]. Taken together, bio-ceramics could act as ideal carriers for nucleic acid and/or biomolecular delivery, with enhanced therapeutic potential providing innovation of elegant methods for synthesis and new ideas of surface functionalization.

7.3.2 Ceramic-Based Polymeric Nanoparticles for Oral Delivery

To develop an oral administration, careful design of the core and exterior coat is required. Recent developments in bone marrow tissue engineering, dentistry, orthopedics and plastic surgery are making use of these kinds of composites in the macroscale; utilizing the same principle in nanoscale will ensure more promising drug development [54]. Ceramic nanoparticles are being designed in order to assist protein adsorption on them and various layers of biodegradable polymers are then coated on their surface as discussed in the succeeding text, in order to protect them from lytic or gastrointestinal enzymes and for obtaining sustained-release kinetics of the loaded drug. Although there is an increasing list of publications on DNA transfection using ceramic nanoparticles, only a few studies with respect to protein loading onto the nanocarriers have been reported so far.

7.3.2.1 Natural Biodegradable Polymers

Chitosan and alginate were extensively studied as surface modifiers for calcium phosphates; effects of long-term reactions between OHyAp and chitosan acetates were analyzed to predict surface and interfacial chemistry [55,56]. In an interesting study, protein drug delivery has been tested with porous hydroxyapatite nanoparticles entrapped in alginate matrix. Oral administration of insulin via these ceramic nanoparticles suggests more efficient and controlled release [57]. However, to the best of our knowledge, no one has used these alginate-coated nanocarriers for pDNA or RNAi delivery for gene targeting in any disease *in vitro* or *in vivo* in human or animal systems. It is tempting to expand the utility of nanoparticles for delivery of therapeutic enzymes, because particles larger than 20 μm are prone to be washed out, being inefficient for mucosal delivery [58], and it is difficult to maintain their native structures [59]. In this regard, innovative techniques, as discussed earlier, have put forward the use of ceramics in high-tech applications like delivering chemicals and biologicals effectively *in vitro* and *in vivo* [60]. We recently for the first time used the covalently cross-linked cell-permeable peptides (CPPs—R9 and Tat peptides) complexed with siRNA to survivin, an inhibitor of apoptosis, which is overexpressed in tumors and inflammation (unpublished observations). We designed covalently cross-linked CPP (R9–siRNA to survivin and Tat peptides–siRNA to survivin peptides) complexed with siRNA to survivin for designing alginate-enclosed chitosan ceramic (ACSC) for effective loading and protection of active acid-labile and alkaline-labile large nanocarriers for oral administration. These loaded targeted nanoparticles have been successfully tested in human colon cancer cells (Caco-2 and HT29) in trans-well assays. As an attempt toward increasing the bioavailability, sustained release, mucoadhesiveness and conformational stability, we have loaded covalently complexed CPP with survivin siRNA into ACSC nanocarriers [61]. The prolonged activity was obtained due to slow release of the protein or RNA from the ACSC and the intact structure without denaturation or dehydration during delivery and storage. The encouraging results obtained in this study could propose ACSC for future *in vivo* studies, especially in the delivery of biomolecular drugs including RNA, DNA, protein, and peptides. These nanocarriers were found to be promising in terms of exhibiting special qualities like better therapeutic effect and enhanced protection to the drug. But further studies in relation to pharmacokinetics, toxicology, and animal studies are required for validating clinical utility of the formulation. More recently, cell-permeable dominant-negative survivin R9 (DNSurR9) [62–65] and survivin and HSP-90 antagonists, shepherdin, have been loaded on alginate gel-encapsulated, chitosan ceramic nanocarriers (ACSC-NPs) and are able to induce apoptosis and disintegrate the mitochondria of colon and breast cancer cell lines (but not normal control cells) more efficiently in in vitro cell-based assays. In this study, we loaded noncovalently cross-linked CPP (R9–siRNA to survivin and Tat peptides–siRNA to survivin peptides) complexed with siRNA to survivin and oncogenic antisense microRNA-27a (as-miR-27a) on ACSC-NPs and transferred to human breast cancer MDA-MB-231 and MCF-7 cell lines by endocytosis, as outlined (Figure 7.1). Our results show that both R9 and Tat CPPs covalently complexed with as-miR-27a–loaded ACSC-NPs exhibited oncogenic activity. Suppression of miR-27a inhibits breast cancer cell growth and invasion. Simultaneous covalently complexed CPP (R9 and Tat peptide) with siRNA to survivin and oncogenic as-miR-27a–loaded ACSC-NPs results in down regulation of genes that are important for cell survival and angiogenesis, which is faster and more efficient than monotherapy. In addition, these responses were accompanied by decreased expression of survivin and angiogenic genes, including survivin isoforms, vascular endothelial growth factor (VEGF), and VEGF receptor 1 (VEGFR1). Downregulation of survivin expression in Western blot in the covalently complexed CPP (R9 and Tat peptide) with siRNA to survivin and oncogenic as-miR-27a–loaded ACSC-NP-treated cells was observed in this study. Terminal deoxynucleotidyl transferase dUTP nick end labeling (TUNEL) assay, caspase activity assay, and changes in mitochondrial membrane potential reveal that cell death was mainly through intrinsic apoptosis pathway. Oral delivery of siRNA to survivin and oncogenic as-miR-27a–loaded ACSC-NPs induced apoptosis, necrosis, and cytotoxicity in the xenograft breast cancer model. Oral administration of covalently complexed CPP (R9 and Tat peptide) with siRNA to survivin and oncogenic as-miR-27a–loaded ACSC-NPs in combination regresses tumor growth faster and inhibits angiogenesis in the xenograft breast cancer mouse model as compared to monotherapy [66]. We also compared our results with the doxorubicin- and taxol-loaded ACSC-NPs. Taken together, our results are highly encouraging for the development of bio-ceramic-based technology for combination and nano-therapeutic strategies that combine gene silencing and drug delivery to obtain more potent and targeted therapeutic.

A list of patented bio-ceramic particles has been provided that was developed to achieve controlled release of drugs (6.4.2). Various ceramic materials have been used so far for delivery of several types of drugs such as silica for effective encapsulation of

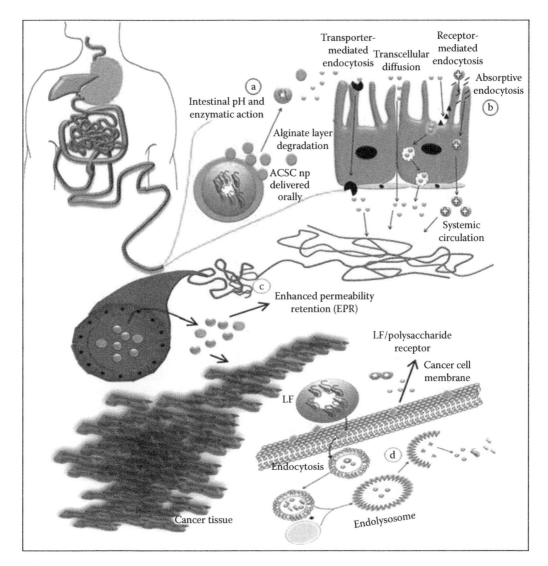

FIGURE 7.1
(See color insert.) Schematic illustration of the proposed mechanism of alginate-enclosed chitosan-coated calcium phosphate bio-ceramic nanoparticle internalization. The alginate coating gets degraded in the alkaline and enzymatic environment in the intestine and the remaining chitosan-coated ceramic particles enter the blood circulation via endocytosis and/or transcytosis. Passive delivery of these ceramic cores at the cancer site occurs due to the EPR effect. Further uptake of the NC in the cancer cells is based on oligosaccharide receptor–mediated endocytosis. Ceramic core further helps in controlled release and biodegradation of the formulation.

2-divinyl-2-(1-hexyloxyethyl)pyropheophorbide (HPPH) [67]; calcium phosphate for achieving controlled release of implanted drug into the porous ceramic, which was to be embedded into the human body [68]; iridium oxide and titanium nitrate for gene transfer into smooth muscle cells using platinum iridium alloy as the underlying layer [69]; titanium oxide, zirconium oxide, scandium oxide, cerium oxide, and yttrium oxide particles for oral and sustained delivery of a combination of drugs [70]; silica fibers and colloidal silica for synthesizing implants, which were to be used for controlled delivery of drugs [71];

hydroxyapatite, fluorapatite, iodapatite, and carbonate-apatite as hybrid imaging particles for enhanced diagnostics and echogenic characteristics using nuclear magnetic resonance (NMR) and x-ray [72]; oxides of titanium, zirconium, scandium, cerium, and yttrium for sustained release of opioid agonists such as oxycodone [73]; and calcium aluminate and calcium silicate as chemically bonded or sintered ceramics to produce microporous structures for drug encapsulation [74].

7.4 Conclusion

The last few decades have seen extensive management of ceramic/polymeric nanomaterials for the treatment of specific diseases including tumor regression and immune modulation. Many of the active drugs developed to treat these diseases, being biopharmaceuticals, developed with the immense advancements in the genetic engineering and biotechnology sector. The formulation of peptide/protein and DNA/RNA-based molecules is certainly problematic and it is likely that vehicles combining controlled release with passive or active targeting will be a necessary requirement. In this regard, microporous and/or mesoporous ceramic materials with enhanced biological applications could be considered as better carriers in the form of nanoparticles. For instance, mesoporous silica nanoparticles have been shown to have improved applications in gene delivery. Bio-inspired calcium phosphate ceramic nanoparticles combined with biodegradable polymers have promising applications for oral delivery of vulnerable and tender anticancer peptides. Combinatorial delivery of synergistic drugs can be made easy with the use of such nano-therapeutics with additional polymeric component for better protection from *in vitro* challenges. Future work in the field of bio-ceramics with improved porosity, bioactivity, and biodegradation along with polymer chemistry can possibly lead to new innovation in active/passive macromolecular delivery in the biopharma sector.

References

1. Williams DF (2009) On the nature of biomaterials. *Biomaterials* 30: 5897–5909.
2. Dorozhkin SV (2010) Bioceramics of calcium orthophosphates. *Biomaterials* 31: 1465–1485.
3. Jandt KD (2007) Evolutions, revolutions and trends in biomaterials science—A perspective. *Advanced Engineering Materials* 9: 1035–1050.
4. Best SM, Porter AE, Thian ES, Huang J (2008) Bioceramics: Past, present and for the future. *Journal of the European Ceramic Society* 28: 1319–1327.
5. Rawat M, Singh D, Saraf S (2006) Nanocarriers: promising vehicle for bioactive drugs. *Biological & Pharmaceutical Bulletin* 29: 1790–1798.
6. Salinas AJ, Vallet-Regí M (2007) Evolution of ceramics with medical applications. *Zeitschrift für Anorganische und Allgemeine Chemie* 633: 1762–1773.
7. Vallet-Regí M, González-Calbet JM (2004) Calcium phosphates as substitution of bone tissues. *Progress in Solid State Chemistry* 32: 1–31.
8. Guzmán Vázquez C, Piña Barba C, Munguía N (2005) Stoichiometric hydroxyapatite obtained by precipitation and sol gel processes. *Revista Mexicana de Fisica* 51: 284–293.
9. Doi Y, Shibutani T, Moriwaki Y, Kajimoto T, Iwayama Y (1998) Sintered carbonate apatites as bioresorbable bone substitutes. *Journal of Biomedical Materials Research* 39: 603–610.

10. Zapanta-Legeros R (1965) Effect of carbonate on the lattice parameters of apatite [18]. *Nature* 206: 403–404.

11. Alauzun J, Mehdi A, Reyé C, Corriu RJP (2005) Hydrophilic conditions: A new way for self-assembly of hybrid silica containing long alkylene chains. *Journal of Materials Chemistry* 15: 841–843.

12. Okamoto K, Goto Y, Inagaki S (2005) Self-organization of crystal-like aromatic-silica hybrid materials. *Journal of Materials Chemistry* 15: 4136–4140.

13. Ragel CV, Vallet-Regi M, Rodriguez-Lorenzo LM (2002) Preparation and in vitro bioactivity of hydroxyapatite/solgel glass biphasic material. *Biomaterials* 23: 1865–1872.

14. Padilla S, Roman J, Sanchez-Salcedo S, Vallet-Regi M (2006) Hydroxyapatite/SiO(2)-CaO-P(2)O(5) glass materials: In vitro bioactivity and biocompatibility. *Acta Biomater* 2: 331–342.

15. Behnamghader A, Bagheri N, Raissi B, Moztarzadeh F (2008) Phase development and sintering behaviour of biphasic HA-TCP calcium phosphate materials prepared from hydroxyapatite and bioactive glass. *Journal of Materials Science: Materials in Medicine* 19: 197–201.

16. Hoexter DL (2002) Bone regeneration graft materials. *Journal of Oral Implantology* 28: 290–294.

17. Nilsson M, Fernandez E, Sarda S, Lidgren L, Planell JA (2002) Characterization of a novel calcium phosphate/sulphate bone cement. *Journal of Biomedical Materials Research* 61: 600–607.

18. Li Y, Kong F, Weng W (2009) Preparation and characterization of novel biphasic calcium phosphate powders (alpha-TCP/HA) derived from carbonated amorphous calcium phosphates. *Journal of Biomedical Materials Research Part B: Applied Biomaterials* 89B: 508–517.

19. Jones JR, Ehrenfried LM, Hench LL (2006) Optimising bioactive glass scaffolds for bone tissue engineering. *Biomaterials* 27: 964–973.

20. Salinas AJ, Martin AI, Vallet-Regí M (2002) Bioactivity of three $CaO-P_2O_5-SiO_2$ sol-gel glasses. *Journal of Biomedical Materials Research* 61: 524–532.

21. Balamurugan A, Balossier G, Kannan S, Michel J, Rebelo AHS et al. (2007) Development and in vitro characterization of sol-gel derived $CaO-P_2O_5-SiO_2-ZnO$ bioglass. *Acta Biomaterialia* 3: 255–262.

22. Olmo N, Martin AI, Salinas AJ, Turnay J, Vallet-Regi M et al. (2003) Bioactive sol-gel glasses with and without a hydroxycarbonate apatite layer as substrates for osteoblast cell adhesion and proliferation. *Biomaterials* 24: 3383–3393.

23. Meseguer-Olmo L, Ros-Nicolas M, Vicente-Ortega V, Alcaraz-Banos M, Clavel-Sainz M et al. (2006) A bioactive sol-gel glass implant for in vivo gentamicin release. Experimental model in Rabbit. *Journal of Orthopaedic Research* 24: 454–460.

24. Singh RK, Srinivasan A, Kothiyal GP (2009) Evaluation of $CaO-SiO_2-P_2O_5-Na_2O-Fe_2O_3$ bioglass-ceramics for hyperthermia application. *Journal of Material Science: Materials in Medicine* 20 (Suppl 1): S147–S151.

25. Ebisawa Y, Miyaji F, Kokubo T, Ohura K, Nakamura T (1997) Bioactivity of ferrimagnetic glass-ceramics in the system $FeO-Fe_2O_3-CaO-SiO_2$. *Biomaterials* 18: 1277–1284.

26. Luciani G, Costantini A, Silvestri B, Tescione F, Branda F et al. (2008) Synthesis, structure and bioactivity of $pHEMA/SiO_2$ hybrids derived through in situ sol–gel process. *Journal of Sol-Gel Science and Technology* 46: 166–175.

27. Tanaka K, Kozuka H (2005) Preparation of acetylcellulose/silica composites by sol-gel method and their mechanical properties. *Journal of Materials Science* 40: 5199–5206.

28. Yang P, Quan Z, Hou Z, Li C, Kang X et al. (2009) A magnetic, luminescent and mesoporous core-shell structured composite material as drug carrier. *Biomaterials* 30: 4786–4795.

29. Slowing, II, Vivero-Escoto JL, Wu CW, Lin VS (2008) Mesoporous silica nanoparticles as controlled release drug delivery and gene transfection carriers. *Advanced Drug Delivery Reviews* 60: 1278–1288.

30. Xia W, Chang J (2006) Well-ordered mesoporous bioactive glasses (MBG): A promising bioactive drug delivery system. *Journal of Controlled Release* 110: 522–530.

31. Matsumura Y, Maeda H (1986) A new concept for macromolecular therapeutics in cancer chemotherapy: mechanism of tumoritropic accumulation of proteins and the antitumor agent smancs. *Cancer Research* 46: 6387–6392.

32. Bhattarai SR, Kim SY, Jang KY, Lee KC, Yi HK et al. (2008) Laboratory formulated magnetic nanoparticles for enhancement of viral gene expression in suspension cell line. *Journal of Virological Methods* 147: 213–218.
33. Badley RD, Ford WT, McEnroe FJ, Assink RA (1990) Surface modification of colloidal silica. *Langmuir* 6: 792–801.
34. Ulrich DR (1990) Prospects for sol-gel processes. *Journal of Non-Crystalline Solids* 121: 465–479.
35. Luo D, Han E, Belcheva N, Saltzman WM (2004) A self-assembled, modular DNA delivery system mediated by silica nanoparticles. *Journal of Controlled Release* 95: 333–341.
36. Chen Y, Xue Z, Zheng D, Xia K, Zhao Y et al. (2003) Sodium chloride modified silica nanoparticles as a non-viral vector with a high efficiency of DNA transfer into cells. *Current Gene Therapy* 3: 273–279.
37. Radu DR, Lai C-Y, Jeftinija K, Rowe EW, Jeftinija S et al. (2004) A Polyamidoamine dendrimer-capped mesoporous silica nanosphere-based gene transfection reagent. *Journal of the American Chemical Society* 126: 13216–13217.
38. Shen H, Tan J, Saltzman WM (2004) Surface-mediated gene transfer from nanocomposites of controlled texture. *Nature Materials* 3: 569–574.
39. Lee JH, Lee K, Moon SH, Lee Y, Park TG et al. (2009) All-in-one target-cell-specific magnetic nanoparticles for simultaneous molecular imaging and siRNA delivery. *Angewandte Chemie International Edition English* 48: 4174–4179.
40. Fratzl P, Gupta HS, Paschalis EP, Roschger P (2004) Structure and mechanical quality of the collagen-mineral nano-composite in bone. *Journal of Materials Chemistry* 14: 2115–2123.
41. Olton D, Li J, Wilson ME, Rogers T, Close J et al. (2007) Nanostructured calcium phosphates (NanoCaPs) for non-viral gene delivery: Influence of the synthesis parameters on transfection efficiency. *Biomaterials* 28: 1267–1279.
42. Liu T, Tang A, Zhang G, Chen Y, Zhang J et al. (2005) Calcium phosphate nanoparticles as a novel nonviral vector for efficient transfection of DNA in cancer gene therapy. *Cancer Biotherapy & Radiopharmaceuticals* 20: 141–149.
43. Kakizawa Y, Furukawa S, Ishii A, Kataoka K (2006) Organic-inorganic hybrid-nanocarrier of siRNA constructing through the self-assembly of calcium phosphate and PEG-based block aniomer. *Journal of Controlled Release* 111: 368–370.
44. Sokolova V, Kovtun A, Prymak O, Meyer-Zaika W, Kubareva EA et al. (2007) Functionalisation of calcium phosphate nanoparticles by oligonucleotides and their application for gene silencing. *Journal of Materials Chemistry* 17: 721–727.
45. Ho ML, Fu YC, Wang GJ, Chen HT, Chang JK et al. (2008) Controlled release carrier of BSA made by W/O/W emulsion method containing PLGA and hydroxyapatite. *Journal of Controlled Release* 128: 142–148.
46. Mukesh U, Kulkarni V, Tushar R, Murthy RS (2009) Methotrexate loaded self stabilized calcium phosphate nanoparticles: a novel inorganic carrier for intracellular drug delivery. *Journal of Biomedical Nanotechnology* 5: 99–105.
47. Oh JM, Choi SJ, Lee GE, Han SH, Choy JH (2009) Inorganic drug-delivery nanovehicle conjugated with cancer-cell-specific ligand. *Advanced Functional Materials* 19: 1617–1624.
48. Cheng X, Kuhn L (2007) Chemotherapy drug delivery from calcium phosphate nanoparticles. *International Journal of Nanomedicine* 2: 667–674.
49. Rawat M, Singh D, Saraf S (2008) Development and in vitro evaluation of alginate gel-encapsulated, chitosan-coated ceramic nanocores for oral delivery of enzyme. *Drug Development and Industrial Pharmacy* 34: 181–188.
50. Portela A, Vasconcelos M, Branco R, Gartner F, Faria M et al. (2010) An in vitro and in vivo investigation of the biological behavior of a ferrimagnetic cement for highly focalized thermotherapy. *Journal of Materials Science Materials in Medicine* 21: 2413–2423.
51. Hou CH, Hou SM, Hsueh YS, Lin J, Wu HC et al. (2009) The in vivo performance of biomagnetic hydroxyapatite nanoparticles in cancer hyperthermia therapy. *Biomaterials* 30: 3956–3960.
52. Shi D, Cho HS, Chen Y, Xu H, Gu H et al. (2009) Fluorescent polystyrene-Fe_3O_4 composite nanospheres for in vivo imaging and hyperthermia. *Advanced Materials* 21: 2170–2173.

53. Ashokan A, Menon D, Nair S, Koyakutty M (2010) A molecular receptor targeted, hydroxyapatite nanocrystal based multi-modal contrast agent. *Biomaterials* 31: 2606–2616.
54. Kanwar JR, Mahidhara G, Kanwar RK (2009) Recent advances in nanoneurology for drug delivery to the brain. *Current Nanoscience* 5: 441–448.
55. Wilson Jr. OC, Hull JR (2008) Surface modification of nanophase hydroxyapatite with chitosan. *Materials Science and Engineering: C* 28: 434–437.
56. Zhou G, Li Y, Zhang L, Zuo Y, Jansen JA (2007) Preparation and characterization of nano-hydroxyapatite/chitosan/konjac glucomannan composite. *Journal of Biomedical Materials Research Part A* 83: 931–939.
57. Paul W, Sharma CP (2006) Nanoceramic matrices: Biomedical applications. *American Journal of Biochemistry and Biotechnology* 2 41–48.
58. Lameiro MH, Lopes A, Martins LO, Alves PM, Melo E (2006) Incorporation of a model protein into chitosan-bile salt microparticles. *International Journal of Pharmaceutics* 312: 119–130.
59. Cleland JL (1997) Protein delivery from biodegradable microspheres. *Pharmaceutical Biotechnology* 10: 1–43.
60. Cherian AK, Rana AC, Jain SK (2000) Self-assembled carbohydrate-stabilized ceramic nanoparticles for the parenteral delivery of insulin. *Drug Development and Industrial Pharmacy* 26: 459–463.
61. Mahidhara G, Kanwar RK, Kanwar JR (2011) A novel nanoplatform for oral delivery of anti-cancer biomacromolecules. *International Journal of Nanotechnology* 9: 942–960.
62. Baratchi S, Kanwar RK, Kanwar JR (2010) Survivin: A target from brain cancer to neurodegenerative disease. *Critical Reviews in Biochemistry and Molecular Biology* 45(6): 535–554.
63. Cheung CHA, Kanwar JR, Krissansen GW (2006) A cell-permeable dominant-negative survivin protein as a tool to understand how Survivin maintains tumour cell survival. *European Journal of Cancer supplements* 4(12): 149.
64. Kanwar RK, Cheung CHA, Chang JY, Kanwar JR (2010) Recent advances in anti-survivin treatments for cancer. *Current Medicinal Chemistry* 17(15): 1509–1515.
65. Cheung CHA, Sun X, Kanwar J, Bai J.-Z, Cheng L, Krissansen G (2010) A cell-permeable dominant-negative survivin protein induces apoptosis and sensitizes prostate cancer cells to TNF-alpha therapy. *Cancer Cell International* 10(36): 1–12.
66. Bawa R (2009) NanoBiotech 2008: Exploring global advances in nanomedicine. *Nanomedicine* 5: 5–7.
67. Bergey EJ, Ohulchansky TY, Prasad P, Pudavar H, Roy I (2004) Ceramic based nanoparticles for entrapping therapeutic agents for photodynamic therapy and method of using same. Patent Application US20040180096 A1.
68. Hakamatsuka Y, Irie H (1990) Drug-impregnated ceramic. U.S. Patent 5318779.
69. Alt E (2010) Drug-releasing stent with ceramic-containing layer. U.S. Patent 7713297.
70. Moerck RE, Sabacky BJ, Prochazka J (2006) Ceramic structures for controlled release of drugs. U.S. Patent publication US20060165787 A1.
71. Mclaughlin CA, Lyles MB, Halff GA, Mallow WA (2006) Porous drug delivery system. U.S. Patent 7001371.
72. Deutsch EA, Deutsch KF, Cacheris WP, Ralston WH, White DH et al. (1997) Treated apatite particles for medical diagnostic imaging U.S. Patent 5690908.
73. Moerck RE, Sabacky BJ, Spitler TM, Prochazka J, Ellsworth D (2006) Ceramic structures for prevention of drug diversion US 20060127486.
74. Hermansson L, Engqvist H (2009) Carriers for drug delivery. U.S. Patent Application 20090061003, Sweden.

8

Imaging and Triggered Release through Design of Ultrastable Core–Shell Iron Oxide Nanoparticles

Erik Reimhult

CONTENTS

8.1 Introduction

Superparamagnetic iron oxide nanoparticles (NPs), with core diameters between 3 and 15 nm, are used in a rapidly expanding number of research and practical applications in the biomedical field [1]; the most common include magnetic cell labeling [2,3], separation [4] and tracking [4], for therapeutic purposes in hyperthermia [5,6] and drug delivery [7], and for diagnostic purposes, for example, as contrast agents for magnetic resonance imaging (MRI) [8–10]. Superparamagnetic iron oxide NPs are considered to be benign to the body [8,11] mainly because iron oxide is dissolved under acidic conditions. The resulting Fe^{3+} ions can be fed into the natural iron storage, which is 3–5 g iron for an adult human [12,13]. Thus, the additional amount of iron released from dissolved iron oxide NPs is negligible for the iron oxide NP concentrations in the µg/kg body weight range that are typically injected [14]. Most other magnetic materials such as Co have a higher saturation magnetization (M_s) compared to iron oxide [15] and would therefore show a stronger magnetic response but are toxic, which provides iron oxide with a decisive advantage for the majority of biomedical applications [16].

The aforementioned applications put tremendous demand on NP stability at physiological conditions, close control over NP size, and controlled surface presentation of functionalities. This chapter recapitulates recent results by us and others that highlight different aspects of the stability of core–shell superparamagnetic iron oxide NPs, in particular its implementation by molecular design of the dispersant shell around the iron oxide core and how this influences the assembly of NPs into superstructures with applications in, for example, drug delivery. Special attention is given to the selection of molecular anchors for the NP dispersant shell, because of their importance to ensure colloidal and functional stability of sterically stabilized superparamagnetic iron oxide NPs.

In the second part, we will discuss a promising and particularly demanding release system in terms of core–shell NP design and assembly: NP-actuated lipid vesicles. These smart drug delivery capsules promise to control localization, timing, and dose of released cargo when fully explored. Controlling location and timing of the release allows using more potent drugs as the interaction with the right target is ensured. In extension, it also can enable the sequential release of multiple drugs in a defined manner. It is therefore one of the central goals of personalized or "nano"-medicine.

8.2 Core–Shell Structure of Iron Oxide Nanoparticles

To enable that superparamagnetic iron oxide NPs can be dispersed in aqueous media and at physiologic salt concentrations, iron oxide cores are coated with amphiphilic or hydrophilic molecular spacers, typically polymers, which are called dispersants. The dispersants surround the core to shield it from direct interactions with the environment and can have a variety of different structures. This coating surrounding the core is referred to as the NP shell.

NPs will rapidly aggregate without a dispersant shell through interactions between themselves or through interactions with biological molecules. The result of aggregation is eventually loss of function and precipitation out of solution. In particular, for most biomedical and *in vivo* applications, NPs have to resist adsorption of biomacromolecules such as proteins to the NP surface. This property is referred to as "stealth." If plasma proteins such as opsonins are adsorbed on the NP, they can lead to aggregation of NPs, but even more problematic, they induce NP uptake by macrophages, monocytes, and dendritic cells, which remove NPs from the body and break them down [17]. Therefore, the ability to resist protein adsorption strongly correlates with colloidal stability and determines the ability of NPs to achieve targeting and delivery in the body or in other biofluids.

Today, commercially available NPs for MRI contrast enhancement are primarily encapsulated in dextran to achieve water solubility and suppress nonspecific protein interactions. However, the most commonly used dispersant spacer in the research on NPs as well as for nonfouling biointerfaces is poly(ethylene glycol) (PEG) [12,18], which often is referred to as "stealth coating." PEG-modified surfaces exhibit the smallest attractive van der Waals forces to proteins compared to other well-known water-soluble polymers due to the low refractive index of PEG [19] and carry a water shell through acting as both hydrogen donor and acceptor that prevents direct contact with proteins [20,21]. Furthermore, the confinement of the stabilizing polymer chain that would result from protein adsorption results in an energetically unfavorable entropy decrease [22]. This steric repulsive force to colloidal aggregation and protein adsorption increases with polymer grafting density on the NP surface and polymer chain molecular weight (MW). On flat surfaces, it has been demonstrated that these two parameters can be combined through calculating the projected (ethylene glycol) monomer surface density that determines the nonfouling properties of the interface [23,24]. Thus, at least within a certain range, the MW and the grafting

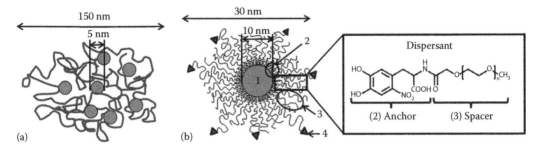

FIGURE 8.1
Steric stabilization of iron oxide NPs. (a) Schematic of commercially available iron oxide magnetic contrast agents such as Feridex™ and Endorem™. The superparamagnetic iron oxide NP cores are coated with physisorbed dispersants such as dextran. Due to the weak affinity and multiple binding sites of the dextran, multiple iron oxide cores are embedded in one cluster. The resulting hydrodynamic diameter is many times larger than the core diameter. (b) Superparamagnetic iron oxide NPs stabilized with low MW dispersants result in a defined core–shell structure. Such NPs can be divided into four components: (1) core, (2) anchor, (3) spacer, and (4) end functional group.

density can be traded off against each other. Furthermore, it is clear from many studies on planar interfaces that protein resistance requires that the PEG conformation is in the brush regime, in which the distance between adjacent chains, D, is smaller than the Flory radius, R_F, of the polymers [25,26]. Superparamagnetic iron oxide NPs have very high surface curvature. This leads to a conically increasing free volume with distance from the particle surface for dispersants grafted to the surface (cf. Figure 8.1b). Thus, even a dense brush at the particle surface will quickly grow out of the brush regime as the polymer expands from the surface. For this reason, the ethylene glycol density on NPs likely has to be even higher than on planar surfaces to render NPs resistant to protein adsorption.

Different types of superparamagnetic iron oxide NPs stabilized with weakly adsorbed, high MW dispersants (Figure 8.1a), such as dextran, are FDA approved and applied for diverse purposes such as magnetic labeling and cell separation and as magnetic resonance contrast agents [27–29]. They are used in clinics as negative magnetic resonance contrast agents to detect lesions mainly in the liver and spleen [28,30]. High MW of the polymer dispersant, typically significantly higher than 10 kDa, makes it possible to physisorb the dispersant on the NPs to stabilize it without further chemical modification of polymer or particle core. However, the utilized weak interactions lead to reversible adsorption of the dispersants and thereby low colloidal stability, for example, at high dilution [31]. These dispersants tend to enwrap and cluster multiple iron oxide NP cores by direct physisorption. The magnetic response of one stabilized core–shell particle is thereby increased compared to single-core particles, but the control over cluster size and thereby also over overall NP size is poor [18,32–34].

Alternatively, superparamagnetic iron oxide NPs can be stabilized with low MW dispersants, that is, dispersants with MWs lower than 10 kDa, which possess a polymer spacer with a covalently bound anchor that has high affinity for the NP surface (Figure 8.1b). By covalently binding a high-affinity anchor to the end of a linear dispersant molecule, a well-defined adsorption geometry to the NP surface can be achieved. The resulting core–shell NPs can be divided into four components: core, anchor, spacer, and optional surface functionalities (Figure 8.1b). Each of these components can independently be adjusted through modular buildup starting from the dispersants grafted to the NP interface. The resulting defined geometry renders such NPs very versatile for a multitude of applications [35,36]. One of the resulting advantages of this modular and controlled buildup is that the hydrodynamic size can be precisely controlled. Furthermore, the well-defined assembly of dispersants at the NP surface enables controlled surface presentation of functionalities. These factors (size, stability, dispersant shell thickness, and control over functionalities presented at the NP surface) are the factors that critically determine NP performance in the demanding environment of a biological fluid through defining the properties of the NP shell [37,38]. The shell shields the NP core and will mediate any interactions with the environment. In particular, the achieved grafting density of the polymer and the desorption rate of the anchor will determine the colloidal stability of the NP.

8.3 Nanoparticle Stability

The term "stability" is used with very different meaning in the literature on NPs, mostly without explicit acknowledgment of these differences. Generally, stability can refer to the stability of the NP structure or of the colloidal suspension of NPs. The two are closely dependent, since a structurally unstable NP is unlikely to form a stable colloidal solution. We therefore highlight the colloidal stability and its dependence on the structural design of the NP.

NPs are often imprecisely considered stable if they do not visibly precipitate over a finite period of time [39,40]. A thorough characterization of NP dispersions, for example, with scattering techniques, allows us to define NP stability more precisely in experimental terms. While colloidal stability can be achieved by many means and the most common method to do so is electrostatic repulsion, biomedical applications in practice limit the applicable methods to only steric stabilization. Steric stabilization is achieved through a highly solvated organic spacer layer, which prohibits biomolecules or other particles to come into contact with the particle core and induce aggregation [41,42]. For example, the aforementioned strategy for creating nonfouling, grafted polymer films on planar surfaces translates into strategies to sterically stabilize NP suspensions by the same basic principle of preventing direct core surface access through a polymer film coating.

We recently demonstrated the importance of the techniques and conditions used to characterize NP stability. This was exemplified by investigations of aqueous suspensions of superparamagnetic iron oxide NPs stabilized by grafting of poly(ethylene glycol)–hydroxydopamine (PEG–hydroxydopamine) shells. PEG–hydroxydopamine-stabilized iron oxide NPs were stable for more than a year if stored and analyzed at room temperature after dialysis [35]. However, PEG–hydroxydopamine-stabilized iron oxide NPs were later shown to agglomerate if they instead were subjected to multiple filtrations (Figure 8.2) [36]. This is likely explained by equilibrium of free, reversibly adsorbed dispersants with dispersants bound to the NP surface, which is sufficient to ensure colloidal stability as long as the dispersants are not efficiently and repeatedly removed as in the case of repeated filtration. NPs start to agglomerate if the dispersant packing density drops below a critical value. Such instability might not easily be demonstrated by a long-time storage at high NP concentration where dispersant surface density might only extremely slowly change over time. In contrast, NPs with reversibly adsorbed dispersants will always agglomerate upon the high dilution they experience in any application (Figure 8.2b). This should prompt experimental investigations of stability to always include investigation of conditions of high dilution, as well as rapid and complete removal of reversibly desorbed dispersants. Furthermore, many applications of magnetic NPs to smart biomedical materials

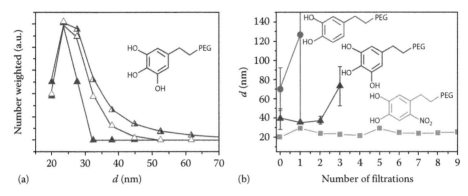

FIGURE 8.2

(See color insert.) Characterization of the stability of superparamagnetic iron oxide NPs. The stability of superparamagnetic iron oxide NPs was measured with DLS at 25°C. (a) The hydrodynamic diameter of PEG(0.55 kDa)–hydroxydopamine-stabilized iron oxide NPs as stabilized (-▲-), after storage for 1 year in PBS (-▲-) and after storing them for 20 months in HEPES (-△-). (From Amstad, E. et al., Surface functionalization of single superparamagnetic iron oxide nanoparticles for targeted magnetic resonance imaging. *Small.* 5. 1334–1342. Copyright Wiley-VCH Verlag GmbH & Co. KGaA. Reproduced with permission.) (b) The stability of NPs evaluated after repeated filtrations performed to remove excessive dispersants from iron oxide NPs stabilized with PEG(5 kDa)–nitrodopamine (-■-), PEG(5 kDa)–hydroxydopamine (-▲-), and PEG(5 kDa)–dopamine (-●-) (Reproduced with permission from Amstad, E., Gillich, T., Bilecka, I., Textor, M., and Reimhult. E. Ultrastable iron oxide nanoparticle colloidal suspensions using dispersants with catechol-derived anchor groups. *Nano Lett.* 9. 4042–4048. Copyright 2009, American Chemical Society.). While PEG–hydroxydopamine-stabilized iron oxide NPs were stable at room temperature for more than 20 months, they started to agglomerate after excessive dispersants were removed by more than two filtrations.

applications that are currently under investigation make use of NPs exposed to temperature higher than body temperature. Applications that include hyperthermia or thermally actuated release induced, for example, by alternating magnetic fields (AMFs), expose the NPs to locally much higher temperatures, which can put extreme demands on the NP shell structural stability. A useful "stress test" for relevant NP colloidal stability is therefore to perform, for example, DLS measurements as function of temperature. Irreversible aggregation at increased temperature implies low dispersant anchor affinity to the NP surface or too low grafting density for such novel applications [36].

The criteria for full colloidal stability of NPs especially for biomedical applications should therefore include NP characterization under dilute conditions and elevated temperature and in the presence of flocculants or sticky macromolecules that are biomedically relevant to give a reasonable guide to their applicability as biomedical or smart materials.

8.4 Steric Stabilization of Iron Oxide Nanoparticles

We will now in more detail study the steric stabilization of NPs through the grafting of polymer shells to the NP surface to produce individually stabilized NPs with a defined architecture. Due to their defined structure, they can be optimized for any application, and they have thereby become the state of the art in the research to develop novel superparamagnetic iron oxide NPs for imaging, drug delivery, and other biomedical smart material applications.

Steric stabilization relies on polymers that densely surround the NP core to overcome attractive van der Waals and magnetic interactions under the relevant application environmental conditions. If two sterically stabilized cores approach each other, the volumes of the respective polymer shells are confined. This reduces the entropy of dispersants and increases the osmotic pressure between NPs to push them apart. The resulting repulsive potential critically depends on the dispersant density profile, packing density [43], binding reversibility, and solvent quality with respect to the dispersants [44]. The polymers used to stabilize NPs are effectively limited to low MW (MW <10 kDa) hydrophilic dispersants due to packing constraints at the core surface and the desired small hydrodynamic radius of the stabilized NP.

8.5 Anchored Low MW Dispersants

Low MW dispersants can be bound to NP surfaces either through the "grafting to" or the "grafting from" technique (Figure 8.3). For the latter approach (Figure 8.3a), initiators are covalently bound to the NP surface. Spacers can subsequently be grown

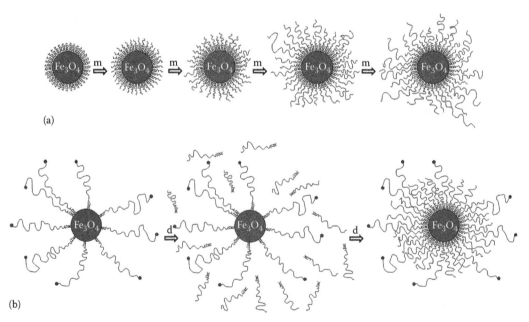

FIGURE 8.3
Grafting methods applied to stabilization of NPs with low MW dispersants. Low MW dispersants can be (a) "grafted from" and (b) "grafted to" the NP surface. By grafting from, (a) the polymer shell is polymerized, typically by radical polymerization, from initiators first bound to the NP surface. The shell benefits from very high dispersant packing density. By grafting to, (b) presynthesized dispersants with high affinity anchor groups for the NP surface are adsorbed to the particle core. The resulting dispersant packing density is below that of dispersants grafted from the NP surface. However, the grafting to approach allows for close control over the dispersant shell thickness and density of functionalities presented at the NP surface by simple mixing of different dispersants for the adsorption step.

in situ, for example, by radical polymerization, from initiators that have been attached to NP surfaces [45,46]. This approach results in high dispersant packing densities and therefore good NP stability, because the brush density is set by the density of anchors from which the grafting from is initiated. Although it leads to a desirable high shell density, the "grafting from" technique has some inherent drawbacks. Dispersant characterization and control over the dispersant polydispersity and layer thickness are difficult, and they have to be performed per finished synthesized NP batch. Furthermore, functionalization of stabilized NPs with different ligands or other functional units and controlling the density of functional groups presented at the NP interface are challenging [47].

Low MW dispersants can also be synthesized and fully characterized before application including a reactive group to anchor it to the NP surface. The finished dispersants can then be "grafted to" the NP surface using the anchor moiety on the polymer chain by a coupling reaction where both NPs and dispersants are present in solution (Figure 8.3b). This self-assembly approach has the advantage that it is cost-effective, has high reproducibility, and is easy to scale up. Furthermore, the dispersant layer thickness can be controlled by the dispersant packing density, polymer MW, and conformation. A particular advantage of the "grafting to" approach is that the surface density of one or multiple surface-presented (bio)ligands can be tailored by co-adsorbing differently functionalized dispersants in one or several subsequent assembly steps [35,48]. The polymer grafting density achieved using the "grafting to" approach is not as easy to control as for the "grafting from" approach. On flat surfaces, it has been shown to be significantly lower [47]. This difference in polymer grafting density is a result of the steric repulsion of adjacent polymer chains that defines the density achievable for the grafting to approach [49,50], that is, as the full MW polymer dispersants try to bind their anchor groups to the substrate, they cannot do so closer to each other than a distance defined by the radius of gyration of the free particle under the solvent conditions at which the grafting is performed. However, the rapidly increasing free volume available to dispersants on NPs with increasing distance from the core surface (cf. Figure 8.1b) will result in a lower steric repulsion of adjacent dispersants. Therefore, the difference in grafting density between surfaces modified through the "grafting from" and the "grafting to" approach decreases with increasing surface curvature. It will, however, persist and limit the maximum polymer MW that can be grafted into a complete shell around the core. The grafting density is also increased by performing the "grafting to" approach using a poor solvent for the polymer to be grafted and by using an anchor with high reaction rate and core surface affinity.

For both grafting approaches, but in particular for the "grafting to" approach, the purification of free polymers and added functional units is of critical importance for correct characterization and functionality. Purification of superparamagnetic iron oxide NPs is challenging due to the similarity in size and properties of the NPs and the excess free polymer dispersants. For this reason, it is also increasingly difficult to separate free dispersants from stabilized NPs the higher the MW of the dispersants is. As described earlier, using weakly or physisorbed dispersants, as is still mostly the case [1], leads to aggregation of NPs already during the purification step [36]. An illustration of the unexpected challenges in NP purification is that it was even reported that noncomplexed ^{64}Cu added as a radiotracer could not be removed by centrifugation but required purification of iron oxide NPs by column separation [51]. Column separation has repeatedly been shown superior to dialysis, centrifugation, and magnetic washing for dispersant separation [1].

8.6 Anchors

Irrespective of whether low MW dispersants are grafted to or from the surface, they have to firmly adhere to the NP surface through suitable anchors (Figure 8.1b). The binding affinity of anchors should be high and the desorption rate k_{off} low so that anchors can irreversibly bind spacers to uncoated NPs and are able to replace hydrophobic capping agents such as oleic acid often used to synthesize monodisperse superparamagnetic iron oxide NPs [52]. Typical anchors used to surface modify superparamagnetic iron oxide NPs described in the literature are catechol [53–56] or catechol derivatives [35,36,57,58], carboxy groups [59–61], phosphonates [60,62,63], and silanes [64–68]. Despite the central importance to define the stability of the NP shell and to define the density of functional units on a NP, the influence of different anchors on iron oxide NP stability was only recently systematically studied for some of these compounds by Reimhult et al. [36]. This particular study demonstrated an optimal binding affinity and "irreversible" off rate for nitrocatechols in contrast to the other investigated commonly used "high-performance" anchors, which displayed reversible binding affinity and/or led to NP dissolution [36,69,70]. The likely mechanism for the irreversible binding affinity of nitrocatechols to the Fe(III) surface ions of Fe_3O_4 core particles through oxidative electron transfer was supported by recent studies [71,72]. Thorough characterization of iron oxide NP stability and its relation to successful *in vivo* application as a function of choice of anchor still remain scant, but the aforementioned described work further demonstrated that the higher anchor stability directly translated to higher colloidal stability for applications of PEGylated NP systems [36,71,72].

8.7 Multifunctional Iron Oxide Nanoparticles

The magnetic core of iron oxide NPs provides the basic functionality of the NP for both imaging and actuation. For a discussion of the impact of NP stabilization and functionalization on the magnetic properties of the core, we refer to Reference 1. Additional functionalities can be attached to the NP when the core is stabilized by a structurally well-controlled dispersant shell. The addition of further functionalities than the superparamagnetic core is becoming increasingly important for many applications, in particular in the biomedical field [7,30,73]. The foremost such functionality is targeting of cellular- or tissue-specific binding sites. Moieties that have been studied for such targeting are, for example, antibodies [74], peptides [67,75–77], aptamers [78], DNA [79,80], and RNA [81] sequences.

In vivo targeting with superparamagnetic iron oxide NPs is claimed in numerous reports. However, the vast majority of iron oxide NPs are targeted toward the liver, kidney, or lymph nodes, locations they naturally end up during clearance [82]. Alternatively, iron oxide NPs were targeted to tumors and cancer cells, where they naturally accumulate due to the enhanced permeation–retention (EPR) effect [30,83]. Proving specific targeting to such organs is therefore difficult as increased accumulation can occur also without specific binding to a target and the addition of a targeting moiety might be helping only by building on the underlying propensity to accumulate in the target tissue.

A prerequisite for successful targeting of NPs is that the targeting moieties are irreversibly bound to the NP surface. If NPs are stabilized with low MW dispersants, the ligand density presented at the NP surface can be closely controlled by co-adsorbing function-alized and unfunctionalized dispersants to the NP surface at defined molar ratios [35]. This is in stark contrast to NP surfaces coated with physisorbed high MW dispersants where the serpentine, constantly changing conformation, prevents efficient addition and controlled presentation of ligands at the NP interface [84].

Targeting has to be specific. The addition of targeting moieties to the NP surface there-fore cannot be allowed to severely compromise the NP stability or its stealth property. The addition of ligands, in particular large ligands such as proteins and RNA, to a particle surface will always tend to compromise the ability of a NP to remain undetected *in vivo*, and it will add interactions with other molecules in a biological environment. The loss of stealth function is minimized if the number of proteins in the ligand shell is minimized. This can be achieved by covalently linking ligands directly to the stealth PEG disper-sant shell at a controlled density. Superparamagnetic iron oxide NPs that are stabilized with low MW dispersants such as PEG silanes allow controlled presentation of ligands at the NP interface. This was demonstrated, for example, on chlorotoxin-functionalized iron oxide NPs [65]. Their performance and uptake were subsequently studied *in vitro* in cell assays [10,67]. A controlled surface presentation of ligands from the shell segments statistically located with a high probability on the outside of the stabilized NP is thought to increase the targeting efficiency by decreasing the risk that ligands are buried in the dis-persant shell. Furthermore, it allows for a closer control over changes in the hydrodynamic diameter of NPs upon functionalization and enables optimizing the number of ligands bound per NP to ensure sufficient binding affinity while minimizing nonspecific interac-tions. Because the hydrodynamic size of superparamagnetic iron oxide NPs significantly influences NP uptake by cells [85], control over the hydrodynamic diameter of NPs upon functionalization is crucial.

It is also often desirable, especially for research purposes, to add multiple labels (imag-ing modalities), such as fluorophores or radiotracers to iron oxide NP surfaces. A second modality helps to uniquely locate such contrast agents and differentiate them from imag-ing artifacts. Multiple labels and functionalities are most easily added to the shell through incorporation in the organic shell, for which a wealth of polymer coupling strategies can be applied. Clearly strategies that irreversibly attach both the dispersants in the shell to the core and the functional moiety to the dispersants in the shell are required [13].

For fluorophores, it should additionally be considered that fluorophores in proximity also to oxide NPs can suffer significant quenching by direct excitation transfer to elec-tronic states of the NP core. For example, iron oxide has been reported to quench CY5.5 and FITC if on average less than two fluorophores were attached per iron oxide core [86]. Combining multiple labels on superparamagnetic iron oxide NPs can generally lead to undesirable crosstalk. Also superparamagnetic iron oxide NPs functionalized with near-infrared (NIR) fluorescent molecules have shown reduced functionality due to the absorp-tion of light caused by the iron oxide cores. Thus, also fluorophores might have to be coupled to the outer segments of the shell in order to be functional, and the introduction of dual imaging modalities by addition of a radioactive tracer to the shell might be a safer choice. As functional units have to be co-localized with the cores as well as survive the NP purification step, we once again emphasize the importance of that the anchor chem-istry yields sufficiently low-dispersant desorption rate. An advantage of the "grafting to" approach in the context of adding functionalities to the NP shell is that using irreversibly binding anchors makes it possible to add any number of *different* functional groups for imaging, tracing, targeting, or therapeutics, as well as detailed control over the number of functional groups per NP [35].

8.8 Magnetic Nanoparticle–Actuated Vesicle Release Systems

We will now discuss recent developments of the application of superparamagnetic NPs to control drug release from vesicles, which build on the development of the above discussed tools for controlling the architecture, colloidal stability, and functionality of iron oxide NPs. The assembly and characterization of drug delivery systems has been the subject of many studies [87–92]. This tremendous scientific interest is closely related to the industrial demand for stable, smart capsules that can easily and cost-effectively be assembled in a versatile way with controlled release properties. Encapsulating drugs and vaccines has been shown to be highly beneficial for multiple reasons. The encapsulated cargo is protected from enzymatic degradation [93]. Furthermore, the surface functionality of the delivery vehicle can be modified to target them to certain locations [88,89,93,94]. Decreased degradation and in particular a high degree of localization of released cargo allow the use of much lower overall concentrations of the drug compared to systemic release. With this decrease in the injected drug dose, the risk of adverse side effects, for example, toxicity or immune system reaction, induced by free drugs is lowered [93,95]. Ideally, delivery vehicles should accumulate and release cargo only at the targeted location. Additionally, they should ideally allow for external control over timing and dose of the released cargo at the target location.

A particularly elegant way to realize the release from vesicles is to embed NPs, serving as actuators, into a vesicular release vehicle [96,97], as is schematically shown in Figure 8.4. Magnetic fields penetrate and are relatively benign to tissue but can interact strongly with magnetic NPs [96]. This gives magnetic NPs an advantage over metallic counterparts as only NIR light has a significant penetration depth in tissue, but such long wavelengths require particles with at least one dimension larger than 100 nm. Encoding a magnetic antenna function through encapsulation of superparamagnetic NPs in a vesicle that can change the release rate of compounds from the vesicle in response to NP-induced changes to the local environment thereby provides an unusually good opportunity to construct systems for biomedical release applications. Such constructs can, for example, be used to circumvent the typical problem of having to trade off low passive permeability against high

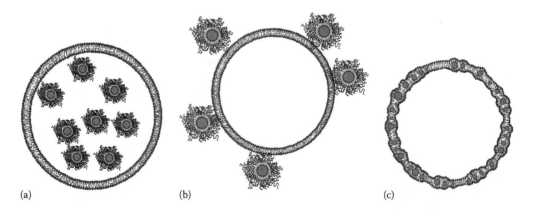

(a)   (b)   (c)

FIGURE 8.4
Schematic of NP-functionalized liposomes. Three different composite vesicles containing NPs for actuation can be envisioned for a unilamellar lipid bilayer vesicle separating an aqueous interior from the surroundings. The NPs used as actuators to trigger drug release can be (a) encapsulated in the vesicle lumen and (b) attached to the vesicle surface and/or (c) in the hydrophobic core of the lipid membrane.

release efficiency for thermally activated release vehicles. If subjected to an AMF, superparamagnetic NPs locally generate heat mainly through Brownian and Néel relaxations [98]. Local delivery of drugs through thermally actuated release has been one of the main research directions for triggered drug release, since it can be controlled by external stimuli. Magnetically responsive NPs can be as small as a few nanometers in diameter [12] and can therefore be encapsulated into the nanoscale structures of drug delivery vehicles, such as membranes. Through magnetic heating, NP-enhanced thermoresponsive vesicles can be locally heated to temperatures far exceeding the body temperature. Thus, the requirements of low leakage at body temperature, high cargo release efficiency, and minimal harm to surrounding tissue by global heating can be simultaneously met. This is achieved through that the temperature-controlled transition of the vesicle permeability can be placed at a temperature far higher than body temperature.

An efficient permeability barrier that can be modified for use in vesicular drug delivery vehicles is the cell mimetic membrane that has a hydrophobic membrane core surrounded by hydrophilic faces. A vesicle surrounded by an amphiphilic membrane allows incorporation of hydrophobic and amphiphilic drugs into the membrane, while hydrophilic drugs can be stored in the aqueous interior [99]. The most considered vesicular drug delivery vehicles are liposomes (Figure 8.5), which have membranes comprising synthetic or natural lipids. The biocompatibility and ease of assembly of liposomes render them especially attractive for biomedical applications [95,100–102]. By choosing the appropriate assembly protocol, the size of liposomes can be varied between 50 nm and several micrometers [101]. Furthermore, incorporation of PEGylated lipids into the liposome bilayer not only increases their blood circulation time *in vivo* [103] but was also shown to significantly reduce the leakiness of liposomes if stored at temperatures below T_m [104].

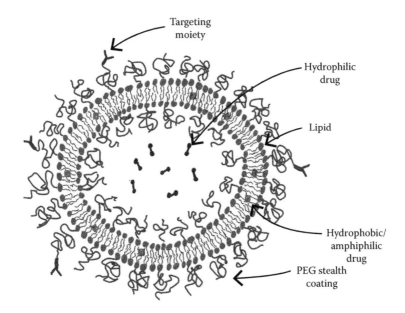

FIGURE 8.5
(See color insert.) Schematic of drug delivery liposome. The basic constituents of a drug delivery liposome are lipids forming a liposome membrane in the 100 nm size range, thereby encapsulating small hydrophilic drugs in the lumen or hydrophobic and amphiphilic drug in the membrane. Optionally the lipids can tether, for example, PEG to increase *in vivo* circulation time and charged lipids or specific binding groups to target the liposome to the release site.

The permeability of liposomes drastically increases around the membrane melting temperature (T_m) [105,106]. Through the choice of high-T_m lipid compositions, the temperature at which leakage occurs can be designed to be significantly above body temperature, ensuring low passive leakage of cargo.

Liposomes have been functionalized with NPs localized in their lumen, membrane, or attached to the membrane surface (Figure 8.4) to actuate release from liposomes with high T_m. These strategies lead to differences in design criteria and release properties that will be discussed in the following sections.

8.9 Nanoparticles in the Liposome Lumen and Associated with the Membrane

Liposomes loaded with iron oxide NPs in their lumen (Figure 8.4a) [107–112] or in which iron oxide NPs were synthesized in the lumen through aqueous precipitation methods using Fe^{3+} and Fe^{2+} salts are known for more than 20 years [113]. Additionally, liposomes containing iron oxide NPs occur naturally in magnetotactic bacteria, the so-called magnetosomes, which have also been proposed for and applied to biomedical applications [114]. In many cases discussed in the literature, the NPs are coated in such a way, for example, unfunctionalized or with ascorbic or citric acid, that they also strongly interact with the membrane and even on purpose decorated on the vesicle surface (Figure 8.4b) [115–119].

Several studies have successfully shown AMF-triggered content release from liposomes loaded with iron oxide NPs in their lumen or bound to the membrane surface [99,107–112,115,116,120]. Superparamagnetic iron oxide NPs as small as 5 nm in diameter were shown to convert energy from an AMF into heat sufficiently effectively to trigger content release [12,104]. However, while the cargo release efficiency depends on the total concentration of iron oxide NPs contained in the liposome dispersion, it was shown to be insensitive to the ratio of the number of iron oxide NPs localized in the liposome lumen to that of NPs freely dispersed in solution [110]. The insensitivity of cargo release efficiency on the ratio of encapsulated to freely dispersed NPs is a consequence of the unchanged specific absorption rate (SAR) of iron oxide NPs upon encapsulation in the liposome lumen [112] and the fast heat dissipation caused by the high thermal conductivity of water [110]. It has been shown that for liposomes heated by NPs incorporated in the lumen, the bulk water temperature is the same as the temperature in the bilayer [110]. If the majority of iron oxide NPs are freely dispersed, the decreased concentration of freely dispersed NPs upon dilution of the liposome dispersion significantly decreases the cargo release efficiency. Freely dispersed NPs will diffuse away from the vicinity of the liposomes when diluted, and this will significantly decrease the local concentration of magnetically heated particles around the liposomes. Such dilution will be the case in real applications, and close attention should therefore be paid to the actual encapsulation of NPs inside the vesicles when characterizing such delivery vehicles.

Recently, it was shown that cargo release from liposomes encapsulating uncoated 6 nm $CoFe_2O_4$ NPs can be triggered with low-frequency magnetic fields [115]. Release through low-frequency magnetic fields does not rely on heating but likely has a mechanical origin and therefore circumvents some of the potential problems with AMF-induced release, such as nonspecific heating of surrounding tissue.

Iron oxide NPs loaded into the liposome lumen of the vast majority of the reported magneto-liposomes agglomerated due to poor steric stabilization [99,110–112,116]. The NP encapsulation efficiency has been shown to decrease if NPs agglomerate prior to or during encapsulation, for example, for starch-coated iron oxide NPs that were reported to be primarily located outside liposomes rather than in the liposome lumen [112]. Iron oxide NPs sterically stabilized with reversibly adsorbing dispersants such as starch, dextran, or oleic acid interact with the liposome membrane due to the high affinity of the phosphate lipid head groups to partly exposed iron at the NP core surface [121]. The interaction of iron oxide NPs sterically stabilized by physisorbed carboxydextran with liposome membranes led to an unknown amount of iron oxide NPs that were bound to the outer surface of the liposomes [122].

The influence of NPs interacting strongly with the lipid membrane on the liposome permeability is not quantitatively known, but several examples from the literature indicate that it is substantial. It has been shown that 12%–50% of the agglomerated carboxydextran-functionalized iron oxide NPs initially encapsulated in the lumen of soy phosphocholine liposomes leaked out within 48 h after liposome preparation [122]. This indicates that liposomes became unstable within 48 h. Additionally, increasing the loading of dextran-coated iron oxide NPs in the liposome lumen resulted in a reduced temperature at which increased liposome permeability was observed [99]. Thus, to gain control over the concentration of encapsulated NPs and their magnetic properties and to retain the liposome membrane properties upon loading iron oxide NPs in their lumen, it is crucial to sterically stabilize NPs with dispersants that adsorb irreversibly to the NP surface and that completely cover it. In summary, the control over both the loading and the release efficiency gets lost if the incorporated NPs are not sufficiently colloidally stable. Also the dispersion of the vesicles risk faster degradation with storage time, which can particularly be an issue with vesicles that employ membrane surface interacting NPs for actuation.

8.10 Nanoparticles in the Membrane

Incorporating NPs directly into the vesicle membrane (Figure 8.4c) has intrinsic advantages over delivery vehicles where NPs are loaded into the lumen or associated with the membrane surface. However, a major challenge is to incorporate NPs without compromising membrane integrity and cause leaking of encapsulated compounds. NPs that are not embedded in the membrane require heating of bulk water to increase the membrane temperature and trigger release [97]. NPs incorporated in the membrane can directly transfer the locally generated heat into the membrane. Thus, externally triggered release from vesicles incorporating NPs in the membrane has been shown more efficient compared to release from vesicles containing NPs in the lumen [104,123].

Liposome membranes are only 4–6 nm thick, matching the maximum thickness of a lipid bilayer. Theoretical studies revealed that it is energetically favorable to embed hydrophobic uncharged NPs with diameters below 6.5 nm into liposome bilayers whereas larger hydrophobic NPs are surrounded by a phospholipid monolayer resulting in micelles [124,125]. Only NPs with hydrophobic shells insert into the hydrophobic core of the membrane, as demonstrated for liposomes prepared with hexanethiol and mercaptosuccinic acid (MSA)-coated Au NPs, respectively [123].

The theoretical prediction that individually stabilized NPs larger than 6.5 nm in diameter cannot be incorporated into liposome membranes is in good agreement with experimental studies. For example, tri-*n*-octylphosphine oxide (TOPO)-stabilized CdSe quantum dots (QDs) with a diameter of 5 nm could be incorporated into giant 1,2-dimyristoyl-sn-glycero-3-phosphatidylcholine liposomes in contrast to 8 nm diameter QDs [126]. Thus, irreversible steric stabilization of hydrophobic NPs is absolutely necessary for embedding them in the hydrophobic part of the phospholipid bilayer. Agglomerated hydrophobic NPs are larger than 6.5 nm and form micelles, while individually stabilized NPs insert into the liposome membrane, as was experimentally observed in several studies [104,127].

The issue of how NPs insert into the membrane is still contested in the literature. Experimentally, both the close packing of NPs [128], attributed to a reduction of the energy related to the imposed membrane curvature and reduced packing density of the lipids, and the observation of well-dispersed NPs in the membrane have been observed [104]. The difference between these two results might stem from the use of much smaller NPs in the first study (NP diameter ~2 nm) than in the second study (NP diameter ~5 nm).

Paasonen et al. were the first ones to show triggered release of liposomes containing hexanethiol-stabilized 2.5 nm Au NPs in the lipid bilayer [123]. Triggered cargo release from liposomes containing superparamagnetic NPs in the membrane interior by applying low- [129] and high-frequency AMFs [104,127] was only demonstrated recently. However, the significant passive leakage observed for most of these liposomes reduces their applicability [127,129,130]. Again, it has been shown that the stability of the NP coating is central to maintaining the particle and vesicle structure, which is necessary to suppress passive leakage. For example, no significant leakage was measured for liposomes functionalized with palmityl-nitroDOPA-stabilized NPs [104] in contrast to liposomes containing oleic acid–stabilized NPs in the membrane [104,127,129]. The high stability and precise localization of palmityl-nitroDOPA surface-modified NPs in the membrane made it possible to control cargo release by pulsing the applied AMF.

The use of AMFs thus allows effective actuation of thermoresponsive vesicles through localized heating with NPs internalized in the membrane wall or closely associated with it. There is still some controversy about the exact mechanism responsible for the enhanced permeability of the liposome membrane subject to this direct actuation. Mechanical distortion of the capsule wall or membrane is generally believed to give the main contribution to increased permeability for release triggered by low-frequency AMFs applied to membrane-associated NPs [115,116]. For high-frequency AMFs, the reports claim that local heat generation is the predominant factor leading to increased membrane permeability [96,97,104,127], although it has not been directly demonstrated and it has been argued that the heat diffusion is too fast to explain the observed release [97]. It should be emphasized that the permeability change observed for liposomes containing NPs within the lipid membrane under AMF actuation has been monitored at global temperatures lower than the T_m of the lipid composition [104,127] and with the liposome structure remaining intact through repeated actuations [104]. Thus, the question whether mechanical distortions add to the enhanced release warrants further studies.

For encapsulation and release mechanisms relying on the melting transition of the lipid membrane or distortions to the order of the lipid membrane, the effect on T_m and the lipid ordering of NP incorporation into the membrane is of importance. SAXS analysis of lipid membranes with embedded 2.5 nm hexanethiol-stabilized Au NPs did not reveal any change in the ordering of the lipids in the bilayers [131]. This might indicate that NPs are localized in the middle of the liposome bilayer and therefore do not affect the spacing of the phospholipid head groups. This interpretation is supported by the fact that the T_m of liposome bilayers remained unaltered after incorporating palmityl-nitroDOPA stabilized 5 nm core diameter iron oxide NPs in the lipid membrane interior [104]. An alternative explanation to the absence of changes in the distance of adjacent phospholipid head groups upon loading liposomes with Au NPs would be that the loading of Au NPs in the liposome membrane was too low to detect small changes in the spacing of just a few affected phospholipid head groups around embedded NPs [131]. A similar argument could be made for the absence of a shift in T_m.

8.11 Conclusions

The increasingly demanding and expanding range of requirements imposed on super-paramagnetic iron oxide NPs intended for biomedical applications require close control over the NP size, structure, and surface properties. The key requirement is colloidal stability under physiological conditions that can only be met if iron oxide NPs are sterically stabilized with dispersants that firmly and for practical purposes irreversibly bind to the NP surface. Dispersants consisting of a suitable anchor covalently linked to a spacer have been shown to meet this stringent requirement. This, however, is only possible if iron oxide NPs are stabilized with optimized dispersants that consist of an irreversibly binding anchor covalently linked to a spacer long enough to provide good steric stability but still small enough to allow high dispersant packing density. If these requirements are fulfilled, individually stabilized, core–shell, superparamagnetic iron oxide NPs can be used as highly stable, well-dispersed NPs for a multitude of biomedical applications ranging from imaging to drug delivery and smart materials. End-grafted and irreversibly bound dispersants further allow for the addition at controlled surface densities of multiple different functionalities to the shell of individually stabilized NPs. This modular approach greatly enhances the versatility of applying NPs to the various biomedical applications they are now being designed for, including their use as actuators in drug delivery systems.

Core–shell magnetic NPs can be incorporated into vesicles for externally triggered drug release using AMFs. Control over the release from NP-functionalized vesicles, for example, liposomes incorporating superparamagnetic iron oxide NPs, requires exquisite control over the colloidal stability of the NPs within the vesicular structure. The physicochemical properties of the NP shell determine the NP location within a vesicle. For example, individually stabilized NPs smaller than 6.5 nm in diameter self-assemble into the hydrophobic part of liposome bilayers. Hydrophilic NPs are localized in the vesicle lumen or outside the vesicle. When the NP shell is not well controlled, it might nonspecifically interact with the liposome membrane, which leads to reduced control over structure and actuation. The efficiency of release has been shown to be much higher for vesicular structures where NPs are stably incorporated into the vesicle membrane and actuates the permeability through localized heating, compared to other structures where heating of the entire bulk liquid is required to achieve permeability change and release. Control over both timing and dose has been demonstrated for optimized NP-actuated vesicles.

In summary, recent years have seen tremendous advances in the understanding of the parameters necessary to control for advanced biomedical applications of superparamagnetic iron oxide NPs. These are now being translated into new protocols for synthesis and assembly of such NPs as well as serving to expand the range of possible applications.

References

1. Amstad, E., M. Textor, and E. Reimhult. 2011. Stabilization and functionalization of iron oxide nanoparticles for biomedical applications. *Nanoscale* 3: 2819–2843.
2. Lewin, M. et al. 2000. Tat peptide-derivatized magnetic nanoparticles allow in vivo tracking and recovery of progenitor cells. *Nature Biotechnology* 18: 410–414.
3. Pittet, M. J., F. K. Swirski, F. Reynolds, L. Josephson, and R. Weissleder. 2006. Labeling of immune cells for in vivo imaging using magnetofluorescent nanoparticles. *Nature Protocols* 1: 73–79, doi:10.1038/nprot.2006.11.
4. Wang, D. S., J. B. He, N. Rosenzweig, and Z. Rosenzweig. 2004. Superparamagnetic Fe_2O_3 Beads-CdSe/ZnS quantum dots core-shell nanocomposite particles for cell separation. *Nano Letters* 4: 409–413, doi:10.1021/nl035010n.
5. Halbreich, A. et al. 1998. Biomedical applications of maghemite ferrofluid. *Biochimie* 80: 379–390.
6. Pankhurst, Q. A., N. K. T. Thanh, S. K. Jones, and J. Dobson. 2009. Progress in applications of magnetic nanoparticles in biomedicine. *Journal of Physics D—Applied Physics* 42.
7. Namdeo, M. et al. 2008. Magnetic nanoparticles for drug delivery applications. *Journal of Nanoscience and Nanotechnology* 8: 3247–3271.
8. Weissleder, R. et al. 1987. Mr imaging of splenic metastases—Ferrite-enhanced detection in rats. *American Journal of Roentgenology* 149: 723–726.
9. Weissleder, R. et al. 1990. Ultrasmall superparamagnetic iron-oxide—Characterization of a new class of contrast agents for Mr imaging. *Radiology* 175: 489–493.
10. McCarthy, J. R. and R. Weissleder. 2008. Multifunctional magnetic nanoparticles for targeted imaging and therapy. *Advanced Drug Delivery Reviews* 60: 1241–1251.
11. Lewinski, N., V. Colvin, and R. Drezek. 2008. Cytotoxicity of nanoparticles. *Small* 4: 26–49, doi:10.1002/smll.200700595.
12. Krishnan, K. M. 2010. Biomedical nanomagnetics: A spin through possibilities in imaging, diagnostics, and therapy. *IEEE Transactions on Magnetics* 46: 2523–2558, doi:10.1109/tmag.2010.2046907.
13. Louie, A. 2010. Multimodality imaging probes: Design and challenges. *Chemical Reviews* 110: 3146–3195, doi:10.1021/cr9003538.
14. Wang, Y. X. J., S. M. Hussain, and G. P. Krestin. 2001. Superparamagnetic iron oxide contrast agents: Physicochemical characteristics and applications in MR imaging. *European Radiology* 11: 2319–2331.
15. Hutten, A. et al. 2004. New magnetic nanoparticles for biotechnology. *Journal of Biotechnology* 112: 47–63, doi:10.1016/j.jbiotec.2004.04.019.
16. Duran, J. D. G., J. L. Arias, V. Gallardo, and A. V. Delgado. 2008. Magnetic colloids as drug vehicles. *Journal of Pharmaceutical Sciences* 97: 2948–2983, doi:10.1002/jps.21249.
17. Owens, D. E. and N. A. Peppas. 2006. Opsonization, biodistribution, and pharmacokinetics of polymeric nanoparticles. *International Journal of Pharmaceutics* 307: 93–102, doi:10.1016/j.ijpharm.2005.10.010.
18. Basiruddin, S. K., A. Saha, N. Pradhan, and N. R. Jana. 2010. Advances in coating chemistry in deriving soluble functional nanoparticle. *Journal of Physical Chemistry C* 114: 11009–11017, doi:10.1021/jp100844d.
19. Jeon, S. I., J. H. Lee, J. D. Andrade, and P. G. Degennes. 1991. Protein surface interactions in the presence of polyethylene oxide.1. simplified theory. *Journal of Colloid and Interface Science* 142: 149–158.
20. Feldman, K., G. Hahner, N. D. Spencer, P. Harder, and M. Grunze. 1999. Probing resistance to protein adsorption of oligo(ethylene glycol)-terminated self-assembled monolayers by scanning force microscopy. *Journal of the American Chemical Society* 121: 10134–10141.
21. Wang, R. L. C., H. J. Kreuzer, and M. Grunze. 1997. Molecular conformation and solvation of oligo(ethylene glycol)-terminated self-assembled monolayers and their resistance to protein adsorption. *Journal of Physical Chemistry B* 101: 9767–9773.

22. Bhat, R. and S. N. Timasheff. 1992. Steric exclusion is the principal source of the preferential hydration of proteins in the presence of polyethylene glycols. *Protein Science* 1: 1133–1143.
23. Pasche, S., S. M. De Paul, J. Voros, N. D. Spencer, and M. Textor. 2003. Poly(L-lysine)-graft-poly(ethylene glycol) assembled monolayers on niobium oxide surfaces: A quantitative study of the influence of polymer interfacial architecture on resistance to protein adsorption by ToF-SIMS and in situ OWLS. *Langmuir* 19: 9216–9225.
24. Dalsin, J. L. et al. 2005. Protein resistance of titanium oxide surfaces modified by biologically inspired mPEG-DOPA. *Langmuir* 21: 640–646.
25. Degennes, P. G. 1980. Conformations of polymers attached to an interface. *Macromolecules* 13: 1069–1075.
26. Kenworthy, A. K., K. Hristova, D. Needham, and T. J. McIntosh. 1995. Range and magnitude of the steric pressure between bilayers containing phospholipids with covalently attached poly(ethylene glycol). *Biophysical Journal* 68: 1921–1936.
27. Jung, C. W. and P. Jacobs. 1995. Physical and chemical-properties of superparamagnetic iron-oxide Mr contrast agents—Ferumoxides, ferumoxtran, ferumoxsil. *Magnetic Resonance Imaging* 13: 661–674.
28. Sosnovik, D. E., M. Nahrendorf, and R. Weissleder. 2008. Magnetic nanoparticles for MR imaging: Agents, techniques and cardiovascular applications. *Basic Research in Cardiology* 103: 122–130.
29. Jin, H. Z. and K. A. Kang. 2007. Application of novel metal nanoparticles as optical/thermal agents in optical mammography and hyperthermic treatment for breast cancer. *Oxygen Transport to Tissue Xxviii* 599: 45–52.
30. Laurent, S. et al. 2008. Magnetic iron oxide nanoparticles: Synthesis, stabilization, vectorization, physicochemical characterizations, and biological applications. *Chemical Reviews* 108: 2064–2110, doi:10.1021/cr068445e.
31. Lin, M. M., D. K. Kim, A. J. El Haj, and J. Dobson. 2008. Development of superparamagnetic iron oxide nanoparticles (SPIONS) for translation to clinical applications. *IEEE Transactions on Nanobioscience* 7: 298–305.
32. Cengelli, F. et al. 2006. Interaction of functionalized superparamagnetic iron oxide nanoparticles with brain structures. *Journal of Pharmacology and Experimental Therapeutics* 318: 108–116, doi:10.1124/jpet.106.101915.
33. Pardoe, H., W. Chua-anusorn, T. G. St Pierre, and J. Dobson. 2001. Structural and magnetic properties of nanoscale iron oxide particles synthesized in the presence of dextran or polyvinyl alcohol. *Journal of Magnetism and Magnetic Materials* 225: 41–46.
34. Bautista, M. C. et al. 2004. Comparative study of ferrofluids based on dextran-coated iron oxide and metal nanoparticles for contrast agents in magnetic resonance imaging. *Nanotechnology* 15: S154–S159.
35. Amstad, E. et al. 2009. Surface functionalization of single superparamagnetic iron oxide nanoparticles for targeted magnetic resonance imaging. *Small* 5: 1334–1342.
36. Amstad, E., T. Gillich, I. Bilecka, M. Textor, and E. Reimhult. 2009. Ultrastable iron oxide nanoparticle colloidal suspensions using dispersants with catechol-derived anchor groups. *Nano Letters* 9: 4042–4048, doi:10.1021/nl902212q.
37. Lu, A. H., E. L. Salabas, and F. Schuth. 2007. Magnetic nanoparticles: Synthesis, protection, functionalization, and application. *Angewandte Chemie—International Edition* 46: 1222–1244, doi:10.1002/anie.200602866.
38. Jun, Y. W., J. H. Lee, and J. Cheon. 2008. Chemical design of nanoparticle probes for high-performance magnetic resonance imaging. *Angewandte Chemie-International Edition* 47: 5122–5135, doi:10.1002/anie.200701674.
39. Xiao, Z. P., K. M. Yang, H. Liang, and J. Lu. 2010. Synthesis of magnetic, reactive, and thermoresponsive Fe_3O_4 nanoparticles via surface-initiated RAFT copolymerization of N-isopropylacrylamide and acrolein. *Journal of Polymer Science Part A—Polymer Chemistry* 48: 542–550, doi:10.1002/pola.23752.

40. Somaskandan, K., T. Veres, M. Niewczas, and B. Simard. 2008. Surface protected and modified iron based core-shell nanoparticles for biological applications. *New Journal of Chemistry* 32: 201–209.

41. Thanh, N. T. K. and L. A. W. Green. 2010. Functionalisation of nanoparticles for biomedical applications. *Nano Today* 5: 213–230, doi:10.1016/j.nantod.2010.05.003.

42. Verma, A. and F. Stellacci. 2010. Effect of surface properties on nanoparticle-cell interactions. *Small* 6: 12–21, doi:10.1002/smll.200901158.

43. Fritz, G., V. Schadler, N. Willenbacher, and N. J. Wagner. 2002. Electrosteric stabilization of colloidal dispersions. *Langmuir* 18: 6381–6390, doi:10.1021/la015734j.

44. Gast, A. P. 1996. Structure, interactions, and dynamics in tethered chain systems. *Langmuir* 12: 4060–4067.

45. Du, B. Y. et al. 2009. Poly[N-isopropylacrylamide-co-3-(trimethoxysilyl)-propylmethacrylate] coated aqueous dispersed thermosensitive Fe$_3$O$_4$ nanoparticles. *Journal of Physical Chemistry C* 113: 10090–10096, doi:10.1021/jp9016536.

46. Wang, S. X., Y. Zhou, W. Guan, and B. J. Ding. 2008. One-step copolymerization modified magnetic nanoparticles via surface chain transfer free radical polymerization. *Applied Surface Science* 254: 5170–5174, doi:10.1016/j.apsusc.2008.02.021.

47. Zhao, B. and W. J. Brittain. 2000. Polymer brushes: Surface-immobilized macromolecules. *Progress in Polymer Science* 25: 677–710.

48. Bae, K. H. et al. 2010. Bioinspired synthesis and characterization of gadolinium-labeled, magnetite nanoparticles for dual contrast T-1- and T-2-weighted magnetic resonance imaging. *Bioconjugate Chemistry* 21: 505–512, doi:10.1021/bc900424u.

49. Nagase, K., J. Kobayashi, and T. Okano. 2009. Temperature-responsive intelligent interfaces for biomolecular separation and cell sheet engineering. *Journal of the Royal Society Interface*: S293-S309, doi:10.1098/rsif.2008.0499.focus.

50. Knoll, W. and R. C. Advincula. 2011. *Functional Polymer Films*. Vol. 2 Wiley-VCH, Weinheim, Germany.

51. Jarrett, B. R., B. Gustafsson, D. L. Kukis, and A. Y. Louie. 2008. Synthesis of Cu-64-labeled magnetic nanoparticles for multimodal imaging. *Bioconjugate Chemistry* 19: 1496–1504, doi:10.1021/bc800108v.

52. Sun, S. H. et al. 2004. Monodisperse MFe$_2$O$_4$ (M = Fe, Co, Mn) nanoparticles. *Journal of the American Chemical Society* 126: 273–279.

53. Xie, J. et al. 2006. Linking hydrophilic macromolecules to monodisperse magnetite (Fe$_3$O$_4$) nanoparticles via trichloro-s-triazine. *Chemistry of Materials* 18: 5401–5403.

54. Xie, J. et al. 2008. Ultrasmall c(RGDyK)-coated Fe$_3$O$_4$ nanoparticles and their specific targeting to integrin αvβ3-rich tumor cells. *Journal of the American Chemical Society* 130: 7542–7543.

55. Xu, C. J. et al. 2004. Dopamine as a robust anchor to immobilize functional molecules on the iron oxide shell of magnetic nanoparticles. *Journal of the American Chemical Society* 126: 9938–9939.

56. Gu, H. W., Z. M. Yang, J. H. Gao, C. K. Chang, and B. Xu. 2005. Heterodimers of nanoparticles: Formation at a liquid-liquid interface and particle-specific surface modification by functional molecules. *Journal of the American Chemical Society* 127: 34–35.

57. Amstad, E., L. Isa, and E. Reimhult. 2011. Nitrocatechol dispersants to tailor superparamagnetic Fe$_3$O$_4$ nanoparticles. *CHIMIA* 64: 826–826, doi:10.2533/chimia.2010.826.

58. Isa, L., E. Amstad, M. Textor, and E. Reimhult. 2010. Self-assembly of iron oxide-poly(ethylene glycol) core-shell nanoparticles at liquid-liquid interfaces. *CHIMIA* 64: 145–149, doi:10.2533/chimia.2010.145.

59. Yu, S. and G. M. Chow. 2004. Carboxyl group (-CO$_2$H) functionalized ferrimagnetic iron oxide nanoparticles for potential bio-applications. *Journal of Materials Chemistry* 14: 2781–2786.

60. White, M. A., J. A. Johnson, J. T. Koberstein, and N. J. Turro. 2006. Toward the syntheses of universal ligands for metal oxide surfaces: Controlling surface functionality through click chemistry. *Journal of the American Chemical Society* 128: 11356–11357, doi:10.1021/ja064041s.

61. Song, H. T. et al. 2005. Surface modulation of magnetic nanocrystals in the development of highly efficient magnetic resonance probes for intracellular labeling. *Journal of the American Chemical Society* 127: 9992–9993.

62. Basly, B. et al. 2010. Dendronized iron oxide nanoparticles as contrast agents for MRI. *Chemical Communications* 46: 985–987.
63. Lalatonne, Y. et al. 2008. Bis-phosphonates—Ultra small superparamagnetic iron oxide nanoparticles: A platform towards diagnosis and therapy. *Chemical Communications* 2008: 2553–2555, doi:10.1039/b801911h.
64. Zhou, Y., S. X. Wang, B. J. Ding, and Z. M. Yang. 2008. Modification of magnetite nanoparticles via surface-initiated atom transfer radical polymerization (ATRP). *Chemical Engineering Journal* 138: 578–585, doi:10.1016/j.cej.2007.07.030.
65. Forge, D. et al. 2011. An original route to stabilize and functionalize magnetite nanoparticles for theranosis applications. *Journal of Magnetism and Magnetic Materials* 323: 410–415, doi:10.1016/j.jmmm.2010.09.031.
66. Sun, C. R. et al. 2010. PEG-mediated synthesis of highly dispersive multifunctional superparamagnetic nanoparticles: Their physicochemical properties and function in vivo. *ACS Nano* 4: 2402–2410, doi:10.1021/nn100190v.
67. Veiseh, O. et al. 2009. Inhibition of tumor-cell invasion with chlorotoxin-bound superparamagnetic nanoparticles. *Small* 5: 256–264, doi:10.1002/smll.200800646.
68. Larsen, E. K. U. et al. 2009. Size-dependent accumulation of PEGylated silane-coated magnetic iron oxide nanoparticles in murine tumors. *ACS Nano* 3: 1947–1951, doi:10.1021/nn900330m.
69. Chen, Z. P. et al. 2008. Stability of hydrophilic magnetic nanoparticles under biologically relevant conditions. *Journal of Nanoscience and Nanotechnology* 8: 6260–6265, doi:10.1166/jnn.2008.343.
70. Kim, M., Y. F. Chen, Y. C. Liu, and X. G. Peng. 2005. Super-stable, high-quality Fe_3O_4 dendron-nanocrystals dispersible in both organic and aqueous solutions. *Advanced Materials* 17: 1429, doi:10.1002/adma.200401991.
71. Amstad, E. et al. 2011. The influence of electronegative substituents on the binding affinity of catechol-derived anchors to Fe_3O_4 nanoparticles. *Journal of Physical Chemistry C* 115: 683–691.
72. Amstad, E., H. Fischer, A. U. Gehring, M. Textor, and E. Reimhult. 2011. Magnetic decoupling of surface Fe^{3+} in magnetite nanoparticles upon nitrocatechol anchored dispersant binding. *Chemistry—A European Journal* 17: 7369–7398.
73. Latham, A. H. and M. E. Williams. 2008. Controlling transport and chemical functionality of magnetic nanoparticles. *Accounts of Chemical Research* 41: 411–420.
74. Tsourkas, A. et al. 2005. In vivo imaging of activated endothelium using an anti-VCAM-1 magnetooptical probe. *Bioconjugate Chemistry* 16: 576–581, doi:10.1021/bc050002e.
75. Kelly, K. A. et al. 2005. Detection of vascular adhesion molecule-1 expression using a novel multimodal nanoparticle. *Circulation Research* 96: 327–336.
76. Montet, X., M. Funovics, K. Montet-Abou, R. Weissleder, and L. Josephson. 2006. Multivalent effects of RGD peptides obtained by nanoparticle display. *Journal of Medicinal Chemistry* 49: 6087–6093, doi:10.1021/jm060515m.
77. Martin, A. L. et al. 2010. Synthesis of bombesin-functionalized iron oxide nanoparticles and their specific uptake in prostate cancer cells. *Journal of Nanoparticle Research* 12: 1599–1608, doi:10.1007/s11051–009–9681–3.
78. Yigit, M. V. et al. 2007. Smart "Turn-on" magnetic resonance contrast agents based on aptamer-functionalized superparamagnetic iron oxide nanoparticles. *ChemBioChem* 8: 1675–1678, doi:10.1002/cbic.200700323.
79. Josephson, L., J. M. Perez, and R. Weissleder. 2001. Magnetic nanosensors for the detection of oligonucleotide sequences. *Angewandte Chemie-International Edition* 40: 3204–3206.
80. Cutler, J. I., D. Zheng, X. Y. Xu, D. A. Giljohann, and C. A. Mirkin. 2010. Polyvalent oligonucleotide iron oxide nanoparticle "Click" conjugates. *Nano Letters* 10: 1477–1480, doi:10.1021/nl100477m.
81. Veiseh, O. et al. 2010. Chlorotoxin bound magnetic nanovector tailored for cancer cell targeting, imaging, and siRNA delivery. *Biomaterials* 31: 8032–8042.
82. Cho, E. C., C. Glaus, J. Y. Chen, M. J. Welch, and Y. N. Xia. 2010. Inorganic nanoparticle-based contrast agents for molecular imaging. *Trends in Molecular Medicine* 16: 561–573, doi:10.1016/j.molmed.2010.09.004.

83. Gindy, M. E. and R. K. Prud'homme. 2009. Multifunctional nanoparticles for imaging, delivery and targeting in cancer therapy. *Expert Opinion on Drug Delivery* 6: 865–878, doi:10.1517/17425240902932908.

84. Sun, C., J. S. H. Lee, and M. Q. Zhang. 2008. Magnetic nanoparticles in MR imaging and drug delivery. *Advanced Drug Delivery Reviews* 60: 1252–1265, doi:10.1016/j.addr.2008.03.018.

85. Huang, J. et al. 2011. Effects of nanoparticle size on cellular uptake and liver MRI with polyvinyl-pyrrolidone-coated iron oxide nanoparticles. *Acs Nano* 4: 7151–7160, doi:10.1021/nn101643u.

86. Josephson, L., M. F. Kircher, U. Mahmood, Y. Tang, and R. Weissleder. 2002. Near-infrared fluorescent nanoparticles as combined MR/optical imaging probes. *Bioconjugate Chemistry* 13: 554–560, doi:10.1021/bc015555d.

87. Kim, K. T., S. A. Meeuwissen, R. J. M. Nolte, and J. C. M. van Hest. 2010. Smart nanocontainers and nanoreactors. *Nanoscale* 2: 844–858, doi:10.1039/b9nr00409b.

88. Huynh, N. T., C. Passirani, P. Saulnier, and J. P. Benoit. 2009. Lipid nanocapsules: A new platform for nanomedicine. *International Journal of Pharmaceutics* 379: 201–209, doi:10.1016/j.ijpharm.2009.04.026.

89. Zha, L. S., B. Banik, and F. Alexis. 2011. Stimulus responsive nanogels for drug delivery. *Soft Matter* 7: 5908–5916, doi:10.1039/c0sm01307b.

90. Strong, L. E. and J. L. West. 2011. Thermally responsive polymer-nanoparticle composites for biomedical applications. *Wiley Interdisciplinary Reviews—Nanomedicine and Nanobiotechnology* 3: 307–317, doi:10.1002/wnan.138.

91. Timko, B. P., T. Dvir, and D. S. Kohane. 2010. Remotely triggerable drug delivery systems. *Advanced Materials* 22: 4925–4943, doi:10.1002/adma.201002072.

92. Date, A. A., N. Desai, R. Dixit, and M. Nagarsenker. 2010. Self-nanoemulsifying drug delivery systems: Formulation insights, applications and advances. *Nanomedicine* 5: 1595–1616, doi:10.2217/nnm.10.126.

93. Igarashi, E. 2008. Factors affecting toxicity and efficacy of polymeric nanomedicines. *Toxicology and Applied Pharmacology* 229: 121–134, doi:10.1016/j.taap.2008.02.007.

94. Ruoslahti, E., S. N. Bhatia, and M. J. Sailor. 2010. Targeting of drugs and nanoparticles to tumors. *Journal of Cell Biology* 188: 759–768, doi:10.1083/jcb.200910104.

95. Malam, Y., M. Loizidou, and A. M. Seifalian. 2009. Liposomes and nanoparticles: Nanosized vehicles for drug delivery in cancer. *Trends in Pharmacological Sciences* 30: 592–599, doi:10.1016/j.tips.2009.08.004.

96. Amstad, E. and E. Reimhult. 2012. Nanoparticle actuated hollow drug delivery vehicles. *Nanomedicine* 7: 145–164.

97. Preiss, M. R. and G. D. Bothun. 2011. Stimuli-responsive liposome-nanoparticle assemblies. *Expert Opinion on Drug Delivery* 8: 1025–1040, doi:10.1517/17425247.2011.584868.

98. Shliomis, M. I., A. F. Pshenichnikov, K. I. Morozov, and I. Y. Shurubor. 1990. Magnetic-properties of ferrocolloids. *Journal of Magnetism and Magnetic Materials* 85: 40–46.

99. Tai, L. A. et al. 2009. Thermosensitive liposomes entrapping iron oxide nanoparticles for controllable drug release. *Nanotechnology* 20: 135101–135109.

100. Fattahi, H. et al. 2011. Magnetoliposomes as multimodal contrast agents for molecular imaging and cancer nanotheragnostics. *Nanomedicine* 6: 529–544, doi:10.2217/nnm.11.14.

101. Maherani, B., E. Arab-Tehrany, M. R. Mozafari, C. Gaiani, and M. Linder. 2011. Liposomes: A review of manufacturing techniques and targeting strategies. *Current Nanoscience* 7: 436–452.

102. Sawant, R. R. and V. P. Torchilin. 2010. Liposomes as 'smart' pharmaceutical nanocarriers. *Soft Matter* 6: 4026–4044, doi:10.1039/b923535n.

103. Maruyama, K. 2011. Intracellular targeting delivery of liposomal drugs to solid tumors based on EPR effects. *Advanced Drug Delivery Reviews* 63: 161–169, doi:10.1016/j.addr.2010.09.003.

104. Amstad, E. et al. 2011. Triggered release from liposomes through magnetic actuation of iron oxide nanoparticle containing membranes. *Nano Letters* 11: 1664–1670, doi:10.1021/nl2001499.

105. Papahadjopoulos, D., K. Jacobson, S. Nir, and T. Isac. 1973. Phase-transitions in phospholipid vesicles—Fluorescence polarization and permeability measurements concerning effect of temperature and cholesterol. *Biochimica et Biophysica Acta* 311: 330–348.

106. Inoue, K. 1974. Permeability properties of liposomes prepared from dipalmitoyllecithin, dimyristoyllecithin, egg lecithin, rat-liver lecithin and beef brain sphingomyelin. *Biochimica Et Biophysica Acta* 339: 390–402.
107. Bulte, J. W. M. et al. 1993. Selective MR imaging of labeled human peripheral-blood mononuclear-cells by liposome mediated incorporation of dextran-magnetite particles. *Magnetic Resonance in Medicine* 29: 32–37.
108. Decuyper, M. and M. Joniau. 1988. Magnetoliposomes—Formation and structural characterization. *European Biophysics Journal with Biophysics Letters* 15: 311–319.
109. Menager, C. and V. Cabuil. 1994. Synthesis of magnetic Ddab vesicles. *Colloid and Polymer Science* 272: 1295–1299.
110. Bothun, G. D. and M. R. Preiss. 2011. Bilayer heating in magnetite nanoparticle-liposome dispersions via fluorescence anisotropy. *Journal of Colloid and Interface Science* 357: 70–74, doi:10.1016/j.jcis.2011.01.089.
111. Nappini, S., T. Al Kayal, D. Berti, B. Norden, and P. Baglioni. 2011. Magnetically triggered release from giant unilamellar vesicles: Visualization by means of confocal microscopy. *Journal of Physical Chemistry Letters* 2: 713–718, doi:10.1021/jz2000936.
112. Pradhan, P. et al. 2010. Targeted temperature sensitive magnetic liposomes for thermo-chemotherapy. *Journal of Controlled Release* 142: 108–121, doi:10.1016/j.jconrel.2009.10.002.
113. Mann, S. and J. P. Hannington. 1988. Formation of iron-oxides in unilamellar vesicles. *Journal of Colloid and Interface Science* 122: 326–335.
114. Jogler, C. and D. Schuler.2009.Genomics, Genetics, and cell biology of magnetosome formation. *Annual Review of Microbiology* 63: 501–521.
115. Nappini, S., F. B. Bombelli, M. Bonini, B. Norden, and P. Baglioni. 2010. Magnetoliposomes for controlled drug release in the presence of low-frequency magnetic field. *Soft Matter* 6: 154–162, doi:10.1039/b915651h.
116. Nappini, S. et al. 2011. Controlled drug release under a low frequency magnetic field: Effect of the citrate coating on magnetoliposomes stability. *Soft Matter* 7: 1025–1037, doi:10.1039/c0sm00789g.
117. Mart, R. J., K. P. Liem, and S. J. Webb. 2009. Magnetically-controlled release from hydrogel-supported vesicle assemblies. *Chemical Communications*: 2287–2289, doi:10.1039/b901472a.
118. Qin, G. T. et al. 2011. Partially polymerized liposomes: Stable against leakage yet capable of instantaneous release for remote controlled drug delivery. *Nanotechnology* 22, 155605, doi: 10.1088/0957–4484/22/15/155605.
119. Wu, G. H. et al. 2008. Remotely triggered liposome release by near-infrared light absorption via hollow gold nanoshells. *Journal of the American Chemical Society* 130: 8175, doi:10.1021/ja802656d.
120. Babincova, M., P. Cicmanec, V. Altanerova, C. Altaner, and P. Babinec. 2002. AC-magnetic field controlled drug release from magnetoliposomes: Design of a method for site-specific chemotherapy. *Bioelectrochemistry* 55: 17–19, doi:10.1016/s1567–5394(01)00171–2.
121. Aliouane, N. et al. 2010. Novel polydentate phosphonic acids: Protonation and stability constants of complexes with Fe(III) and Cu(II) in aqueous medium. *Heteroatom Chemistry* 21: 51–62, doi:10.1002/hc.20575.
122. Cintra, E. R. et al. 2009. Nanoparticle agglomerates in magnetoliposomes. *Nanotechnology* 20, 045103, doi: 10.1088/0957–4484/20/4/045103.
123. Paasonen, L. et al. 2007. Gold nanoparticles enable selective light-induced contents release from liposomes. *Journal of Controlled Release* 122: 86–93, doi:10.1016/j.jconrel.2007.06.009.
124. Wi, H. S., K. Lee, and H. K. Pak. 2008. Interfacial energy consideration in the organization of a quantum dot-lipid mixed system. *Journal of Physics-Condensed Matter* 20, doi:10.1088/0953–8984/20/49/494211.
125. Ginzburg, V. V. and S. Balijepailli. 2007. Modeling the thermodynamics of the interaction of nanoparticles with cell membranes. *Nano Letters* 7: 3716–3722, doi:10.1021/nl0720531.
126. Gopalakrishnan, G. et al. 2006. Multifunctional lipid/quantum dot hybrid nanocontainers for controlled targeting of live cells. *Angewandte Chemie—International Edition* 45: 5478–5483.

127. Chen, Y. J., A. Bose, and G. D. Bothun. 2010. Controlled release from bilayer-decorated magn-etoliposomes via electromagnetic heating. *ACS Nano* 4: 3215–3221, doi:10.1021/nn100274v.

128. Rasch, M. R. et al. 2010. Hydrophobic gold nanoparticle self-assembly with phosphatidyl-choline lipid: Membrane-loaded and janus vesicles. *Nano Letters* 10: 3733–3739, doi:10.1021/nl102387n.

129. Nappini, S., M. Bonini, F. Ridi, and P. Baglioni. 2011. Structure and permeability of magnetoli-posomes loaded with hydrophobic magnetic nanoparticles in the presence of a low frequency magnetic field. *Soft Matter* 7: 4801–4811, doi:10.1039/c0sm01264e.

130. An, X. Q., F. Zhang, Y. Y. Zhu, and W. G. Shen. 2010. Photoinduced drug release from ther-mosensitive AuNPs-liposome using a AuNPs-switch. *Chemical Communications* 46: 7202–7204, doi:10.1039/c0cc03142a.

131. Paasonen, L. et al. 2010. Gold-embedded photosensitive liposomes for drug delivery: Triggering mechanism and intracellular release. *Journal of Controlled Release* 147: 136–143, doi:10.1016/j.jconrel.2010.07.095.

9

Natural and Synthetic Nanoporous Membranes for Cell Encapsulation Therapy

Sachin Kadam and David H. Gracias

CONTENTS

9.1 Introduction

A number of diseases are caused when cells are damaged or destroyed either due to trauma or a variety of genetic or immune disorders. This damage results in the lack of the body's ability to produce critical molecules such as hormones, enzymes, or factors. A classic example is type 1 diabetes [1] that is a multifactorial disease wherein pancreatic β-cells are typically attacked by the immune system, leading to their destruction. Pancreatic β-cells are responsible for the production of insulin, a hormone that regulates glucose levels in the body. When there is significant destruction of β-cells, patients are at risk for hyperglycemia (high blood sugar) and eventual diabetic ketoacidosis, which can lead to severe illness or death. Similarly, several neurodegenerative diseases such as Parkinson's disease or amyotrophic lateral sclerosis (ALS) are caused by damage to nerve cells. In Parkinson's disease, for example, there is considerable loss of dopaminergic neurons in the substantia nigra leading to a dopamine-deficient state [2,3]. Since these diseases are associated with the destruction of only specific cell types, the idea of replacing the defective or damaged cells with transplanted cells has been around for over 50 years. It has been argued that cells could be derived, multiplied, and transplanted from a variety of sources including the same individual (autotransplantation), a non-genetically identical individual of the same species (allotransplantation), or from other species (xenotransplantation). Although, in principle, autologous cells do not elicit an immune response, they are often compromised due to congenital factors, their retrieval requires additional surgery, and there is a challenge with the time required expanding these cells [4]. Hence, there is a demonstrated need for transplantation of allogenic or xenogenic cells. Xenotransplantation offers the possibility to overcome the acute shortage of human donors and could provide an effective therapy for several diseases (Figure 9.1). For example, animal insulin is fully active in humans and has been used for over 80 years. Pigs, for example, have glucose levels that are similar to humans, and porcine insulin is effective in humans [5]. Recent reports suggest that the transplantation of porcine fetal neuronal cells has resulted in some positive outcomes for patients with Parkinson's or Huntington's disease [6]. One major concern of transplantation of xenotransplantation is the cross-species transmission of infectious agents such as the porcine endogenous retrovirus (PERV) [7], but the magnitude and manageability of the risk are active areas of debate [8,9].

In contrast to transplanting free cells, encapsulation of transplanted cells has been proposed as a means of protecting the cells from the external immune response while at the same time allowing the bidirectional diffusion of nutrients toward and therapeutic biochemicals away from the encapsulated cells to the external environment (Figure 9.2) [10–13]. Cell encapsulation therapy (CET) poses major materials and engineering challenges since the encapsulant needs to be biocompatible, enable long-term encapsulated cell viability, be mechanically and chemically stable, adequately protect transplanted cells from immune and foreign cell attack, and provide efficient mass transport of small molecules. Nanotechnological approaches can enable solutions to these challenges since thin yet mechanically sturdy membranes can enable short diffusion paths, thereby enhancing encapsulated cellular response times. Further, nanoporosity of these membranes can facilitate size-exclusion-based immunoisolation. Moreover, nanomaterials can be engineered with tailored surface properties, which could provide anti-fouling and protein-resistant surfaces, thereby minimizing foreign cell responses and fibrosis. It is noteworthy that nanoporous membranes must encapsulate a finite volume of cells, so they need to be

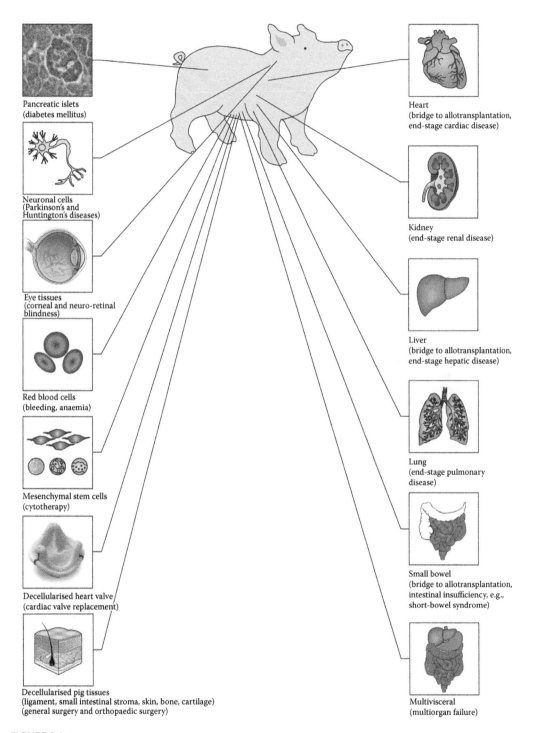

Pancreatic islets
(diabetes mellitus)

Neuronal cells
(Parkinson's and
Huntington's diseases)

Eye tissues
(corneal and neuro-retinal
blindness)

Red blood cells
(bleeding, anaemia)

Mesenchymal stem cells
(cytotherapy)

Decellularised heart valve
(cardiac valve replacement)

Decellularised pig tissues
(ligament, small intestinal stroma, skin, bone, cartilage)
(general surgery and orthopaedic surgery)

Heart
(bridge to allotransplantation,
end-stage cardiac disease)

Kidney
(end-stage renal disease)

Liver
(bridge to allotransplantation,
end-stage hepatic disease)

Lung
(end-stage pulmonary
disease)

Small bowel
(bridge to allotransplantation,
intestinal insufficiency, e.g.,
short-bowel syndrome)

Multivisceral
(multiorgan failure)

FIGURE 9.1
(See color insert.) Disorders for which xenotransplantation is a potential therapy. (Reprinted from *The Lancet*, 379, Ekser, B. et al., Clinical xenotransplantation: The next medical revolution? 672–683, Copyright 2012, with permission from Elsevier.)

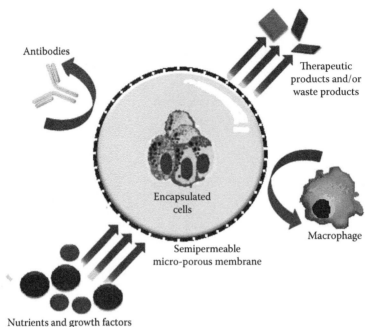

Antibodies

Therapeutic
products and/or
waste products

Encapsulated
cells

Macrophage

Semipermeable
micro-porous membrane

Nutrients and growth factors

FIGURE 9.2
(See color insert.) Conceptual representation of the encapsulation of cells within nanoporous membranes. The
nanoporous membranes permit selective diffusion of nutrients toward cells and of therapeutic and waste products away from the encapsulated cells and to the outer side of membrane. At the same time, the small size of
pores in the membrane prevents passage of immune components such as macrophages and antibodies.

shaped in capsular geometries, which open up a variety of synthesis, fabrication, and processing challenges. In this chapter, we focus on nanomaterials that have been utilized for
CET and discuss advantages and disadvantages of both natural and synthetic nanoporous
membranes.

9.2 Requirements of Nanoporous Membranes for CET Applications

There are a number of structural and chemical requirements of nanoporous membranes
for CET applications, and these are discussed in the following text.

9.2.1 Pore Characteristics

The premise for CET is that the encapsulant should function like a semipermeable membrane and effectively block off any harmful immune components from entering the encapsulant and damaging the transplanted cells while enabling bidirectional flow of nutrients
and therapeutic molecules. This is by no means a trivial task since the immune system has
a large number of cellular and molecular components with a wide range of sizes. A variety
of parameters that need to be optimized in the ideal membrane encapsulant are discussed
in the following text.

9.2.1.1 Pore Size

The challenge in the choice of pore size is due to the fact that relevant CET molecules have a wide range of molecular weights. The feasibility of utilizing nanoporous membranes for CET stems from the fact that nutrients and therapeutic molecules are smaller than most biomolecular and cellular components of the immune system. For example, therapeutic molecules such as dopamine and the insulin have molecular weights of approximately 153 Da and 5.8 kDa, respectively, while the size of antibody components of the immune system are; IgG = 150 kDa, IgA = 160 kDa, IgD = 185 kDa, IgE = 200 kDa, and IgM = 900 kDa. Cellular immune components such as macrophages or neutrophils are significantly larger. The typical size-exclusion pore sizes in commonly utilized alginate capsules have cutoff sizes of 100 kDa as measured by the dextran viscosity radius [14,15], and pore sizes in the range of 15–80 nm are relevant for CET applications. It is noteworthy that apart from antibodies, the immune system also features smaller immunomodulating and signaling molecules, smaller cytokine molecules such as TNF-α (25 kDa) and interleukins that could have deleterious effects on encapsulated cells. However, the importance of smaller immune molecules such as cytokines in determining the immune response to xenogenic encapsulated cells remains unclear. For example, it has been observed that while human cytokines significantly decrease viability of human islets, they do not significantly impact viability of animal-derived islets [16]. This observation has been attributed to the fact that xenogenic islet cells have a lower affinity to bind to human cytokines due to receptor variations on the cell surface. And even when cytokine-mediated damage to encapsulated islet cells was observed, it was greatly reduced in low-permeability membranes (while maintaining adequate insulin) [17] and medium-sized (400–500 μm) capsules [18]. This reduction may be due to the fact that many of these small immune mediator molecule species are short lived and consequently effective only over short distances. Hence, the ability to delay the entry of these components by manipulating the pore geometry and depth can greatly minimize their effect. Alternatively, surface modification with molecular cytokine traps [19,20] could also capture these smaller immune components.

9.2.1.2 Pore Density

High pore densities are required to enhance mass transport by increasing the effective area over which molecules can diffuse. It is well known that the insulin secretion characteristics of β-cells are strongly correlated to their metabolic activity and also that the metabolic demand of islet cells is high, about million-fold greater than chondrocytes, for example [21–23]. It is also known that a human pancreas is composed of 10^6 islets or approximately 10^9 β-cells. Hence, any cell encapsulation diabetes therapy needs to accommodate large numbers of cells while providing adequate nutrients to them. This is a major challenge given that the native tissues are vascularized whereas these cell encapsulants are not, and it is known that cells in the body are no further than 150–200 μm from a capillary [24]. It is also not possible to utilize a nanoporous encapsulant and enable vascularization within its interior since any vascular conduits would be larger than the small size of the pores.

9.2.1.3 Pore Fouling

When nanoporous membrane devices are implanted, they will end up being coated with a variety of biomolecules including proteins, which can block the pores. Hence, the physical and chemical characteristics of the surface of the membrane are critical to reduce fouling that leads to immune reactions, necrosis of transplanted cell due to inadequate diffusion of nutrients, and device failure. Physical characteristics include roughness, morphology, defects, and geometry of the pores and the membrane. For example, it has been suggested that tortuous porosity is more prone to biofouling (and consequently blockage) as compared to straight pores [25].

9.2.2 Biocompatibility and Surface Properties

The material composition of a cell encapsulant is important since it is expected to reside *in vivo* over extended periods of time. And while the immunogenicity of the encapsulant can be modulated to some extent by surface coating, it is important that encapsulants are made using bioinert and nontoxic materials. While gels such as alginate are widely used for CET, it can be challenging to purify alginate to be endotoxin-free, and impure alginate is known to increase islet necrosis, splenocyte proliferation, and TNF-α secretion resulting in increased alginate immunogenicity and decreased encapsulated islet viability [26]. Hence, the source of the material as well as all contaminants such as organic solvents that are sometimes needed to aid processing needs to be carefully evaluated for toxicity. To improve biocompatibility and foreign cell response, capsules can be modified by a number of surface coatings that have been developed to resist protein adhesion, which is generally considered to be the first step in foreign cell adhesion [27,28].

9.2.3 Capsule Size and Geometry

It is important that the nanoporous membranes are structured in 3D geometries so that they can encapsulate cells. Typical capsule sizes are 500 μm in diameter ensuring that encapsulated cells are no further than 250 μm from the external environment. When CET devices are utilized with flow (such as in a shunt), the size can be larger due to enhanced convective mass transport. In addition to size, the pore distribution is important. For example, Kalinin et al. observed that cells encapsulated in cubic capsules with pores arranged on only one surface showed larger hypoxic or necrotic regions (Figure 9.3) as compared to those with 3D porosity [29]. They experimentally confirmed that cubic capsules with 3D porosity showed an increased insulin response and cells transplanted in these capsules were viable for longer periods. Further, the density of encapsulated cells within each capsule is also important to ensure adequate cell viability as can be seen from the graphs in Figure 9.3.

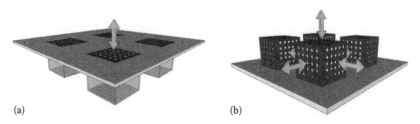

(a)   (b)

FIGURE 9.3
Importance of membrane pore placement and density on encapsulated cell viability. (a,b) Conceptual schematic of microwell cell encapsulant arrays featuring membrane surfaces and porosity in either (a) one face, 2D, or (b) five porous faces, 3D.

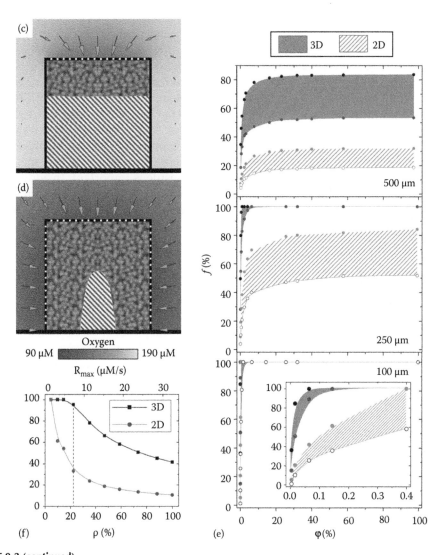

FIGURE 9.3 (continued)

Importance of membrane pore placement and density on encapsulated cell viability. (c–f) Numerical simulations and comparison of cell viability in a single 2D versus 3D porous microwell with cylindrical geometry. (c,d) Spatial variation of viable (speckles) and necrotic (lines) cells within a microwell with (c) one porous face and (d) a microwell with porosity on all faces except the one at the bottom. The O_2 concentration outside the microwell is color coded as per the legend in the figure, and the arrows represent the diffusive flux of O_2 in the medium surrounding the microwell. (e) Plots of the fraction f of the volume of the microwell where the O_2 concentration is larger than the threshold concentration c_{cr} (0.1 µM) required for viable β-TC-6 cells versus the porosity of each face. The three panels correspond to cylindrical microwells with heights (or diameters) of 500, 250, and 100 µm. The regions shown in solid (3D microwell) and hatched (2D microwell) are bounded by low and high literature values of consumption rates. (From Rydgren, T. et al., *Diabetes*, 55, 1407, 2006; Malaisse, W.J., *Acta Diabetologica*, 33, 173, 1996.) (f) Plot of f versus both the cell density and O_2 consumption rate within a microwell with a height (or diameter) of 500 µm and wall porosity φ = 2.3%. The intersection of the dotted line in the plot corresponds to the parameters used in determining the spatial variation of viable cells shown in panels (c) and (d). (From Randall, C.L., Kalinin, Y.V., Jamal, M., Manohar, T., and Gracias, D.H., Three-dimensional microwell arrays for cell culture, *Lab on a Chip*, 11, 127–131, 2011. Reprinted with permission of The Royal Society of Chemistry.)

9.2.4 Mechanical and Chemical Stability

Long-term stability of the capsules is important to limit wear or premature device rupture. If the membrane ruptures, it exposes the encapsulated cells to the host immune system, which can result in an immune attack of the unprotected, transplanted cells or worse still a systemic immune response. A straightforward means to increase mechanical stability is to increase the capsular wall thickness, but this is not an ideal solution since thin walls result in a more rapid encapsulated cellular response and enhanced mass transport. In addition to mechanical strength, the encapsulant is exposed to fluids that contain a number of enzymes that can degrade a number of biological materials; hence, the chemical composition of the material used to structure the capsule is important. In addition to the material composition, the technique or process by which the capsules are formed is also critical since any defects formed within the capsule during synthesis or processing can lead to catastrophic failure.

9.3 Natural Nanoporous Membranes

Naturally occurring materials such as alginate, chitosan, and collagen are well studied and are often used for cell encapsulation. One possible reason behind this is that they are composed of similar classes of biochemicals (e.g., polypeptides and polysaccharides) that are also present in human tissues. Moreover, these materials feature porosity, which can be tuned to some extent by processing or cross-linking. However, as these materials are often extracted from algal or animal tissues, there is considerable batch variation during their production and purification. Also, due to the similarity in the constituent molecular blocks of these materials as compared to human tissues, they can be immunogenic. In spite of these drawbacks, natural gelatinous polymers are still very popular, and many researchers are working on different encapsulation methods using either pure natural materials or in combination with other synthetic polymers.

9.3.1 Natural Materials

The most commonly utilized natural materials for CET are briefly discussed as follows. The reader is cautioned that the list is by no means exhaustive and numerous combinations of gels and biopolymers have also been utilized for cell encapsulation.

9.3.1.1 Alginate

Alginate is a polysaccharide-based polymer derived from bacterial cell walls, plants, algae, and seaweed and is the most commonly used material for cell encapsulation [30]. An aqueous solution of alginate becomes a gel upon cross-linking with divalent cations like Ca^{2+} and Ba^{2+}. Gelation occurs due to the binding of G block of alginate chains with Ca^{2+} ions that forms interchain ionic bridges [31]. The extent of cross-linking and manipulation of porosity of alginate matrix with respect to its composition mannuronic acid/guluronic acid units and divalent cation concentration has been well studied [32]. Alginate capsules are typically coated with poly-L-lysine (PLL) [33]. PLL–alginate capsules have reduced pore sizes improving immunoprotection, and the polycationic–polyanionic interaction stabilizes the capsule against osmotic swelling and consequently rupture. However, PLL is cell adherent, and despite efforts to add another coat of alginate on the capsules, PLL–alginate capsules have been observed to enhance host cell interaction resulting in compromised cell encapsulation [34–37].

9.3.1.2 Chitosan

Chitosan is a cationic linear polysaccharide polymer obtained from chitin, a major unit of the crustacean shell, fungal cell walls, and insect cuticles. Chitin is made up of copolymers of D-glucosamine and N-acetyl-D-glucosamine and is widely utilized in tissue engineering since it is biocompatible [38], nonantigenic, and nontoxic [39]. Despite its attractive biofunctionality, chitosan is susceptible to enzymatic degradation [40] and it is biodegradable, and hence, it is more relevant as a matrix material within cell encapsulation devices. However, due to the high mechanical strength of chitosan, complexes of chitosan with other materials such as alginate are being explored for CET [41].

9.3.1.3 Chondroitin Sulfate

Chondroitin sulfate is an important structural component of cartilage and lubricating fluids of joints in body. It is a sulfated glycosaminoglycan (GAG) consisting of repeating units of N-acetylgalactosamine and glucuronic acid [31]. These polysaccharides are often covalently linked to proteins to form proteoglycans [42]. Since chondroitin sulfate is readily water soluble, its application in CET requires cross-linking with other materials such as chitosan [43], hyaluronan [44], and collagen [45] where it is often used to regulate pore structure and porosity.

9.3.1.4 Collagen

Collagen is an extracellular matrix protein that gives support to fibrous connective tissues such as skin, cartilage, and bone [46–48]. Collagen contains interactive cellular peptides, which are important for cell adhesion giving rise to superior cell viability and biocompatibility. However, this major advantage can be detrimental when one is trying to limit foreign cell adhesion of a cell encapsulant. Of the 27 identified collagen types, type 1 collagen is widely used in major tissue engineering constructs. Positively charged collagen has also been coacervated with negatively charged polymer to form microcapsules that can be formed without exposing the encapsulated cells to organic solvents, and these are being explored for CET applications [49].

9.3.1.5 Fibrin

Fibrin is a biocompatible fibrinogen-derived polymer that has the ability to improve cellular adhesion due to the presence of RGD motifs that are present within its network; hence, fibrin is used in many tissue engineering applications, including cell delivery. The gelation of fibrin occurs during enzymatic degradation of fibrinogen by thrombin followed by polymerization. After implantation, proteolytic enzymes secreted by mammalian cells rapidly degrade fibrin; hence, stabilizing additives such as aprotinin and factor XIII are required to improve structural characteristics over extended periods [50]. Studies have also showed that such composite fibrin scaffolds and cell encapsulants facilitate higher cell viability and longevity [51].

9.3.1.6 Hyaluronic Acid

Hyaluronic acid (HA) is a naturally occurring non-sulfated GAG found in connective tissues. HA is a linear polysaccharide composed of α-1,4-D-glucuronic acid and β-1,3-N-acetyl-D-glucosamine [52]. HA is typically present as its polyanion called hyaluronan under *in vivo* conditions and has many attractive properties for CET including it being biocompatible, non-immunogenic, and nonadhesive [53,54]. However, HA is enzymatically degraded and typically would need to be cross-linked or combined with other polymers to be stable over long time scales [55].

9.3.2 Methods for Cell Encapsulation with Natural Materials

The naturally occurring biomaterials discussed previously are nanoporous as a consequence of their native molecular structure or intentional cross-linking. Cells can be encapsulated within these membranous materials in a variety of architectures, and we summarize the important methods in the following text.

9.3.2.1 Encapsulation in Capsules

Cell encapsulation in spherical capsules involves the inclusion of cells within a thin shell with a preferred diameter of 0.3–1.5 mm, typically 0.5 mm [56]. Although gelatin coacervation was introduced as a form of microencapsulation by the colloidal chemist Barrett Green in 1930, it was Ennis and James who developed the "droplet sizer," a device used to obtain single uniformly sized droplets [57]. Chang was the first to propose the microencapsulation of biological reagents in the 1960s [58,59]. Microencapsulation of cells in spherical capsules is typically achieved by first mixing cells with the liquid polymer (e.g., alginate) and extruding the mixture through a needle or catheter into a beaker containing a cross-linking agent (e.g., $CaCl_2$) with stirring (Figure 9.4). After capsule formation and cell encapsulation, additional coating steps may be applied by successive washes to improve capsule biocompatibility. An accurate control over the capsule size, pore size, pore density, and wall thickness requires careful control over the relative concentrations and processing parameters, and this can be challenging. Further, it is well documented that the islet cell number variation in capsules can be large with many capsules being empty [60]. More recently, microfluidic approaches are being applied to overcome the aforementioned challenges [61,62]. Some of these methods even provide encapsulation of single cells, which could also be useful in cell screening and single-cell analysis in small volumes.

9.3.2.2 Encapsulation in Fibers

The use of hollow fiber membranes (HFMs) for cell encapsulation goes back to the 1970s, and a variety of materials such as Millipore membranes [63], XM-50 [64], and AN-69 [65] hollow fibers have been utilized (Figure 9.5). Typically, HFMs are prepared by extruding a polymer solution or melt through a spinneret [66]. The polymer is injected

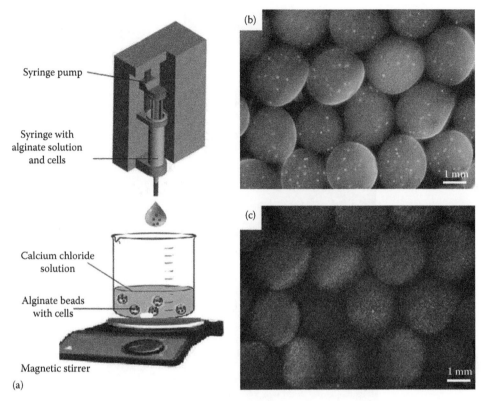

FIGURE 9.4
(See color insert.) Encapsulation of β-TC-6 cells in sodium alginate–based capsules. (a) Schematic representation of a typical setup required for encapsulating cells in alginate capsules. (b) Bright-field image of encapsulated β-TC-6 cells within alginate capsules. (c) Fluorescence image of Calcein AM–stained β-TC-6 encapsulated within alginate capsules. Scale bar 1 mm.

through a programmable infusion pump, while saline is flowed through the co-extrusion needle. HFMs can be prepared over a wide range of diameters (50 μm–3 mm) and exhibit cross-sectional thicknesses between 100 and 400 μm. Hollow fibers having diameters ranging from 50 to 200 μm are usually called *fine hollow fibers* and can withstand high hydrostatic pressures (up to 1000 psi or more). HFMs are packed into bundles and potted into tubes to form hollow fiber modules or cartridges. The key features and advantages of HFMs over other membrane modules are its packing density (surface area per module volume), which can be as high as 10,000 m²/m³. They are also relatively inexpensive and compact. Owing to these advantages, hollow fiber–based modules have led to high commercial interest and are being widely used in a range of diverse applications such as biomedical devices (hemodialyzers, artificial lungs, bioartificial livers, and pancreas), bioseparations (removal of viruses from blood), water purification (reverse osmosis), ultrafiltration, microfiltration, membrane bioreactors, liquid separations (pervaporation), gas separations (purification of natural gas, oxygen enrichment), and membrane contactors (alternatives to gas absorbers and liquid/liquid extraction columns) [67–69]. Recently, a technique known as cell electrospinning has been developed that enables the spinning of live cells encapsulated within polymeric fibers [70].

FIGURE 9.5
Cell encapsulation in HFMs. (a) Macroscopic view of a hollow fiber. (b) Fibers filled with cells. (c) Peritoneal implantation. (From Bessis, N. et al.: Encapsulation in hollow fibers of xenogeneic cells engineered to secrete IL-4 or IL-13 ameliorates murine collagen-induced arthritis (CIA). *Clin. Exp. Immunol.* 1999. 117. 376–382. Copyright Wiley-VCH Verlag GmbH & Co. KGaA. Reprinted with permission.)

9.3.2.3 Encapsulation in Polymer Sheets

Encapsulation of cells in alginate sheets has been utilized to create the so-called islet sheets. These sheets seek to overcome some of the limitations of microcapsule encapsulation by allowing better processing, uniform barriers, higher volume fractions of loaded islets, and the possibility for intact retrieval, if required [71]. Sheets can be created by a variety of means such as casting from solutions or more recently by electrospinning mats. Electrospinning involves the use of electrical fields to draw out fibers from a thin nozzle and can be used to prepare fibrous sheets from natural and/or synthetic polymers, ceramic, and their blends. The sheets generated by this method can be used for the fabrication of highly porous nanofibrous scaffolds, which resemble the extracellular matrix of host tissue. Electrospinning allows the generation of scaffolds with nanosized features, large surface area, and highly porous interconnected network. When transplanted in tissue engineering techniques, these features facilitate cell adhesion, proliferation, migration, and efficient nutrient flux at the implant site. Interconnected pores enhance the mineral deposition rate as well as the depth of infiltration and provide space for the development of new blood vessels [72].

9.3.2.4 Encapsulation in Single Tubes

Polymer tube encapsulation can be categorized as a macroencapsulation technique and is typically achieved by molding (Figure 9.6). The size of the tube can vary from 10 mm to a few centimeters. The curing of the polymer is achieved either by controlling temperature [73] or by treatment with cross-linkers or enzymes [74]. A typical mold could consist of two concentric glass tubes, as has been used to mold a fibrin tube (Figure 9.6a and b).

Tubes can also be formed by simply rolling up a sheet by hand as we have demonstrated in our own laboratory with nanoporous polycarbonate membranes. Polycarbonate membranes have been used as a 3D matrix for cell culture for a long time, and its biocompatibility has been evaluated [75,76]. We rolled the polycarbonate membranes with the help of a glass capillary to form the tubes, and we then sealed one end while we loaded live cells through the other end (Figure 9.6c). We measured cell viability up to 21 days after encapsulation and the cells were viable (Figure 9.6d). The attractive feature of this approach is that polycarbonate membranes are commercially available with a range of pore sizes in the range of 15–100 nm. One disadvantage, however, is that although the entire dimensions of the tube can be well controlled, the large millimeter-sized diameter of the tubes makes encapsulated cells susceptible to necrosis due to hypoxic cores.

9.4 Synthetic Nanoporous Membranes

Polycarbonate is a synthetic material, and the formation of tubes with nanoporous membranes of this material highlights the value in utilizing materials in which nanoporosity is engineered by a physical or chemical process. For example, pores in polycarbonate membranes are made by track etching, which is a process of etching that is achieved during irradiation of thin polymeric films by energetic heavy ions (e.g., Ar^{9+}, 220 MeV) [77,78]. The highlight of this approach is that membrane thickness as well as pore size can be readily tuned, offering considerable precision and versatility to the end user. One limitation,

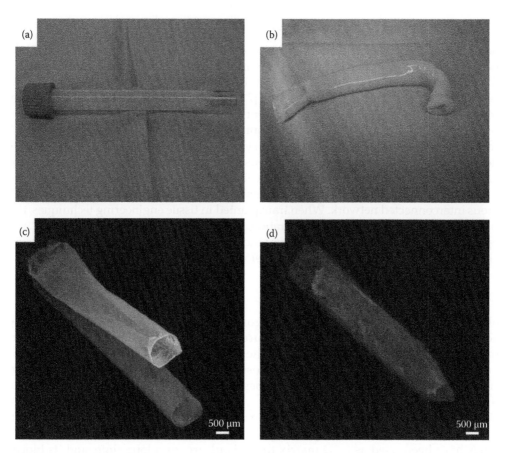

FIGURE 9.6
(See color insert.) Cell encapsulation in tubes. (a,b) Fibrin tubes and (a) image of a custom-built casting mold used to mold a fibrin tube. The mold consisted of two 15 cm long glass tubes, which were closed and fixed at one end. The internal tube had an outside diameter of 5 mm, and the inside diameter of the external tube was 11 mm with a 3 mm broad interspace among both tubes, in which the fibrin preparation was filled. Due to shrinkage during the hardening of the fibrin, preparation segments with a wall thickness of 1.5–2.5 mm were obtained. (b) Image of a segment after 10 days of incubation *in vitro* before placing it in the bioreactor. (Reprinted from *Eur. J. Vasc. Endovasc. Surg.*, 33, Aper, T., Schmidt, A., Duchrow, M., and Bruch, H.P., Autologous blood vessels engineered from peripheral blood sample, 33–39, Copyright 2007, with permission from Elsevier.) (c,d) Manually rolled up nanoporous membranes. (c) Nanoporous (0.03 μm pore size) polycarbonate membrane rolled into tubelike structures with an inner diameter of 1 mm and sealed at one end. (d) Image of green fluorescence indicating viability of β-TC-6 cells with 0.5% agar that were packed in these tubes and stained with Calcein AM after 7 days.

however, is that ion etch tracks are randomly distributed in the membrane and it can be challenging to control their ordering, spacing, and density.

9.4.1 Lithographically Structured Nanoporous Membranes

In the late 1990s, silicon (Si)-based micro- and nanofabrication was suggested as an alternative route for engineering robust, reproducible, and nanoporous biocapsules for CET [79]. The motivation was primarily to overcome the limitations of gel-based materials notably limited precision in terms of capsule size and shape, geometry, and arrangement of pores. It was argued that Si-based micro- and nanofabrication had high reproducibility in terms of engineering systems with identical sizes and shapes, as is evident in the fabrication of integrated circuits. Si-based devices also have excellent chemical and mechanical stability. Since then, alternate material compositions have been used, but inherently similar geometric structures have been demonstrated [80]. They all feature a capsule, which is made of a macroscale or microfabricated recess that is covered with a lithographically patterned nanoporous membrane. While it was shown that these devices can feature significant advantages as compared to gel-based encapsulants, the critical limitation of this approach is that since lithographically fabricated, nanoporous membranes are typically attached to only one or at most two faces. Hence, the diffusion is very limited resulting in hypoxic zones, which becomes more pronounced for larger devices. This diffusion limitation is compounded by the fact that the membranes are attached serially, sometimes by hand, and as a result cannot be easily scaled up. Consequently, the researchers have often utilized large centimeter scale capsules, which suffer from the presence of hypoxic zones as discussed earlier.

9.4.2 Self-Assembled Nanoporous Membranes

Self-assembly can be used to transform the precision of lithography to the third dimension and has been used to create polyhedral capsules in a variety of sizes and shapes and with precisely patterned porosity (Figure 9.7) [81,82]. The self-assembly mechanisms allow one to spontaneously roll up or fold a 2D template by utilizing forces derived from differential stress or surface tension. So, for the first time now, it is possible to precisely structure hollow capsules, in a highly parallel manner, using the precision of planar lithographic methods. The method can be utilized with a variety of materials including metals, silicon, and polymers, with sizes ranging from 100 nm to 1 cm and with precise wall porosity with a resolution of 15 nm. Due to high precision of the fabrication and assembly methodology, polyhedral and tubular capsules can be first designed using computer-aided design (CAD), which allows one to optimize mass transport as well as immunoisolation. Further, the capsules can be made with inert coatings such as gold that can be further functionalized to resist foreign cell adhesion. Surface coatings are important for a number of reasons as is evident in the PLL–alginate system described previously. Unlike materials like alginate that are extracted from algal sources and are challenging to purify, materials like gold, silicon, or synthetic polymers are available with high purity and well-defined surfaces and are amenable to cell nonadhesive polyethylene glycol (PEG) surface coatings using methods such as atom transfer radical polymerization (ATRP) [83]. The highlight of the self-assembly process for capsule formation with lithographically patterned nanoporous membranes is the high precision in terms of precise pore size, shape, and placement in all three dimensions. Randall et al. demonstrated *in vitro* feasibility of this approach by encapsulating islet cells within cubic self-assembled capsules with a variable range of pore sizes from 78 nm to 2 μm (Figure 9.8). It is noteworthy that the pore characterization was done on the external surface and the pores are probably smaller along their length. It was observed that the self-assembled capsules with the smallest pore sizes could successfully limit diffusion of antibodies such as IgG while permitting diffusion of insulin [84].

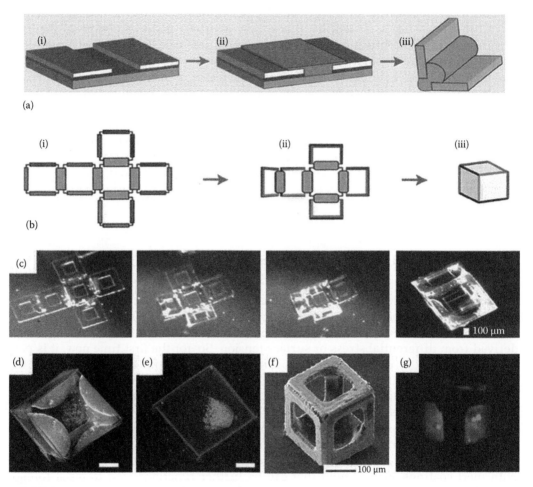

FIGURE 9.7

(See color insert.) Self-folding process for creating 3D capsules. (a) Schematic representation of the fabrication process for self-folding of polymeric containers. (i) Si wafer containing spin-coated sacrificial and photolithographically patterned SU-8 panels. (ii) PCL deposition in hinge gaps. (iii) 2D templates were lifted off via dissolution of the PVA layer in water and self-assembly occurred on heating above 58°C. (b, i–iii) Schematic demonstrating self-folding of a cubic container. External "locking" hinges are colored in pairs to denote corresponding meeting edges. (c) Video capture sequence (over 15 s) showing a 1 mm sized, six-windowed polymeric container self-folding at 60°C. (d) Bright-field and (e) fluorescence z-plane stack image of stained fibroblast cells encapsulated within a nonporous polymeric container. Green fluorescence indicates live cells. (Reprinted from *Biomed. Microdev.*, 13, Azam, A., Laflin, K.E., Jamal, M., Fernandes, R., and Gracias, D.H., Self-folding micropatterned polymeric containers, 51–58, Copyright 2011, with permission from Elsevier.) (f) Scanning electron microscopy image of a hollow, open-surfaced biocontainer and (g) fluorescence microscopy images of a metallic biocontainer loaded with cell–ECM–agarose with the cell viability stain, Calcein AM. (Reprinted from *Biomed. Microdev.*, 7, Gimi, B. et al., Self-assembled three-dimensional radio frequency (RF) shielded containers for cell encapsulation, 341–345, Copyright 2005, with permission from Elsevier.)

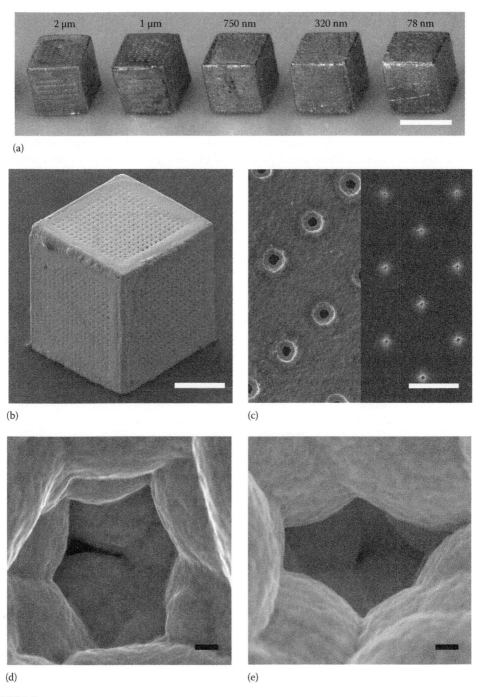

FIGURE 9.8
Self-assembling synthetic nanoporous capsules. (a) Bright-field image of Au-coated 500 μm sized containers with the decreasing pore sizes indicated. (b–e) SEM images of the porous containers. (b) SEM of a container illustrating 3D porosity. (c) SEM of pores on the face of containers of different pore diameters. (d,e) Zoomed-in SEM focusing on a single 320 nm (d) and 78 nm (e) pore. Scale bars: (a) = 500 μm, (b) = 200 μm, (c) = 10 μm, and (d,e) = 200 nm. (Reprinted from *Nanomedicine*, 7, Randall, C.L., Kalinin, Y.V., Jamal, M., Shah, A., and Gracias, D.H., Self-folding immunoprotective cell encapsulation devices, 686–689, Copyright 2011, with permission from Elsevier.)

9.5 Summary

The list of materials utilized for CET as well as the pros and cons are listed in Tables 9.1 and 9.2. While much has been achieved, the promise of widespread applicability of CET still remains a major challenge mainly due to the need to achieve high biocompatibility while eluding attack from the various components of the immune system that results in implant rejection. Still, the ability to transplant cells remains a holy grail of medicine, and future advances in engineering nanomaterials promise to enable this methodology for a range of diseases. Hence, there are a number of challenges as well as opportunities for nanotechnologists.

TABLE 9.1

Materials Used for Cell Encapsulation and the Targeted Diseases

Type of Material	Encapsulation Material	Cells Encapsulated	Targeted Disease	Examples
Natural materials	Alginate	Pancreatic islets, stem cells, chondrocytes	Diabetes, nervous disorders, myocardial defects, cancer	[10,85–90]
	Chitosan	Pancreatic islets, epithelial cells, stem cells	Diabetes, orthopedics, wound healing, cartilage tissue engineering, neuronal disorders	[91–96]
	Chondroitin sulfate	Cartilage cells	Arthritis	[97,98]
	Heparin	Hepatocytes/ stem cells	Liver disease	[99,100]
	Gelatin	Mammalian cells	Cancer, gene therapy, diabetes	[101–104]
	Fibrin	Keratinocytes	Diabetes, hemophilia, cardiovascular diseases	[50,105]
	Collagen	Pancreatic islets, fibroblast, stem cells, endothelial cells, epithelial cells	Diabetes, bone tissue regeneration, wound healing	[106,107]
	HA	Stem cells, epithelial cells	Diabetes, neuronal disease, cancer, bone defects	[55,108]
Synthetic materials	PEG	Polypeptides, polyesters, polyphosphazenes, hemoglobin	Diabetes	[109,110]
	Polyacrylates	Fibroblasts, hemoglobin, spleen cells	Diabetes	[111,112]
	Polyfumarates	Stromal cells, epithelial cell	Osteoarthritis, diabetes	[113,114]
	Silicon	Beta-cells	Diabetes	[115]
	Epoxy (SU8)	Fibroblasts, beta-cells	Diabetes	[116,117]
	Metal (gold coated)	Beta-cells	Diabetes	[27,83,116]

TABLE 9.2

Pros and Cons of Different Nanomaterial Encapsulants

Encapsulation Material	Advantage	Disadvantage
Sodium alginate	Naturally and readily available; biocompatible; bioinert; economical.	Unstable upon contact with phosphate of potassium or sodium ions, limited to culture time of 10–15 days.
Chitosan	Naturally available; biocompatible; shows less evidence of foreign body reaction and no fibrosis.	Needs organic solvents, lengthy and complex preparation process.
Chondroitin sulfate	Immobilizing chondroitin sulfate to synthetic scaffolds stimulates matrix synthesis.	Material properties not easily tunable; Possibility of immune response.
Heparin	Innate physical characteristics of ECM.	Material properties not easily tunable; Degradation products lead to immune response; possibility of immune response.
Gelatin	Cells readily interact with the material.	Material properties not easily tunable.
Fibrin	Biocompatible; can be delivered with growth factors; performs direct cellular function.	Material properties not easily tunable and too weak and degrade rapidly.
Collagen	High content in tripeptides increasing its bioavailability; highly biocompatible, contains several cell adhesive sites including RGD; relatively resistant to proteases.	Material properties not easily tunable, difficult to control cell adhesion and inadequate mechanical strength for many applications.
HA	Provides excellent microenvironment for proliferation and migration of functional cells; enzymatically degradable.	Can elicit immune response; properties are difficult to manipulate.
PEG	Can fine-tune properties; easily produced in large quantities with little variation; use of photolithography or microfluidic processing allows fabrication of microarchitectures that potentially recapitulate key aspects of tissue architecture to guide behavior of cells with respect to morphology, cytoskeletal structure, and functionality.	Little cell interaction; must be tailored to direct cellular function.
Polyacrylates	Can fine-tune properties; easily produced in large quantities with little variation; network structure can be easily modified to mimic critical aspects of the original microenvironments.	Little cell interaction; must be tailored to direct cellular function.
Polyfumarates	Can fine-tune properties; easily produced in large quantities with little variation.	Little cell interaction; must be tailored to direct cellular function.
Silicon	Compatible with micro- and nanofabrication processes; relatively inert; surface can be modified.	Relatively expensive; high-throughput loading can be challenging.
Epoxy (SU8)	Machinable; photopatternable; FDA approved in shunts.	Relatively expensive; high-throughput loading can be challenging; possibly toxicity concerns from leaching of small cross-linking molecules.
Metal (gold coated)	Highly precise; high mechanical strength.	Relatively expensive; high-throughput loading can be challenging; metal toxicity/allergy concerns.

Acknowledgment

We acknowledge funding from the Iacocca Family Foundation.

References

1. Daneman, D. 2006. Type 1 diabetes. *Lancet* 367: 847–858, doi:10.1016/S0140-6736(06)68341-4.
2. Lang, A. E. and A. M. Lozano. 1998. Parkinson's disease. First of two parts. *The New England Journal of Medicine* 339: 1044–1053, doi:10.1056/NEJM199810083391506.
3. Lang, A. E. and A. M. Lozano. 1998. Parkinson's disease. Second of two parts. *The New England Journal of Medicine* 339: 1130–1143, doi:10.1056/NEJM199810153391607.
4. Poncelet, A. J., D. Denis, and P. Gianello. 2009. Cellular xenotransplantation. *Current Opinion in Organ Transplantation* 14: 168–174.
5. Yang, Y. G. and M. Sykes. 2007. Xenotransplantation: Current status and a perspective on the future. *Nature Reviews Immunology* 7: 519–531, doi:10.1038/nri2099.
6. Fink, J. S. et al. 2000. Porcine xenografts in Parkinson's disease and Huntington's disease patients: Preliminary results. *Cell Transplantation* 9: 273–278.
7. Patience, C., Y. Takeuchi, and R. A. Weiss. 1997. Infection of human cells by an endogenous retrovirus of pigs. *Nature Medicine* 3: 282–286.
8. Dinsmore, J. H., C. Manhart, R. Raineri, D. B. Jacoby, and A. Moore. 2000. No evidence for infection of human cells with porcine endogenous retrovirus (PERV) after exposure to porcine fetal neuronal cells. *Transplantation* 70: 1382–1389.
9. Tan, P. L. 2010. Company profile: Tissue regeneration for diabetes and neurological diseases at Living Cell Technologies. *Regenerative Medicine* 5: 181–187, doi:10.2217/rme.10.4.
10. Lim, F. and A. M. Sun. 1980. Microencapsulated islets as bioartificial endocrine pancreas. *Science* 210: 908–910.
11. Lanza, R. P., J. L. Hayes, and W. L. Chick. 1996. Encapsulated cell technology. *Nature Biotechnology* 14: 1107–1111.
12. Orive, G., R. M. Hernandez, A. R. Gascon, M. Igartua, and J. L. Pedraz. 2002. Encapsulated cell technology: From research to market. *Trends Biotechnology* 20: 382–387.
13. Ekser, B. et al. 2012. Clinical xenotransplantation: The next medical revolution? *Lancet* 379: 672–683.
14. Wang, T. et al. 1997. An encapsulation system for the immunoisolation of pancreatic islets. *Nature Biotechnology* 15: 358–362.
15. Brissova, M., M. Petro, I. Lacik, A. C. Powers, and T. Wang. 1996. Evaluation of microcapsule permeability via inverse size exclusion chromatography. *Analytical Biochemistry* 242: 104–111.
16. de Vos, P. and P. Marchetti. 2002. Encapsulation of pancreatic islets for transplantation in diabetes: The untouchable islets. *Trends in Molecular Medicine* 8: 363–366.
17. de Groot, M. et al. 2001. Microcapsules and their ability to protect islets against cytokine-mediated dysfunction. *Transplantation Proceedings* 33: 1711–1712.
18. Basta, G. et al. 2004. Optimized parameters for microencapsulation of pancreatic islet cells: An in vitro study clueing on islet graft immunoprotection in type 1 diabetes mellitus. *Transplant Immunology* 13: 289–296, doi:10.1016/j.trim.2004.10.003.
19. Economides, A. N. et al. 2003. Cytokine traps: Multi-component, high-affinity blockers of cytokine action. *Nature Medicine* 9: 47–52, doi:10.1038/nm811.
20. Rydgren, T., D. Bengtsson, and S. Sandler. 2006. Complete protection against interleukin-1beta-induced functional suppression and cytokine-mediated cytotoxicity in rat pancreatic islets in vitro using an interleukin-1 cytokine trap. *Diabetes* 55: 1407–1412.

21. Malaisse, W. J. 1996. Regulation, perturbation, and correction of metabolic events in pancreatic islets. *Acta Diabetologica* 33: 173–179.
22. Wang, W., L. Upshaw, D. M. Strong, R. P. Robertson, and J. Reems. 2005. Increased oxygen consumption rates in response to high glucose detected by a novel oxygen biosensor system in non-human primate and human islets. *Journal of Endocrinology* 185: 445–455.
23. Fort, A., N. Fort, C. Ricordi, and C. L. Stabler. 2008. Biohybrid devices and encapsulation technologies for engineering a bioartificial pancreas. *Cell Transplantation* 17: 997–1003.
24. Thomlinson, R. H. and L. H. Gray. 1955. The histological structure of some human lung cancers and the possible implications for radiotherapy. *British Journal of Cancer* 9: 539–549.
25. Desai, T. A., D. J. Hansford, L. Leoni, M. Essenpreis, and M. Ferrari. 2000. Nanoporous anti-fouling silicon membranes for biosensor applications. *Biosensors and Bioelectronics* 15: 453–462.
26. De Vos, P., B. J. De Haan, G. H. Wolters, J. H. Strubbe, and R. Van Schilfgaarde. 1997. Improved biocompatibility but limited graft survival after purification of alginate for microencapsulation of pancreatic islets. *Diabetologia* 40: 262–270.
27. Clubb, F. J., Jr. et al. 1999. Surface texturing and coating of biomaterial implants: Effects on tissue integration and fibrosis. *ASAIO Journal* 45: 281–287.
28. Li, L., S. Chen, J. Zheng, B. D. Ratner, and S. Jiang. 2005. Protein adsorption on oligo (ethylene glycol)-terminated alkanethiolate self-assembled monolayers: The molecular basis for nonfouling behavior. *Journal of Physical Chemistry B* 109: 2934–2941, doi:10.1021/jp0473321.
29. Randall, C. L., Y. V. Kalinin, M. Jamal, T. Manohar, and D. H. Gracias. 2011. Three-dimensional microwell arrays for cell culture. *Lab on a Chip* 11: 127–131, doi:10.1039/c0lc00368a.
30. Smidsrod, O. and G. Skjakbraek. 1990. Alginate as immobilization matrix for cells. *Trends in Biotechnology* 8: 71–78.
31. Malafaya, P. B., G. A. Silva, and R. L. Reis. 2007. Natural-origin polymers as carriers and scaffolds for biomolecules and cell delivery in tissue engineering applications. *Advanced Drug Delivery Reviews* 59: 207–233, doi:DOI 10.1016/j.addr.2007.03.012.
32. Simpson, N. E., C. L. Stabler, C. P. Simpson, A. Sambanis, and L. Constantinidis. 2004. The role of the CaCl2-guluronic acid interaction on alginate encapsulated beta TC3 cells. *Biomaterials* 25: 2603–2610, doi:DOI 10.1016/j.biomaterials.2003.09.046.
33. Goren, A., N. Dahan, E. Goren, L. Baruch, and M. Machluf. 2010. Encapsulated human mesenchymal stem cells: a unique hypoimmunogenic platform for long-term cellular therapy. *FASEB Journal* 24: 22–31, doi:Doi 10.1096/Fj.09-131888.
34. Strand, B. L. et al. 2001. Poly-L-lysine induces fibrosis on alginate microcapsules via the induction of cytokines. *Cell Transplantation* 10: 263–275.
35. Clayton, H. A., N. J. London, P. S. Colloby, P. R. Bell, and R. F. James. 1991. The effect of capsule composition on the biocompatibility of alginate-poly-l-lysine capsules. *Journal of Microencapsulation* 8: 221–233.
36. Vandenbossche, G. M. R. et al. 1993. Host-reaction against empty alginate-polylysine microcapsules–influence of preparation procedure. *Journal of Pharmacy and Pharmacology* 45: 115–120.
37. DeVos, P., B. DeHaan, and R. VanSchilfgaarde. 1997. Effect of the alginate composition on the biocompatibility of alginate-polylysine microcapsules. *Biomaterials* 18: 273–278.
38. VandeVord, P. J. et al. 2002. Evaluation of the biocompatibility of a chitosan scaffold in mice. *Journal of Biomedical Materials Research* 59: 585–590.
39. Felt, O., P. Buri, and R. Gurny. 1998. Chitosan: A unique polysaccharide for drug delivery. *Drug Development and Industrial Pharmacy* 24: 979–993.
40. Drury, J. L. and D. J. Mooney. 2003. Hydrogels for tissue engineering: Scaffold design variables and applications. *Biomaterials* 24: 4337–4351, doi:Doi 10.1016/S0142-9612(03)00340-5.
41. Baruch, L. and M. Machluf. 2006. Alginate-chitosan complex coacervation for cell encapsulation: Effect on mechanical properties and on long-term viability. *Biopolymers* 82: 570–579 doi:Doi 10.1002/Bip.20509.
42. Pieper, J. S., A. Oosterhof, P. J. Dijkstra, J. H. Veerkamp, and T. H. van Kuppevelt. 1999. Preparation and characterization of porous crosslinked collagenous matrices containing bioavailable chondroitin sulphate. *Biomaterials* 20: 847–858.

43. Park, Y. J. et al. 2000. Controlled release of platelet-derived growth factor-BB from chondroitin sulfate-chitosan sponge for guided bone regeneration. *Journal of Controlled Release* 67: 385–394.
44. Chang, C. H., H. C. Liu, C. C. Lin, C. H. Chou, and F. H. Lin. 2003. Gelatin-chondroitin-hyaluronan tri-copolymer scaffold for cartilage tissue engineering. *Biomaterials* 24: 4853–4858 doi:Doi 10.1016/S0142-9612(03)00383-1.
45. Flanagan, T. C. et al. 2006. A collagen-glycosaminoglycan co-culture model for heart valve tissue engineering applications. *Biomaterials* 27: 2233–2246, doi:DOI 10.1016/j.biomaterials.2005.10.031.
46. Eyre, D. R. 1980. Collagen: Molecular diversity in the body's protein scaffold. *Science* 207: 1315–1322.
47. Kemp, P. D. 2000. Tissue engineering and cell-populated collagen matrices. *Methods in Molecular Biology* 139: 287–293, doi:10.1385/1-59259-063-2:287.
48. Lee, C. H., A. Singla, and Y. Lee. 2001. Biomedical applications of collagen. *International Journal of Pharmaceutics* 221: 1–22.
49. Chia, S. M. et al. 2002. Multi-layered microcapsules for cell encapsulation. *Biomaterials* 23: 849–856.
50. Hunt, N. C. and L. M. Grover. 2010. Cell encapsulation using biopolymer gels for regenerative medicine. *Biotechnology Letters* 32: 733–742, doi:DOI 10.1007/s10529-010-0221-0.
51. Park, S. H., J. H. Cui, S. R. Park, and B. H. Min. 2009. Potential of fortified fibrin/hyaluronic acid composite gel as a cell delivery vehicle for chondrocytes. *Artificial Organs* 33: 439–447 doi:10.1111/j.1525-1594.2009.00744.x.
52. Laurent, T. C., U. B. G. Laurent, and J. R. E. Fraser. 1996. Serum hyaluronan as a disease marker. *Annals of Medicine* 28: 241–253.
53. Liao, Y. H., S. A. Jones, B. Forbes, G. P. Martin, and M. B. Brown. 2005. Hyaluronan: Pharmaceutical characterization and drug delivery. *Drug Delivery* 12: 327–342, doi:Doi 10.1080/10717540590952555.
54. Morra, M. 2005. Engineering of biomaterials surfaces by hyaluronan. *Biomacromolecules* 6: 1205–1223.
55. Khademhosseini, A. et al. 2006. Micromolding of photocrosslinkable hyaluronic acid for cell encapsulation and entrapment. *Journal of Biomedical Materials Research Part A* 79A: 522–532 doi:Doi 10.1002/Jbm.A.30821.
56. Uludag, H., P. De Vos, and P. A. Tresco. 2000. Technology of mammalian cell encapsulation. *Advanced Drug Delivery Reviews* 42: 29–64.
57. Ennis, W. B. and D. T. James. 1950. A simple apparatus for producing droplets of uniform size from small volumes of liquids. *Science* 112: 434–436.
58. Chang, T. M. S. 1964. Semipermeable microcapsules. *Science* 146: 524–525.
59. Chang, T. M. S., Macintosh, F. C., and S. G. Mason. 1966. Semipermeable aqueous microcapsules. I. Preparation and properties. *Canadian Journal of Physiology Pharmacology* 44: 115–128.
60. Freimark, D. et al. 2010. Use of encapsulated stem cells to overcome the bottleneck of cell availability for cell therapy approaches. *Transfusion Medicine and Hemotherapy [Offizielles Organ der Deutschen Gesellschaft fur Transfusionsmedizin und Immunhamatologie]* 37: 66–73 doi:10.1159/000285777.
61. Kim, M. S., J. H. Yeon, and J. K. Park. 2007. A microfluidic platform for 3-dimensional cell culture and cell-based assays. *Biomedical Microdevices* 9: 25–34, doi:DOI 10.1007/s10544-006-9016-4.
62. Brouzes, E. et al. 2009. Droplet microfluidic technology for single-cell high-throughput screening. *Proceedings of the National Academy of Sciences of the United States of America* 106: 14195–14200 doi:DOI 10.1073/pnas.0903542106.
63. Chick, W. L. et al. 1977. Artificial pancreas using living beta cells–effects on glucose homeostasis in diabetic rats. *Science* 197: 780–782.
64. Knazek, R. A., P. O. Kohler, P. M. Gullino, and R. L. Dedrick. 1972. Cell-culture on artificial capillaries–approach to tissue growth in-vitro. *Science* 178: 65–67.
65. Bessis, N. et al. 1999. Encapsulation in hollow fibres of xenogeneic cells engineered to secrete IL-4 or IL-13 ameliorates murine collagen-induced arthritis (CIA). *Clinical and Experimental Immunology* 117: 376–382.
66. Mckinney, R. 1987. An experimental approach to the preparation of hollow fiber membranes. *Desalination* 62: 37–47.

67. Gabelman, A. and S. T. Hwang. 1999. Hollow fiber membrane contactors. *Journal of Membrane Science* 159: 61–106.
68. Scott, K. 1998. Introduction to membrane separations. *Handbook of Industrial Membranes*, 2nd edn., Elsevier, Oxford, U.K.
69. Legallais, C., B. David, and E. Dore. 2001. Bioartificial livers: Current technological aspects and future developments. *Journal of Membrane Science* 181: 81–95.
70. Townsend-Nicholson, A. and S. N. Jayasinghe. 2006. Cell electrospinning: A unique biotechnique for encapsulating living organisms for generating active biological microthreads/scaffolds. *Biomacromolecules* 7: 3364–3369, doi:Doi 10.1021/Bm060649h.
71. Storrs, R., R. Dorian, S. R. King, J. Lakey, and H. Rilo. 2001. Preclinical development of the Islet Sheet. *Annals of the New York Academy of Sciences* 944: 252–266.
72. Shin, M., H. Yoshimoto, and J. P. Vacanti. 2004. In vivo bone tissue engineering using mesenchymal stem cells on a novel electrospun nanofibrous scaffold. *Tissue Engineering* 10: 33–41.
73. Muthyala, S., V. R. R. Raj, M. Mohanty, P. V. Mohanan, and P. D. Nair. 2011. The reversal of diabetes in rat model using mouse insulin producing cells—A combination approach of tissue engineering and macroencapsulation. *Acta Biomaterialia* 7: 2153–2162, doi:DOI 10.1016/j.actbio.2011.01.036.
74. Aper, T., A. Schmidt, M. Duchrow, and H. P. Bruch. 2007. Autologous blood vessels engineered from peripheral blood sample. *European Journal of Vascular Endovascular Surgery* 33: 33–39 doi:DOI 10.1016/j.ejvs.2006.08.008.
75. Lee, S. J. et al. 2004. Response of MG63 osteoblast-like cells onto polycarbonate membrane surfaces with different micropore sizes. *Biomaterials* 25: 4699–4707, doi:DOI 10.1016/j.biomaterials.2003.11.034.
76. Biagini, G. et al. 1994. Fibroblast proliferation over dialysis membrane–An experimental-model for tissue biocompatibility evaluation. *International Journal of Artificial Organs* 17: 620–628.
77. Pra, L. D. D., E. Ferain, R. Legras, and S. Demoustier-Champagne. 2002. Fabrication of a new generation of track-etched templates and their use for the synthesis of metallic and organic nanostructures. *Nuclear Instruments and Methods in Physics B* 196: 81–88.
78. Price, P. B. and R. M. Walker. 1962. Chemical etching of charged-particle tracks in solids. *Journal of Applied Physics* 33: 3407.
79. Desai, T. A. et al. 1998. Microfabricated immunoisolating biocapsules. *Biotechnology and Bioengineering* 57: 118–120.
80. Gimi, B. et al. 2009. Cell encapsulation and oxygenation in nanoporous microcontainers. *Biomedical Microdevices* 11: 1205–1212, doi:10.1007/s10544-009-9338-0.
81. Randall, C. L., E. Gultepe, and D. H. Gracias. 2012. Self-folding devices and materials for biomedical applications. *Trends in Biotechnology* 30: 138.
82. Fernandes, R. and D. H. Gracias. 2012. Self-folding polymeric containers for encapsulation and delivery of drugs. *Advance Drug Delivery Reviews* 64(14): 1579–1589.
83. Matyjaszewski, K. and N. V. Tsarevsky. 2009. Nanostructured functional materials prepared by atom transfer radical polymerization. *Nature Chemistry* 1: 276–288.
84. Randall, C. L., Y. V. Kalinin, M. Jamal, A. Shah, and D. H. Gracias. 2011. Self-folding immunoprotective cell encapsulation devices. *Nanomedicine* 7: 686–689, doi:10.1016/j.nano.2011.08.020.
85. Gattas-Asfura, K. M., C. A. Fraker, and C. L. Stabler. 2012. Perfluorinated alginate for cellular encapsulation. *Journal of Biomedical Materials Research A* 100: 1963–1971, doi:10.1002/jbm.a.34052.
86. Shinde, U. and M. Nagarsenker. 2011. Microencapsulation of eugenol by gelatin-sodium alginate complex coacervation. *Indian Journal of Pharmaceutical Sciences* 73: 311–315, doi:10.4103/0250-474X.93524.
87. Maguire, T., E. Novik, R. Schloss, and M. Yarmush. 2006. Alginate-PLL microencapsulation: Effect on the differentiation of embryonic stem cells into hepatocytes. *Biotechnology and Bioengineering* 93: 581–591, doi:10.1002/bit.20748.
88. Schagemann, J. C. et al. 2009. Cell-laden and cell-free biopolymer hydrogel for the treatment of osteochondral defects in a sheep model. *Tissue Engineering Part A* 15: 75–82, doi:10.1089/ten.tea.2008.0087.

89. Karoubi, G., M. L. Ormiston, D. J. Stewart, and D. W. Courtman. 2009. Single-cell hydrogel encapsulation for enhanced survival of human marrow stromal cells. *Biomaterials* 30: 5445–5455 doi:10.1016/j.biomaterials.2009.06.035.

90. Leor, J. et al. 2000. Bioengineered cardiac grafts: A new approach to repair the infarcted myocardium? *Circulation* 102: III56-61.

91. Zhi, Z. L., A. Kerby, A. J. King, P. M. Jones, and J. C. Pickup. 2012. Nano-scale encapsulation enhances allograft survival and function of islets transplanted in a mouse model of diabetes. *Diabetologia* 55: 1081–1090, doi:10.1007/s00125-011-2431-y.

92. Xu, K. et al. 2012. Thiolene-based biological/synthetic hybrid biomatrix for 3-D living cell culture. *Acta Biomaterialia* 8: 2504–2516, doi:10.1016/j.actbio.2012.03.049.

93. Yang, K. C. et al. 2010. The cytoprotection of chitosan based hydrogels in xenogeneic islet transplantation: An in vivo study in streptozotocin-induced diabetic mouse. *Biochemical and Biophysical Research Communication* 393: 818–823, doi:10.1016/j.bbrc.2010.02.089.

94. Yang, K. C. et al. 2008. Chitosan/gelatin hydrogel as immunoisolative matrix for injectable bioartificial pancreas. *Xenotransplantation* 15: 407–416, doi:10.1111/j.1399–3089.2008.00503.x.

95. Hardikar, A. A., M. V. Risbud, and R. R. Bhonde. 1999. A simple microcapsule generator design for islet encapsulation. *Journal of Biosciences* 24: 371–376.

96. Suh, J. K. and H. W. Matthew. 2000. Application of chitosan-based polysaccharide biomaterials in cartilage tissue engineering: A review. *Biomaterials* 21: 2589–2598.

97. Reis, A. V. et al. 2009. Nanometer-and submicrometer-sized hollow spheres of chondroitin sulfate as a potential formulation strategy for anti-inflammatory encapsulation. *Pharmaceutical Research* 26: 438–444, doi:10.1007/s11095-008-9732-y.

98. Brown, K. E. et al. 1998. Gelatin/chondroitin 6-sulfate microspheres for the delivery of therapeutic proteins to the joint. *Arthritis and Rheumatism* 41: 2185–2195, doi:10.1002/1529-0131(199812)41:12<2185::AID-ART13>3.0.CO;2-C.

99. Yu, S., B. Han, and C. Peng. 2012. A preliminary study of alginate, heparin-chitosan-alginate and heparin microencapsulated hepatocytes system. *Hepato-Gastroenterology* 59: 1234–1240, doi:10.5754/hge11638.

100. Kim, M., J. Y. Lee, C. N. Jones, A. Revzin, and G. Tae. 2010. Heparin-based hydrogel as a matrix for encapsulation and cultivation of primary hepatocytes. *Biomaterials* 31: 3596–3603 doi:10.1016/j.biomaterials.2010.01.068.

101. Imani, R., S. H. Emami, P. R. Moshtagh, N. Baheiraei, and A. M. Sharifi. 2012. Preparation and characterization of agarose-gelatin blend hydrogels as a cell encapsulation matrix: An in-vitro study. *Journal of Macromolecular Science B* 51: 1606–1616, doi:Doi 10.1080/00222348.2012.657110.

102. Peran, M. et al. 2012. Functionalized nanostructures with application in regenerative medicine. *International Journal of Molecular Sciences* 13: 3847–3886, doi:Doi 10.3390/Ijms13033847.

103. Xu, J. and M. Amiji. 2012. Therapeutic gene delivery and transfection in human pancreatic cancer cells using epidermal growth factor receptor-targeted gelatin nanoparticles. *Journal of Visualized Experiments* e3612, doi:10.3791/3612.

104. Munarin, F., P. Petrini, S. Fare, and M. C. Tanzi. 2010. Structural properties of polysaccharide-based microcapsules for soft tissue regeneration. *Journal of Materials Science Materials in Medicine* 21: 365–375, doi:10.1007/s10856-009-3860-8.

105. Mauth, C. et al. 2009. Cell-seeded polyurethane-fibrin structures—A possible system for intervertebral disc regeneration. *European Cells and Materials* 18: 27–38; discussion 38–39.

106. Chan, B. P., T. Y. Hui, M. Y. Wong, K. H. Yip, and G. C. Chan. 2010. Mesenchymal stem cell-encapsulated collagen microspheres for bone tissue engineering. *Tissue Engineering Part C, Methods* 16: 225–235, doi:10.1089/ten.tec.2008.0709.

107. Yuan, M., K. W. Leong, and B. P. Chan. 2011. Three-dimensional culture of rabbit nucleus pulposus cells in collagen microspheres. *The Spine Journal: Official Journal of the North American Spine Society* 11: 947–960, doi:10.1016/j.spinee.2011.07.004.

108. Bae, K. H., J. J. Yoon, and T. G. Park. 2006. Fabrication of hyaluronic acid hydrogel beads for cell encapsulation. *Biotechnology Progress* 22: 297–302, doi:10.1021/bp050312b.

109. Teramura, Y., L. N. Minh, T. Kawamoto, and H. Iwata. 2010. Microencapsulation of islets with living cells using polyDNA-PEG-lipid conjugate. *Bioconjugate Chemistry* 21: 792–796 doi:10.1021/bc900494x.
110. Phelps, E. A. et al. 2012. Maleimide cross-linked bioactive PEG hydrogel exhibits improved reaction kinetics and cross-linking for cell encapsulation and in situ delivery. *Advanced Materials* 24: 64–70, 62, doi:10.1002/adma.201103574.
111. Boag, A. H. and M. V. Sefton. 1987. Microencapsulation of human fibroblasts in a water-insoluble polyacrylate. *Biotechnology and Bioengineering* 30: 954–962, doi:10.1002/bit.260300806.
112. Gharapetian, H., M. Maleki, G. M. O'Shea, R. C. Carpenter, and A. M. Sun. 1987. Polyacrylate microcapsules for cell encapsulation: Effects of copolymer structure on membrane properties. *Biotechnology and Bioengineering* 30: 775–779, doi:10.1002/bit.260300610.
113. Temenoff, J. S. et al. 2004. Thermally cross-linked oligo(poly(ethylene glycol) fumarate) hydrogels support osteogenic differentiation of encapsulated marrow stromal cells in vitro. *Biomacromolecules* 5: 5–10, doi:10.1021/bm030067p.
114. Zhang, M. W. et al. 2010. Adapting biodegradable oligo(poly(ethylene glycol) fumarate) hydrogels for pigment epithelial cell encapsulation and lens regeneration. *Tissue Engineering Part C, Methods* 16: 261–267, doi:10.1089/ten.TEC.2009.0162.
115. Desai, T. A. et al. 2000. Microfabricated immunoisolating biocapsules. *Biotechnology and Bioengineering* 57: 118–120.
116. Kwon, J. et al. 2009. SU-8-based immunoisolative microcontainer with nanoslots defined by nanoimprint lithography. *Journal of Vacuum Science and Technology B, Microelectronics and Nanometer Structures: Processing, Measurement, and Phenomena: An Official Journal of the American Vacuum Society* 27: 2795–2800, doi:10.1116/1.3258146.
117. Azam, A., K. E. Laflin, M. Jamal, R. Fernandes, and D. H. Gracias. 2011. Self-folding micropatterned polymeric containers. *Biomedical Microdevices* 13: 51–58.
118. Gimi, B. et al. 2005. Self-assembled three-dimensional radio frequency (RF) shielded containers for cell encapsulation. *Biomedical Microdevices* 7: 341–345.

10

Lipid and Cyclodextrin Nanocarriers Loading Bioactive Agents: Stabilization on Polymeric Supports

Jesús Hierrezuelo, Laura Peláez, Juana Benavente, Juan Manuel López-Romero, Rodrigo Rico, and Ricard Armengol

CONTENTS

10.1 Introduction

Encapsulation technologies are nowadays commonly used in pharmaceutical, cosmetic, and medical industries for the development of controlled-release delivery systems [1–4]. Many therapeutically active molecules need to be encapsulated in a drug-carrying system to prevent the loaded drug from degradation but also to ensure its effective delivery in the biological media. Nanoencapsulation may also protect the active compound from oxidation, evaporation, and off-flavor [5]. The nanomaterials used for encapsulation include micelles, liposomes, micro- and nano-polymeric particles [6–9], and, more recently, molecular carriers like cyclodextrins, solid lipid nanoparticles (SLNs), nanostructured lipid carriers (NLCs), and lipid nanocapsules (LNCs) [9–12].

In this context, lipid nanoparticles (LNs) together with β-cyclodextrin (CD) inclusion complex nanoparticles represent a promising class of particulate carriers for bioactive organic molecules, since they improve their solubility in aqueous media and chemical stability, preventing degradation of the loaded bioactive compound. Other advantages of LNs over other carrier systems include good-to-high loading capacity, ease of preparation, low production costs, and the possibility of large-scale industrial production [9]. Moreover, they can be prepared from natural sources by using biocompatible components, and their properties can be tuned by controlling the nanoparticle shape, size, and state of aggregation [9,13]. Due to their good compatibility with different ingredients, both lipid and cyclodextrin nanoparticles can be added to existing formulations without problems.

Particularly, cyclodextrins are a group of natural cyclic oligosaccharides with a hydrophobic cavity that allows the inclusion of organic and inorganic guest molecules, altering their physicochemical behavior and reducing their undesirable effects, like odor and color. In the pharmaceutical, cosmetics, and food industries, cyclodextrins have been used as complexing agents to increase the water solubility of various compounds, such as drugs, vitamins, and food colorants. It has been demonstrated that complexation can also considerably increase the chemical stability and bioavailability of the guest molecule [5].

However, the use of these nanomaterials as delivery systems is still limited due to stability problems, mainly when biological fluids are involved (e.g., drug expulsion, gelation, or modification of particle size) [14] but also during storage or administration [15]. To overcome some of these limitations, polymeric nanosphere gels and hybrid materials have been proposed for the topical delivery of lipophilic molecules [16–18].

This chapter shows the preparation of lecithin-tristearin nanoparticles and also β-cyclodextrin inclusion complexes loaded with dihydromyricetin (DMy-LNs and DMy-CD) or 2,4-dihydroxybenzophenone (DHB-LNs), as well as their inclusion in a cellulosic film support. We chose these active molecules due to the UV-filter application in sunscreen for skin protection of DHB [19], while DMy exhibits antitumor effects against bladder carcinoma, showing also antialcohol intoxication properties [20,21]. The morphologies and chemical compositions of nanoparticles were analyzed by atomic force microscope (AFM), scanning electron microscopy (SEM), micro-Raman, and NMR. The surface coverage and the presence of the DHB-LNs in the cellulose film were obtained by XPS, while changes in the film elastic properties are an indication of its presence in the bulk support structure. Results showing DHB release from LNs and the cellulosic engineering tissue are also presented.

10.2 Materials and Methods

10.2.1 Chemicals

DMy (85%, Provital Group Spain, Barberá del Vallés, Spain, catalogue # 1950) was purified by crystallization from ethyl acetate. DHB (99%, Sigma-Aldrich, Madrid, Spain, catalogue # 126217), CD (99%, Sigma-Aldrich, # W402826), Tween® 80 (TW, Sigma-Aldrich, # P1754), sodium taurocholate (ST) (Sigma-Aldrich, # T4009), glyceryl tristearate (GT) (Fluka, Madrid,

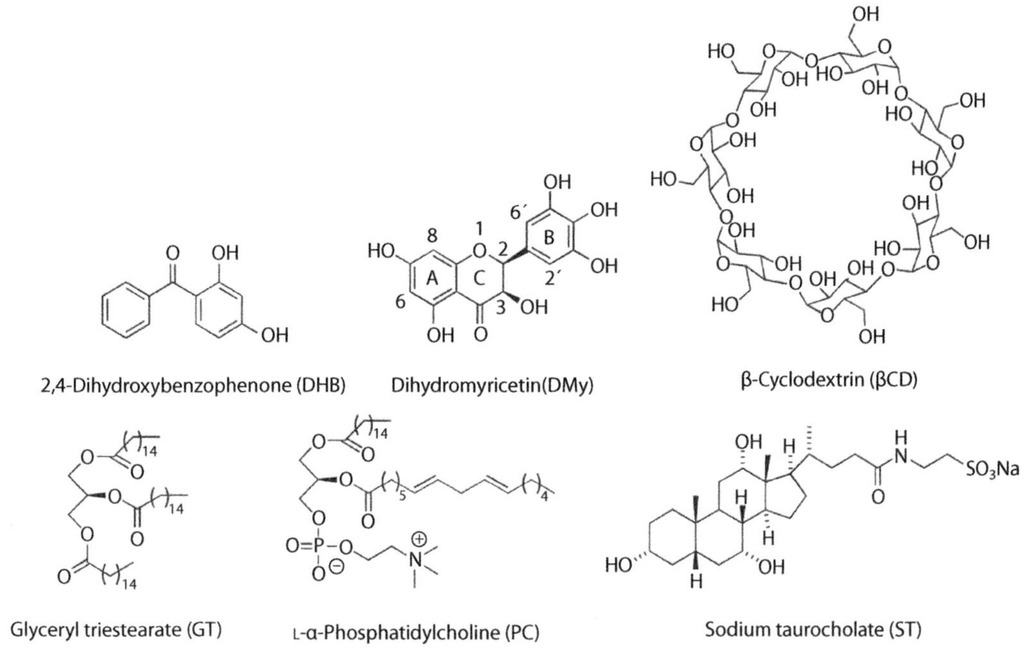

FIGURE 10.1
Chemical structure of actives (DHB and DMy), lipids (GT), β-cyclodextrin (CD), and surfactants (PC and ST).

Spain, # 69498), and lecithin 60% from egg yolk (PC, L-α-phosphatidylcholine, Fluka, # 61755) were used as purchased without any purification (Figure 10.1). If water was needed, deionized water was used in all experiments.

A highly hydrophilic flat regenerated cellophane membrane from Cellophane Española S.A. (Burgos, Spain) with 0.03 kg/m² of regenerated cellulose (RC-3) was used as the polymeric support.

10.2.2 Preparation of Lipid Nanoparticles Loaded with Actives

LNs were obtained by combining ultrasound (Bandelin Sonopuls, Berlin, Germany) and high-pressure homogenization (Nano DeBee, Bee International Inc., South Easton, Massachusetts, USA) methods, using L-α-phosphatidylcholine (6 wt%), ST (4 wt%), and Tween 80 (4 wt%) as surfactants and n-butanol (2 wt%) and GT (12 wt%) as the main lipid component [22]. DHB and DMy were the active organic components used. In a typical experiment, a mixture of GT (1.2 g) and PC (0.6 g) was vigorously stirred at about 75°C for 15 min. Then, preheated at 70°C, TW (0.4 g), ST (0.4 g), n-butanol (0.2 g), and DHB (0.04 g) or DMy (0.03 g) were added. After 45 min stirring, the mixture was dispersed in water (7.16 g), and the o/w microemulsion obtained was sonicated (50 W) for 15 min. The milky suspension obtained was filtered off (ceramic filter by Sigma-Aldrich, with an average diameter pore size of 70 ± 30 μm) to obtain a clear solution, which was then high-pressure homogenized (2 cycles). The final suspension, kept at 5°C and protected from light, was used for further physicochemical characterization.

10.2.3 Preparation of the Inclusion Complexes

DMy-CD samples were prepared by adaptation of two methods: sonication and precipitation. In a typical experiment, CD (2.06 g, 1.82 mmol) was dissolved at 45°C by stirring in deionized water (35 mL). The clear solution was left to reach room temperature, and then DMy (0.58 g, 1.81 mmol, 1:1 mol ratio active: CD) was added.

The mixture was sonicated for a 3 min interval during 30 min (Bandelin, Sonopuls) to complete solubilization. After this period, the mixture was stirred at 25°C for 16 h, and within this period, a white solid appeared. The solid was filtered off, washed with ethyl acetate, and dried in a desiccator. The ethyl acetate phase was concentrated to dryness under vacuum.

10.2.4 Entrapment Efficiency (EE%)

To determine the amount of incorporated drug into the LNs, 1 mL of the nanoparticle dispersion was ultracentrifugated (15,000 g). The aqueous layer (0.5 mL) was decanted and extracted with dichloromethane (3 mL) and the solid was freeze-dried. For the quantitative determination of DHB or DMy in the solvent, a reverse-phase HPLC method was used (Hewlett-Packard 1050, Los Angeles, California, USA). The apparatus was equipped with an RP-18 column. Analyses were performed at 24°C eluting with a mixture of methanol to water 75:25 volume% as mobile phase, at a flow rate of 0.8 mL/min, and using a UV detector at 330 nm. The freeze-dried LNs were analyzed by HPLC directly after dissolution in CH_2Cl_2.

Entrapment efficiency (EE%) was determined as follows [22,23]:

$$EE\% = \left[\frac{(W_{initial\,drug} - W_{free\,drug})}{W_{initial\,drug}} \right] \times 100\% \qquad (10.1)$$

10.2.5 Micro-Raman Spectroscopy

DMy-CD samples used for the Raman study were prepared by depositing the white solid onto a freshly cleaved glass plate. Samples were kept at 20°C overnight in a desiccator ($CaCl_2$). The samples were analyzed at room temperature using a micro-Raman spectrophotometer (Renishaw inVia Reflex Raman, Madrid, Spain). The excitation source was an Ar^+ laser with an excitation wavelength of 514.5 nm. The spectra were recorded with a spatial resolution of 2 µm and the 514.5 nm exciting line of the Ar^+ laser of the Raman microscope.

10.2.6 SEM and AFM Imaging

Nanoparticle morphology was characterized by SEM and AFM. Dried nanospheres were coated with gold for 3 min by sputtering (JEOL JCC-1100, JEOL Ltd., Madrid, Spain) before SEM was recorded. The accelerating voltage ranged from 5 to 15 kV during scanning. SEM of the electrospun samples was performed on a JEOL JSM-840 JE (JEOL Ltd., Madrid, Spain) electron microscope.

For AFM analyses, an AFM (Veeco Instruments, Multimode AFM with Nanoscope V controller, Mannheim, Germany) was equipped with a 12 µm EV scanner. A tapping Mode™ tip in air (42 N/m, 320 kHz) was selected to scan the surface. WSxM 4.0 (Nanotec Electrónica, S.L., Madrid, Spain) software was used for the AFM image processing.

10.2.7 ¹H NMR Studies

All NMR experiments were performed on a 400 MHz Bruker ARX-400 spectrometer (Bruker, S.A., Madrid, Spain) using the residual solvent peak in $CDCl_3$ (∂_H 7.24 ppm for ^1H),

D$_2$O (∂_H 4.45 ppm for ^{1}H), or DMSO-d$_6$ (∂_H 2.50 ppm for ^{1}H). The spectra were acquired in an inversed probe head at 298 K in 5 mm diameter tubes. All chemical shifts were relative to the DOH signal at 4.70 ppm. The NMR measurements were recorded with standard Bruker pulse sequences.

10.2.8 Controlled Release

Quantification of DHB released from LNs through time was carried out by HPLC analysis in a Thermo Fisher instrument (Thermo Fisher Scientific, Madrid, Spain). For analyses, the nanoparticles were immersed in cyclohexane and aliquots were taken at different times.

10.2.9 Support Film Modification with LNs and CNs

Pieces of the RC-3 cellophane support film were immersed for 15 min in a slowly stirred water dispersion of DHB-loaded LNs or CNs at room temperature. The samples were then removed, washed with water, and dried at room temperature for 24 h in a desiccator, and samples of the RC-3/DHB-LN- and RC-3/DMy-CD-engineered materials were obtained.

A piece of the RC-3/DHB-LN film was immersed in cyclohexane for 30 min and analyzed by HPLC in order to determine the effect of the cellulosic support on the DHB release.

10.2.10 X-Ray Photoelectron Spectroscopy Analysis

Chemical surface characterization of the RC-3/DHB-LN- and RC-3/DMy-CD-modified membranes was carried out by x-ray photoelectron spectroscopy or XPS analysis; for comparison, results previously obtained for the DHB-loaded LNs and the RC-3 cellulosic support are also presented [22,24]. XPS spectra were recorded with a Physical Electronics PHI 5700 spectrometer with a multichannel hemispherical electroanalyzer. An Mg Kα x-ray was used as the excitation source (hν = 1253.6 eV). Accurate ±0.1 eV binding energies were determined with respect to the position of the adventitious C 1s peak at 284.8 eV. High-resolution spectra were recorded at a takeoff angle of 45° by a concentric hemispherical analyzer operating in the constant pass energy mode at 29.35 eV, using a 720 µm diameter analysis area. The residual pressure in the analysis chamber was maintained below 10^{-9} torr during data acquisition. A PHI ACCESS ESCA-V6.0F software package was used for acquisition and data analysis. The RC-3-modified tissue was irradiated in less than 20 min to minimize x-ray-induced sample damage [25]. Atomic concentration (AC, %) percentages of the characteristic membrane elements were determined after subtraction of a Shirley-type background taking into account the corresponding area sensitivity factor [26] for the different measured spectral regions.

10.2.11 Elastic Measurements

Elastic properties of the RC-3 and the RC-3/DHB-LNs' cellulosic-based solid structures were determined from tensile stress versus elongation curves. These curves were obtained by applying different tensions to samples of 2 cm initial length and measuring the length changes by using a force digital gauge (Mark-T, ES20 model) with a maximum tension of 100 N, while the length accurate was ±0.01 mm.

10.3 Results and Discussion

10.3.1 Drug Incorporation and Loading Capacity

Encapsulation percentage was analyzed by HPLC of the ultracentrifugated samples. This technique shows a 12 wt% of DHB and 10 wt% of DMy in the freeze-dried DHB-LN or DMy-LN solids, respectively. These values are similar to those reported by different authors; particularly, Alukda et al. give a value of 8.3% ± 0.7% of encapsulation percentage of tenofovir loaded into SLNs made of Softisan 100 lipids for HIV prevention applications [27], while Ito et al., dealing with polymeric nanoparticles loaded with the model hydrophilic drug isoniazid, indicate an encapsulation percentage around 8% by using conventional techniques, which is improved up to 20% by using a frozen inner water phase [28]. An accurate HPLC analysis of the washing water shows 5% of free DHB, meaning 95% of EE% for fresh samples. The high EE% can be explained by the lattice defects of the lipid structure that offers space to accommodate the drug, even when DHB does not have a high solubility in the lipid melt [22]. Similar results were obtained for DMy-LN samples, reaching 11% of free DMy and 89% of EE%. The EE% value for DHB-LNs improves those obtained when only homogenization technique is used [18].

10.3.2 Morphological Characterization of LNs and CNs

Morphology of the LNs was already studied by AFM [18]. The topographic analysis shows particle diameters in the range of 400–500 nm, as can be observed in Figure 10.2a with a height average of 100 nm. Taking into account the AFM results, an average value of

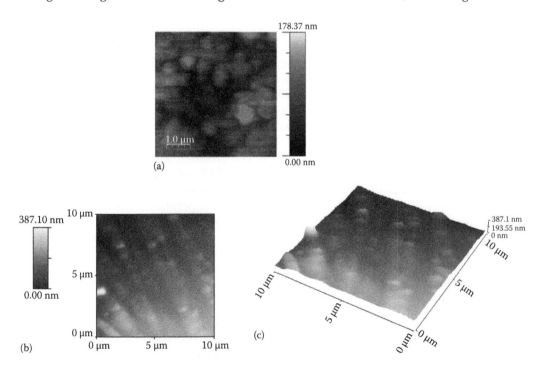

FIGURE 10.2
Morphology of DHB-LNs (a) and DMy-LNs (b and c) by AFM.

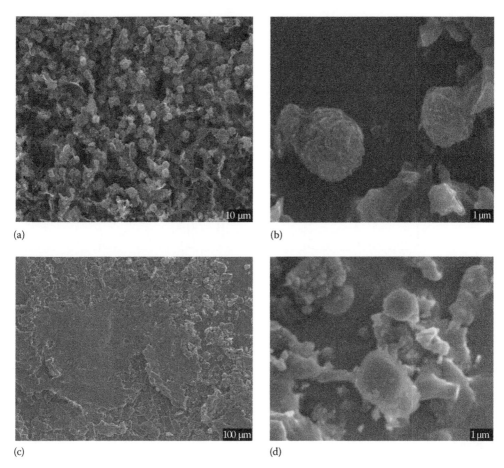

(a) (b)

(c) (d)

FIGURE 10.3
Morphology of DHB-LNs (a and b), the single component DMy (c), and the binary system DMy-CD (d) by SEM.

230 ± 30 nm for the DHB-LN radius was estimated, while for DMy-LNs, the AFM analysis shows an average ratio of 350 nm (Figure 10.2b and c).

Spherical morphology for both LNs and CDs was confirmed by SEM, as can be observed in Figure 10.3. Figure 10.3a shows the morphology of DHB-LNs in a 10 μm scale, while in Figure 10.3b one can clearly observe the spherical shape of these nanoparticles (1 μm scale). On the other hand, Figures 10.3c and d show the shape of pure DMy material and the binary system DMy-CD, respectively. DMy exists in a crystal-like form, whereas CD is usually seen as amorphous spherical or pieces of spherical particles [29]. Inclusion complex appeared in the form of irregular particles (Figure 10.3d) in which the original morphology of the active component disappeared and tiny aggregates of amorphous pieces of irregular size were present.

10.3.3 Characterization of DMy-CDs

The formation of inclusion complexes can be proven from the changes of chemical shifts of DMy and CD in ^{1}H NMR spectra [30]. ^{1}H NMR data for DMy dissolved in $CDCl_3 + MeOD$ (Figure 10.4a) show well-resolved signals: at 4.19 and 4.38 ppm,

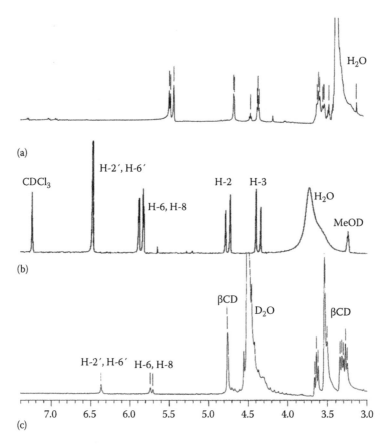

FIGURE 10.4
^{1}H NMR spectra of pure CD (a, DMSO-d_6), DMy (b, CDCl$_3$), and DMy-CD complex (c, D$_2$O).

two doublets (J = 8.5 Hz) corresponding to methine protons, H-2 and H-3; two double doublets at 5.82 and 5.90 ppm corresponding to H-6 and H-8 protons (J = 1.3 Hz); and a singlet signal integrating by 2H centered at 6.46 ppm due to the methine nuclei H-2′ and H-6′.

Figure 10.4 illustrates that most of the protons of DMy are influenced by the presence of CD. Upfield shifts are observed for all the aromatic protons of the A- and B-rings. The broadening of the proton signals of DMy in the presence of CD suggests that the B-ring, the C-ring, and the A-ring are included in the cavity, and hence, the motions of these protons are restricted indicating that DMy molecule fitted tightly.

In Figure 10.5 are shown the Raman spectra for CD, DMy, and their guest–host complex DMy-CD in the spectral range between 200 and 2000 cm^{-1}. As can be seen in Figure 10.5c, the characteristic bands of CD were obtained in the following regions: at 1450 cm^{-1} assigned to δ(CH$_2$) motions; a broad signal between 1400 and 1300 cm^{-1} attributed to δ(OH); the bands falling in the 1145–1000 cm^{-1} region, which are attributed to ν(CO); the bands around 955 cm^{-1}, which are attributed to ν(CC) motions; and the bands falling in the 865–700 cm^{-1} region due to ρ(CH$_2$) and to δ(OH). The bands below these regions are attributed to molecular skeletal vibrations. It is important to point out that the most intense CD band appears at 477 cm^{-1}.

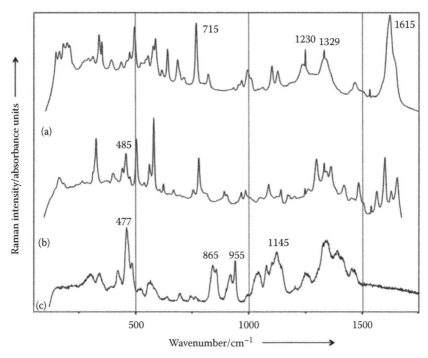

FIGURE 10.5
Micro-Raman spectra representing the band intensity obtained at different wave numbers (cm^{-1}) corresponding to DMy (a), DMy-CD complex (b) and CD (c).

Similarly, the bands of DMy are shown in Figure 10.5a. Specifically, one intense signal at 1615 cm^{-1} and a weaker band at 715 cm^{-1} are typical of CH stretching and CH twist vibrations and ring out-of-plane deformations of the aromatic rings, respectively. Other weaker bands are also found at 1230 and 1329 cm^{-1}, which are usually assigned to C–H in-plane bending and aromatic C–C ring stretching. Important signals are found between 200 and 250 cm^{-1} attributed to structural deformations of the molecule.

Regarding the Raman spectrum of the CD and DMy complex (Figure 10.5b), it is worth mentioning that signals corresponding to DMy are encountered. The lower Raman signal seen in the complex with CD can be attributed to a lower resonance Raman effect of DMy in the presence of this CD due to the stronger structural change induced in the flavonol. One can notice the absence of the Raman bands given by the CD molecule vibrations in the following spectral ranges: 200–250 and 1500–1650 cm^{-1}. These spectral regions can be analyzed in detail in order to evidence the changes caused by the guest–host interaction concerning the positions and widths of the DMy Raman bands. In the 200–250 cm^{-1} range of the Raman spectra of DMy-CD, broadened signals attributed to DMy can be clearly observed. On the other hand, in the 1500–1650 cm^{-1} range of the DMy-CD Raman spectra, the intense signal at 1615 cm^{-1} is now observed to be split into four signals at 1560, 1615, 1625, and 1630 cm^{-1}.

The interaction with CD seems to also affect the carbohydrate structure [31]. In fact, the bands observed in the Raman spectrum at 948 cm^{-1} shift downward to 925 cm^{-1}, while those at 477 cm^{-1} are shifted upward to 485 cm^{-1}, thus indicating that a structural change involving the molecular skeleton and the CH$_2$ groups is taking place. In contrast, the corresponding bands of CD in the physical mixture remain practically unchanged.

10.3.4 Dissolution and Stability Studies

It has been reported that DMy solubility in water is notably affected by temperature. The pH of solution has little effect on the solubility, but when pH > 6, degradation of the active is observed [29]. We checked the stability and solubility of DMy once it was included into the cyclodextrin skeleton in solutions at pH 5 and at temperatures similar to that of the skin.

The solubility was determined by adding excess amounts of DMy or DMy-CD to water at pH = 5 and at a temperature of 40°C. DMy solubility in water is 218 µg/mL at 25°C and the pH is 4.6. The suspensions formed were stirred for 120 min and then filtrated through a 0.45 µm membrane filter to obtain a clear solution for HPLC assay. Each sample was determined in duplicate. The percentage of drug dissolved was plotted against time (Figure 10.6).

The dissolution curves of DMy and DMy-CD complex at different times are shown in Figure 10.7. The saturation concentration of DMy-CD in water reaches the 3 mg/mL

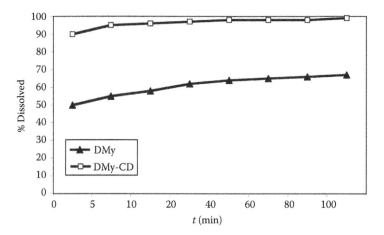

FIGURE 10.6
Dissolution curves of DMy alone and DMy-CD at pH = 5 and 40°C.

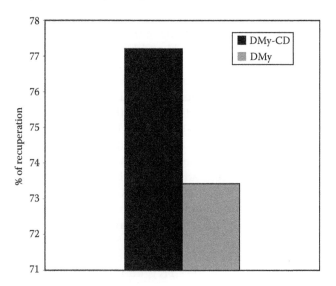

FIGURE 10.7
Oxidation response of pure DMy and DMy-CD.

concentration at 40°C, being around 14 times more soluble than pure DMy. It is also evident that the complex exhibits faster dissolution rates than pure DMy. Similar observations were reported [29].

The structure of flavonols is the determinant of their antioxidants properties [32] but also of its lability. Under the previously mentioned conditions, we studied the stability against the oxidation for DMy and its inclusion complex. When DMy and DMy-CD samples were treated with potassium chlorate as oxidant (0.1 mol/L) over a period of 6 h at 40°C, the product oxidation rate was reduced by 4%.

10.3.5 Chemical and Elastic Characterization of the Engineered Materials

Chemical surface changes in the RC-3 support and the modified as a consequence of the inclusion of the DHB-loaded LNs and the DMy cyclodextrin (RC-3/DHB-LNs and RC-3/DMy-CD) were studied by analyzing the XPS spectra, which allow the determination of the AC percentages of the different chemical elements present on the material surfaces. Table 10.1 shows the obtained AC values for the different samples, where the results for RC-3 and RC-3/DHB-LNs represent the average values of the two surfaces of both membranes (error lower than 7%); however, due to the significant differences in the values obtained for both faces of the RC-3/DMy-CD cyclodextrin–modified membrane, these results are indicated individually in Table 10.1. These results show different coverage on both membrane surfaces by the DMy-CD, and this fact was already observed with similar cellulosic supports and dendrimer quantum dots. Moreover, for comparison, the individual results obtained for the DHB-loaded LNs is also indicated.

The DHB-LNs' previous results indicated that the location of the DHB active is located in the nanoparticle core and surrounded by the lipids and surfactants' derivatives [22], according to the very small percentages for phosphorous and nitrogen detected, but the P/N atomic ratio corresponds to the theoretical value ($[P/N]T = 1$). This point confirms the previous results related to spherical geometry obtained by SEM (Figure 10.3). Moreover, the increase of carbon and nitrogen percents in the RC-3/DHB-LN–engineered material when compared with the RC-3 cellulosic support and the decrease in oxygen percentage values clearly show the surface coverage of the RC-3 by the DHB-LNs.

However, information on changes in the bulk material characteristic parameters is necessary to unambiguously establish the inclusion of the DHB-LNs into the cellulosic chains to check the mechanic stability of this engineered material or tissue. For this reason, the

TABLE 10.1

Atomic Concentration Percentages of the Elements Found on the Samples' Surfaces

Sample	C 1s (%)	O 1s (%)	P 2p (%)	N 1s (%)
DHB-LNs	85.6	13.6	0.4	0.4
RC-3	61.7	37.6	—	0.2
RC-3/DHB-LNs	72.2	25.9	0.4	1.2
RC-3/DMy-CD(A)	65.2	34.6	—	0.2
RC-3/DMy-CD(B)	77.4	22.5	—	0.1

Note: Small percentages obtained for non-characteristic material elements or impurities (AC < 0.3) are not indicated and, consequently, the total% might be lower than 100%.

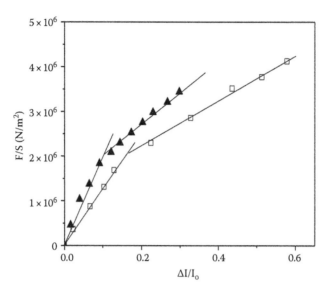

FIGURE 10.8
Membrane elongation as a function of the tensile stress for samples RC-3 (□) and RC-3/DHB-LNs (▲).

elastic behavior of both RC-3 and RC-3/DHB-LN solid structures was studied by comparing their normal stress–strain curves. This sample was chosen due to its higher homogeneity according to the XPS results. Figure 10.8 shows normal tensile stress (F/S) versus elongation or strain ($\Delta L/L$) for membrane RC-3 and composite RC-3/DHB-LNs, where differences in the values and shape of the curves depending on the system considered can be observed.

Both curves show two different linear behaviors depending on the applied tensile strength, with significantly higher slopes for tensile stress lower than 2 MN/m^2, but the slopes of both samples decrease at higher stress values; a similar behavior was already reported in the literature for different cellulose-based membranes [33,34]. As can be observed, the inclusion of the LNs in the structure of the cellulose support clearly modified the slope of the linear part of the elastic curve at low deformations, which represents the Young or elasticity modulus, from 10 to 20 MPa, but it also reduces the membrane maximum elongation in around 44%. The increase of the Young modulus seems to indicate a more rigid character for the RC-3/DHB-LN structure, which is associated to the presence of the LNs, and it might also affect the transport of solute/molecules across the modified composite material.

10.3.6 Drug Delivery Results

The DHB delivery from the loaded LNs at different times was obtained by HPLC analysis using cyclohexane and these results are shown in Figure 10.9. As can be observed, a significant increase of DHB was detected at the beginning of the experiments (t < 20 min), but a plateau was already reached.

To estimate the effect of the cellulosic support on the DHB release, a comparison of the values obtained at a given time (30 min) from both DHB-LN and RC-3/DHB-LN systems

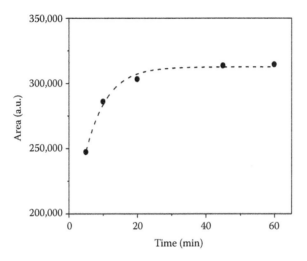

FIGURE 10.9
DHB delivery from the loaded LNs at different times.

was performed and the results obtained indicate that the DHB realized from the modified film or engineered material is around 5% of that determined for the LNs.

10.4 Conclusions

Preparation of DHB- and DMy-loaded GT–PC nanoparticles (DHB-LNs and DMy-LNs) by combining the ultrasound and high-pressure homogenization methods was successfully performed. An active drug loading of around 10 wt% in both kinds of LNs was reached, while an EE% of 95% and 89% for DHB-LNs and DMy-LNs, respectively, was achieved. Physicochemical characterization of LNs and CNs was performed by Raman, AFM, and SEM. Nanoparticles show quasi-spherical morphology, while a higher chemical stability against oxidation was obtained for DMy.

DHB-LNs and DMy-CDs were embedded in a highly hydrophilic support by a simple immersion method and the resulting engineering material was chemically and mechanically analyzed to establish the LNs' effect on the support structure.

DHB release from free LNs was studied as well as the effect of the LNs' entrapment into the solid support on the drug liberation. These results show the possibility of applying this biocompatible tissue for drug delivery under skin-mimic conditions.

Acknowledgments

We thank MINECO, Spain (Projects CTQ/2010-17633 and CTQ/2011-27770) and Junta de Andalucía (Project P10-FQM-6778) for financial support.

References

1. Müller, R. H., Mäder, K., and Gohla, S. 2000. Solid lipid nanoparticles (SLN) for controlled drug delivery—A review of the state of the art. *Eur. J. Pharm. Biopharm.* 50:161–177.
2. Siepmann, J. and Peppas, N. A. 2001. Modeling the drug release from delivery systems based on hydroxypropyl methylcellulose (HPMC). *Adv. Drug Deliv. Rev.* 48:139–157.
3. Yener, G., Dal, Ö., and Üner, M. 2009. Effect of vehicles on release of meloxicam from various topical formulations. *Open Drug Deliv. J.* 3:19–23.
4. Goodsell, D. S. 2004. *Bionanotechnology*. Hoboken, NJ: Wiley–Liss.
5. Bilensoy, E. 2011. *Cyclodextrins in Pharmaceutics, Cosmetics, and Biomedicine*. Hoboken, NJ: John Wiley & Sons, Inc.
6. Amsden, B. G. and Goosen, M. F. A. 1997. An examination of factors affecting the size, distribution and release characteristics of polymer microbeads made using electrostatics. *J. Control. Release* 43:183–196.
7. Xie, J., Marijnissen, J. C. M., and Wang, C. 2006. Microparticles developed by electrohydrodynamic atomization for the local delivery of anticancer drugs to treat C6 glioma in vitro. *Biomaterials* 27:3321–3332.
8. Xie, J., Lim, L. K., Phua, Y., Hua, J., and Wang, C. 2006. Electrohydrodynamic atomization for biodegradable polymeric particle production. *J. Colloid Interface Sci.* 302:103–112.
9. Gupta, R. B. and Kompella, U. B. 2006. *Nanoparticle Technology for Drug Delivery*. New York: Taylor & Francis, CRC Press.
10. Muchow, M., Maincent, P., and Müller, R. H. 2008. Lipid nanoparticles with a solid matrix (SLN, NLC, LDC) for oral drug delivery. *Drug Dev. Ind. Pharm.* 34:1394–1405.
11. Radtke, M., Souto, E. B., and Müller, R. H. 2005. Nanostructured lipid carriers: A novel generation of solid lipid drug carriers. *Pharm. Technol. Eur.* 17:45–50.
12. Le-Jiao, I., Zhang, D.-R., Li, Z.-Y., Feng, F.-F., Wang, Y.-Ch., Dai, W.-T., Duan, C.-X., and Zhang, Q. 2010. Preparation and characterization of silybin-loaded nanostructured lipid carriers. *Drug Deliv.* 17:11–18.
13. Huynh, N. T., Passirani, C., Saulnier, P., and Benoit, J. P. 2009. Lipid nanocapsules: A new platform for nanomedicine. *Int. J. Pharm.* 379:201–219.
14. Korting, H. C. and Schäfer-Korting, M. 2010. Carriers in the topical treatment of skin disease. *Handb. Exp. Pharmacol.* 197:435–468.
15. Martins, S., Sarmento, B., Ferreira, D. C., and Souto, E. B. 2007. Lipid–based colloidal carriers for peptide and protein delivery—Liposomes versus lipid nanoparticles. *Int. J. Nanomed.* 2:595–607.
16. Feng, S.-S. and Huang, G. 2001. Effects of emulsifiers on the controlled release of paclitaxel (Taxol®) from nanospheres of biodegradable polymers. *J. Control. Release* 71:53–69.
17. Benavente, J., Vázquez, M. I., Hierrezuelo, J., Rico, R., López-Romero, J. M., and López-Ramírez, M. R. 2010. Modification of a regenerated cellulose membrane with lipid nanoparticles and layers. Nanoparticle preparation, morphological and physicochemical characterization of nanoparticles and modified membranes. *J. Membr. Sci.* 355:45–52.
18. Vázquez, M. I., Peláez, L., Benavente, J., Hierrezuelo, J., Rico, R., López-Romero, J. M., Guillén, E., and López-Ramírez, M. E. 2011. Functionalized lipid nanoparticles-cellophane hybrid films for molecular delivery. *J. Pharm. Sci.* 100:4815–4822.
19. Barret, C. M., Ross, T., Guerrero, R., Hood, D. K. and Fares, H. 2011. Sun–care compositions. PCT Int. WO 2011017491 (A1).
20. Zhang, B., Dong, S., Cen, X., Wang, X., Liu, X., Zhang, H., Zhao, X., and Wu, Y. 2012. Ampelopsin sodium exhibits antitumor effects against bladder carcinoma in orthotopic xenograft models. *Anti-Cancer Drugs* 23:590–596.
21. Shen, Y., Lindemeyer, K., González, C., Shao, X. M., Spigelman, I., Olsen, R. W., and Liang, J. 2012. Dihydromyricetin as a novel anti-alcohol intoxication medication. *J. Neurosci.* 32:390–401.

22. Hierrezuelo, J., Benavente, J., López–Romero, J. M., Martínez de Yuso, M. V., and Rodríguez-Castellón, E. 2012. Preparation, chemical and electrical characterization of lipid nanoparticles loaded with dihydroxybenzophenone. *Med. Chem.* 8:541–548.

23. Hou, D., Xie, C., Huang, K., and Zhu, C. 2003. The production and characteristics of solid lipid nanoparticles (SLNs).*Biomaterials* 24:1781–1785.

24. Vázquez, M. I., de Lara, R., and Benavente, J. 2008. Chemical surface, diffusional, electrical and elastic characterizations of two different dense regenerated cellulose membranes. *J. Colloid Interface Sci.* 328:331–337.

25. Ariza, M. J., Benavente, J., and Rodríguez-Castellón, E. 2009. The capability of X-ray photoelectron spectroscopy in the characterization of membranes: Correlation between surface chemical and transport properties in polymeric membranes. In: *Handbook of Membranes: Properties, Performance and Applications*, ed. S. V. Gorley, pp. 257–290. New York: Nova Science Publishers, Inc.

26. Moulder, J. F., Stickle, W. F., Sobol, P.E., and Bomben, K. D. 1992. *Handbook of X-Ray Photoelectron Spectroscopy*. Minneapolis, MN: Perkin-Elmer Corporation.

27. Alukda, D., Sturgis, T., and Youan B.-B. C. 2011. Formulation of tenofovir-loaded functionalized solid lipid nanoparticles intended for HIV prevention. *J. Pharm. Sci.* 100:3345.

28. Ito, I., Fujimori, H., Kawakami, H., Kanamura, K., and Makino, K. 2011. Technique to encapsulate a low molecular weight hydrophilic drug in biodegradable polymeric particles in a liquid–liquid system. *Colloids Surf. A: Physicochem. Eng. Asp.* 384:368–373.

29. Ruan, L.-P., Yu, B.-Y., Fu, G.-M., and Zhu, D.-N. 2005. Improving the solubility of ampelopsin by solid dispersions and inclusion complexes. *J. Pharm. Biomed. Anal.* 38:457–464.

30. Jullian, C., Miranda, S., Zapata-Torres, G., Mendizábal, F., and Olea-Azar, C. 2007. Studies on inclusion complexes of natural and modified cyclodextrin with (+)catechin by NMR and molecular modeling. *Bioorg. Med. Chem.* 15:3217–3224.

31. Blanch, G. P., Ruiz del Castillo, M. L., Caja, M. M., Pérez–Méndez, M., and Sánchez Cortés, S. 2007. Stabilization of all-trans-lycopene from tomato by encapsulation using cyclodextrins. *Food. Chem.* 105:1335–1341.

32. Lucas-Abellán, C., Fortea, I., Gabaldón, J. A., and Nuñez-Delicado, E. 2008. Encapsulation of quercetin and myricetin in cyclodextrins at acidic pH. *J. Agric. Food. Chem.* 56:255–259.

33. Liang, S., Zhang, L., and Xu, J. J. 2007. Morphology and permeability of cellulose/chitin blend membranes. *J. Membr. Sci.* 19:287–291.

34. Ramos, J. D., Milano, C., Romero, V., Escalera, S., Alba, M. C., Vázquez, M. I., and Benavente, J. 2010. Water effect on physical-chemical and elastic parameters for a dense cellulose regenerated membrane. Transport of different aqueous electrolyte solutions. *J. Membr. Sci.* 352:153–159.

11

Silver Nanoparticles: Nanotoxicity Testing and Bioapplications

R. Sharma, S. Kwon, and C.J. Chen

CONTENTS

11.1 Introduction

Nanomaterials in general have emerged as potential tools in almost every field from space to environment and health to robotics. With increasing demand of silver nanomaterials, it is necessary to evaluate toxicity carefully before accepting silver nanomaterial in wider anti-inflammatory or wound healing, drug delivery bacterial systems, and several other implant bioapplications. In this chapter, technical developments on silver nanomaterials are described with account of historical development, experimental models, and potential applications. Chapter 11 is divided in subsections on preparation of silver-impregnated cellulose (SIC); antibacterial use of silver nanocomposites; characteristics and mechanism of silver-induced cytotoxicity with emphasis on cytomorphological changes in bacterial membrane, nucleus, and gene with account of metabolic integrity loss; oxidative stress; DNA damage; and cell cycle arrest induced by silver nanoparticles (Ag-NPs). The next section describes the future of possible bioapplications of impregnated Ag-NPs with major focus on proper selection of sensitive, reproducible, accurate, and user-friendly techniques used in characterization and testing cytotoxicity of nanocomposites at nanoscale to capitalize the silver nanocomposites to design possible long-lasting anti-inflammatory safe delivery systems to target and treat inflammation and bacterial infections. Authors put forth a concept of "optimal use of right element or particle size in enough amount is best choice" and believe that bacteria or tissue cells have a program to use their energy and metabolic resources in defense such as bacteria and cells in tissues experience initially metabolic integrity loss, apoptosis, respiratory burst, enzymatic regulation, cell cycle changes, proliferation, metabolic arrest, genomic, cytological toxicity, and cell death in a sequence depending on influence of silver particle–specific concentration, elemental silver oxidation state, surface, and shape.

11.1.1 Nanotoxicity of Silver Nanomaterials

The concept of "silver nanotoxicity" was imbibed long time back since World War I by showing healing power of silver paste. However, later silver paste was found to be toxic and antibacterial. In recent past, rapid progress and early acceptance of nanobiotechnology increased further the potential use of silver nanomaterials in spite of their adverse effects due to nitric oxide production in alveolar cells [1]. However, the environmental impact of silver nanomaterials remained a controversial hot topic of increased multiple usage of silver nanomaterials in future. In particular, the behavior of Ag-NPs sitting inside the mesenchymal stem cells to cause metabolic consequences and immunological and inflammatory responses in osteogenic cells induced by Ag-NPs still remains poorly understood [2]. Moreover, nanotoxicity testing of Ag-NPs takes up this challenge to explore the molecular events that regulate bioaccumulation of toxic products and events of silver cytotoxicity as a consequence of Ag-NP-induced toxicity. Toxicity testing of silver nanomaterials serves as a testing criterion of designing and characterizing new biocompatible silver nanomaterials or silver nanocomposites suitable for bioapplications shown in Table 11.1. The following sections describe the cytotoxicity of metallic silver complexes.

11.1.2 Motivation: Combining Nanotechnology and Silver Nanoparticle Surface Science with Growing Bioapplications

We describe our "right choice is the best choice" concept and review the newly explored applications of silver nanocomposites for their new medical applications and toxicity testing. Initially, silver was considered active antibacterial and anti-inflammatory ingredient for external applications. Since ages, silver has been in use as dental replacement material. Recently, applications of nanosilver were reported as antimicrobial, anticancer, antibacterial, anti-inflammatory, antiseptic, or wound healing based on the Ag-NPs' action on DNA, gene, and cell organelles as outlined in Table 11.1. This growing knowledge on silver

TABLE 11.1

Toxicity Testing of Silver Nanoparticles and Bioapplications

Author	Action	Application	Reference
Xiu et al.	Antibacterial	Experimental study	[3]
Du et al.	Antibacterial	Experimental study	[4]
Jena et al.	Antibacterial	Experimental study	[5]
Araújo et al.	Antibacterial	Preclinical study	[6]
Jain et al.	Antimicrobial	Experimental study	[7]
Gutherie et al.	Wound healing	Experimental study	[8]
De Lima et al.	Cytotoxicity	Clinical study	[9]
Sintubin et al.	Biomedical	Preclinical study	[10]
Sathishkumar et al.	Photosynthesis	Experimental study	[11]
Lemcke et al.	Antimicrobial	Clinical trial	[12]
Keen et al.	Wound healing	Silver dressing	[13]
VerdúSoriano et al.	Wound healing	ALEA clinical trial	[14]
Miller et al.	Ulcer healing	Clinical trial	[15]
Gravante et al.	Burn healing	Clinical trial	[16]
You et al.	Antibacterial	Feasibility study	[17]
Kim et al.	Cytotoxicity	Mechanistic study	[18]
Nirmala et al.	Cytotoxicity	Hydroxyapatite nanofiber	[19]
Böhmert et al.	Cytotoxicity	Human study	[20]
Song et al.	Cytotoxicity	Experimental study	[21]
Christen et al.	Cytochrome P4501A	ER stress response	[22]
Stoehr et al.	Cytotoxicity	Experimental study	[23]
Kang et al.	Cytotoxicity	Heme oxygenase-1 pathway	[24]
Hackenberg et al.	Cytotoxicity	Experimental study	[25]
Eom et al.	Cytotoxicity	Experimental cancer study	[26]
Gopinath et al.	Cytotoxicity	Gene therapy study	[27]
Sanpui et al.	Cytotoxicity	Experimental cancer study	[28]
Austin et al.	Cytotoxicity	Experimental study	[29]
Sriam et al.	Antitumor activity	Experimental study	[30]
Sur et al.	Cytotoxicity	Experimental study	[31]
Samberg et al.	Skin toxicity	Experimental study	[32]
Kokura et al.	Skin cosmetics	Human study	[33]
Panacek et al.	Antibacterial	Clinical study	[34]
Asharani et al.	Genotoxicity	Clinical study	[35]
Boca et al.	Anticancer	Experimental study	[36]
Piao et al.	Antioxidant	Experimental study	[37]
Ma et al.	Skin cytotoxicity	Human study	[38]
Garcia-Contreas et al.	Dental stabilization	Clinical study	[39]
Hernandez-Sierraet	Dental stabilization	Clinical study	[40]

metal or nanoparticle physical and chemical behavior and silver–tissue cell interactions motivated us to focus on right choice of silver particle physical state, size, surface, and shape in enough amounts to achieve specific interaction with tissue cell metabolism as desired bioapplication. In present time, more promising applications of Ag are emerging in genotoxicity, cell cytotoxicity in cancer, skin and wound healing, and replacement of hard implants such as bones and teeth.

11.2 Silver Nanoparticles and Toxicity Testing

Ag-NPs have become most commonly used in electronics and medical fabrics with future potentials in usage as Ag nanoprobes in cancer research and drug delivery in pharmaceuticals. Increased usage of Ag nanoprobes necessitates investigating the cytotoxicity and genotoxicity of Ag nanoprobes and dependence of Ag reactivity on structure, size, and chemical composition. Ag nanoprobes are prepared by coating 15–25 nm sized nanoparticles by hydroxyl-terminated polyethylene glycol (PEG). A most common cytotoxicity test is the reduction of 3-(4,5-dimethylthiazol-2-yl)-2,5-diphenyltetrazolium bromide (MTT), a yellow tetrazole, to formazan-based MTT assay. MTT assay is based on mitochondrial dehydrogenase reaction to form formazan crystals and dissolution by adding MTT reagent (Sigma, St Louis) in presence of added Ag-NPs to measure survival rates of cells in culture. Another most common genotoxicity test is cytokinesis-block micronucleus (CBMN) assay.

In the following sections, Ag cytotoxicity and genotoxicity are described with detailed account of SIC nanocomposite for its antibacterial and anti-inflammatory properties.

Cytotoxicity of Ag in various forms was used for centuries as an antimicrobial agent. Ancient Greeks used Ag coins as a way of disinfecting stored liquids [41]. In the late 1800s, Ag nitrate solutions were used to prevent ocular infections in neonates, typhoid, and anthrax bacilli [42]. In the nineteenth century, colloidal Ag was approved by the FDA for use in wound treatment and through the first half of the twentieth century was used in applications such as decreasing the stuffiness in nonhealing wounds and controlling infection in burn wounds [43]. Approximately 30 years later, Ag began to receive renewed clinical attention as cytotoxic antimicrobial agent [44]. In present time, Ag-impregnated catheters and wound dressings were used in therapeutic applications. In spite of the wider usage of Ag-NP in wound dressings, which can cause easy entry into the cells, very few reports on the toxicity of Ag-NPs at cellular level are available. The larger surface area and smaller size of the nanoparticles are expected to increase the *in vivo* activity. Although a few research groups have investigated the toxicity of Ag nanocomposites and nanoparticles in cell lines to estimate viability and reactive oxygen species (ROS) generation, little is known about the mechanisms of Ag-NP toxicity, primary targets, and distribution patterns. Recent reports have established involvement of mitochondria-dependent c-Jun N-terminal kinase (JNK) pathway in Ag-NP toxicity [45]. *In vivo* experiments in rats have established lung function changes and inflammation. Ag-NPs stabilized with starch and BSA may induce distinct development defects in zebra fish embryos. Various tests of Ag-NP toxicity were explored based on the effect of starch-coated Ag-NP on cell viability, ATP production, DNA damage, chromosomal aberrations, and cell cycle. It was expected that the biokinetics of nanoparticles can be measured as the rate of nanoparticle uptake, intracellular distribution based on the fact that exocytosis contributes to nanoparticle toxicity [46]. The nanoparticle size, surface area, and surface functionalization are other major factors that influence biokinetics and toxicity. In the following section, antimicrobial action and cytotoxicity of Ag-NPs are described.

The characteristics of silver particles that contribute to its increasing use are considerable as follows. First, there are evidences that silver is an effective antimicrobial agent. It has been found that yeasts, fungi, viruses, and a broad spectrum of anaerobic, aerobic, Gram-positive, and Gram-negative bacteria are susceptible to the effects of ionic silver [47]. As an advantage for its use in clinical applications, silver has also been categorized to be "oligodynamic," in that large bactericidal effects can be achieved by very small quantities [42,44]. Regarding silver's antibiotic effects, it is largely accepted that silver is only active in its soluble form, Ag^+ or Ag^0 [45]. In this active form, three proposed mechanisms of action in bacteria were reported: (1) silver interfered with DNA replication, inhibits electron transport, and/or causes conformational changes on the cell membrane [46]; (2) if silver was used in optimal concentrations, it was not toxic to mammalian cells [47–49]; and (3) silver showed the relatively low propensity for bacteria to develop a tolerance to silver-induced effects [50].

In the following section, we describe the techniques developed over the years for SIC made of ionic silver with water-soluble polymers by introducing the Ag^+–polymer complex into the cellulose matrix. The focus was to test silver cytotoxicity in *Candida albicans*, *Micrococcus luteus*, *Pseudomonas putida*, and *Escherichia coli* for establishing antimicrobial property of Ag-NPs. In our protocol, all microbial cells were allowed to grow in the presence of SIC at concentrations below 0.0035 Ag w/v% [50]. The major outcome was as follows: (1) At low concentration of 0.00035 Ag w/v%, *Pseudomonas aeruginosa* failed to grow, but *C. albicans*, *Staphylococcus epidermidis*, and *E. coli* grew partially at low concentration; (2) ionic silver showed cytotoxicity to human fibroblast cells at five different concentrations of SIC, while no cytotoxicity to human fibroblasts was observed in SIC of 0.035 Ag w/v% or lower. Our lab established that the present approach of silver cytotoxicity has novelty and serves the purpose of new human fibroblast cell model of cytotoxicity to test antimicrobial action using SIC. The study has wider implications both in microbiology and industry. In the following section, we describe the preparation of silver nanocomposites, antimicrobial testing methods, and cellular toxicity measurement protocols used in the lab for interested readers [50].

11.2.1 Qualitative Elemental Analyses of Silver Nanoparticles

Elemental analyses of Ag-NPs were conducted using inductively coupled plasma (ICP, PerkinElmer Optima 3000 DV). ICP can detect silver at an optimized wavelength of 328.068 nm, with relative detection limit of 0.9 parts per billion Ag^+ ion. In brief, the method of digestion was as follows: SIC pellet was shaken in 50 mL of double ionized water at 150 rpm for 24 h. The supernatant was tested via ICP for residual silver concentration. The pellet was immersed in a digestion apparatus containing a boiling solution of nitric acid and hydrogen peroxide and left for 24 h. The resulting solution of digested pellet sample was diluted to 50 mL and tested via ICP for residual silver concentration. To analyze the characteristics of silver coating to cellulose matrix, an Electro Scan environmental scanning electron microscope (ESEM, model E3) was used as described elsewhere [48].

11.2.2 Silver-Impregnated Cellulose

The effect of silver-coated cotton fibers on human epithelial cells serves the purpose of developing a user-friendly cytotoxicity method in isolated cells in culture. The intriguing issue is why silver ions do not exhibit cytotoxicity to human cells but they show significant antimicrobial effects. The answer may lie in the difference in culture media, growth conditions, membrane structure, and enzyme properties between human cells and bacterial cells. It may also depend on the concentration of ionic silver and the structure of silver–polymer complex with cellulose. Furthermore, SIC has no particle leaching problems and no particle formation during molecular coating within a defined concentration range. Such property of SIC has increased beneficial effects of SIC on human cells with increased antimicrobial activity. Human fibroblasts may be seeded into silver-coated cotton (cellulose matrices), or fibroblast cells could attach and grow on both plastic- and silver-coated cotton surfaces. Such culture configuration is suitable to study the effects of silver-coated cotton on human cell viability and cytotoxicity. Silver coating method was developed by our team [50]. In brief, Ag cation was dissociated from sodium chloride (AgCl) in water, and the Ag–polymer complex was formed by adding a water-soluble polymer (carboxymethyl cellulose) to the solution. Then the Ag–polymer complex was absorbed into a cellulose matrix (cotton pellet) by adding cotton pellets to the Ag–polymer complex in solution and was incubated and dried. In order to avoid the cotton pellet's dispersion in water, pellets were treated with a trivalent aluminum solution before coating the cotton pellets with silver and soaked in solution under vacuum pressure until sufficient absorption had occurred. The pellets were then rinsed, until no more aluminum effluent was observed, and subsequently dried. Finally, the SIC pellets were dried and kept in a desiccator cabinet [50]. ESEM pictures of a silver-coated cotton pellet prepared from the following concentrations: 0, 0.00035, 0.0035, 0.035, 0.35, and 0.7 Ag w/v% indicated peculiar behavior of silver crystals (particles) in the treated cotton pellets at the following concentration range: 0.035–4.5 Ag w/v%. However, no silver particles were observed at the concentration of 0.0035 Ag w/v% or lower. This concentration range for particle forming was confirmed by energy dispersive x-ray analysis system. In this x-ray analysis, silver was evenly distributed over all the surfaces. Therefore, silver may be chemically bonded to cellulose by the mechanism indicated in Figure 11.1. In the following section, observations of cultures and silver toxicity are described using leaching test and digestion results of various cell media containing bacteria and human fibroblast cells.

Other studies also indicated that the silver–polymer composite makes the complex with the cellulose matrix as SIC. The complex comprised the associating ionic silver (Ag^+) with water-soluble polymers by introducing the Ag^+–polymer complex into the cellulose matrix by possible binding between Ag^+ and celluloses. Investigators established that ionic silver can have up to four ligands, and it is well known that hydroxyl groups from C-2 to C-3 can bind to metal ions in a bidentate fashion [51]. The C-3 hydroxyl group has the highest acidity among the three hydroxyl groups [51], so it is reasonable to believe that one of the C-3 hydroxyl groups binds to the silver in alkoxylate form. These hydroxyl groups effectively chelate to the metal Ag^+ ion and prevent its possible leaching in aqueous solution. In the following section, we describe the silver-induced cytotoxicity in bacteria.

11.2.3 Silver Nanoparticle Cytotoxicity in Bacteria

Several bacteria including *C. albicans*, *S. epidermidis*, *P. aeruginosa*, and *E. coli* are choice of Ag-NP cytotoxicity testing mainly available from American Type Culture Collection (ATCC). For Ag-NP toxicity experiments, bacterial cells may be cultivated in Luria broth (LB) media for 24 h at 37°C, spun down, and washed twice in phosphate buffered saline (PBS). In our lab, bacterial cells in PBS were incubated with silver-coated cotton pellets at six different concentrations of silver in shaking flasks at starting concentration of microbial cells ~10^4 cells/mL or ~10^9 cells/mL. Kim et al. reported the concentration of microbial cells from 100 μL of liquid sample of each flask monitored by spectrophotometer set at 600 nm after the incubation of microbial cells with silver-coated cotton pellets at six different concentrations for 48 h [50]. Investigators indicated that all microbial cells failed to grow in the presence of SIC at concentrations above 0.0035 Ag w/v%. Even at a concentration of 0.00035 Ag w/v%, *P. aeruginosa* failed to grow, but *C. albicans*, *S. epidermidis*, and *E. coli* grew partially at this concentration. These observations indicated that cotton pellets

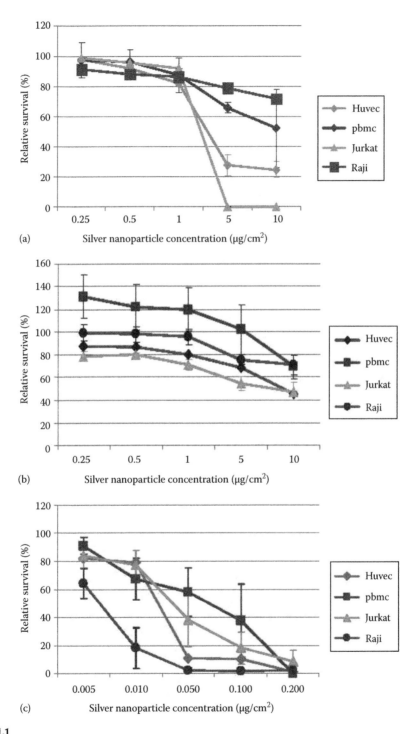

FIGURE 11.1
(a) Cytotoxicity of dialyzed silver 15 nm nanoparticles in different human cells following 24 h treatment by MTT assay. (b) Cytotoxicity of PEG-dialyzed Ag-NP in different human cells following 24 h treatment by MTT assay. (c) Cytotoxicity of nanorods in different human cells following 24 h treatment by MTT assay.

(continued)

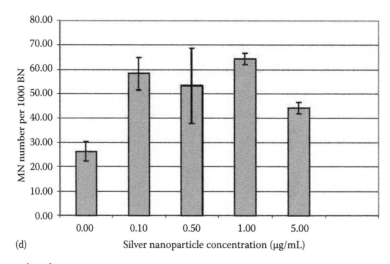

(d)

FIGURE 11.1 (continued)
(d) Genotoxicity of 15 nm Ag-NP in different human peripheral blood cells following >20 h treatment by micronuclei formation determined by CBMN assay.

treated with a trivalent aluminum solution before coating with silver were not dispersed in water and not allowed bacterial cell adsorption to cotton pellets. In the following section, we describe working protocol on silver-induced toxicity in fibroblasts.

11.2.4 Silver Toxicity on Human Fibroblasts

Silver toxicity on human fibroblasts is commonly available from ATCC (CCD-25Sk, CRL-1474) resource. These fibroblast cell lines originate from human skin. In our lab, Kwon et al. developed a technique of silver-coated cotton pellets (silver-coated cotton fibers) over years and applied on the surface of each well of 24 well plates to establish the silver toxicity on fibroblasts. Fibroblast cell lines can be seeded at 2×10^4 cells/cm^2 onto each well with the media containing 1.0 mL minimum essential medium (MEM) supplemented with nonessential amino acids, sodium pyruvate (1 mM), 10% fetal bovine serum (FBS), lactalbumin hydrolysate (0.1%), amphotericin B (5.6 mg/L), and penicillin–streptomycin (100 U/mL). Fibroblast cells grow in silver-coated cotton fibers with different concentrations of Ag-NPs.

The effect of silver-coated cotton fibers on human epithelial cells was established by observing fibroblast growth on SIC as shown in Figure 11.1. Cells were attached onto both a plastic surface and SIC. Cytotoxicity was monitored at day 1, day 2, and day 7. Live/dead assay kit (Molecular Probes, Eugene, OR) was used to identify the effect of silver-coated cotton fibers on cell viability and cytotoxicity. Five different concentrations of SIC tested with human fibroblasts showed viable cells (green) predominant in SIC of 0.035 Ag w/v% or lower, and an increased portion of dead cells (red) was detected in SIC of 0.35 Ag w/v% or higher, at both day 2 and day 7. These observations did not show cytotoxicity to human fibroblasts observed in SIC concentrations of 0.035 Ag w/v% or lower (Figure 11.2).

The following section describes the current information with new observations to speculate mechanism of silver cytotoxicity in cultured cells and cellular toxicity at organelle level to alter the metabolic integrity and control in eukaryotic cells. The details are given in the interest of readers. The employed Ag-NPs varied 6–20 nm in size with an absorption maximum at 400 nm. The calculated size distribution histogram is a standard method to

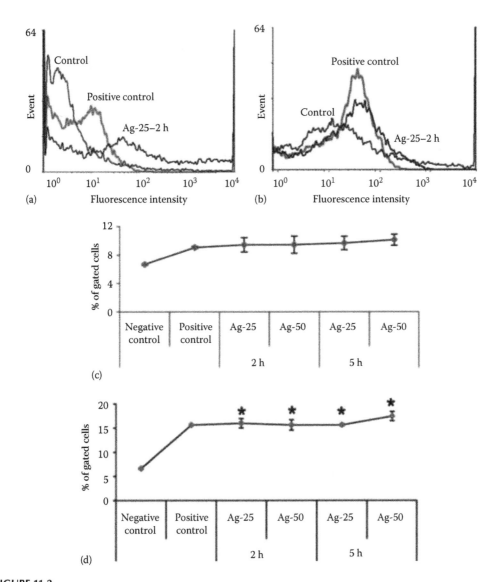

FIGURE 11.2
(a) Histogram represents data from DCF-DA staining for detecting hydrogen peroxide production in the Ag-NP-treated fibroblasts (25 g/mL). Shift was independent of the time of incubation starting from 2 h of incubation. (b) DHE staining of the cells suggests superoxide production and increased ROS generation. The x-axis represents the fluorescence intensity, and the y-axis represents the number of cells collected (10,000 cells). The graph represents the percent of gated cells for (c) DCF-DA staining and (d) DHE staining as obtained from the statistics generated by WinMDI 2.8 software. For DCF-DA staining along with untreated control, H_2O_2-treated cells were used as positive control. DDC was used as positive control for detecting superoxide production; *represents $P < 0.05$. (Reproduced with permission from AshaRani, P.V., G. Low Kah Mun, M.P. Hande, S. Valiyaveettil. Cytotoxicity and genotoxicity of silver nanoparticles in human cells. *Nano* 3(2): 279–90. Copyright 2009, American Chemical Society.)

confirm the size distribution of nanoparticles. The Ag-NPs showed their stability in water, concentration-dependent cytotoxicity (low metabolic activity), genotoxicity (DNA damage and chromosomal aberrations), and cell cycle arrest in Ag-NP-treated cells. The electron micrographs of tissue cells showed the presence of endosomes with Ag-NPs in the cytosol, suggesting a receptor-mediated endocytosis mechanism.

For cytotoxicity, dose–response study demonstrated poorly significant toxicity of Ag-NP treatment for 24 h transition time below 5 µg/cm^2 or lower in Jurkat cells, human umbilical vein endothelial cells (HUVEC), peripheral blood mononuclear cells (PBMC), and RAJI cells as shown in Figure 11.1a–c for Ag-NPs, dialyzed silver–PEG, and silver nanorods. Dialyzed PEG-coated Ag-NPs show specific dose–response to the survival of T cell leukemia Jurkat, B lymphoblastoid cells Raji, HUVEC, and human PBMC depending on types, concentrations, treatment times, and target cell types.

At increased transition times, cells showed more toxicity and less survival at lower dosage of nanoparticles. Polyethylene glycated Ag-NPs exhibit less toxicity, while silver nanorods exhibit highest cytotoxicity. For genotoxicity, human peripheral blood cells treated with 15 nm Ag-NPs at different concentrations showed dose–response graph of chromosome breakage or micronuclei (MN) loss per 100 binucleate cells as shown in Figure 11.1d.

Observations of MTT assays suggest that cytotoxic effects are dependent on the physiochemical properties, NP concentrations, duration of nanoparticle exposure, and target cell types. Ag-NPs clearly show the genotoxic effect in human blood cells especially chromosomal damage in cells. Moreover, it is alarming that the use of nanoparticles is possibly a public health hazard with issues of environmental impacts induced by engineered Ag-NPs and their products. In the following section, we describe the silver cytotoxicity in bacteria and its antimicrobial application.

11.2.5 Mechanism of Antimicrobial Action and Cellular Cytotoxicity

The mechanism of the antimicrobial action of silver ions is not well characterized yet even after continued efforts over three decades. Several investigators explored different mechanisms. Major emphasis was to indentify changes in cell morphology and intracellular, molecular, biochemical, and cytological events. With technical development, in present time, different investigators approached to define induced effects of silver particle size, shape, concentration, surface properties, atomic or elemental interactions in lattice, and tissue binding molecules with silver particles on induced cell damage at different levels of cell membrane, cell viability, mitochondria, nucleus, DNA, cell proliferation, cell cycle, and chemical reactions. The most widely accepted mechanism was closely related to the silver ion interaction with thiol groups [52–54], although other target sites were also well known as other possibilities [55,56]. Other findings also suggest that interaction of Ag$^+$ with thiol groups in enzymes and proteins may play an essential role in bacterial inactivation, although other previously mentioned cellular components may be involved. Hydrogen bonding, effects of hydrogen bond–breaking agents, and specificity of Ag$^+$ for thiol groups were discussed in greater detail [57]. Virucidal properties were explained by binding to thiol groups [58]. Recently, a study proposed that silver salts act by binding to key functional groups of fungal enzymes. Ag$^+$ causes the release of K$^+$ ions from microorganisms [59]; the enzymes associated with microbial plasma or cytoplasmic membrane were important target sites for Ag$^+$ activity [50,60]. In addition to its effects on enzymes, Ag$^+$ produces other changes in microorganisms. Silver nitrate causes marked inhibition of *Cryptococcus neoformans* growth and is deposited in the vacuole and cell wall as granules [61]. Ag$^+$ also inhibits cell division and damages the cell envelope and contents of *P. aeruginosa*. Bacterial cells increase in size, and the cytoplasmic membrane, cytoplasmic contents, and outer cell layers all exhibit structural abnormalities, although without any blebs (protuberances) [61]. In the following section, we describe the silver-induced changes in cell morphology.

11.2.6 Silver-Induced Effect on Cell Morphology

The effect on cell morphology is the first visible effect on cell shape or morphology in a monolayer culture. In previous report, microscopic observations of treated cells showed distinct morphological changes indicating the unhealthy cells, while control cells appeared normal [62]. Investigators described the Ag-NP-treated cells appeared as clustered with a few cellular extensions, and cell spreading patterns were restricted as compared to control cells due to disturbances in cytoskeletal functions as a consequence of nanoparticle treatment. Similar results were observed by other groups in dermal fibroblast cells treated with citrate-coated gold nanoparticles. Dark orange patches seen on the cell surface were due to the adsorption of Ag-NPs on the cell surface. However, only a few floating cells were observed under the microscope to suggest the absence of widespread cell death due to necrosis [61]. In the following section, we describe the silver-induced changes in cell viability.

11.2.7 Silver-Induced Effects on Cell Viability

Cell viability–based cytotoxicity assays indicate the cellular response to a nanoparticle toxicant with information of cell death, survival, and metabolic activities. Other more sensitive luminescence-based assays, fluorescent-based assays, and ATP assays can assess the toxicity of Ag-NPs more accurately in concentration- and time-dependent drop in luminescence intensity in cancer cells and normal cells, signifying time- and dose-dependent toxicity. The drop in ATP content of the cells in the presence of nanoparticles represents as adverse effects of nanoparticles in concentration-dependent manner (reduced nanoparticle agglomeration and subsequent precipitation, reduced uptake rate of nanoparticles), which could be observed as a decrease in ATP depletion and higher cytotoxicity. For example, starch as control does not show any significant cytotoxicity in cells. Such inert behavior or biocompatibility of starch serves as capping agent in nanoparticles. Cytotoxicity of nanorods, gold, and silver on isolated cell cultures has been a robust research area in recent years. Gold nanoparticles and nanorods showed no significant toxicity in HeLa cells [63,64]. However, several challenges are important in using nanoparticles such as purity of nanoparticles, free from reactants used in the synthetic steps, and presence of any toxic materials left over from the synthesis. Similar outcome was reported in previous report with no evidence of toxicity in supernatant liquid comparable to control [65]. Authors advocated the high cell viability or low cytotoxicity with microscopic observation showing no indication of massive cell death, low ATP levels due to metabolic arrest, or low mitochondrial activity by cell titer assay (Figure 11.1d). In our view, structural and functional damage of the mitochondria or cell viability assays play pivotal role in metabolic arrest (not cell death) followed by a decrease in ATP and NADPH yield. To observe cell death, cell cycle, kinetics of apoptosis–necrosis, and genotoxicity are the choices. However, low ATP measure or poor mitochondrial activity may not always represent cell death but leads to metabolic inhibition in cells. In other studies of testing Ag-NP, starch controls did not show toxicity. It indicated that the observed toxicity was due to Ag-NP. Size-dependent toxicity of Ag-NPs was observed in fibroblast, epithelial cells, and melanoma cells [65]. Investigators emphasized that Ag-NPs show different degrees of *in vitro* cytotoxicity limited by time of exposure to Ag-NP and surface functionalization [65]. In the following section, we describe the silver-induced oxidative stress.

11.2.8 Role of Ag-Nanoparticles in Oxidative Stress

Oxidative stress in cells plays a role in cells due to nanoparticle toxicity. Oxidative stress has specific effects in the cells, including oxidative damage to protein and DNA [66]. In previous study, investigators established the role of oxidative stress as a decisive factor in starch-capped Ag-NP toxicity using dichlorofluorescein diacetate (DCF-DA) and dihydro-ethidium (DHE) staining methods. Authors described the presence of ROS by measuring fluorescent intensity of the cells stained with dyes increased or right shift of the emission maximum [67]. Untreated cells or control standards of ROS production showed the percentage of cells with increased fluorescence intensity due to increase in hydrogen peroxide and superoxide production in cells treated with 25 and 50 g/mL of Ag-NP. The percent of gated cells from DCF-DA stained and DHE staining indicated the ROS production. Authors described this effect due to exchange interactions between the unpaired electrons of the free radicals and the conduction band electrons of the Ag-NPs. Such effect has been reported for gold nanoparticles. It is possible that activation of a cellular antioxidant network had counterbalanced the effect of ROS [66]. In the following section, we describe the silver-induced changes in respiratory burst and radical formation.

11.2.9 Role of Ag-Nanoparticles in Mitochondrial Respiratory Chain, Synthesis of ATP, and ROS

Mitochondrial respiratory chain, synthesis of ATP, and ROS production with low cellular ATP content may be the effect of cell damage or slow mitochondrial respiratory chain indicated by the reduced dehydrogenase activity (measured by the reduction of resazurin to resofurin by CellTiter-Blue viability assay). The root cause of mitochondrial dysfunction in toxicology is ROS production and subsequent oxidative stress. Oxidative stress is a common mechanism for the cell damage induced by nano- and ultrafine particles [66]. Other reasons of cytotoxicity may be mechanical injury caused by accumulated nanoparticle deposits in mitochondria or mitochondrial damage. Such concept is based on nanoparticles of various sizes, and nanoparticle chemical compositions localized in mitochondria may induce major structural mitochondrial damage and may contribute to oxidative stress [67]. The majority of other nanomaterials such as zinc oxide, carbon nanotubes (CNTs), and silicon dioxide also exert their toxic effects through oxidative stress and ROS to cause metabolic disturbance [68], similar to titanium dioxide nanoparticles reported earlier [69].

From a biochemical standpoint, mitochondria are the major sites of ROS production in the cell. During oxidative phosphorylation, oxygen is reduced to water by addition of electrons in a controlled manner through the respiratory chain. Some of these electrons occasionally escape from the chain and are accepted by molecular oxygen to form the extremely reactive superoxide anion radical $\left(O_2^{-\cdot}\right)$, which gets further converted to hydrogen peroxide (H_2O_2) and in turn may be fully reduced to water or partially reduced to hydroxyl radical (OH•), one of the strongest oxidants in nature [68]. Toxic agents increase the rate of superoxide anion production, either by blocking the electron transport or by accepting an electron from a respiratory carrier and transferring it to molecular oxygen without inhibiting the respiratory chain. The inhibition of respiratory chain is expected to cause a decrease in ATP synthesis. The deposition of Ag-NP in mitochondria can alter normal functioning of mitochondria by disrupting the electron transport chain, ultimately resulting in ROS and low ATP yield. ROS are highly reactive and result in oxidative damage to proteins and DNA. Hence, it is indispensable to investigate genome stability in cells with significantly higher ROS production. It is possible that surface oxidation of Ag-NP, upon contact with cell culture medium or proteins in the cytoplasm, liberates Ag+ ions that could amplify the toxicity. Reactions between H_2O_2 and Ag-NPs are presumed to be one of the factors causing Ag+ ions to release *in vivo*. Another study has reported similar activity of cobalt and nickel nanoparticles to induce the release of corresponding ions due to enhanced toxicity [70]. A possible chemical reaction involves

$$2Ag + H_2O_2 + 2H^+ \rightarrow 2Ag^+ + 2H_2O \; E^0 = 0.17 \text{ V}$$

Half reaction

$$H_2O_{2(aq)} + 2H^+ + 2e^- \rightarrow 2H_2O_{(1)} \; E^0 = +1.77 \text{ V}$$

$$2Ag^+_{(s)} \rightarrow 2Ag^+_{(aq)} + 2e^- E^0 = 2(-0.8) \text{ V}$$

where H+ ions are present in abundance inside the mitochondria, where H+ efflux is the main event (proton motive force) in ATP synthesis.

Another study of toxicity of silver ions on respiratory chain in *E. coli* and other microbial cells [71] suggested mechanisms for the action of Ag+ ion in the respiratory chain to inhibit phosphate uptake and exchange in *E. coli*, which causes efflux of accumulated phosphate [72]. Thiols reversed the effect or enhanced phosphate uptake due to the reversal of binding of Ag+ with thiol-containing proteins in the respiratory chain. Our view is that it is one possibility out of other possibilities such as Ag+ ions may cause a leakage of protons through the membranes of *cells* to cause the collapse of proton motive force, possibly by binding to membrane proteins. Other possibilities may be NADH:ubiquinone reductase complex (Complex I) in high metabolic cells such as *E. coli* containing two types of NADH dehydrogenases, both containing cysteine residues with high affinity for silver. Both hydrogenases may appear as possible sites of Ag+ ion recognition [73]. Binding of Ag+ to these low-potential enzymes of the bacterial respiratory chain may result in an inefficient passage of electrons to oxygen at the terminal oxidase to cause the production of large quantities of ROS, thus explaining the toxicity of nanometal ions to *cells* at submicromolar concentrations [74]. A consequence of interaction of Ag+ ions with enzymes of the respiratory chain may be sudden stimulation of respiration followed by cell death, due to uncoupling of respiratory control from ATP synthesis. However, prokaryotic cells and eukaryotic cells have entirely different respiratory functions, which determine sensitivity and survival rate upon exposure to nanoparticles. Eukaryotic cells have a prominent nucleus, a complex DNA repair mechanism, and cell cycle pathway to control cell death and survival, which are absent in prokaryotic cells. Yamanaka et al. (2005) [75] studied the effect of Ag+ ions on expression of various proteins in *E. coli* by proteomic analysis. Silver ions were assumed to penetrate through ion channels in the cell without causing damage to the membrane surface. Proteomic analysis of cells treated with Ag+ ions further showed a reduction in expression of ribosomal subunit S2, succinyl coenzyme (CoA) synthetase, and maltose transporter [75]. Conceptually, the reduction in expression of ribosomal subunit S2 impairs the synthesis of other proteins, whereas the reduction in synthesis of succinyl CoA synthetase and maltose transporter causes the suppression of intracellular production of ATP, resulting in death of the cell. We believe that nanoparticle toxicity is multifactorial, where several factors of nanoparticle size, shape, surface functionalization, and potential to release the corresponding metal ions could play pivotal roles. In the following section, we describe the silver-induced changes in cell cycle status.

11.2.10 Effect of Ag-Nanoparticles on Cell Cycle

The effect of Ag-NPs on cell cycle can be attributed as oxidative stress in Ag-NP-treated cells indicating the possibility of DNA damage where the early effect may evidence the changes in cell cycle progression. Conceptually, cells with damaged DNA accumulate in gap1 (G1), DNA synthesis (S), or in gap2/mitosis (G2/M) phase. Cells with irreversible damage most of the times undergo apoptosis, giving rise to the accumulation of cells in subG1 phase [74]. Toxicity studies seem feasible by cell cycle analysis to detect parameters such as apoptosis, cell cycle arrest, and evidence of DNA damage [74]. The influence of nanoparticles on the cell cycle was analyzed by subjecting the nanoparticle-treated cells to flow cytometry. Authors reported statistical data from raw histograms using WinMDI software and the percent of cells in each phase of the cell cycle compared with that of controls. Both cell types showed a concentration-dependent G2 arrest, which was observed as an increase in cell population in G2/M phase compared to control. The lowest concentration of nanoparticles tested (25 g/mL) marked the onset of G2/M arrest. As the concentration of Ag-NPs was increased to 400 g, there was a massive increase (approximately 30%) in G2 population. In controls, major cell population was observed in G1 phase, whereas in Ag-NP-treated cells, a decrease in G1 population accompanied by an increase in G2/M population was detected [74]. The proportion of cells in S phase was less affected as compared to the G2/M population. Interestingly, no significant apoptosis was observed, as indicated by the absence of cell population in subG1.

11.2.11 Role of Ag-Nanoparticles in Apoptosis and Necrosis

Apoptosis and necrosis of cells may be used to assess the extent and mode of cell death. Annexin-V staining is a choice to confirm apoptosis [76]. In a previous study, statistical data extracted from the dot plots using WinMDI software showed different colors for the percentages of unstained cells (viable cells), red fluorescent labels (necrotic cells), green labels (apoptotic cells), and dual stained cells (late apoptosis) to indicate that only a small percentage of cells was undergoing apoptosis and necrosis at higher concentrations of Ag–starch nanoparticles [76]. Investigators reported an increase (5%–9% with respect to control) in the apoptotic cell population from 25 to 100 g/mL for fibroblasts, due to the observed ROS production, while 16% of cell death observed was due to late apoptosis and necrosis [76]. The induction of apoptosis specifically in low doses of nanoparticles accompanied by proliferation arrest at high concentrations suggested differential sensitivity of nanoparticle concentrations. Investigators interpreted it a situation where cells sustain DNA damage and gain resistance to cell death. Investigators further reported a concentration-dependent increase in DNA damage and G2/M arrest and established that DNA damage is increasing with concentration. Our recent report also has identified apoptosis as a major mechanism of cell death in exposure to nanomaterials [1]. However, conflicting results suggest on additional parameters in nanoparticle-mediated cell death, and we believe that it requires detailed study. Further experiments should be designed in detail to identify the molecular mechanisms underlying nanoparticle-mediated cell death such as DNA fragmentation analysis to study DNA fragmentation characteristic of late apoptosis in cells. Similar study showed no laddering patterns observed in the gel, which confirmed the absence of late apoptosis where nuclear fragmentation occurs. The absence of massive apoptosis and necrosis at higher concentrations of Ag-NP accompanied by G2/M arrest indicated a retarded cell proliferation. This inference was also supported by the cell cycle and genotoxicity data [77]. The following section describes the silver-induced genotoxicity and DNA damage.

11.2.12 Genotoxicity of Ag-Nanoparticles

The genotoxicity of Ag-NPs or DNA damage is measured by routine comet assay. Recent study on Ag-NPs using comet assay and cytokinesis-blocked micronucleus assay confirmed the chromosome rearrangement abnormalities as a direct consequence of DNA damage such as double-strand breaks and misrepair of strand breaks in DNA [78]. Investigators reported the MN formed in dividing cells from chromosome fragments or whole chromosomes unable to engage with the mitotic spindle during mitosis. Investigators observed an extensive and dose-dependent damage to DNA after treatment of the cells with Ag-NPs. Comet assay of Ag-NP-treated cells also showed a concentration-dependent increase in comet tail as compared to control cells that gave the extent of DNA damage as shown in Figure 11.3. Conceptually, comet-like tail implies the presence of a damaged DNA strand that lags behind if electrophoreses is done with an intact nucleus. The length of the tail increases with the extent of DNA damage as shown in Figure 11.3. Investigators reported different tail momentums of control DNA compared with nanoparticle-treated cells and evaluated the extent of DNA damage. We reported an increase in DNA damage with

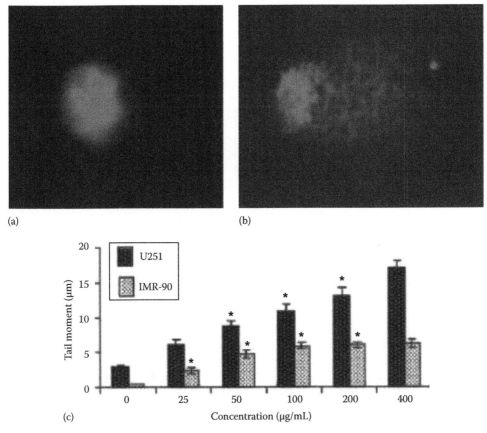

FIGURE 11.3
(See color insert.) Comet analysis: (a) untreated and (b) Ag-NP-treated cancer cells stained by SYBR green (conc. 400 g/mL). (c) Representation of the tail moments of DNA (m). Fibroblasts exhibited a concentration-dependent increase in DNA damage up to 100 g, above which the values remained constant, whereas cancer cells showed a steady increase; *represents $P < 0.05$. (Reproduced from Asharani, P.V. et al., *Adv. Funct. Mater.*, 20, 1233, 2010. With permission.)

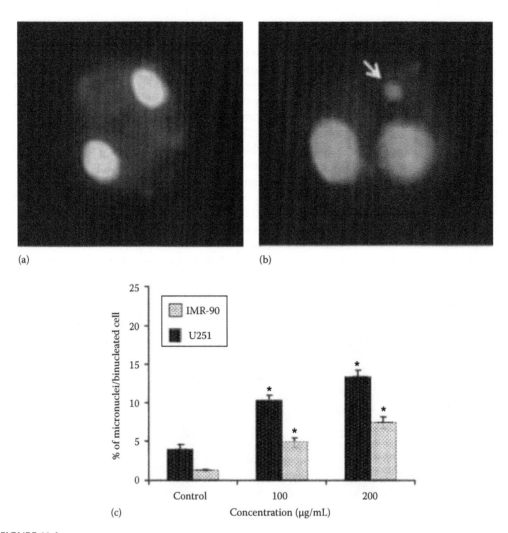

(a)   (b)

(c)

FIGURE 11.4
(See color insert.) Micronucleus analysis of (a) untreated and (b) Ag-NP (100 g/mL)-treated fibroblasts showing binucleated cell. (b) White arrow indicates the micronucleus formed among the binucleated cells. (c) Data from MNA show chromosomal aberrations. The data represent 1000 binucleated cells for U251 and 700 binucleated cells for fibroblasts; *represents $P < 0.05$. (Reproduced from Asharani, P.V. et al., *Adv. Funct. Mater.*, 20, 1233, 2010. With permission.)

increase in nanoparticle concentration observed in cancer cells, whereas the fibroblasts showed no increase in DNA damage beyond a nanoparticle concentration of 100 g/mL. Another study reported, in addition, the cytokinesis-blocked micronucleus assay results in corroborating the chromosomal breaks in Ag-NP-treated cells as compared to the untreated cells as shown in Figure 11.4. The extent of DNA damage was much higher in cancer cells as compared to fibroblasts, and significant numbers of MN were formed in cancer cells than fibroblasts. Apoptotic or necrotic cells were also observed by Coomassie blue micronuclei (CBMN) analysis, and annexin-V staining showed only a few apoptotic and necrotic cells.

As described earlier, the presence of Ag-NPs may cause the formation of ROS and reduction in ATP content. ROS are considered to be the major source of spontaneous damage to DNA. Oxidative attack on the DNA results in mutagenic structures such as 8-hydroxyadenine and 8-hydroxyguanine, which may induce instability of repetitive sequences. Earlier study reported the chemical reactions that bring about such mutations based on the formation of highly reactive and short-lived hydroxyl radical (OH•) in close proximity to DNA [77]. ROS-mediated genotoxicity has been previously observed for metal oxide nanoparticles [76]. This first report was published on quantitative measurements of DNA damage and chromosomal aberrations in Ag-NP-treated cells indicating the DNA damage through single- and double-strand breaks, inter- and intrastrand cross-linking, etc. On the other hand, silver ions have been shown to interact with DNA and RNA under *in vitro* conditions [79]. Ag ions usually form a type I complex by binding to N7 of guanine or adenine, and in a type II complex, it forms interstrand AT and GC adducts without causing much change in the conformation of DNA. Liu et al. [80] proposed initially the stabilization of DNA by Ag ions by studying the binding of Ag ions to plasmid and chromosomal DNA in *in vitro* condition. However, they found that, in the presence of ascorbate, Ag^+ ions cause significantly more damage to DNA than the ascorbate alone. Conceptually, Ag^+ ion-catalyzed oxidation of ascorbate anion by molecular oxygen may cause the formation of free radicals, which could further damage the DNA.

11.2.13 Role of Ag-Nanoparticles in DNA Damage, Reduced Cellular ATP Content, and Cell Cycle Arrest

DNA damage, reduced cellular ATP content, and cell cycle arrest in cells are major events. DNA damage causes the arrests of cell cycle progression at the G2/M boundary, allowing cells enough extra time to repair the DNA damage prior to segregation of chromosomes. Conceptually, DNA repair machinery makes access in the nucleosome in order to carry out the DNA repair using two classes of enzymes that are involved in regulating the accessibility to chromatin. One class enzyme modifies the core group histone amino acids and the other class consisting of large multisubunit complexes known as "chromatin remodelers," which may make use of the energy from ATP hydrolysis to weaken the interactions between histones and the surrounding DNA. The reduction in ATP content after Ag-NP treatment could affect the DNA repair, as ATP is required for a cascade of events requiring phosphorylation of several proteins taking part in repair of DNA damage [81]. The role of ATP in cell cycle arrest was studied through specifically inhibiting the mitochondrial production of ATP. It was shown that a small reduction in the cellular level of ATP induced a significant increase in the G1 cell population, while further decrease (up to 35%) elicited a G2/M accumulation followed by the onset of cytotoxicity [81]. It suggested that the checkpoints regulating passage through cell cycle events are sensitive to alteration in the ATP status of the cell. Investigators observed the extensive damage of DNA measured by comet assay and CBMN assay reflected into the arrest of the IMR-90 and U251 cells in the S and G2/M phases [82]. The number of cells in the G2/M phase also increased with increasing dose of the Ag-NP. Cell cycle arrest provided enough time for the cells to repair the damaged DNA. Similar results were reported for carbon-black nanoparticles [83]. Investigators demonstrated that cells treated with carbon-black nanoparticles suffered with DNA damage, which led to cell cycle arrest [83]. Recently, Ag-NP unveiled the effect of Ag-NP on the cell cycle and DNA damage caused by Ag-NP to the U251 cells and IMR-90 cells in terms of steeper increase in the number of cells in the G2/M phase with concentration of Ag-NP. The increased sensitivity of U251 cells to DNA damage was due to impaired repair pathways. In fact, current cancer therapy relies heavily on DNA damaging agents to induce programmed cell death in cancer cells.

With advancements in technical development, it is possible to visualize the size, shape, aggregation, and surface characteristics of Ag-NP in tissues and cells with a detail of cytological changes by transmission electron microscopy (TEM). In the following section, we describe present awareness on silver-induced morphology changes.

11.2.14 Transmission Electron Microscopy

Cell sections may visualize the biodistribution of the Ag-NP. Nanoparticles get deposited in endosomes or nucleus and membrane. TEM analysis of the cells treated with 100 g/mL of nanoparticles was a choice to confirm cellular abnormalities [84]. Ag-NP-treated cells showed endosomes near the cell membrane with a large number of nanoparticles inside. The nanoparticles were found to distribute throughout the cytoplasm, inside lysosomes, and nucleus. Clumps of nanoparticles found inside endosomes and in cytoplasm were similar to nanoaggregates found in cells. However, magnified images showed the presence of individual nanoparticles within the clump. Investigators observed large endosomes with nanoparticles in the cytoplasm of the cells and near the cell and nuclear membrane, which suggested that nanoparticles were entering the cells through endocytosis rather than diffusion [83]. Cytoplasm of the cells showed multiple endosomes with engulfed nanoparticles, and such endosomes were also observed near the nuclear membrane. Nanoparticles accumulate, and deposits of nanoparticles were observed in the nucleus and nucleolus. The nuclear envelope has multiple pores (nuclear pore complexes) with an effective diameter of 9–10 nm, through which transport of proteins takes place. Owing to their small size, Ag-NPs could be readily diffused into the nucleus through the pores. The Ag-NPs or some of the Ag+ ions inside the cell nucleus may bind with DNA and augment the DNA damage caused by the ROS radicals. Small vesicles carrying nanoparticles were observed to be in contact with invaginations of nuclear membrane. The nanoparticles were also seen deposited inside other organelles such as mitochondria [83]. Investigators reported the cytoplasm of the cells with heavy deposits of nanoparticles, outside the vesicles. We believe a possible reason could be the damage to the heavy nanoparticle-loaded microsomes, lysosomes, and endosomes, resulting in phagocytosis due to the deposition of the particles in cytoplasm. The cells with a small number of nanoparticles are believed to survive longer perhaps due to less phagocytosis. Recent reports have established a similar mechanism whereby gold nanoparticles were taken up by the cells through clathrin- and caveolae-mediated endocytosis [84]. The report established the influence of surface chemistry where different surface functionalizations resulted in distinct uptake pathways. Similar properties can be expected for Ag-NP. However, no nanoparticle deposition in nucleus was observed in unmodified gold nanoparticles, but Ag-NPs showed up intracellular distribution of Ag-NP due to more toxic than gold nanoparticles. The tendency of the Ag-NP to accumulate in the nuclei of the cells can be assumed to be associated with the small size, which allows them to diffuse freely through a nuclear pore complex as reported for gold and silica nanoparticles [84]. However, detailed investigation needs to be done to study if small vesicles carrying nanoparticles lodged in the nuclear invaginations can play a role in transferring nanoparticles to the nucleus. As of today, the mechanism of deposition of nanoparticles in mitochondria remains unknown. The current evidence from electron micrography sheds light on the endocytic pathway of nanoparticle uptake. There are different types of active endocytic pathways such as receptor-mediated endocytosis (clathrin or caveolae mediated) and macropinocytosis [85]. Investigators described a detailed study conducted to unravel the mechanism involved in Ag-NP uptake. Elemental mapping of cell sections using scanning transmission electron microscope (STEM) confirmed the distribution of Ag-NP within the cell.

The embedded Ag-NP locations represented red color dots. Scanning transmission electron micrographs and elemental mapping of the cell sections further confirmed the TEM observations. Ag-NP was found to be toxic to both human lung fibroblast (e.g., IMR-90) and the human glioma (e.g., U251) cell lines used in the study. A change in morphology of the cells was observed upon Ag-NP treatment as the first indication of toxicity. Electron micrographs confirmed a significant number of nanoparticles in vital organs such as mitochondria and nucleus. A significant decrease in cell viability was observed, probably as a result of reduction in ATP production, generation of ROS, and damage to the mitochondrial respiratory chain. The ROS production is believed to be the trigger for DNA damage, followed by cell cycle arrest at G2/M. The cells arresting at G2/M are not undergoing massive apoptosis or necrosis, and no fragmented nuclei or necrotic cells were observed in CBMN analysis. The comet and CBMN assays demonstrated extensive DNA damage to both cell lines, the U251 cells being much more vulnerable than the IMR-90 cells. We believe that nanoparticle-induced accumulation of cells at the G2/M interface is associated with DNA repair, which could lead to cell death or survival at a later stage. It opens a window to speculate that Ag-NP at the range of optimized concentrations used may result in G2/M arrest in the cells, which might lead to cell death if repair pathways cannot correct DNA repair.

11.3 Future Developments in Toxicology/Safety of Nanoparticles

Nanoparticles pose toxicity concern that nanoscale particles may have negative health and environmental impacts [86]. There is paucity of information on the possibility of risk of free available nanoparticles without any adverse inhaling effect. In the future, injectable nanometals will be introduced into the body for disease detection, imaging, and treatment. However, there will be public concerns about potential toxicity of such nanomaterials [87]. In the last decade, exceptional properties of carbon nanoparticles have evidenced adverse effects, and it may lead to unique health hazards in the coming years [88]. This is because the size of the nanomaterials is comparable to human cell pores and large proteins. Such similarity may result that regular human immune system may not work against CNT [89]. Both carbon- and nanosilver-based nanoparticles may enter the body by inhalation, skin contact, or ingestion. However, the major concern is inhalation and the use and reuse of these ultrafine nanoparticles to get involved with metabolic reactions. Humans continuously come into contact with small nanoparticles in daily life, and they are generally not hurt seriously by being exposed to the nanoparticles. However, inhalation of micron-sized quartz particles or of asbestos nanofibers has proven quite harmful to health [90]. Similarly, carbon nanoparticles resulting from combustion processes such as forest fires or industrial pollution have a detrimental effect [90]. In addition, studies have been conducted to observe the pulmonary toxicity of nanosized particles in rats [91]. It has been found that toxicity was related to inhalation of nanoparticles, which can penetrate into deep lung tissue. It has been suggested that the surface properties such as surface chemistry and the area of nanoparticles and free-radical generation with cell interaction play a significant role in the increased toxicity of nanoparticles [92]. In the future, the small optimized size of the nanoparticles and the nanocoating on these nanoparticles will be both important considerations [89]. The adverse effects of CNTs exposed to human epidermal keratinocytes (HaCaT) were evaluated as detrimental effect in animals [91]. Shvedova et al. [91] established that multiwalled nanotubes (MWNTs) can localize within cells and may initiate an irritation response in a target epithelial cell (human epidermal keratinocytes). These investigators observed the exposure of HaCaT cells to CNT resulted in accelerated oxidative stress and a reduction of total antioxidant reserve and vitamin E. In addition, investigators showed ultrastructural and morphological changes in cultured skin cells followed by CNT exposure. In the future, polymer-based nanocomposites will be developed to possess biocompatible, less toxic properties. However, one needs to be aware of the possibility of nanosized particles being released into the atmosphere as a result of production and then being inhaled by humans [93]. Major issues still remain to solve such as toxicity of sunscreen made of TiO_2, increased particle emission, decreased particle size distribution, and industrial safety are not very well understood. The introduction of nanomaterials into the industry requires safety evaluation as well as an understanding of the impact on the environment and human health or other biological species [94]. Fiorito et al. investigated highly purified carbon nanomaterials (single-walled carbon nanotubes [SWNTs], C fullerenes) and concluded that they possess a very low toxicity against human macrophage cells [95]. To establish the toxicity of silver nanoparticles, it is quite complicated, and more work is needed on all of the new silver nanomaterials to assess their toxicity and health risks [96]. Another concern is the need for caution when interpreting the data obtained using these highly complex bio-/nanomaterial systems. Hurt et al. have emphasized upon the more sensitive short-term bioassays and subchronic exposures of Ag-NP [97]. In general, nanoparticles must be tested in well-characterized *in vitro* (macrophage and/or epithelial cell cultures) and *in vivo* (intratracheal or inhalation in rats, mice, or hamsters) experimental animal models and calibrated using validated negative and positive controls [97]. There is an ongoing debate among toxicologists, industry, and regulatory agencies about the utility of *in vitro* cellular toxicity assay versus subchronic animal testing in screening for potential adverse health effects of carbon nanomaterials [97]. The same group of Hurt et al. described the overall status and trends of toxicology of carbon-related nanomaterials [97]. Investigators mentioned that instead of supporting either extreme viewpoint—that nanomaterials pose no health risks or that nanotechnology presents extreme risks that warrant cessation of development activities—it is better to continue research on toxicity and safety with a wide range of new material formulations.

A worthy goal for toxicologists and material scientists should be the joint development of "green" nanomaterial formulations—those co-optimized for function and minimal health impact. Occupational and health effect data on workers involved with nanotechnology are limited because of their incipient nature, the relatively small number of records on workers potentially exposed to date, and the lack of time for chronic disease to develop and be detected in short duration. The most relevant human experience deals with exposures to ultrafine particles (which include particles with diameters <100 nm) and fine particles (particles with diameters <2.5 μm). Ultrafine and finer nanoparticles have been assessed in epidemiologic air pollution studies and in studies of occupational cohorts exposed to mineral dusts, fibers, welding fumes, combustion products, and poorly soluble, low-toxicity particulates such as titanium dioxide and carbon black [98,99]. The hazards of these exposures and extreme exposures to new engineered carbon nanoparticles can be identified in animal studies [100]. Recent reports strongly suggest the significant role of surface area, oxidative stress, and pro-inflammatory effects of carbon and silver nanoparticles in the lung with the suggestion of more detailed studies on the oxidative stress as more likely risk of inflammation and cytotoxicity at both molecular and cellular levels [101]. The findings from experimental animal studies ultimately need to be interpreted in terms of the exposure (dose) that humans might receive. Although there is still some debate, the evidence from air pollution studies associates the increased particulate air pollution (the finer carbon particulate matter fraction, PM2.5, with an aerodynamic diameter <2.5 μm) with adverse health effects in susceptible members of the population—particularly the elderly with respiratory and cardiovascular diseases at high risk [102]. Moreover, the concentrations associated with measurable effects on the health of populations may be quite low.

Silver nanocomposites have emerged for various applications in photochemistry, thermochemistry, sonochemistry, wet chemistry, biochemistry, and electrochemistry such as Ag-diamond-like carbon (DLC), oxygen absorbents, silver-based textiles and paints, prevention of HIV-1 binding to host cells, photoinduced reorientation of liquid crystalline polymers in photonics, optoelectronics, nonlinear optics, optical detectors, laser-induced shape transformation, sensor, imaging, display, solar cell, photocatalysis, photoelectrochemistry, and biomedicine. In the future, silver nanocomposite toxicity will be capitalized for bioapplications such as bacterial-resistant paints, prevention of virus in body, impregnated cellulose, and controlled drug delivery systems. Synthesized Ag-NP added to water paints or cotton fabrics will show better pronounced antibacterial/antifungal effect. Oxygen atoms react preferentially with silver preventing the reaction of oxygen with carbon from the DLC films. DLC film may be useful in space research. The surface-modified Ag-NPs have proved to have improved biocompatibility and intracellular uptake for drug delivery. The biomolecules conjugated with nanoparticles have received the most attention in the field of clinical diagnosis and drug delivery because of their high stability, good biocompatibility, and high affinity for biomolecule. Silver Nano (Silver Nano Health System) is a trademark name of an antibacterial Samsung technology that uses Ag-NP in washing machines, refrigerators, air conditioners, air purifiers, and vacuum cleaners.

In occupational studies, the populations that are repeatedly exposed to hazardous mineral dusts and nanofibers in the inhalable range (e.g., carbon, quartz, and asbestos, respectively) have well-known health effects related to the dose inhaled [103]. With asbestos, the critical risk factors for developing respiratory diseases are fiber length, diameter, and biopersistence. For poorly soluble, low-toxicity dusts such as titanium dioxide, smaller in the nanometer size range may appear to cause an increase in risk for lung cancer in animals on the basis of particle size and surface area. Although the findings are not conclusive, various studies of engineered carbon and silver nanoparticles in animals raised concerns about the existence and severity of hazards posed to exposed workers. Possible adverse effects include the development of fibrosis and other pulmonary effects after short-term exposure to CNTs [104], the translocation of nanoparticles to the brain via the olfactory nerve, the ability of nanoparticles to translocate into the circulation, and the potential for nanoparticles to activate platelets and enhance vascular thrombosis [105]. None of these findings are conclusive about the nature and extent of the hazards, but they may be sufficient to support the precautionary action [106]. Ultimately, the significance of hazard information depends on the extent to which workers are exposed to the hazard. Hazard information is a very crucial criterion of nanomaterial risk for the probability that an exposed worker will become ill. There is a real need to identify Ag-NP-specific risk assessments. Interestingly, risk assessments that use the most appropriate dose metrics rather than typical mass of nanoparticles will be unique to nanotechnology development as advocated by NAE 2004 [107]. Risk assessment has been widely used to manage the uncertainty of risks posed to humans by newly introduced nanochemicals or processes. However, nanotechnology offers a diverse range of CNT and silver nanocomposite compositions, structures, and applications, so a single risk assessment and management strategy may not be appropriate [108].

11.4 Conclusions

Ag-NPs may show nanotoxicity or potential use in bioapplications depending on particle size, concentration, and tissue cell interactions as a proposed concept of "use right choice" to get right outcome. Ag-NPs serve as potential antimicrobial agents on human fibroblasts using *C. albicans, S. epidermidis, P. aeruginosa,* and *E. coli* in the presence of variable SIC concentrations. Silver nanocomposites show cytotoxicity at threshold Ag concentrations to human fibroblasts. Ag-NP-induced toxicity is possibly due to the accumulation of cells at the G2/M interface associated with insufficient DNA repair. It could lead to cell death or survival at a later stage if repair pathways cannot correct DNA repair depending on the right choice of silver particle selection. In fact, silver element chemical state matters not actually nanoparticle properties. In the future, more studies will be available on details of silver element chemistry, silver–tissue cell interactions, mechanisms of toxicity, beneficial optimized use of silver metal or particles to regulate intermediary cell metabolisms, and mechanisms of Ag-NP for biomedical use, and it will be explored in context with cell damage, cell cycle, and tissue engineering in repair.

Acknowledgments

The authors acknowledge the experimental assistance in their graduate work, technical know-how, and their published data in different journals included in this chapter from both students Mrs. Stoker and Mr. Forrest at the Department of Biological Engineering in the Utah State University. The authors also acknowledge the permissions to reproduce figures and results from previously published data in literature as cited references in the text. The authors appreciate the encouragement of Dean Dr. Ching J Chen, former director of Center for Nanomagnetics and Biotechnology at Florida State University, while preparing this manuscript.

References

1. Kwon, S., R. Sharma. 2009. Role of effectors on hypoxia due to nitric oxide production in human alveolar epithelial cells and oxygen depletion in human hepatocytes. *Int J Med Biol Front* 15(5): 425–442.
2. George, J., Y. Kuboki, T. Miyata. 2006. Differentiation of mesenchymal stem cells into osteoblasts on honeycomb collagen scaffolds. *Biotechnol Bioeng* 95(3): 404–411.
3. Xiu, Z., Q. Zhang, H.L. Puppala, V.L. Colvin, P.J. Alvarez. 2012. Negligible particle-specific antibacterial activity of silver nanoparticles. *Nano Lett* 12(8): 4271–4275.
4. Du, H., T.M. Lo, J. Sitompul, M.W. Chang. 2012. Systems-level analysis of *Escherichia coli* response to silver nanoparticles: The roles of anaerobic respiration in microbial resistance. *Biochem Biophys Res Commun* 424(4): 657–662.
5. de Lima, R., A.B. Seabra, N. Durán. 2012. Silver nanoparticles: A brief review of cytotoxicity and genotoxicity of chemically and biogenically synthesized nanoparticles. *J Appl Toxicol* 32(11): 867–879.

6. Sintubin, L., W. Verstraete, N. Boon. 2012. Biologically produced nanosilver: Current state and future perspectives. *Biotechnol Bioeng* 109(10): 2422–2436.

7. Jena, P., S. Mohanty, R. Mallick, B. Jacob, A. Sonawane. 2012. Toxicity and antibacterial assessment of chitosan-coated silver nanoparticles on human pathogens and macrophage cells. *Int J Nanomedicine* 7: 1805–1818.

8. Guthrie, K.M., A. Agarwal, D.S. Tackes, K.W. Johnson, N.L. Abbott, C.J. Murphy, C.J. Czuprynski et al. 2012. Antibacterial efficacy of silver-impregnated polyelectrolyte multilayers immobilized on a biological dressing in a murine wound infection model. *Ann Surg* 256(2): 371–377.

9. Araújo, E.A., N.J. Andrade, L.H. da Silva, P.C. Bernardes, A.V. de C Teixeira, J.P. de Sá, J.F. Fialho Jr., P.E. Fernandes. 2012. Antimicrobial effects of silver nanoparticles against bacterial cells adhered to stainless steel surfaces. *J Food Prot* 75(4): 701–705.

10. Jain, J., S. Arora, J.M. Rajwade, P. Omray, S. Khandelwal, K.M. Paknikar. 2009. Silver nanoparticles in therapeutics: Development of an antimicrobial gel formulation for topical use. *Mol Pharm* 6(5): 1388–1401.

11. Sathishkumar, G., C. Gobinath, K. Karpagam, V. Hemamalini, K. Premkumar, S. Sivaramakrishnan. 2012. Phyto-synthesis of silver nanoscale particles using Morinda citrifolia L. and its inhibitory activity against human pathogens. *Colloids Surf B Biointerfaces* 95: 235–240.

12. Lemcke, J., F. Depner, U. Meier. 2012. The impact of silver nanoparticle-coated and antibiotic-impregnated external ventricular drainage catheters on the risk of infections: A clinical comparison of 95 patients. *Acta Neurochir Suppl* 114: 347–350.

13. Keen, J.S., P.P. Desai, C.S. Smith, M. Suk. 2012. Efficacy of hydrosurgical debridement and nanocrystalline silver dressings for infection prevention in type II and III open injuries. *Int Wound J* 9(1): 7–13.

14. Verdú Soriano J, A. Nolasco Bonmati. ALEA study. 2010. Treatment of chronic wounds infected by the application of silver dressings nanocrystalline combined with dressings hydrocellular. *Rev Enferm* 33(10): 6–14.

15. Miller, C.N., N. Newall, S.E. Kapp, G. Lewin, L. Karimi, K. Carville, T. Gliddon, N.M. Santamaria. 2010. A randomized-controlled trial comparing cadexomer iodine and nanocrystalline silver on the healing of leg ulcers. *Wound Repair Regen* 18(4): 359–367.

16. Gravante, G., R. Caruso, R. Sorge, F. Nicoli, P. Gentile, V. Cervelli. 2009. Nanocrystalline silver: A systematic review of randomized trials conducted on burned patients and an evidence-based assessment of potential advantages over older silver formulations. *Ann Plast Surg* 63(2): 201–205.

17. You, C., C. Han, X. Wang, Y. Zheng, Q. Li, X. Hu, H. Sun. 2012. The progress of silver nanoparticles in the antibacterial mechanism, clinical application and cytotoxicity. *Mol Biol Rep* 39(9) 9193–9201.

18. Kim, A.S., C.H. Chae, J. Kim, J.Y. Choi, S.G. Kim, G. Băciult. 2012. Silver nanoparticles induce apoptosis through the Toll-like receptor 2 pathway. *Oral Surg Oral Med Oral Pathol Oral Radiol* 113(6): 789–798.

19. Nirmala, R., H.S. Kang, H.M. Park, R. Navamathavan, I.S. Jeong, H.Y. Kim. 2012. Silver-loaded biomimetic hydroxyapatite grafted poly (epsilon-caprolactone) composite nanofibers: A cytotoxicity study. *J Biomed Nanotechnol* 8(1): 125–132.

20. Böhmert, L., B. Niemann, A.F. Thünemann, A. Lampen. 2012. Cytotoxicity of peptide-coated silver nanoparticles on the human intestinal cell line Caco-2. *Arch Toxicol* 86(7): 1107–1115.

21. Song, X.L., B. Li, K. Xu, J. Liu, W. Ju, J. Wang, X.D. Liu, J. Li, Y.F. Qi. 2012. Cytotoxicity of water-soluble mPEG-SH-coated silver nanoparticles in HL-7702 cells. *Cell Biol Toxicol* 28(4): 225–237.

22. Christen, V., K. Fent. 2012. Silica nanoparticles and silver-doped silica nanoparticles induce endoplasmatic reticulum stress response and alter cytochrome P4501A activity. *Chemosphere* 87(4): 423–434.

23. Stoehr, L.C., E. Gonzalez, A. Stampfl, E. Casals, A. Duschl, V. Puntes, G.J. Oostingh. 2011. Shapematters: Effects of silver nanospheres and wires on human alveolar epithelial cells. *Part Fibre Toxicol* 8: 36.

24. Kang, S.J., I.G. Ryoo, Y.J. Lee, M.K. Kwak. 2012. Role of the Nrf2-hemeoxygenase-1pathwayin silver nanoparticle-mediated cytotoxicity. *Toxicol Appl Pharmacol* 258(1): 89–98.

25. Hackenberg, S., A. Scherzed, M. Kessler, S. Hummel, A. Technau, K. Froelich, C. Ginzkey, C. Koehler, R. Hagen, N. Kleinsasser. 2011. Silver nanoparticles: Evaluation of DNA damage, toxicity and functional impairment in human mesenchymal stem cells. *Toxicol Lett* 201(1): 27–33.

26. Eom, H.J., J. Choi. 2010. p38 MAPK activation, DNA damage, cell cycle arrest and apoptosis as mechanisms of toxicity of silver nanoparticles in Jurkat T cells. *Environ Sci Technol* 44(21): 8337–8342.

27. Gopinath, P., S.K. Gogoi, A. Chattopadhyay, S.S. Ghosh. 2008. Implications of silver nanoparticle induced cell apoptosis for in vitro gene therapy. *Nanotechnology* 19(7): 75–104.

28. Sanpui, P., A. Chattopadhyay, S.S. Ghosh. 2011. Induction of apoptosis in cancer cells at low silver nanoparticle concentrations using chitosan nanocarrier. *ACS Appl Mater Interfaces* 3(2): 218–228.

29. Austin, L.A., B. Kang, C.W. Yen, M.A. El-Sayed. 2011. Plasmonic imaging of human oral cancer cell communities during programmed cell death by nuclear targeting silver nanoparticles. *J Am Chem Soc* 133(44): 17594–17597.

30. Sriram, M.I., S.B. Kanth, K. Kalishwaralal, S. Gurunathan. 2010. Antitumor activity of silver nanoparticles in Dalton's lymphoma ascites tumor model. *Int J Nanomedicine* 5: 753–762.

31. Sur, I., D. Cam, M. Kahraman, A. Baysal, M. Culha. 2010. Interaction of multi-functional silver nanoparticles with living cells. *Nanotechnology* 21(17): 175104.

32. Samberg, M.E., S.J. Oldenburg, N.A. Monteiro-Riviere. 2010. Evaluation of silver nanoparticle toxicity in skin in vivo and keratinocytes in vitro. *Environ Health Perspect* 118(3): 407–413.

33. Kokura, S., O. Handa, T. Takagi, T. Ishikawa, Y. Naito, T. Yoshikawa. 2010. Silver nanoparticles as a safe preservative for use in cosmetics. *Nanomedicine* 6(4): 570–574.

34. Panácek, A., M. Kolár, R. Vecerová, R. Prucek, J. Soukupová, V. Krystof, P. Hamal, R. Zboril, L. Kvítek. 2009. Antifungal activity of silver nanoparticles against Candida spp. *Biomaterials* 30(31): 6333–6340.

35. AshaRani, P.V., G. Low Kah Mun, M.P. Hande, S. Valiyaveettil. 2009. Cytotoxicity and genotoxicity of silver nanoparticles in human cells. *ACS Nano* 3(2): 279–290.

36. Boca, S.C., M. Potara, A.M. Gabudean, A. Juhem, P.L. Baldeck, S. Astilean. 2011. Chitosan-coated triangular silver nanoparticles as a novel class of biocompatible, highly effective photo thermal transducers for in vitro cancer cell therapy. *Cancer Lett* 311(2): 131–140.

37. Piao, M.J., K.A. Kang, I.K. Lee, H.S. Kim, S. Kim, J.Y. Choi, J. Choi, J.W. Hyun. 2011. Silver nanoparticles induce oxidative cell damage in human liver cells through inhibition of reduced glutathione and induction of mitochondria-involved apoptosis. *Toxicol Lett* 201(1): 92–100.

38. Ma, J., X. Lü, Y. Huang. 2011. Genomic analysis of cytotoxicity response to nanosilver in human dermal fibroblasts. *J Biomed Nanotechnol* 7(2): 263–275.

39. García-Contreras R, L. Argueta-Figueroa, C. Mejía-Rubalcava, R. Jiménez-Martínez, S. Cuevas-Guajardo, P.A. Sánchez-Reyna, H. Mendieta-Zeron. 2011. Perspectives for the use of silver nanoparticles in dental practice. *Int Dent J.* 61(6): 297–301.

40. Hernández-Sierra, J.F., F. Ruíz, J.P. Castanedo-Cázares, V. Martinez-Ruiz, P. Mandeville, M. Pierdant-Pérez, A. Gordillo-Moscoso, J. Pozos-Guillén Ade. 2010. In vitro determination of the chromatic effect of a silver nanoparticles solution linked to the gantrez S-97 copolymer on tooth enamel. *J Clin Pediatr Dent* 35(1): 65–68.

41. Fong. J., F. Wood. 2006. Nanocrystalline silver dressings in wound management: A review. *Int J Nanomedicine* 1(4): 441–449.

42. Dunn, K., V. Edwards-Jones. 2004. The role of Acticoat with nanocrystalline silver in the management of burns. *Burns* 30 (Suppl 1): S1–S9.

43. Klasen, H.J. 2000. A historical review of the use of silver in the treatment of burns. II. Renewed interest for silver. *Burns* 26(2): 131–138.

44. Wright, J.B., K. Lam, D. Hansen, R.E. Burrell. 1999. Efficacy of topical silver against fungal burn wound pathogens. *Am J Infect Control* 27(4): 344–350.

45. Percival, S.L., P.G. Bowler, D. Russell. 2005. Bacterial resistance to silver in wound care. *J Hosp Infect* 60(1): 1–7.

46. Lansdown, A.B. 2002. Silver. I: Its antibacterial properties and mechanism of action. *J Wound Care* 11(4): 125–130.
47. Ewald, A., S.K. Gluckermann, R. Thull, U. Gbureck. 2006. Antimicrobial titanium/silver PVD coatings on titanium. *Biomed Eng Online* 5: 22.
48. Lansdown, A.B. 2002. Toxicity in mammals and how its products aid wound repair. *J Wound Care* 11(5): 173–177.
49. Burd, A., C.H. Kwok, S.C. Hung, H.S. Chan, H. Gu, W.K. Lam, L. Huang. 2007. A comparative study of the cytotoxicity of silver-based dressings in monolayer cell, tissue explant, and animal models. *Wound Repair Regen* 15(1): 94–104.
50. Kim, J., S. Kwon, E. Ostler. 2009. Antimicrobial effect of silver-impregnated cellulose: Potential for antimicrobial therapy. *J Biol Eng* 3: 20. doi: 10.1186/1754-1611-3-20.
51. Klemm, D., B. Philipp, U. Heinze, W.Wagenknecht. 1998. *Comprehensive Cellulose Chemistry*, Vol. 2, Wiley-VCH, Weinheim, Germany.
52. Eom, H.J., J. Choi. 2010. p38 MAPK activation, DNA damage, cell cycle arrest and apoptosis as mechanisms of toxicity of silver nanoparticles in Jurkat T cells. *Environ Sci Technol* 44 (21): 8337–8342. doi: 10.1021/es1020668.
53. Asghari, S., S.A. Johari, J.H. Lee, Y.S. Kim, Y.B. Jeon, H.J. Choi, M.C. Moon, I.J. Yu. 2012. Toxicity of various silver nanoparticles compared to silver ions in *Daphnia magna*. *J Nanobiotechnol* 10: 14. doi:10.1186/1477-3155-10-14.
54. Park, S.W., S.B. Hyun, C.X. Zhi, O.H. Kwon, M.W. Huh, I.K. Kang. 2009. Preparation and properties of silver containing nylon 6 nanofibers formed by electrospinning. *J Appl Polym Sci* 112(4): 2320–2326.
55. Park, J.P., J. Yi, Y. Kim, K. Choi, K. Park. 2010. Silver nanoparticles induce cytotoxicity by a Trojan-horse type mechanism. *Toxicol In Vitro* 24: 872–878.
56. Yang, X., A.P. Gondikas, S.M. Marinakos, M. Auffan, J. Liu, H. Hsu-Kim, J.N. Meyer. 2012. Mechanism of silver nanoparticle toxicity is dependent on dissolved silver and surface coating in caenorhabditis elegans. *Environ Sci Technol* 46(2): 1119–1127.
57. Xiu, Z., Q. Zhang, H. L. Puppala, V.L. Colvin, P.J. J. Alvarez. 2012. Negligible particle-specific antibacterial activity of silver nanoparticles. *Nano Lett* 12(8): 4271–4275.
58. Thurmann, R.B., C.P. Gerba. 1989. The molecules mechanisms of copper and silver ion disinfection of bacteria and viruses. *Crit Rev Environ Control* 1: 295–315.
59. Fabrega, J., S.R. Fawcett, J.C. Renshaw, J.R. Lead. 2009. Silver nanoparticle impact on bacterial growth: Effect of pH, concentration, and organic matter. *Environ Sci Technol* 43 (19): 7285–7290.
60. Schreurs, W.J.A., H. Rosenburgh. 1982. Effect of silver ions on transport and retention of phosphate by Escherichia coli. *J Bacteriol* 152: 7–13.
61. Richards, R.M.E., H.A. Odelola, B. Anderson. 1984. Effect of silver on whole cells and spheroplasts of a silver resistant Pseudomonas aeruginosa. *Microbios* 39: 151–258.
62. Hussain, S.M., A.K. Javorina, A.M. Schrand, H.M. Duhart, S.F. Ali, J.J. Schlager. 2006. The interaction of manganese nanoparticles with PC-12 cells induces dopamine depletion. *Toxicol Sci* 92: 456–463.
63. Braydich-Stolle, L., S. Hussain, J.J. Schlager, M.C. Hofmann. 2005. In vitro cytotoxicity of nanoparticles in mammalian germline stem cells. *Toxicol Sci* 88: 412–419.
64. Kone, B.C., M. Kaleta, S.R. Gullans. 1988. Silver ion (Ag+) induced increases in cell membrane K^+ and Na^+ permeability in renal proximal tubule: Reversal by thiol reagents. *J Membr Biol* 102: 11–19.
65. Lee, K.S., M.A. El-Sayed. 2006. Gold and silver nanoparticles in sensing and imaging: sensitivity of plasmon response to size, shape, and metal composition. *J Phys Chem* 110: 19220–19225.
66. Oberdorster, G., E. Oberdorster, J. Oberdorster. 2005a. Nanotoxicology: An emerging discipline evolving from studies of ultrafine particles. *Environ Health Perspect* 113: 823–839.
67. Oberdorster, G., A. Maynard, K. Donaldson, V. Castranova, J. Fitzpatrick, K. Ausman, J. Carter et al. 2005b. Principles for characterizing the potential human health effects from exposure to nanomaterials: Elements of a screening strategy. *Part Fibre Toxicol* 2: 8. http://www.particleandfibretoxicology.com/content/pdf/1743-8977-2-8.pdf

68. Donaldson, K., V. Stone, C.L. Tran, W. Kreyling, P.J.A. Borm. 2004. Nanotoxicology. *Occup Environ Med* 61: 727–728.
69. Donaldson, K., C.L. Tran. 2002. Inflammation caused by particles and fibers. *Inhal Toxicol* 14: 5–27.
70. Jose, R.M., L.E. Jose, C. Alejandra, H. Katherine, B.K. Juan, T.R. Jose, J.Y. Miguel. 2005. The bactericidal effect of silver nanoparticles. *Nanotechnology* 16: 2346–2353.
71. Nel, A. 2005. Air pollution-related illness: Effects of particles. *Science* 6 (308): 804–806.
72. Elechiguerra, J.L., J.L. Burt, J.R. Morones, A. Camacho-Bragado, X. Gao, H.H. Lara, M.J. Yacaman. 2005. Interaction of silver nanoparticles with HIV-1. *J Nanobiotechnol* 3: 6.
73. Almofti, M.R., T. Ichikawa, K. Yamashita, H. Terada, Y. Shinohara. 2003. Silver ion induces a cyclosporine a-insensitive permeability transition in rat liver mitochondria and release of apoptogenic cytochrome C. *J Biochem* 134: 43–49.
74. Trop, M., M. Novak, S. Rodl, B. Hellbom, W. Kroell, W. Goessler. 2006. Silver-coated dressing Acticoat caused raised liver enzymes and argyria-like symptoms in burn patient. *J Trauma* 60: 648–652.
75. Yamanaka, M., K. Hara, J. Kudo. 2005. Bactericidal actions of a silver ion solution on *Escherichia coli*, studied by energy-filtering transmission electron microscopy and proteomic analysis. *Appl Environ Microbiol* 71: 7589–7593.
76. Xia, T., M. Kovochich, J. Brant, M. Hotze, J. Sempf, T. Oberley, C. Sioutas, J.I. Yeh, M.R. Wiesner, A.E. Nel. 2006. Comparison of the abilities of ambient and manufactured nanoparticles to induce cellular toxicity according to an oxidative stress paradigm. *Nano Lett* 6: 1794–1807.
77. Wright, J.B., K. Lam, A.G. Buret, M.E. Olson, R.E. Burrell. 2002. Early healing events in a porcine model of contaminated wounds: effects of nanocrystalline silver on matrix metalloproteinases, cell apoptosis, and healing. *Wound Repair Regen* 10: 141–151.
78. Sayes, C.M., R. Wahi, P.A. Kurian, Y. Liu, J.L. West, K.D. Ausman, D.B. Warheit, V.L. Colvin. 2006. Correlating nanoscale titania structure with toxicity: A cytotoxicity and inflammatory response study with human dermal fibroblasts and human lung epithelial cells. *Toxicol Sci* 92: 174–185.
79. Nel, A., T. Xia, L. Madler, N. Li. 2006. Toxic potential of materials at the nanolevel. *Science* 311: 622–627.
80. Liu, X., M.J. Keane, J.C. Harrison, E.V. Cilento, T. Ong, W.E. Wallace. 1998. Phospholipid surfactant adsorption by respirable quartz and in vitro expression of cytotoxicity and DNA damage. *Toxicol Lett* 96–97: 77–84.
81. Allain, L.R., T. Vo-Dinh. 2002. Surface-enhanced Raman scattering detection of the breast cancer susceptibility gene BRCA1 using a silver-coated microarray platform. *Anal Chim Acta* 469: 149–154.
82. Barlow, P.G., A. Clouter-Baker, K. Donaldson, J. Maccallum, V. Stone. 2005. Carbon black nanoparticles induce type II epithelial cells to release chemotaxins for alveolar macrophages. *Part Fibre Toxicol* 2: 11. http://www.particleandfibretoxicology.com/content/pdf/1743-8977-2-11.pdf
83. Bogle, K.A., S.D. Dhole, V.N. Bhoraskar. 2006. Silver nanoparticles: Synthesis and size control by electron irradiation. *Nanotechnology* 17: 3204–3208.
84. Lee, K.S., M.A. El-Sayed. 2006. Gold and silver nanoparticles in sensing and imaging: Sensitivity of plasmon response to size, shape, and metal composition. *J Phys Chem* 110: 19220–19225.
85. Asharani, P.V., S. Sethu, S. Vadukumpully, S. Zhong, C.T. Lim, M.P. Hande, S. Valiyaveettil. 2010. Investigations on the structural damage in human erythrocytes exposed to silver, gold and platinum nanoparticles in human cells. *Adv Funct Mater* 20: 1233–1245.
86. Chengang, C., C. David. 2003. Processing and morphological development of montmorillonite epoxy nanocomposites. *Nanotechnology* 14: 643–648.
87. Toensmeier, P.A. 2004. Nanotechnology faces scrutiny over environment and toxicity. *Plastic Eng* 60: 14–17.
88. Borm, P.J.A. 2002. Particle toxicology: From coal to nanotechnology. *Inhal Toxicol* 14: 311–324.

89. Agarwal, S., E. Tatli, N.N. Clark, R. Gupta. 2005. Nanoscale science and technology research. In: *International Symposium on Polymer Nanocomposites Science and Technology*, Boucherville, Canada, pp. 401.

90. Koyama, S., M. Endo, Y.A. Kim, T. Hayashi, T. Yanagisawa, K. Osaka, H. Koyama, H. Haniu, N. Kuroiwa. 2006. Role of systemic T-cells and histopathological aspects after subcutaneous implantation of various carbon nanotubes in mice. *Carbon* 44(6): 1079–1092.

91. Shvedova, A.A., E.R. Kison, R. Mercer. 2005. Unusual inflammatory and fibrogenic pulmonary responses to single walled carbon nanotube in mice. *Am J Physiol Lung Cell Mol Physiol* 289(5): L698–L708.

92. Lademann, J., H.J. Weigmann et al. 1999. Penetration of titanium dioxide micro particles in a sunscreen formulation into the horny layer and the follicular orifice. *Skin Pharmacol Appl Skin Physiol* 12: 247–256.

93. Dunford, R., A. Salinaro, L. Cai, N. Serpone, S. Horikoshi, H. Hidaka, J. Knowland. 1997. Chemical oxidation and DNA damage catalyzed by inorganic sunscreen ingredients. *FEBS Lett* 418(1–2): 87–90.

94. Smart, S.K., A.I. Cassady, G.Q. Lu, D.J. Martin. 2006. The biocompatibility of carbon nanotubes. *Carbon* 44(6): 1034–1047.

95. Fiorito, S., A. Serafino, F. Andreola, P. Bernier. 2006. Effects of fullerenes and single wall carbon nanotubes on marine and human macrophages. *Carbon* 44(6): 1100–1105.

96. Donaldson, K., R. Aitken, L. Tran, V. Stone, R. Duffin, G. Forrest. 2006. Carbon nanotubes: A review of their properties in relation to pulmonary toxicology and workplace safety. *Toxicol Sci* 92: 5–22.

97. Hurt, R.H., M. Monthioux, A. Kane. 2006. Toxicology of carbon nanomaterials: Status, trends, and perspectives on the special issue. *Carbon* 44(6): 1028–1033.

98. Elder A, R. Gelein, V. Silva, T. Feikert, L. Opanashuk, J. Carter. 2006. Translocation of inhaled ultrafine manganese oxide particles to the central nervous system. *Environ Health Perspect* 114: 1172–1178.

99. Aitken, R.J., K.S. Creely, C.L. Tran. 2004. *Nanoparticles: An Occupational Hygiene Review*. Health safety executive, research report 274, HSE Books, London, U.K.

100. Kipen, H.M., D.L. Laskin. 2005. Smaller is not always better: Nanotechnology yields nanotoxicology. *Am J Physiol Lung Cell Mol Physiol* 289(5): L696–L697.

101. Radomski, A., P. Jurasz, P. Alonso-Escolano, M. Drew, M. Morandi, M. Tadeusz. 2005. Nanoparticle-induced platelet aggregation and vascular thrombosis. *Br J Pharmacol* 146: 882–893.

102. Warheit, D.B. 2004. Nanoparticles: Health impacts? *Mater Today* 7(2): 32–35.

103. Monteiro-Riviere, N.A., R.J. Nemanich, A.O. Inman, Y.Y. Wang, J.E. Riviere. 2004. Multi-walled carbon nanotube interactions with human epidermal keratinocytes. *Toxicol Lett* 155: 377–384.

104. Lam, C.W., J.T. James, R. McCluskey, R.L. Hunter. 2004. Pulmonary toxicity of single-wall carbon nanotubes in mice 7 and 90 days after intratracheal instillation. *Toxicol Sci* 77: 126–134.

105. Mark, D. 2004. *Nanomaterials—A Risk to Health at Work?* Health and Safety Laboratory, Buxton, UK: SK17 9 JN. Available: www.hsl.gov.uk/capabilities/nanosymrep_

106. Samuels, S. 2003. Occupational medicine and its moral discontents. *J Occup Environ Med* 45: 1226–1233.

107. National Academy of Engineering. 2004. Emerging Technologies and Ethical Issues in Engineering: Papers in NAE Workshop on 14–15 October 2003, http://newton.nap.edu/books/

108. Wardak, A., D. Rejeski. 2003. Nanotechnology and regulation: A case study using the Toxic Substances Control Act (TSCA) Washington, DC, Woodrow Wilson International Center for Scholars, Foresight and Governance Project. Publication 6.

12

Nanoparticle Production for Biomedical Applications via Laser Ablation Synthesis in Solution

---

Reza Zamiri, Hossein Abbastabar Ahangar, and Gregor P.C. Drummen

CONTENTS

12.1 Introduction

Over the past two decades, various fields of science and engineering are increasingly investigating engineered nanoparticles. These nanometer-sized particles, typically smaller than 100 nm in at least one dimension, have become important materials in a myriad of applications, for example, as imaging agents in biological and (bio)medical research; as drug delivery vehicles in (bio)medicine; as constituents of paints, coatings, and pastes; in electronics and photovoltaic applications; in light emitting diode (LEDs) and lasers; and many other advanced technologies.

At the mesoscopic scale, nanoparticles display unique and in most cases unexpected divergent optical, electronic, thermal, magnetic, and structural properties compared with their bulk material counterparts [1–5]. The divergent behavior of gold is a good example of how the nanoscale size affects a material's properties. Gold in its bulk form is a material that is perceived as a shiny yellow and relatively inert material but with good electrical and thermal conduction properties. Nanogold, however, has dramatically different properties: depending on the size and shape, the gold nanoparticles show a red to blue–gray color, become good catalysts, and no longer behave as a noble metal. Additionally, recent research has shown that not only nanogold but also palladium nanoparticles exhibit ferromagnetic properties in a particular size range [6]. These properties are largely size and shape dependent. Additionally, because nanoparticles are so small and their surface area is extremely large, the majority of their atoms are located at the surface, which is especially a favorable property in chemical catalysis. Collectively, these extraordinary properties and the large surface-to-volume ratios open up new areas and opportunities for their utilization, in particular in biomedical, chemical, optical, and electronic applications. Recently, Roduner intelligibly and comprehensively described why size and shape matter and why nanoparticles are so distinctly different [4].

Metal nanoparticles (MNPs) are an important class of materials with extraordinary properties and are particularly interesting because they display various plasmonic phenomena (a quantum of plasma oscillations, which occurs when light radiation interacts with free electrons in a plasmonic nanoparticle [1,2]), which can be exploited particularly for diagnostic biomolecule detection, (bio)imaging, and other optical applications. Especially the dimensional tunability is highly attractive. In silver nanoparticles (AgNPs), for instance, the surface plasmon resonance (SPR) is size and shape dependent, and its energy is located far from the interband transition energy, which is a major advantage compared with other MNPs [7]. Furthermore, noble MNPs display SPR in the visible region of the electromagnetic spectrum and redshift their wavelengths as the particle size increases [8,9]. These properties not only offer the opportunity to tailor the nanoparticle for particular applications in bioassays, even allowing a multiple detection scenario, but also allows close monitoring of their synthesis by spectroscopic methods.

There are a number of methods to synthesize nanoparticles, both physical and chemical, which can broadly be subdivided into wet chemical methods, ion implantation, and chemical or physical vapor deposition, each of which has its own distinct advantages and disadvantages. Noble MNPs specifically can be synthesized chemically by reduction of a metal salt in aqueous solution with agents such as borohydrides, hydrazine, alcohols, alkyl sulfate, or sodium citrate. Unfortunately, these synthesis methods are relatively expensive and suffer from a number of drawbacks, such as the use of harmful and toxic materials (they include chemicals that are fatal to electrical reliability, i.e., reducing agents containing sodium or potassium); the risk of surface fouling with residual chemicals is a hazard in (bio)medical applications, and they produce particles with broad size distributions and display low stability because of the propensity of nanoparticles to aggregate, to name but a few. In particular, the tendency of MNPs to agglomerate into larger clusters is a major drawback for a variety of applications and the (semi) product's shelf life. Therefore, a number of groups have started using surfactants to prevent nanoparticle agglomeration and to extend their stability, which has been shown to be relatively successful.

To overcome many of the aforementioned disadvantages, a laser ablation method in solution for the synthesis of MNPs was developed in the groups of Henglein and Cotton

[12–14], analogous to the laser-induced preparation of gas-phase clusters [10]. Laser ablation synthesis in solution (LASiS) is an attractive alternative to produce nanoparticles without any hazardous components, by ablating a solid metal target in solution with a pulsed laser. This is an inexpensive, fast, and straightforward method with an uncomplicated experimental setup and synthesis procedure [11–14]. Furthermore, the preparation of MNPs by LASiS does not require harmful chemicals, and thus, no surface contamination and concomitant potential toxic effects occur [11]. However, nanoparticles that have been synthesized by LASiS still show the same propensity to agglomeration because of their high surface energy and potential chemical reactivity. Therefore, the major challenge in MNP synthesis is to devise methods that allow efficient synthesis without the need for further purification and produce nanoparticles that can directly be utilized or easily functionalized for biomedical applications, with narrow size distributions, and that have significantly reduced tendencies to precipitate or flocculate. Generally, compounds that act as surface passivation reagents effectively prevent nanoparticle agglomeration by neutralizing surface charges or reactivities and by sterically hindering particles from closely approaching and coalescing. Recently, natural materials such as long-chain fatty acids (stearic, palmitic, and lauric acid), lauryl amine [15], and chitosan [16] have been used as stabilizers in the synthesis of MNPs. These stabilizing agents prevent particle agglomeration by physically surrounding the MNPs (capping) resulting in charge and/or steric stabilization, allow easy and direct functionalization [17], and are excellent nontoxic, biocompatible, biodegradable, and environmentally friendly alternatives.

12.2 Laser Ablation Synthesis in Solution

12.2.1 Experimental Setup

In a typical LASiS experiment, a solid target of a pure metal (e.g., Ag, Au) or compound metallic material (e.g., transition metal/silicon) is irradiated with a laser beam of appropriate wavelength and laser power in a nonreactive background. The experimental setup is generally uncomplicated and can readily be built from commercially available off-the-shelf parts. This includes a vibration-free optical table, lenses and mirrors, a glass container with target holder and magnetic stirring device (preferably incorporated in the container's base), and a pulsed laser. Furthermore, the initial investment and start-up costs for a typical LASiS setup are relatively low and below €20,000, of which the laser is the most expensive component.

Two variants of a typical experimental setup for LASiS-based nanoparticle synthesis are schematically depicted in Figure 12.1a, although variations have been reported that include movable or rotatable targets. Surface impurities are removed from the target plate by prior ultrasound treatment in a sonication bath. The plate or foil is placed in the container and subsequently irradiated with a high-power pulsed laser beam of nano-(ns) or femtosecond (fs) duration that is focused onto the target with a lens.

In the left setup in Figure 12.1a, the beam passes through a liquid layer that is only several millimeters in thickness, which has consequences for the nanoparticle formation and the final product (see succeeding text). To prevent loss of the laser's energy due to absorption or scattering, the liquid environment must be transparent and the selected

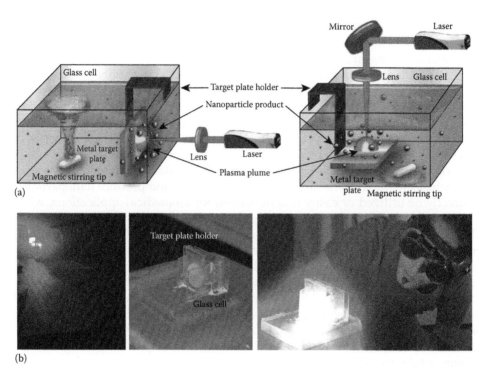

FIGURE 12.1
(See color insert.) Schematic representation of the LASiS experimental setup. (a) Two variants are generally used: (left) The plate is placed close to the wall and irradiated from the same side, with the laser beam passing through a few millimeters of liquid, and (right) the plate is attached to the wall and the laser beam passes through the entire liquid column before reaching the plate. (b) Actual experimental setup and glass cell of the left schematic configuration in (a).

laser wavelength should ideally be outside the absorption range of the liquid medium. These conditions are generally satisfied for most of the available pulsed lasers, because these operate in the near-infrared or far visible range where most liquids, especially water, do not absorb. However, an increase in the number of particles within the liquid medium progressively will cause scattering and/or absorption of the beam and thus reduce the laser's efficiency as LASiS progresses. It is therefore essential that the particles formed near the target be largely dispersed within the bulk through mechanical stirring. Besides the aforementioned phenomena, ablation of the metal plate causes the formation of a plasma plume (Figure 12.2b). This induced plasma can partially absorb or scatter the laser beam and thereby reduce the amount of laser light that effectively reaches the target plate (plasma plume shielding) and thus the ablation efficiency. The loss of laser energy through this mechanism is one order of magnitude higher for ns pulse lasers in comparison with fs pulse lasers and increases with increasing pulse energy [18].

12.2.2 Mechanism of Laser Ablation

The exact mechanism underpinning laser ablation, both in vacuum, gaseous, and liquid environments, is not well understood and is a direct consequence of the complexity of

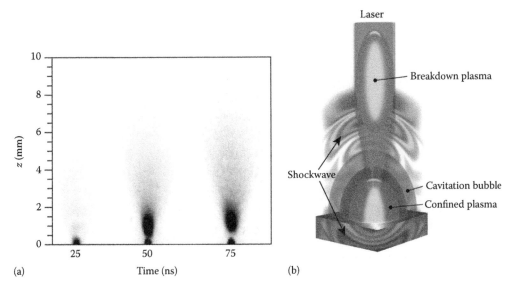

FIGURE 12.2
(See color insert.) Plasma plume formation during laser ablation. (a) High-speed photography of the plasma plume extension during ablation of a silver plate with a 527 nm Nd:glass laser in vacuum. (b) Schematic overview of the mechanism and phenomena involved in plasma plume formation during LASiS.

the processes and physics involved. Nonetheless, with our current understanding, several distinct phenomena can be discerned and at least two mechanisms have been proposed. Interaction of the laser beam with the surface of the target plate results in absorption of photons and is the first step in the laser ablation process. This focusing of an intense laser beam onto the solid target material results in the delivery of a massive power density in a small region of the target material. Consequently, the target is ablated resulting in the ejection of solid fragments, highly excited species, and neutrals, for example, atoms, droplets, clusters, and ions (impact ionization), from its surface or the formation of a rapidly expanding plasma plume. The amount of material removed from the target and the phase of this material both directly depend on the absorbed energy E, and Momma et al. described the following empirical trend [19]:

for

$$\tau_a \gg \tau_l, \quad L_a \propto E^{2/3}, \quad \tau_a \propto E^{1/2}, \quad T_e \propto E^{1/3} \tag{12.1}$$

where
  L_a is the ablation depth
  τ_a is the ablation duration
  T_e is the electronic temperature during ablation
  τ_l is the laser pulse length

At least two mechanisms have been proposed for LASiS: (i) the thermal evaporation mechanism, which is based on the formation of a plasma plume (Figure 12.2), and (ii) the explosive ejection mechanism in which the high-speed ejection of nanodroplets occurs (Figure 12.3).

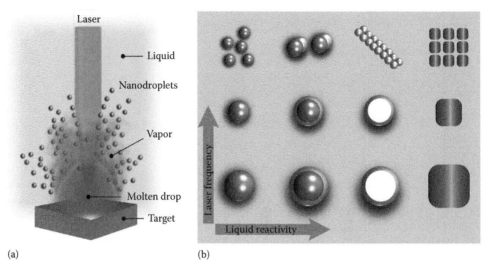

(a)   (b)

FIGURE 12.3
(See color insert.) Schematic overview of the explosive ejection mechanism. (a) Formation of nanodroplets in a millisecond low-energy laser ablation process. (b) Effect of liquid reactivity and laser frequency on morphology of nanostructures when a Pb target was ablated in various liquids (and reactive environments). (Based on Niu, K.Y. et al., *J. Am. Chem. Soc.*, 132, 9814, 2010.)

12.2.2.1 Thermal Evaporation Mechanism

This proposed mechanism involves the formation of a plasma plume from the target plate material, which starts when the laser beam hits the target and transfers its energy to the target. Consequently, the material undergoes a phase transition, and a highly energetic and dense plasma plume that is at nonequilibrium is formed. This plasma expands and travels at supersonic velocity and produces a shock wave that causes an increase in temperature and pressure within the plume and also propagates through the target plate [20], as schematically depicted in Figure 12.2b. Several reports show that in water as the surrounding liquid medium, the plasma pressure and related shock wave directly depend on the laser wavelength, thickness of the water layer, and the focusing conditions [21–23]. Figure 12.2a illustrates the evolution of the plasma plume and ejection of material, albeit in a vacuum; notice the breakdown plasma detaching.

In LASiS, the expansion of the plasma is more confined by the liquid layer, and therefore, the thickness of the plasma above the target's surface is smaller compared with laser ablation in diluted gas or vacuum. This leads to higher pressures and temperatures, as well as higher plasma plume densities compared with gaseous media [24]. Moreover, the lifetime of the plasma plume is at least 10 times shorter for LASiS than similar ablation processes in air [24]. This high-pressure plasma is able to directly interact with the target's surface and causes additional ablation of material, which constitutes a secondary ablation process. This process only occurs during the first 100 ns, because the plasma pressure decreases after this time [25]. Following these ablation events and plasma plume formation, after approximately 1 µs, a cavitation bubble is formed as a result of vaporization of the liquid that is in direct contact with the hot plasma. The bubble expands for 300 µs and subsequently collapses, concomitantly releasing significant amounts of energy, which again causes ablation of the target, termed the third ablation [25,26]. It is noteworthy to mention that the contribution of this process is much smaller than the other two ablation processes,

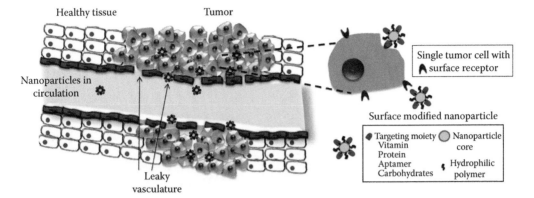

FIGURE 1.3
Schematic representation of passive and active targeting. Nanoparticles are able to extravasate into tumor tissue through the leaky vasculature in tumor tissue and accumulate due to poor lymphatic drainage—also known as the effect of EPR. Once accumulated in the tumor tissue surface, modified nanoparticles can specifically target cancer cells by different ligands and can be internalized by receptor-mediated endocytosis where payload can be released.

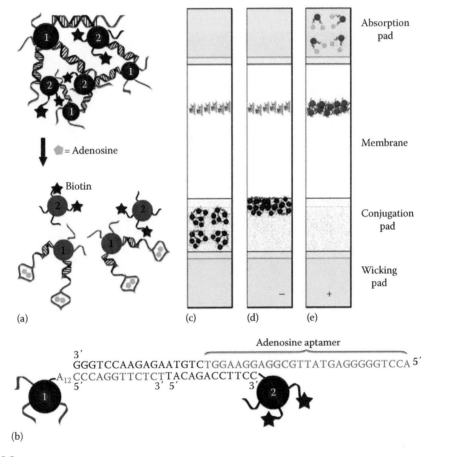

FIGURE 2.2
Aptamer/nanoparticle-based lateral-flow device. (a) Adenosine-induced disassembly of nanoparticle aggregates into red-colored dispersed nanoparticles. Biotin is denoted as black stars. (b) DNA sequences and linkages in nanoparticle aggregates. Lateral flow devices loaded with the aggregates (on the conjugation pad) and streptavidin (on the membrane in cyan color) before use (c) and in a negative (d) or a positive (e) test. (From Liu, J., Mazumdar, D., and Lu, Y. A simple and sensitive "dipstick" test in serum based on lateral flow separation of aptamer-linked nanostructures. *Angew. Chem. Int. Eng. Edn.* 2006. 45. 7955–7959. Copyright Wiley-VCH Verlag GmbH & Co. KGaA. Reproduced with permission.)

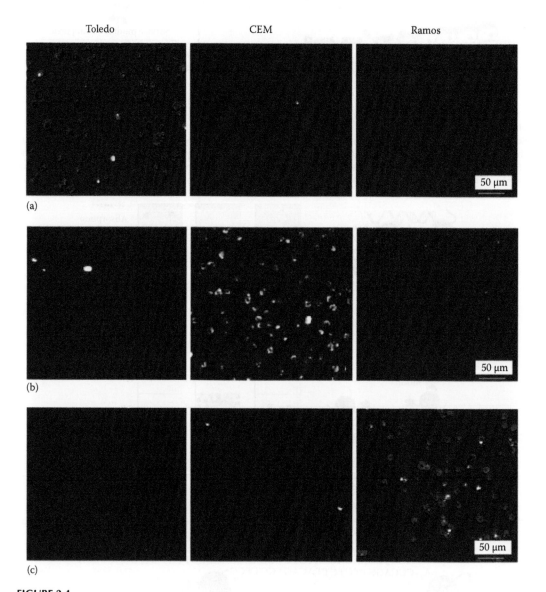

FIGURE 2.4
Confocal microscopy images of individual nanoparticle–aptamer conjugates with the three different cells (Toledo, CCRF-CEM, and Ramos): (a) nanoparticle-(FAM)-aptamer T1, specific for Toledo; (b) nanoparticle-(FAM-R6G)-aptamer sgc8, specific for CCRF-CEM; and (c) nanoparticle-(FAM-R6GROX)-aptamer TDO5, specific for Toledo. (Reprinted with permission from Chen, X. et al., *Anal. Chem.*, 81, 7009. Copyright 2009 American Chemical Society.)

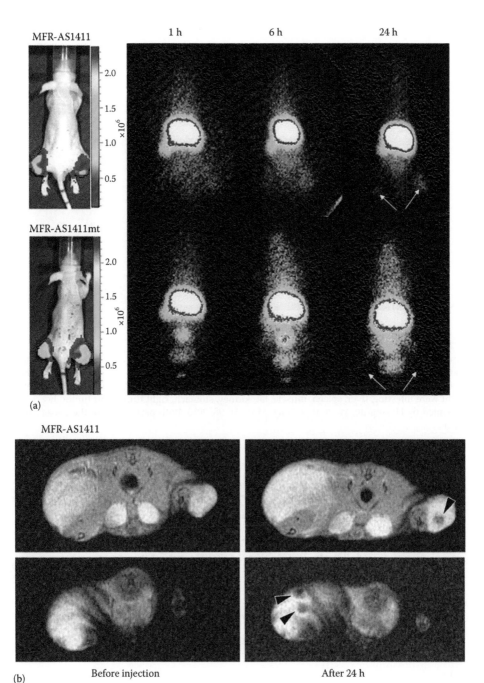

FIGURE 2.5

In vivo multimodal cancer targeting and imaging using MFR-AS1411 particles. (a) MFR-AS1411 particles were intravenously injected into tumor-bearing mice, and radionuclide images were acquired at 1, 6, and 24 h after injection. Scintigraphic images of C6 tumors in mice that received MFR-AS1411 showed that C6 tumors had accumulated MFR-AS1411 at 24 h after injection but did not accumulate MFR-AS1411mt. Tumor growth patterns were followed using bioluminescence signals acquired from luciferase-expressing C6 cells. (b) MR images of tumor-bearing mice before and after injection of MFR-AS1411 were acquired. Dark signal intensities at tumor sites were detected in MFR-AS1411-injected mice (arrowhead).

(continued)

MFR-AS1411mt   MFR-AS1411

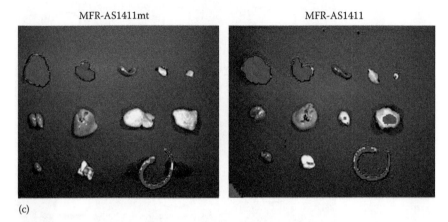

(c)

FIGURE 2.5 (continued)

In vivo multimodal cancer targeting and imaging using MFR-AS1411 particles. (c) Tumors were isolated and their fluorescence verified using IVIS200 system. Fluorescence signal at tumor site injected with MFR-AS1411 was detected, compared with tumors injected with MFR-AS1411mt. Isolated organs in order from upper left to lower right were intestine, liver, spleen, muscle, fat, kidney, stomach, right tumor, left tumor, heart, lung, and tail. (Reprinted by Hwang do, W. et al., *J. Nucl. Med.*, 51, 98, 2010. With permission of the Society of Nuclear Medicine.)

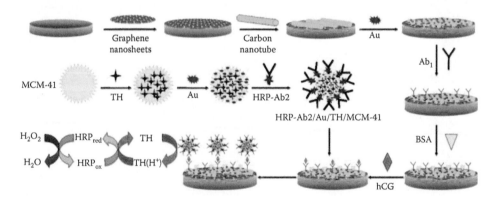

FIGURE 3.6

Fabrication processes of Au/TH/MCM-41 nanomaterials and measurement protocol of the electrochemical immunosensor. (From Lu, J. et al., *Biosens. Bioelectron.*, 33, 1, 29, 2012. With permission.)

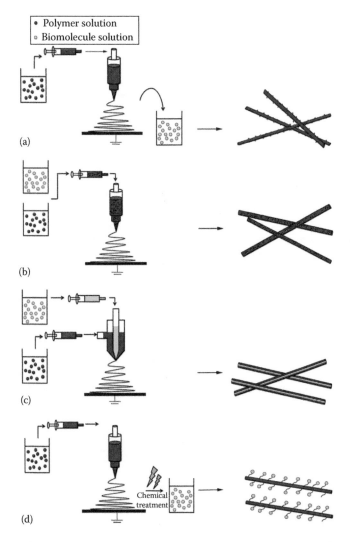

FIGURE 4.5
Schematic approaches to produce nanofiber with bioactive molecules: (a) physical adsorption, (b) blend electrospinning, (c) coaxial electrospinning, and (d) covalent immobilization. (From Ji, W. et al., *Pharmaceutical Research*, 28, 1259, 2011.)

Legend within figure:
- Polymer solution
- Biomolecule solution

Chemical treatment

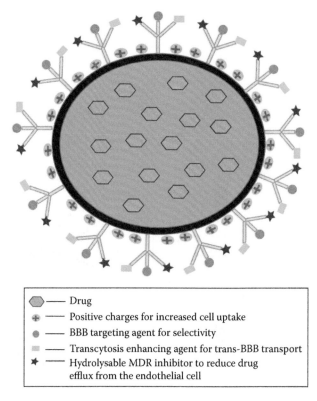

Legend:
- Drug
- Positive charges for increased cell uptake
- BBB targeting agent for selectivity
- Transcytosis enhancing agent for trans-BBB transport
- Hydrolysable MDR inhibitor to reduce drug efflux from the endothelial cell

FIGURE 6.1
A model drug nanoparticle concept for overcoming blood–brain barrier (BBB).

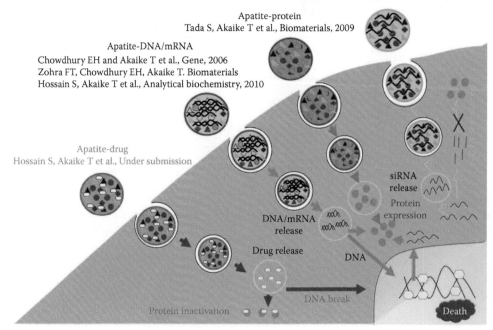

Apatite-siRNA/mRNA
Hossain S, Akaike T et al., J. Control release, 2010

Apatite-protein
Tada S, Akaike T et al., Biomaterials, 2009

Apatite-DNA/mRNA
Chowdhury EH and Akaike T et al., Gene, 2006
Zohra FT, Chowdhury EH, Akaike T. Biomaterials
Hossain S, Akaike T et al., Analytical biochemistry, 2010

Apatite-drug
Hossain S, Akaike T et al., Under submission

siRNA release
Protein expression
DNA/mRNA release
Drug release
DNA
DNA break
Protein inactivation
Death

FIGURE 6.2
Super efficient carbonate apatite as nucleic acids/drug transporter: mode of action.

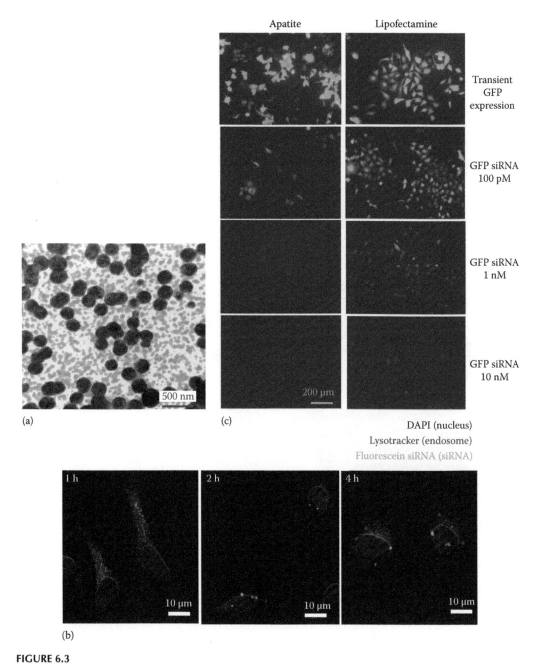

FIGURE 6.3
(a) Desirable nanosize, (b) endosomal entry, time-dependent exit of siRNA, and (c) finally effective knockdown of reporter gene (GFP) mediated by carbonate apatite nanoparticles. Scale bar 500 nm.

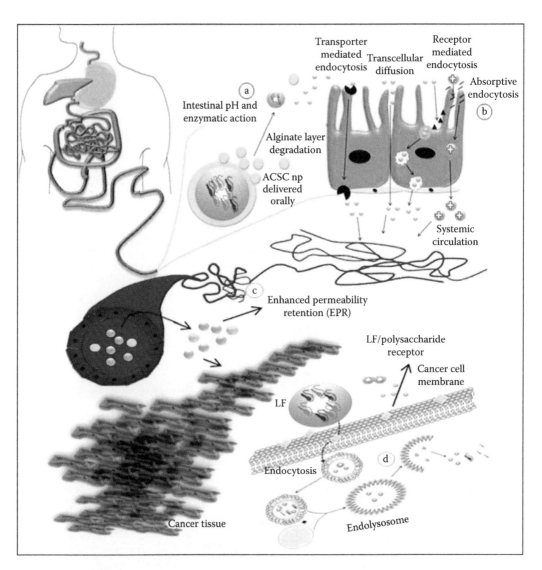

FIGURE 7.1
Schematic illustration of the proposed mechanism of alginate-enclosed chitosan-coated calcium phosphate bio-ceramic nanoparticle internalization. The alginate coating gets degraded in the alkaline and enzymatic environment in the intestine and the remaining chitosan-coated ceramic particles enter the blood circulation via endocytosis and/or transcytosis. Passive delivery of these ceramic cores at the cancer site happens by making use of the EPR effect. Further uptake of the NC will be done in the cancer cells based on oligosaccharide receptor–mediated endocytosis. Ceramic core further helps in controlled release and biodegradation of the formulation.

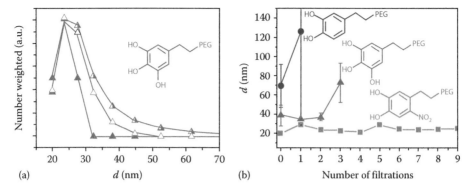

FIGURE 8.2

Characterization of the stability of superparamagnetic iron oxide NPs. The stability of superparamagnetic iron oxide NPs was measured with DLS at 25°C. (a) The hydrodynamic diameter of PEG(0.55 kDa)–hydroxydopa-mine-stabilized iron oxide NPs as stabilized (-▲-), after storage for 1 year in PBS (-△-) and after storing them for 20 months in HEPES (-△-). (From Amstad, E. et al., Surface functionalization of single superparamagnetic iron oxide nanoparticles for targeted magnetic resonance imaging. *Small.* 5. 1334–1342. Copyright Wiley-VCH Verlag GmbH & Co. KGaA. Reproduced with permission.) (b) The stability of NPs evaluated after repeated filtrations performed to remove excessive dispersants from iron oxide NPs stabilized with PEG(5 kDa)–nitrodopamine (-■-), PEG(5 kDa)–hydroxydopamine (-▲-), and PEG(5 kDa)–dopamine (-●-) (Reproduced with permission from Amstad, E., Gillich, T., Bilecka, I., Textor, M., and Reimhult. E. Ultrastable iron oxide nanoparticle colloidal suspensions using dispersants with catechol-derived anchor groups. *Nano Lett.* 9. 4042–4048. Copyright 2009, American Chemical Society.). While PEG–hydroxydopamine-stabilized iron oxide NPs were stable at room temperature for more than 20 months, they started to agglomerate after excessive dispersants were removed by more than two filtrations.

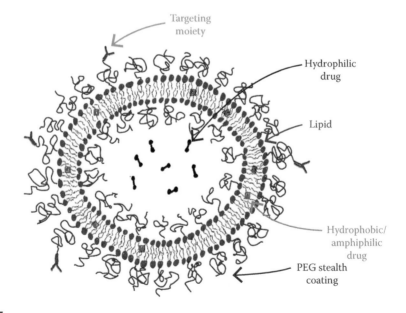

FIGURE 8.5

Schematic of drug delivery liposome. The basic constituents of a drug delivery liposome are lipids forming a liposome membrane in the 100 nm size range, thereby encapsulating small hydrophilic drugs in the lumen or hydrophobic and amphiphilic drug in the membrane. Optionally the lipids can tether, for example, PEG to increase *in vivo* circulation time and charged lipids or specific binding groups to target the liposome to the release site.

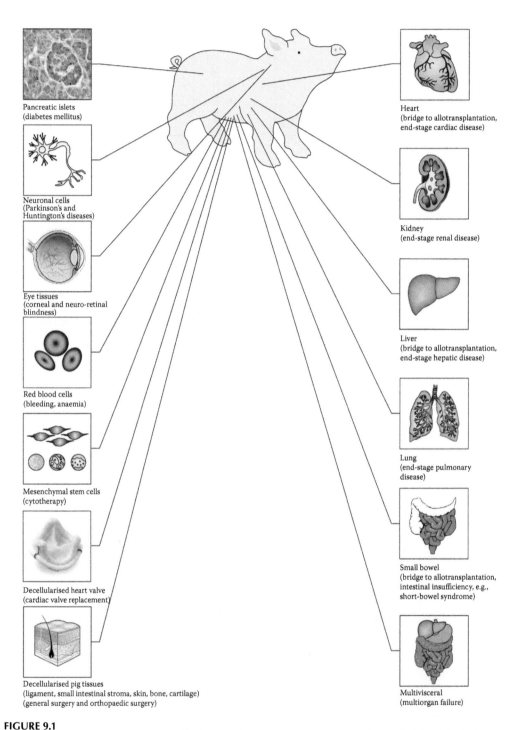

Pancreatic islets
(diabetes mellitus)

Neuronal cells
(Parkinson's and
Huntington's diseases)

Eye tissues
(corneal and neuro-retinal
blindness)

Red blood cells
(bleeding, anaemia)

Mesenchymal stem cells
(cytotherapy)

Decellularised heart valve
(cardiac valve replacement)

Decellularised pig tissues
(ligament, small intestinal stroma, skin, bone, cartilage)
(general surgery and orthopaedic surgery)

Heart
(bridge to allotransplantation,
end-stage cardiac disease)

Kidney
(end-stage renal disease)

Liver
(bridge to allotransplantation,
end-stage hepatic disease)

Lung
(end-stage pulmonary
disease)

Small bowel
(bridge to allotransplantation,
intestinal insufficiency, e.g.,
short-bowel syndrome)

Multivisceral
(multiorgan failure)

FIGURE 9.1
Disorders for which xenotransplantation is a potential therapy. (Reprinted from *The Lancet*, 379, Ekser, B. et al., Clinical xenotransplantation: The next medical revolution? 672–683, Copyright 2012, with permission from Elsevier.)

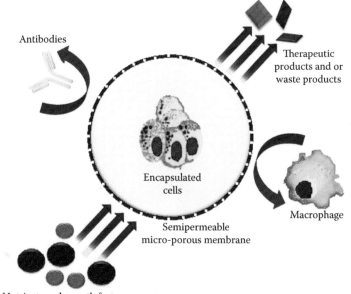

FIGURE 9.2
Conceptual representation of the encapsulation of cells within nanoporous membranes. The nanoporous membranes permit selective diffusion of nutrients toward cells and of therapeutic and waste products away from the encapsulated cells and to the outer side of membrane. At the same time, the small size of pores in the membrane prevents passage of immune components such as macrophages and antibodies.

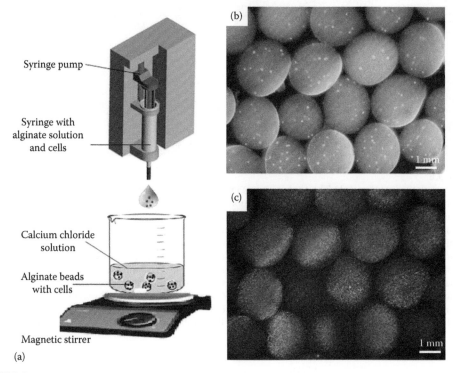

FIGURE 9.4
Encapsulation of β-TC-6 cells in sodium alginate–based capsules. (a) Schematic representation of a typical setup required for encapsulating cells in alginate capsules. (b) Bright-field image of encapsulated β-TC-6 cells within alginate capsules. (c) Fluorescence image of Calcein AM–stained β-TC-6 encapsulated within alginate capsules. Scale bar 1 mm.

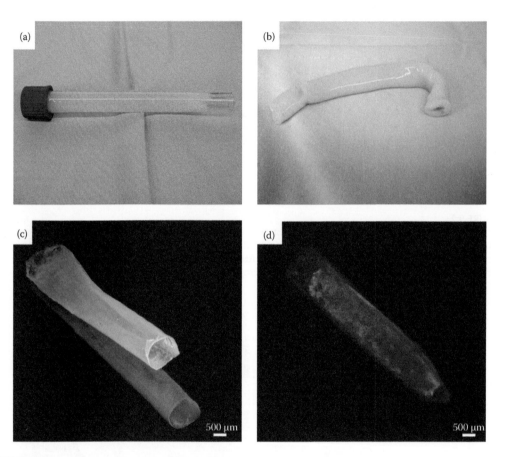

FIGURE 9.6
Cell encapsulation in tubes. (a,b) Fibrin tubes and (a) image of a custom-built casting mold used to mold a fibrin tube. The mold consisted of two 15 cm long glass tubes, which were closed and fixed at one end. The internal tube had an outside diameter of 5 mm, and the inside diameter of the external tube was 11 mm with a 3 mm broad interspace among both tubes, in which the fibrin preparation was filled. Due to shrinkage during the hardening of the fibrin, preparation segments with a wall thickness of 1.5–2.5 mm were obtained. (b) Image of a segment after 10 days of incubation in vitro before placing it in the bioreactor. (Reprinted from *Eur. J. Vasc. Endovasc. Surg.*, 33, Aper, T., Schmidt, A., Duchrow, M., and Bruch, H.P., Autologous blood vessels engineered from peripheral blood sample, 33–39, Copyright 2007, with permission from Elsevier.) (c,d) Manually rolled up nanoporous membranes. (c) Nanoporous (0.03 μm pore size) polycarbonate membrane rolled into tubelike structures with an inner diameter of 1 mm and sealed at one end. (d) Image of green fluorescence indicating viability of β-TC-6 cells with 0.5% agar that were packed in these tubes and stained with Calcein AM after 7 days.

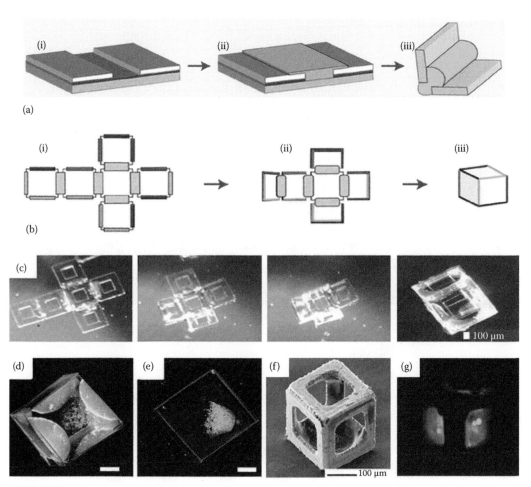

FIGURE 9.7
Self-folding process for creating 3D capsules. (a) Schematic representation of the fabrication process for self-folding of polymeric containers. (i) Si wafer containing spin-coated sacrificial and photolithographically patterned SU-8 panels. (ii) PCL deposition in hinge gaps. (iii) 2D templates were lifted off via dissolution of the PVA layer in water and self-assembly occurred on heating above 58°C. (b, i–iii) Schematic demonstrating self-folding of a cubic container. External "locking" hinges are colored in pairs to denote corresponding meeting edges. (c) Video capture sequence (over 15 s) showing a 1 mm sized, six-windowed polymeric container self-folding at 60°C. (d) Bright-field and (e) fluorescence z-plane stack image of stained fibroblast cells encapsulated within a nonporous polymeric container. Green fluorescence indicates live cells. (Reprinted from *Biomed. Microdev.*, 13, Azam, A., Laflin, K.E., Jamal, M., Fernandes, R., and Gracias, D.H., Self-folding micropatterned polymeric containers, 51–58, Copyright 2011, with permission from Elsevier.) (f) Scanning electron microscopy image of a hollow, open-surfaced biocontainer and (g) fluorescence microscopy images of a metallic biocontainer loaded with cell–ECM–agarose with the cell viability stain, Calcein AM. (Reprinted from *Biomed. Microdev.*, 7, Gimi, B. et al., Self-assembled three-dimensional radio frequency (RF) shielded containers for cell encapsulation, 341–345, Copyright 2005, with permission from Elsevier.)

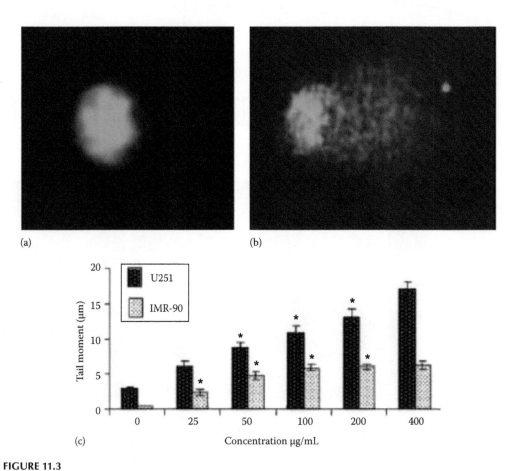

(a)

(b)

(c)

FIGURE 11.3
Comet analysis: (a) untreated and (b) Ag-NP-treated cancer cells stained by SYBR green (conc. 400 g/mL). (c) Represents the tail moments of DNA (m). Fibroblasts exhibited a concentration-dependent increase in DNA damage up to 100 g, above which the values remained constant, whereas cancer cells showed a steady increase; *represents $P < 0.05$. (Reproduced from Asharani, P.V. et al., *Adv. Funct. Mater.*, 20, 1233, 2010. With permission.)

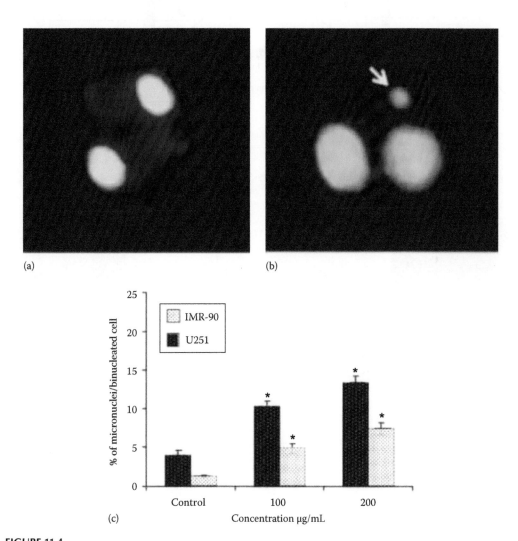

(a)

(b)

(c)

FIGURE 11.4
Micronucleus analysis of (a) untreated and (b) Ag-NP (100 g/mL)-treated fibroblasts showing binucleated cell. (b) White arrow indicates the micronucleus formed among the binucleated cells. (c) Data from MNA show chromosomal aberrations. The data represent 1000 binucleated cells for U251 and 700 binucleated cells for fibroblasts; *represents $P < 0.05$. (Reproduced from Asharani, P.V. et al., *Adv. Funct. Mater.*, 20, 1233, 2010. with permission.)

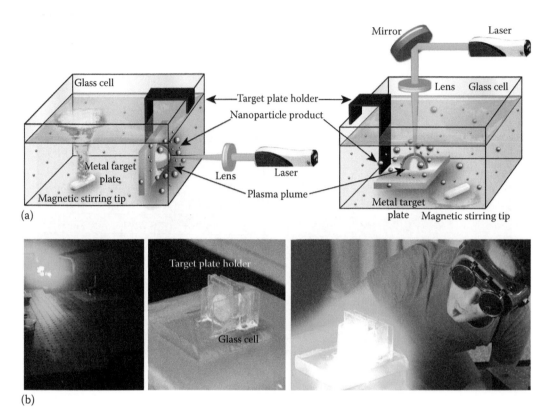

(a)

(b)

FIGURE 12.1
Schematic representation of the LASiS experimental setup. (a) Two variants are generally used: (left) The plate is placed close to the wall and irradiated from the same side, with the laser beam passing through a few millimeters of liquid, and (right) the plate is attached to the wall and the laser beam passes through the entire liquid column before reaching the plate. (b) Actual experimental setup and glass cell of the left schematic configuration in (a).

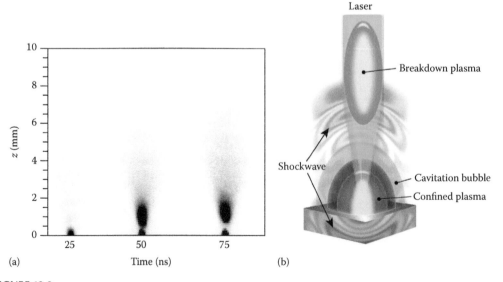

(a)   (b)

FIGURE 12.2
Plasma plume formation during laser ablation. (a) High-speed photography of the plasma plume extension during ablation of a silver plate with a 527 nm Nd:glass laser in vacuum. (b) Schematic overview of the mechanism and phenomena involved in plasma plume formation during LASiS.

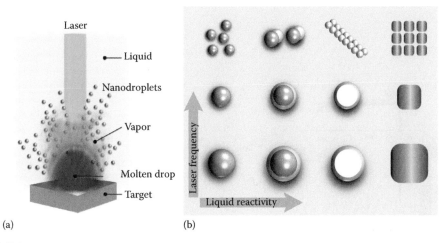

(a)   (b)

FIGURE 12.3
Schematic overview of the explosive ejection mechanism. (a) Formation of nanodroplets in a millisecond low-energy laser ablation process. (b) Effect of liquid reactivity and laser frequency on morphology of nanostructures when a Pb target was ablated in various liquids (and reactive environments). (Based on Niu, K.Y. et al., *J. Am. Chem. Soc.*, 132, 9814, 2010.)

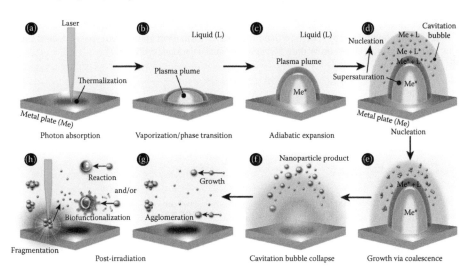

FIGURE 12.4
Mechanism of nanoparticle formation during LASiS.

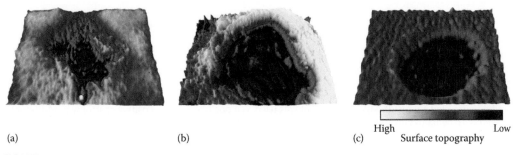

(a)   (b)   (c)

FIGURE 12.6
Scanning electron micrographs showing the topology (CLUT) of a silver target plate (craters) at the point of focus after LASiS in water. (a) Fs laser (120 fs), (b) ns laser (10 ns), and (c) ns LASiS of a gold foil at low laser power (10 ns). (Partially based on Tsuji, T. et al., *Appl. Surf. Sci.*, 206, 314, 2003.)

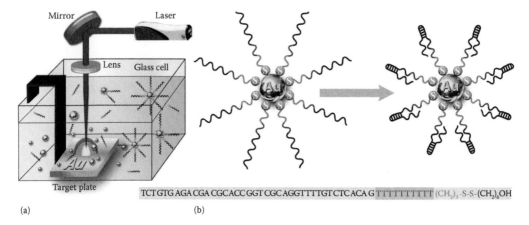

TCT GTG AGA CGA CGCACC GGT CGC AGGTTTTGT CTC ACA G TTTTTTTTTT (CH₂)₃-S-S-(CH₂)₆OH

(a)   (b)

FIGURE 12.7
Generation of aptamer-conjugated AuNPs via *in situ* conjugation. (a) Schematic illustration of *in situ* conjugation of AuNPs with aptamers during laser ablation in an aqueous aptamer solution. (b) Spacer design and resulting mixed monolayer conjugated nanoparticles. Mixed monolayer formation and careful spacer design contributed to correct aptamer folding. (Based on Walter, J.G. et al., *J. Nanobiotechnol.*, 8, 21, 2010.)

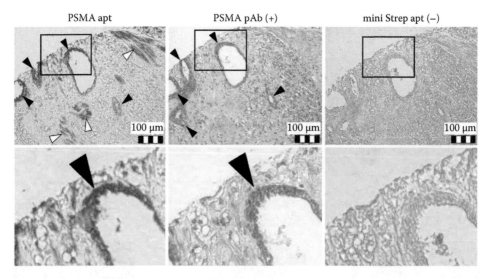

FIGURE 12.8
Detection of PSMA in human prostate cancer tissue. Detection of PSMA-positive structures in prostate cancer tissue sections by immunohistochemical staining using anti-PSMA aptamer (PSMA apt)–conjugated AuNPs. As a negative control, AuNPs conjugated with miniStrep aptamer (miniStrep apt) were used. A polyclonal antibody directed against PSMA (PSMA pAb) was used as a positive control. Positive control was additionally stained with hematoxylin and eosin. Black arrows indicate specific staining, while white arrows flag unspecific binding. (Reproduced from Walter, J.G. et al., *J. Nanobiotechnol.*, 8, 21, 2010.)

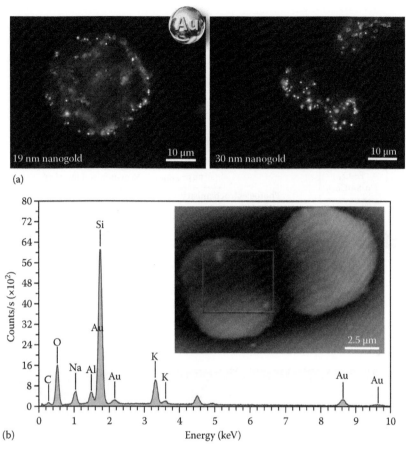

(a)

(b)

FIGURE 12.9
Internalization of nanogold particles produced via LASiS in pancreatic tumor cells. (a) Dark-field imaging of internalized gold nanoparticles in AR42J pancreatic tumor cells. The cells were treated with 19 and 30 nm gold colloids for 6 h (scale bar, 10 µm). (b) EDX spectrum of cells incubated with 30 nm gold nanoparticles. Inset is the SEM image of the cell incubated with 30 nm particle (scale bar, 10 µm). (Based on Sobhan, M.A. et al., *Colloids Surf. B. Biointerfaces*, 92, 190, 2012.)

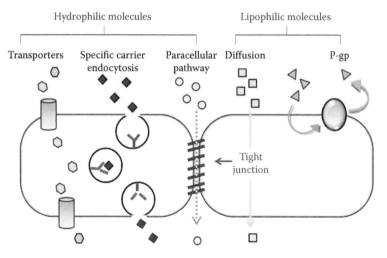

FIGURE 13.1
Molecular transport across the BBB.

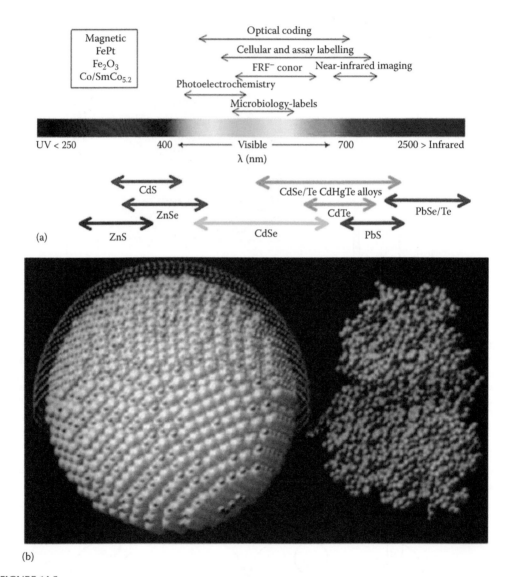

(a)

(b)

FIGURE 14.5
Quantum-dots general features. (a) Representation of the type of applications of semiconductor quantum-dot nanomaterials, their common formulation along with their emission wavelength range (from Reference 126). (b) It shows in molecular scale the size of a quantum-dot made of CdSe/ZnS with 6 nm of diameter in comparison to a protein called MBP (maltose-binding protein, MW 44 KD) with an overall dimensions of 3 x 4 x 6.5 nm.

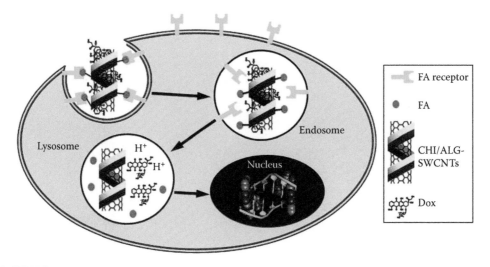

FIGURE 15.3
Scheme of the uptake by endocytosis of SWCNT loading doxorubicin by cancer cells. (Reproduced from *Biomaterials*, 33, Meng, L., Zhang, X., Lu, Q., Fei, Z., and Dyson, P.J., Single walled carbon nanotubes as drug delivery vehicles: targeting doxorubicin to tumors, 1689, Copyright 2012, with permission from Elsevier.)

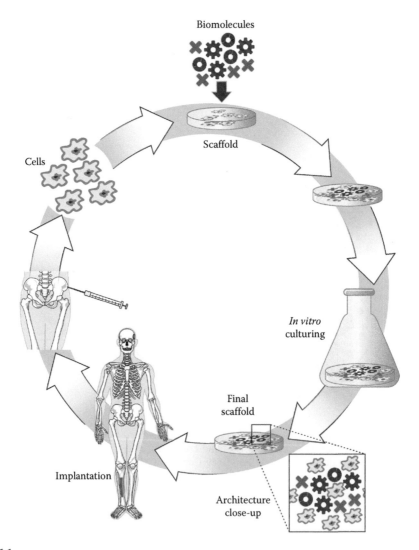

Biomolecules

Cells

Scaffold

In vitro culturing

Final scaffold

Implantation

Architecture close-up

FIGURE 16.1
Bone tissue engineering utilizing 3-D scaffold structures. Cells obtained from healthy bone tissue followed by isolation and expansion in culture. Cells are seeded onto a suitable scaffold with appropriate biomolecules. The construct is cultured *in vitro*. Transplantation of the final scaffold incorporating cells and biomolecules into the defect area.

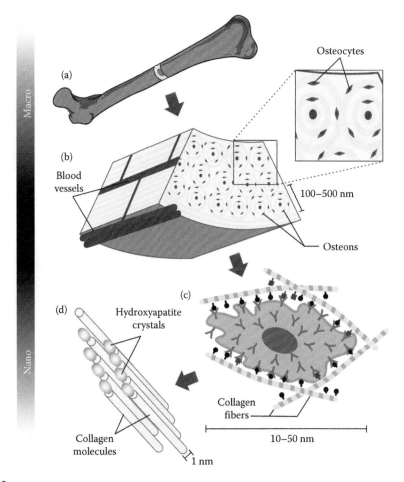

FIGURE 16.2
Hierarchical organization of bone tissue from the macro-(centimeter) scale to the nanoscale structure: (a) At the macrostructural level, bone may be described on the exterior by a hard, calcified, cortical (compact) layer; (b) at the submicron level, osteons (Haversian systems) are evident; (c) cells interact with ligand sites within the ECM through their membrane receptors; (d) and the distinctive nanostructure that comprises the ECM is composed of an inorganic CaP mineral phase and an organic component chiefly composed of type I collagen. (From Stevens, M.M. and George, J.H., *Science*, 310, 1135–1138. Copyright 2005. Redrawn and reproduced with permission of AAAS.)

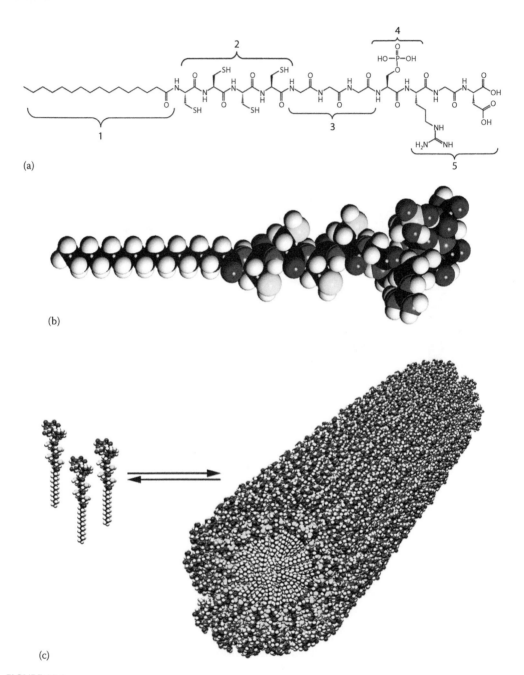

(a)

(b)

(c)

FIGURE 16.4
(a) Peptide amphiphile structure highlighting the distinctive features intrinsic to the self-assembly process. Region 1 represents the alkyl chain that conveys the hydrophobic properties on the molecule. Region 2 highlights four cysteine residues that oxidize and can form disulfide bonds. Region 3 contains three glycine residues that impart flexibility to the hydrophilic head group. Region 4 depicts a phosphorylated serine residue that can interact with Ca ions and help direct mineralization of HA. Region 5 represents the cell adhesion ligand RGD. (b) Molecular model of the PA showing the overall conical shape of the molecule going from the narrow hydrophobic tail to the bulkier peptide region. (c) Schematic showing the cylindrical nanostructure of the self-assembled PA. (From Hartgerink, J.D. et al., *Science*, 294, 1684–1688. Copyright 2001. Reproduced with permission of AAAS.)

because the pressure after collapse of the cavitation bubble is 3–4 orders of magnitude smaller than the pressure during the initial expansion of the plasma.

12.2.2.2 Explosive Ejection Mechanism

Although the thermal evaporation model more or less adequately describes nanoparticle formation during LASiS in a large number of nanoparticle types, it still falls short in other cases. To explain the formation of particular nanostructures, predominantly metal oxides and sulfides, an alternate mechanism was proposed in which, as a consequence of the laser–target interaction, nanodroplets are ejected into the medium at high speed (Figure 12.3a) [27]. This phenomenon is predominantly observed when using millisecond pulsed lasers, which deposit only low-power densities onto the target. Nonetheless, several groups also reported observing the formation of nanodroplets with ns pulsed lasers, which illustrates that the exact mechanism underpinning LASiS is quite complicated. Tsuji et al. observed the formation of Ag, Au, and Si nanodroplets during ns pulsed laser ablation in water using shadowgraphy [28]. They observed two distinct types of ejections at the target–water interface: (1) a straight jet emanating from the center below 100 μs after the laser pulse, consisting of clusters and nanodroplets, and (2) a bubble confining the plasma–gas that developed below 1 μs and expanded up to 200 μs, only to subsequently collapse (Figure 12.4f). The latter led to a "classical" nucleation–growth mechanism. Alternatively, as proposed by Phuoc et al., the laser causes local melting, forming a molten drop (Figure 12.3a) [29]. The liquid adjacent to this drop forms a vapor or plasma with high pressure. This expanding vapor/plasma subsequently explosively causes the molten drop to disintegrate into nanodroplets, which react with the medium to form the final nanoparticle product. Similar considerations were expressed by Yoo et al. during LASiS of silicon [30]. They noted that vapor formed near the superheated silicon mass that nucleated as bubbles. Once a critical size is reached, rapid expansion initiates explosive boiling and mass ejection as particulates.

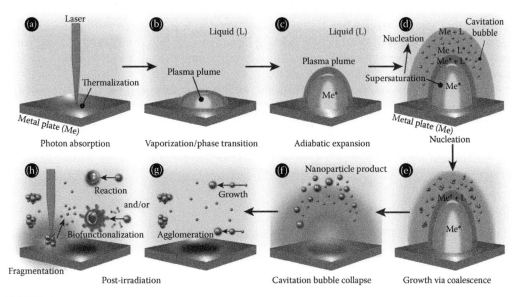

FIGURE 12.4
(See color insert.) Mechanism of nanoparticle formation during LASiS.

In practice, the thermal evaporation and the explosive ejection mechanisms may occur concomitantly or at least some features overlap. Further research will show if a particular measure of control over these mechanisms can be achieved, which would allow more control over the particle's size; thermal evaporation produces a more uniform size, whereas the explosive mechanism produces a broader size distribution but also a significantly larger yield [31–33].

One other feature of the explosive mechanism is the fact that such compact and hot droplets exhibit surface reactions with the surrounding liquid, which depend on the liquid reactivity and laser settings, as observed by Niu and coworkers [27]. In this way, nanoparticle products with various morphologies can be produced, as schematically depicted in Figure 12.3b. Furthermore, it should be possible to exploit this surface reactivity to form more complex hybrid particles, although more research is required in order to obtain more control over the LASiS process.

12.2.3 Mechanism of Nanoparticle Formation

Nanoparticle formation follows a number of steps according to a nucleation, during plasma plume cooling, and subsequent growth mechanism, some of which are shared with more conventional crystallization processes.

As schematically depicted in Figure 12.4, the ablation process starts with an incoming pulse of laser light hitting the target and the energy is absorbed, locally heating the target, and the material melts (a). Subsequently, the material is ejected from the plate and partially evaporated to form a plasma plume (b) that expands adiabatically (c). Within the plasma plume, nucleation occurs (d), for which the driving force is supersaturation, as the plume starts to cool. The formation of the cavitation bubble is a result of vaporization of the liquid that is in contact with the hot plasma and confines the plasma. The embryonic particle grows by attracting emitted material, such as atoms, clusters, and droplets (e), released from the plate during the laser ablation process, which is diffusion limited, to yield the final product (f). Furthermore, clusters of various sizes can be formed by coalescence of individual growing embryonic particles or, as Mafuné et al. determined, particles may continue to grow for days after LASiS (g) [34]. Finally, it is possible to biofunctionalize the particles in a one-pot synthesis by adding ligands to the liquid medium or to perform post-LASiS size manipulations by irradiating the solution with laser pulses of appropriate wavelength, fluence, and pulse duration (h).

The progress of the nanoparticle synthesis can easily be followed via UV/vis spectroscopy. The presence of a sharp plasmon band near 400, 525, and 600 nm for Ag, Au, and Cu shows the presence of the nanoparticles, and the absence of additional peaks demonstrates that the particles have spherical forms. Particle growth postirradiation can be easily detected by observing the position of the absorption maximum. Generally, a redshift indicates an increase in mean particle diameter.

12.2.4 Use of Stabilizers

In general, when LASiS is performed in water, unstable colloidal solutions with relatively broad size distributions are observed. Furthermore, colloids in water tend to agglomerate over time, forming large clusters (Figure 12.5b). Clear differences in the final nanoparticle product can be observed when performing LASiS in organic solvents. In water, particles with 10–40 nm are typically obtained, while particles between 5 and 15 nm are observed for MNPs in organic solvents [11]. Amendola speculates that differences in the thermal

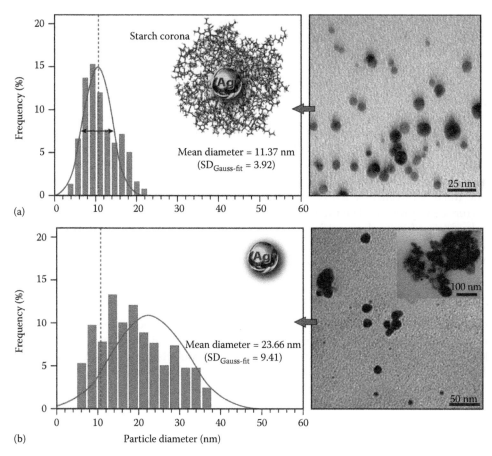

FIGURE 12.5
LASiS-based AgNP formation in aqueous starch solution and neat water. Representative TEM images and size distributions of AgNPs prepared by 532 nm LASiS (60 mJ/pulse; 10 ns; 20 Hz) in an (a) aqueous starch solution (1% wt.) or (b) distilled water: insert shows an example of a large cluster formed postirradiation via agglomeration. (Standard deviations in the normal distribution in parenthesis.)

properties of water compared to organic solvents and/or the strong interaction of organic solvents with the nanoparticle surface, which is competitive with particle growth, might cause these differences.

Because LASiS in pure solvents generally leads to nanoparticle agglomeration, and after weeks of storage flocculation of the product, a number of groups, including ours, turned to using surfactants and natural materials as stabilizers, such as sodium dodecyl sulfate (SDS) [35–37], long-chain fatty acids (stearic, palmitic, and lauric acid), lauryl amine [15], starch [38,39], chitosan [16], and palm and coconut oil [40,41]. These stabilizing agents prevent particle agglomeration by physically surrounding the nanoparticle, which leads to charge and/or steric stabilization (Figure 12.5a). Furthermore, it is speculated that they might serve as synthesis templates, because these materials form complex 3D shapes that form boundaries to the watery environment, such as micelles. In addition, the process in which the stabilizers cover the particle's surface competes with growth processes, such as nuclei coalescence and adsorption of free atoms, and as such, a smaller average size is obtained compared with equivalent LASiS in water.

By utilizing LASiS as a nanoparticle synthesis method, we recently produced stabile nanoparticles in an aqueous starch solution. To this end, LASiS was performed with a pulsed Q-switched Nd:YAG at its second harmonic wavelength of 532 nm (10 ns; 60 mJ/pulse; 20 Hz repetition rate for 15 min). Here, a stabile colloidal solution of AgNPs with an average diameter of ~11 nm was produced and a relatively narrow size distribution (Figure 12.5a). By postirradiating the solutions, the final size could be significantly reduced. The starch coating of the AgNPs not only prevented growth but stabilized the colloids through shielding of surface charges and steric hindrance (Figure 12.5a). The same LASiS in bi-distilled water (Figure 12.5b) illustrates that not only a higher average size and broader size distribution is obtained but also the propensity of nanoparticles in water to agglomerate to larger clusters (see insert).

One major drawback when using carbon-based solvents or stabilizers is the possible occurrence of pyrolysis. The increasing production of degradation products during LASiS may lead to surface fouling, which can affect biocompatibility, stability, and surface accessibility. This should be taken into account when adjusting the laser settings and in the production of medical-grade colloidal solutions.

12.2.5 Influence of Various Laser Parameters

A major advantage of LASiS of a solid target is the fact that adjustment of various laser parameters allows some measure of control over the size and shape of the final nanoparticle product. Conceptually, the tuning of the laser parameters can induce fragmentation, reshaping, and growth of nanoparticles. The influence of various laser parameters is briefly discussed in the succeeding text.

12.2.5.1 Laser Fluence and Beam Size

The effect of laser fluence and beam size on nanoparticles produced by a laser ablation process has been extensively investigated by a number of authors. The majority found that increasing the laser fluence and beam size led to an increase in particle diameter and size distribution [22,25,32,35,42]. Both of these parameters directly affect the thermodynamic properties of the cavitation bubble and the produced plasma. An increase in these parameters not only increases the nanoparticle production rate and concentration but also allows the plasma plume to reach a higher temperature and pressure, which results in a longer existence of the cavitation bubble before collapse and consequently the formation of larger particles.

12.2.5.2 Laser Pulse Duration

LASiS of nanoparticles with fs lasers (10^{-15}) is gaining in popularity, because of the high ablation efficiency of this kind of laser and also because of the enhanced control over the particle size. Compared to laser ablation processes with longer-duration laser pulses, such as picosecond (10^{-12}) and ns (10^{-9}) lasers, significantly smaller particle sizes can be achieved with fs lasers [26–31]. Femtosecond pulses transfer energy to the electrons in the target on a timescale that is faster than electron–phonon thermalization processes. Conversely, picosecond and ns laser pulses transfer energy on a timescale that is comparable with thermal relaxation processes in the target [19,43]. This is essential, since it has consequences for the ablation mechanism and produces different outcomes.

First of all, a number of studies have shown that the plasma plume lasts for tens of nanoseconds [19,44], and thus, for ns pulses, the plasma is present for nearly twice

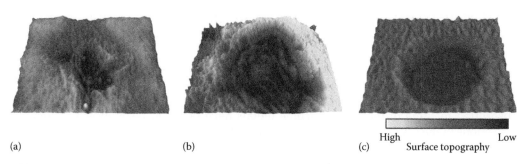

(a) (b) High Low
(c) Surface topography

FIGURE 12.6

(See color insert.) Scanning electron micrographs showing the topology (CLUT) of a silver target plate (craters) at the point of focus after LASiS in water. (a) Fs laser (120 fs), (b) ns laser (10 ns), and (c) ns LASiS of a gold foil at low laser power (10 ns). (Partially based on Tsuji, T. et al., *Appl. Surf. Sci.*, 206, 314, 2003.)

the pulse duration [44]. Consequently, in fs and ps lasers, there is no temporal overlap between the pulses and the ejected material; it was found that for fs laser pulses, the number of ejected atoms per pulse was ~10^2–10^3 times lesser than for ns laser pulses. In ns lasers, the temporal overlap causes the expanding plasma plume to absorb sequential pulses and its temperature rises, favoring atomization and homogenization of the material [11]. At the same time, the plume causes optical shielding of the target, and thus, less laser energy reaches the target. However, the increased plasma temperature leads to a significantly higher interaction of the plasma with the target (secondary ablation). Second, the lifetime of the cavitation bubble is significantly reduced when performing LASiS with fs laser pulses, because the pulse ends before the beginning of the plasma expansion. This phenomenon limits the effective growth time of the nuclei, which leads to the formation of nanoparticles with considerably smaller sizes [24,42–43,45,46].

The effect that the pulse duration has on the LASiS outcome is clearly illustrated when looking at the target focal point, where craters are left behind where the material was ejected and vaporized. As illustrated from the SEM micrographs in Figure 12.6, fs lasers form shallower and less well-defined craters compared with ns laser pulses. Conversely, craters formed with ns laser pulses are deeper and have outer rims, which would be consistent with higher temperatures and increased formation efficiency. Note that the crater in Figure 12.6 virtually has no rim, which may be attributed to the materials used (Au foil), the laser fluence, and a mechanism that favors explosive ejection of ablated material.

12.2.5.3 Laser Wavelength

The used laser wavelength is one of the most important parameters in LASiS and not only determines the efficacy of the ablation but can also be used to control the nanoparticle size. Tsuji and coworkers performed an extensive study of the relationship between the laser wavelength and the particle size for MNPs [48]. In their study, a silver plate was ablated in water with a ns pulsed laser operating at different wavelengths, that is, 1064, 532, or 355 nm. They found that particle size and size distribution directly depend on the laser wavelength and decrease with decreasing laser wavelength: $|\lambda_{laser} = 355$ nm: $d_{particle} < 10$ nm$|$, $|\lambda_{laser} = 532$ nm: $10 < d_{particle} < 30$ nm$|$, and $|\lambda_{laser} = 1064$ nm: $d_{particle} > 30$ nm$|$. Furthermore, from the UV/vis spectra, they found that the nanoparticle formation efficiency at shorter wavelengths was lower than those prepared at longer wavelengths. Most likely, this is a result of self-absorption and/or scattering by already formed nanoparticles in the solution, thereby increasingly preventing the laser light (and energy) from reaching the target as LASiS progresses.

An ambivalent effect of this self-absorption is the occurrence of fragmentation of particles and clusters through thermal excitation of the particle lattice by the absorbed energy. Particle fragmentation by this mechanism must be larger for wavelengths that lie near the absorption maxima of the colloids (some metals have plasmon resonances or interband transitions in the UV/vis part of the spectrum) and results in a significant size reduction at shorter wavelengths. However, even though the particle size decreases, broadening of the size distribution via photoinduced fragmentation occurs.

Generally, the decrease in particle size for shorter wavelengths may be explained by two mechanisms:

1. Absorption by the plasma through inverse Bremsstrahlung decreases at shorter laser wavelengths, and therefore, the lifetime of the plasma and associated cavitation bubble is shorter, which leads to a decrease in the particle growth time [49].

2. Interaction between laser light and the produced nanoparticles, dispersed in solution, increases with decreasing wavelength. As a result of this secondary interaction, nanoparticle fragmentation leads to a decrease in particle size [48,50].

To minimize reabsorption effects, near-infrared wavelengths should be used [36], which significantly reduces nonlinearities between the ablation efficiency and the absorption coefficient or the laser fluence.

12.2.5.4 Laser Irradiation Time

The effect of the laser irradiation time on the size distribution of nanoparticles was investigated in detail by a number of authors, and most of them reported that further fragmentation of particles as a result of the longer irradiation decreased the average nanoparticle size [36,51].

As an illustrative example of these studies, Takami et al. irradiated gold nanoparticles in aqueous solution with a ns pulsed laser at 532 nm (210 mJ cm^{-2}) for 10 min [52]. From TEM analysis, they observed both a reduction in the particle size and a change in their shape. Before irradiation, the size was distributed between 19 and 47 nm, whereas after the irradiation, it was between 5 and 21 nm. Furthermore, before the irradiation, the particle shapes were irregular and spheroid at the best, whereas after the irradiation, they were spherical in shape. Therefore, it may be concluded that the size reduction is a result of heating through absorption via SPR of electrons on the particle surface. This energy is subsequently transferred to the lattice phonon and the temperature of the particle increases. Above the melting point, the particle melts and becomes a liquid droplet that eventually cools to a perfect spherical nanoparticle with reduced size. The size reduction is a result of surface vaporization when the temperature exceeds the boiling point and atoms and small particles are ejected, and thus, the effective amount of material is reduced. The ejected atoms and small particles are highly unstable in solution and tend to agglomerate or are deposited on the surface of the remaining particles. Consequently, the proportion of ejected to deposited atoms determines the final particle size. In addition, an increasing exposure time increases the efficiency of nanoparticle production and reduces the particle size distribution.

12.2.5.5 Laser Repetition Rate

Another parameter of lasers that can afford some control over the mean nanoparticle size is the laser repetition rate. With the aim to obtain nanoparticles with narrow size distributions but with high concentration, Menéndez-Manjón and Barcikowski closely studied the effect of the laser repetition rate on produced gold nanoparticles [53]. In order to assess the aforementioned relationship, the authors defined two parameters for comparison reasons: (1) the nanoparticle productivity as mass ablated per unit of time and (2) the ablation efficiency as ablated mass per pulse. They found that the nanoparticle productivity increased linearly with repetition rates between 100 and 2500 Hz (60 s laser ablation time), but the ablation efficiency remained constant. As the repetition rate approached 5 kHz, the nanoparticle productivity deviated from linearity and the ablation efficiency decreased. Interestingly, when the ablation time was reduced to 30 and 10 s, the productivity and ablation efficiency increased again. This effect is a direct result of increasing shielding by the nanoparticles, not to be confused with shielding due to plasma plume expansion. At longer ablation times, a higher concentration of nanoparticles is produced, which in turn increasingly leads to more absorption of light and consequently decreases the productivity.

In general, the effect of repetition rate during LASiS may be explained in terms of residency time in the focal volume, laser pulse period, and the extent of the fragmentation that consequently occurs. Diffusion of particles or agglomerates through the focal volume generally results in reduction of the colloidal polydispersity. However, if the pulse length is shorter than the diffusion time, multiple pulses will hit a particle, and thus, the laser will affect fewer particles in the total colloidal solution. Consequently, a broader size distribution will be observed. At rates above 500 Hz, the fragmentation efficiency is constant, higher productivities occur, and narrower size distributions will be obtained with higher repetition rates (convective–diffusion from and into the irradiated volume).

Overall, the laser repetition rate is an important parameter, which can be used to steer the polydispersity of the colloidal solution and the final average particle size. However, the exact parameters that influence the repetition rate, especially in terms of shielding, mass transport, focal residency time, and local temperatures, remain to be further investigated. We, for instance, recently noticed that in a particular LASiS setup with starch as a stabilizer (and possibly a synthesis template), the average particle size increased with increasing low repetition rate (unpublished results). Such contradictions illustrate that further research is required.

12.3 Applications of LASiS in Biomedical Science

Nanoparticles that interface with biological systems have attracted widespread interest in the fields of biology, (bio)medicine, and related life sciences over the last few decades and inevitably will propel diagnostic and therapeutic applications beyond current boundaries. Consequently, a significant increase in scientific research effort may be noticed, in which scientists and engineers "boldly go where no man has gone before."

Nanoparticles have unique electrical and optical properties and consequently make excellent semiconductors and imaging agents. Their extremely small size allows them to penetrate cells and interact with (sub)cellular structures and (bio)molecules. This property allows researchers to obtain information on the single-cell or even single-particle level rather than data averaged over an ensemble of cell populations or analytes. Furthermore, it is envisioned that micromanipulations will become possible in the intact organism, for instance, by targeting imageable nanoparticles, carrying photoinducible apoptosis signals to cancer cells, allowing both detection of aberrant cells and site-specific treatment.

As stated previously, a variety of physical and chemical methods are used to synthesize nanoparticles. However, conventional physical and chemical methods carry with them the disadvantage that they often rely on the use of toxic materials, which downstream requires elaborate purification steps to obtain medical-grade end products. Furthermore, biocompatibilization and functionalization require several sequential synthesis and modification steps, all of which increase the probability for nanoparticle destabilization, surface fouling, residual toxins, and other unexpected side effects. Therefore, methods that do not depend on hazardous and toxic chemicals, allow easy and preferably a one-pot synthesis and functionalization, would offer significant advantages over more conventional methods. In this sense, LASiS is an ideal candidate for uncomplicated and green nanoparticle synthesis and biofunctionalization. This paragraph serves to give an impression of what is currently possible with LASiS-based synthesis methods, how biologically relevant nanostructures are produced, but is not meant to be a comprehensive review of the literature.

12.3.1 *In Situ* Biofunctionalization of Gold Nanoparticles

Over the past few years, numerous groups have demonstrated that gold nanoparticle conjugates can be generated by the physical process of laser ablation of a gold target in various liquid environments. Studies by Sylvestre et al. have shown that gold nanoparticles (AuNPs) produced with fs LASiS in water have oxidation states of Au(0), Au(I), and Au(III), as determined by x-ray photoelectron spectroscopy [54]. The latter two oxidation states were a result of the presence of Au–O$^-$ compounds on the nanoparticle surface. Since their surface was partially oxidized, electrostatic repulsion led to stabilization of the colloid solution without the need for chemical stabilizers. The authors further noted that the charge (hydroxylation) can be controlled via the pH, as might be expected. Furthermore, by utilizing different liquids with different ligands, it is possible to directly achieve biologically valuable surface modifications, such as by ablation in *n*-propylamine, which results in the presence of both amine ($-NH_2$) and ammonium ions ($-NH_3^+$) [54].

Such modifications of the nanoparticle surface can be utilized for further surface coating or for making the nanoparticle biocompatible or biofunctional. For instance, cell-penetrating peptides or DNA can be coordinated to surface charges on AuNPs through electron donors like hydroxyl, amino, and thiol groups in the ablation medium [55,56]. Barcikowski's group showed the feasibility of the one-step generation and *in situ* biofunctionalization of AuNP-aptamer conjugates via LASiS in relevant physiological buffers [57], that is, tris(hydroxymethyl)aminomethane (Tris) pH 8.0 or N′-2-hydroxyethylpiperazine-N′-2-ethanesulfonic acid (HEPES) pH 7.4. Aptamers are short, single-stranded DNA or RNA oligonucleotides that recognize targets with high specificity and affinity analogous to antibody–target protein recognition. Because of their properties, aptamers are highly versatile and are suitable for *in vitro* selection via systematic evolution of ligands by exponential enrichment (SELEX) against practically any molecule, including proteins, nucleotides, and small molecules like metal ions [58–60].

The basic LASiS approach is illustrated in Figure 12.7 and involves the direct synthesis of AuNPs by irradiating a gold foil with 120 fs laser pulses at 800 nm (100 µJ pulse energy, 5 kHz repetition rate) in a physiological buffer (Tris or HEPES) containing the aptamer.

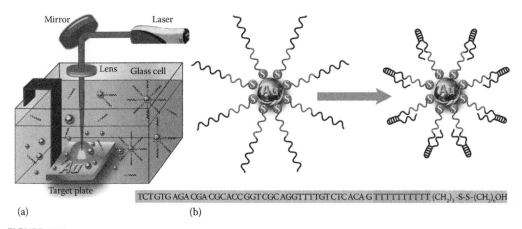

TCT GTG AGA CGA CGC ACC GGT CGC AGG TTTT GT CTC ACA G TTTTTTTTTT (CH$_2$)$_3$-S-S-(CH$_2$)$_6$OH

(a)   (b)

FIGURE 12.7
(See color insert.) Generation of aptamer-conjugated AuNPs via *in situ* conjugation. (a) Schematic illustration of *in situ* conjugation of AuNPs with aptamers during laser ablation in an aqueous aptamer solution. (b) Spacer design and resulting mixed monolayer conjugated nanoparticles. Mixed monolayer formation and careful spacer design contributed to correct aptamer folding. (Based on Walter, J.G. et al., *J. Nanobiotechnol.*, 8, 21, 2010.)

No heat-induced aptamer denaturation occurred during LASiS, as deduced from the high degree of aptamer activity on the AuNP surface.

Several advantages may occur when utilizing LASiS in the synthesis of nucleotide–gold conjugates: (1) AuNPs generated by LASiS based on *in situ* conjugation showed up to five times higher surface coverage in comparison to equivalent particles produced by chemical synthesis and subsequent *ex situ* conjugation [61], and (2) the higher surface coverage in DNA-modified AuNPs efficiently restricts the innate immune response, as deduced from interferon-β expression in mouse macrophages (inversely proportional to the oligonucleotide density), thereby ensuring maximal biocompatibility and circulation time [11]. The latter effect is possibly a result of the impediment of the immune recognition machinery that detects alien nucleic acids, because of steric hindrance and the high surface charge density.

Because aptamer–AuNP conjugates showed promise, Barcikowski et al. provided proof of principle by targeting prostate-specific membrane antigen (PSMA) in prostate cancer (adenocarcinoma) tissue sections [57]. AuNPs conjugated with anti-PSMA aptamer show a staining pattern similar to staining with an anti-PSMA antibody, with a positive staining of acinar epithelial cells (Figure 12.8). Some unspecific binding to muscle cells was observed, which the authors attribute to the presence of a "PSMA-like" target in smooth muscle cells. Indeed it was reported by Kinoshita et al. that the monoclonal PSMA antibody 7E11 also shows this tendency due to possible PSMA analogues [62].

These results revealed the broad applicability of aptamer-conjugated AuNPs, even in highly complex biological matrices and bio-imaging applications.

12.3.2 Cancer Treatment

Heat treatment of aberrant cells by selective uptake of nanoparticles is an attractive strategy for anticancer and antipathogen therapeutic strategies, since nanoparticle-based hyperthermic therapy is not hindered by the limitations of drugs, such as development of antibiotic or multidrug resistance, the difficulty of targeting multiple pathways, and a multitude of possible side effects and potential collateral damage to healthy tissue.

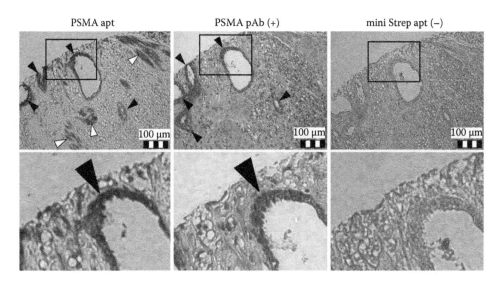

FIGURE 12.8

(See color insert.) Detection of PSMA in human prostate cancer tissue. Detection of PSMA-positive structures in prostate cancer tissue sections by immunohistochemical staining using anti-PSMA aptamer (PSMA apt)–conjugated AuNPs. As a negative control, AuNPs conjugated with miniStrep aptamer (miniStrep apt) were used. A polyclonal antibody directed against PSMA (PSMA pAb) was used as a positive control. Positive control was additionally stained with hematoxylin and eosin. Black arrows indicate specific staining, while white arrows flag unspecific binding. (Reproduced from Walter, J.G. et al., *J. Nanobiotechnol.*, 8, 21, 2010.)

The high optical to thermal conversion efficiency of AuNPs offers an excellent opportunity for the development of hyperthermic-therapy-based anticancer options [63–65]. An intense optical field has the ability to locally increase temperatures in the proximity of the MNP through interaction with free electrons, which leads to plasmon resonance. A recent study by Kessentini and Barchiesi showed that under circular polarization, photothermal effects are largest in optimized hollow gold nanospheres compared with nanorods [66]. When cells are heated beyond physiological values (40°C–45°C), proteins denature, aggregate, and misfold; cytoskeletal and membrane integrity is lost; DNA synthesis, RNA processing, and translation are inhibited; the cell cycle finally arrests; and ultimately, cell death occurs. In order for hyperthermic therapy to deploy its maximal efficacy, the nanoparticles must be internalized by the cells via endocytosis [67].

Endocytic cellular uptake can be divided into (1) nonspecific internalization, a random process without specific biomolecular control, and (2) specific internalization, where the cell selectively internalizes particles often via cell surface receptors. Sobhan et al. reported the spontaneous and nonspecific internalization of three different sizes of pure colloidal gold nanoparticles synthesized by fs LASiS by pancreatic tumor cells [68]. The synthesized AuNPs showed no cytotoxic or disruptive effects, even after 24 h of continuous exposure and independent of their size. Uptake was most efficient and highest for nanoparticles larger than 20 nm, and the authors propose that this might be due to the fact that nanoparticles with a larger surface area are able to bind to the minimum critical number of receptors required for triggering a cellular response responsible for endocytosis. Alternatively, the coordination of random proteins to the nanoparticle surface might facilitate nanoparticle uptake by activation of the cell membrane [69]. Dark-field microscopy (Figure 12.9a)—this is feasible because AuNPs have high scattering cross sections—and energy-dispersive X-ray (EDX) element analysis (Figure 12.9b) showed

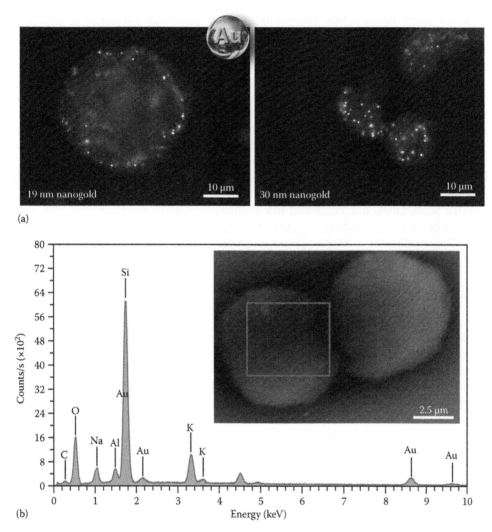

FIGURE 12.9

(See color insert.) Internalization of nanogold particles produced via LASiS in pancreatic tumor cells. (a) Dark-field imaging of internalized gold nanoparticles in AR42J pancreatic tumor cells. The cells were treated with 19 and 30 nm gold colloids for 6 h (scale bar, 10 μm). (b) EDX spectrum of cells incubated with 30 nm gold nanoparticles. Inset is the SEM image of the cell incubated with 30 nm particles (scale bar, 10 μm). (Based on Sobhan, M.A. et al., *Colloids Surf. B. Biointerfaces*, 92, 190, 2012.)

that the AuNPs were predominantly localized in the cytoplasm. Thus, in a dark-field microscope, the AuNPs appear as colorful, bright dots (Figure 12.9a).

12.3.3 Bio-Imaging with Surface-Enhanced Raman Scattering

MNPs absorb and resonantly scatter visible and near-infrared light because of excitation of their surface plasmon oscillations. By changing intrinsic parameters, such as nanoparticle size and shape (spheres, rods, cubes, etc.), but also the material(s) used (bimetallic or hybrid particles), the plasmon resonance band can be tuned over a wide spectral range [70].

In recent years, nanoparticle-based labeling methods with surface-enhanced Raman scattering (SERS) properties have been the focus of attention and development for use in ultrasensitive assays and imaging modalities in biological systems [71–73]. Although Raman scattering generally generates weaker signals compared with Rayleigh scattering or photoluminescence, the adsorption of particular molecules on the surface of metallic nanostructures with plasmonic properties results in an enormous enhancement of the differential Raman scattering cross section [71,74,75]. This enhancement is commonly attributed to an electromagnetic and/or chemical effect. Modulation of the local electric field may occur through oscillations of conduction band electrons at the metal surface (SPR), resulting in a large number of scattered photons or charge transfer processes related to strong electron coupling between the metal and the analyte molecule (chemical enhancement) [71,74].

Metallic SERS labels are valid and interesting alternatives for fluorochromes, such as organic dyes and semiconductor quantum dots, because of their extraordinary properties: (1) The bandwidth of Raman peaks in particular is much narrower than equivalent fluorescence emission bands [74,76]; (2) a multitude of SERS labels can be developed with characteristic spectral signatures due to the distinctive Raman fingerprint of different molecules [76]; (3) unlike fluorescence, Raman measurements are not sensitive to background signals, and therefore, measurements occur against a pitch-black background; (4) SERS labels can be significantly brighter than semiconductor quantum dots [74,76]; and (5) no photobleaching (signal fading due to label destruction) occurs with metallic SERS labels. However, signal intermittency (blinking) does occur, but on the timescale of fractions of seconds, and only might hinder single-particle measurements [71,77]. In SERS, blinking is caused by both a thermo-activated and a light-induced component [77]. Overall, with these properties, higher sensitivity and lower detection limits than conventional fluorescent labels may generally be expected.

To exemplify how LASiS-assisted synthesis of MNPs can be used to produce SERS labels, a brief description of the recent work by Amendola et al. is given here. These authors showed that it is possible to accurately correlate the SERS signal with the concentration of SERS labels based on AuNPs with Raman-active dyes [78]. Gold nanoparticles with average sizes of 19 nm were synthesized via LASiS with a pulsed Nd:YAG laser at 1064 nm (9 ns, 10 J cm^{-2} at 10 Hz repetition rate, 90 min) focused on a gold plate placed at the bottom of a cell (as in Figure 12.1a [right]) with 10^{-5} M NaCl in bi-distilled water. Postirradiation, a Raman-active cationic dye, for example, methylene blue, cresyl violet 670, hexacyanin 3, malachite green, and tetramethylrhodamine isothiocyanate (TRITC), was added to the colloidal gold solution. The dye molecules attached to the AuNP surface via electrostatic interaction with the negative surface charges. Interestingly, the absorption band of the dyes was not observed in the UV/vis spectra of the SERS AuNPs, and the authors therefore estimate that less than 200 dye molecules must have been present on a single AuNP. The particles were stabilized with polyethylene glycol (PEG) to prevent the rapid agglomeration observed in uncoated AuNPs, which yielded particles as schematically depicted in Figure 12.10a.

Electron microscopic evaluation showed well-dispersed spherical particles (Figure 12.10b). By fitting the UV/vis absorption spectra with a Mie–Gans model, the authors were able to correlate the SERS label concentration with the intensity of the Raman signal. Furthermore, multiplexing would be possible because each dye induced an individual Raman fingerprint that could be separated from other Raman signals by deconvolution. To provide proof of principle, the authors used the AuNPs to quantify PMA-differentiated U937 macrophage cells. The assessment of the number of AuNPs internalized by

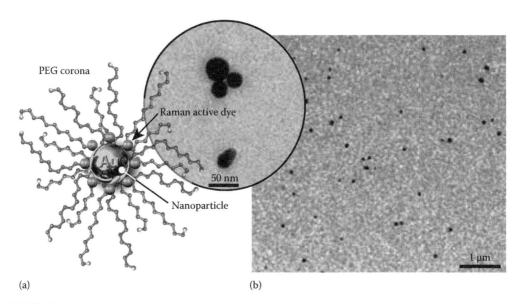

(a)   (b)

FIGURE 12.10
SERS labels for quantitative assays. (a) Sketch of the SERS labels based on AuNP coated with PEG and functionalized with Raman-active dye molecules. (b) Representative TEM images of pegylated AuNP loaded with malachite green dyes. (Based on Amendola, V. et al., *Anal. Method.*, 3, 849, 2011.)

macrophage cells via SERS correlated well with the results from inductively coupled plasma–mass spectrometry (ICP-MS).

Good examples of how both single SERS particle tracking and SERS multiplexing can be applied to basic cell biological research were recently shown by Kawata's and Pezacki's groups. Kawata et al. used 50 nm AuNPs to provide molecular maps of organelle transport and lysosomal accumulation in J774A.1 macrophage cells. This was achieved via simultaneous microscopic particle motion tracking and detection of intracellular molecules at 65 nm spatial resolution and 50 ms temporal resolution by SERS spectroscopy [79]. Pezacki and co-workers functionalized AgNPs with four different organic ligands to produce 25 nm AgNP SERS labels with distinct Raman fingerprints [80]. With these, multiplex imaging through signal deconvolution via Raman microscopy can be achieved to probe receptor colocalization or interaction. In a proof of principle experiment, the authors detected β_2-adrenergic receptors and caveolin-3 on the surface of rat cardiomyocytes and showed that 17% of receptors colocalized.

12.3.4 Transfection Reagents

Transfection is generally used in biological and biomedical research to transfer foreign genetic material into living cells, thereby influencing the expression of particular proteins, or to introduce artificially produced constructs that lead to expression of mutated, engineered, or fusion proteins. None of the currently available techniques, including transfection via viral carriers, electroporation, microinjection, or liposomes, is entirely satisfactory and universally applicable. The major disadvantages involved in transfection include (1) low transfection efficiency *in vivo*, (2) lack of sustained protein expression, (3) possible cell/tissue damage induced by the carriers or methodology, and (4) transfection efficiencies vary with the cell type.

For these reasons, many researchers started to investigate if nanoparticle-based carriers could be effective alternatives to more conventional delivery systems. For instance, Peterson et al. investigated the possible effects of laser-generated AuNPs on the biological functionality of DNA molecules. They tested four differently sized and positively charged AuNPs (14, 24, 59, and 89 nm) by incubating them with recombinant eGFP-C1–HMGB1 DNA expression plasmids that code for eGFP fusion proteins and contain the canine architectural transcription factor HMGB1 and subsequently transfected them with FuGENE HD into the canine mammary cell line MTH53a [81]. The AuNPs were produced via LASiS in bi-distilled water with 120 fs laser pulses at 800 nm (200 μJ at 5 kHz repetition for 12 min). The investigations showed that the presence of the AuNPs did not significantly change protein expression and functionality (DNA binding), while the presence of AuNPs increased the transfection efficiency considerably. The latter effect was size dependent and medium-sized AuNPs enhanced transfection efficiency nearly sixfold.

Baumgart and co-workers reported a fs laser-based high-throughput virus- and fragmentation-free transfection method based on the dispersion of 100 nm out of resonant spherical gold nanoparticles, followed by an irradiation using a low fluence from a weakly focused fs laser to prevent any laser-induced optoperforation [82]. The laser-based transfection was carried out by irradiating WM278 human melanoma cells in the presence of the nanoparticles through the glass bottom dishes from below with a Gaussian beam profile and a spot size of 650 mm in diameter. The irradiation was carried out with a fs laser at 800 nm, a repetition rate of 1 kHz, and maximum output pulse energy of 5 mJ. The results showed that in WM278 human melanoma cells, a very high transfection efficiency of 70%, which is a three times higher value than for conventional lipofection, and very low toxicity (<1%) were observed and that the off-resonance laser excitation effectively prevented nanoparticle fragmentation into possibly toxic DNA intercalating particles. The authors speculate that with such positive and promising effects, even *in vivo* application for cells at the surface of the tissues would be feasible.

12.3.5 Drug Targeting

Nanometer-sized particles have been proposed as platforms for drug and gene delivery, because of their unique functional profiles and easy biocompatibilization, particularly derivatized AuNPs. Furthermore, nanoparticles are in the same size range as biomolecules, such as proteins (albumin $d \approx 5$ nm) and DNA ($d \approx 2$ nm), and are also smaller than organelles, such as mitochondria (500–10,000 nm) and peroxisomes (200–1000 nm). Cellular uptake of engineered nanoparticles can be facilitated by modifying their surface, for example, coating with polymers, induction of particular surface charges, or attachment of substrates for receptors that internalize in receptor-mediated endocytosis, or through external manipulation by exertion of a force, that is, magnetic fields.

Salmaso et al. used AuNPs coated with a thermoresponsive polymer to give temperature-tunable colloidal systems with thermal control of their cellular uptake [83]. This thermoresponsive thiol-terminated poly-N-isopropylacrylamide-co-acrylamide copolymer is extended and hydrophilic at low temperatures and converts to a collapsed and hydrophobic state at higher temperature. The functionalized AuNPs were produced by laser ablation of a solid gold target with Nd:YAG laser pulses at 1064 nm (9 ns; 10 J cm^{-2} at Hz repetition rate for 90 min) in bi-distilled water. Postirradiation, the polymer was added and reacted with the AuNP surface. Cell culture studies in human breast adenocarcinoma MCF7 cells at 40°C showed a more than 80-fold greater uptake of AuNPs compared to cells treated at 34°C. Such systems can cleverly be used for targeting cells in inflamed or malignant cancer tissues, since in these the temperature is higher than normal physiological temperature; it has been shown that the mean temperature in tumorous tissue can be ~1.6°C higher than normal tissue [84]. By chemically linking or adsorbing drugs to the surface of thermoresponsive nanoparticles, a more selective accumulation of the drug in malignant tissue can be achieved or triggered by external temperature control.

Another approach based on multifunctional iron oxide magnetic nanoparticles (FeOxMNPs) was demonstrated by Amendola et al. [85]. FeOxMNPs are promising agents for a multitude of biomedical applications, in particular as theragnostic agents as a combined imaging (MRT) and targeted and controlled drug delivery system. The method is based on LASiS in which a pulsed Nd:YAG laser operating at 1064 nm (9 ns; 10 J cm^{-2} at 10 Hz repetition rate for 90 min) was focused on an iron plate placed at the bottom of a cell containing bi-distilled water. The LASiS produced ultrasmall polycrystalline FeOxMNPs (~75% magnetite) with a saturation magnetization comparable to the bulk equivalent (~80 emu g^{-1}). The functionalization of the FeOxMNPs postirradiation was uncomplicated to perform with a variety of ligands, including carboxylated phosphonates, fluorescent isothiocyanates, fluorescent alkylamines, and bovine serum albumin. The authors went on to characterize if cells loaded with FeOxMNPs can be manipulated with an external magnetic field. To this end, human nonadherent monocyte-like U937 cells were differentiated into macrophages and challenged with the FeOxMNPs. The results showed that loaded macrophages could indeed be manipulated and thus would provide efficient and selective means to transport FeOxMNPs to particular sites within the organism. Applications might include cell sorting, *in situ* cell concentration, wound cicatrization, cancer therapy by magnetic hyperthermia, multimodal imaging of lesions, or theragnostic applications.

12.3.6 Nanonization of Drugs

Nanonization is the process of producing nanometer-sized drug particles from drug powder formulations and is an established method to improve drug absorption and facilitate the intravenous administration of insoluble drugs [86]. Although LASiS techniques have been described for a wide variety of nanoparticles, for example, noble MNPs, quantum dots, silicium-based particles, and organic (e.g., vanadyl phthalocyanine) nanoparticles, laser ablation methods for drug formulation and nanonization are scarce in the literature. Nagare and Senna reported in 2004 on the production of indomethacin-based nanoparticles by a laser ablation technique with a ns Nd:YAG pulsed laser in a dry environment, that is, the absence of a dispersing medium and/or a stabilizing agent [87]. The same group further developed this laser deposition technique to include indomethacin–bovine serum albumin composite nanoparticles with sizes below 50 nm [88].

Femtosecond laser technology is more favorable than the ns laser technology, because with fs laser pulses, the target material is irradiated with less energy, thereby limiting the alterations of the ablated drug nanocrystals and degradation of the drug into either biologically inactive or toxic molecules. Kenth et al. investigated an adjusted fs LASiS technique for the fabrication of paclitaxel nanocrystals in aqueous solution, a normally insoluble anticancer agent isolated from the bark of *Taxus brevifolia* [89]. The particles were produced from a solid paclitaxel tablet placed at the bottom of a glass beaker filled with poloxamer 188 solution, and the target was irradiated with 120 fs pulses from a Ti:sapphire laser at 800 nm (1 kHz repetition rate). A single-step fragmentation was used to obtain a narrowly dispersed colloidal solution of ~400 nm particles. The drug retained its crystalline morphology upon nanonization, but the authors note that the anhydrous crystals were converted to a hydrated form, which is a phenomenon that is also observed during bead milling.

12.4 Conclusions

LASiS is a relatively inexpensive, uncomplicated, and efficient means to produce nanoparticles for biomedical research or medical-grade therapeutics. The method does not require the use of toxic and hazardous chemicals and is thus environmentally friendly. Tuning of a variety of parameters can be used to manipulate the size, shape, and functionalizability; these include the wavelength, repetition rate, pulse frequency, laser power, synthesis environment, reactivity of the environment, presence of ligands, and others. LASiS even allows the synthesis of more complex compound nanoparticles. A major advantage of LASiS is the fact that the process can be carried out in the presence of biocompatible stabilizers that allow further functionalization, and that in principle, the synthesis, stabilization, and biofunctionalization can be carried out in a one-pot synthesis setting.

However, there are a number of distinct drawbacks: (1) limited control over the average size and size distribution, (2) synthesis in organic solvents can induce pyrolysis of organic molecules and consequent contamination of the nanoparticle surface, and (3) mass production by LASiS is only possible with high repetition rate lasers, to name but a few.

Nonetheless, LASIS is a highly promising technology that already is a major asset to nanotechnological and biomedical research. However, in order for this method to become mainstream, the ablation mechanism must be better understood; in particular the surface chemistry of the particles and synthesis procedures as far as laser setting and reaction environments may be concerned should be simplified, automated, and modularized.

Excellent comprehensive reviews on various aspects of laser ablation and its application were recently published by Amendola [11], Yang [44], Besner and Meunier [90], Sasaki and Takada [91], Semaltianos [92], and Zeng [93], and their respective coworkers.

Acknowledgment

The authors are grateful to Dr. Mushtaq Sobhan (MQ BioFocus Research Centre, Macquarie University, Sydney, Australia) for providing the high-quality figures of the AR42J pancreatic tumor cells.

References

1. Henglein, A. 1989. Small-particle research–Physicochemical properties of extremely small colloidal metal and semiconductor particles. *Chem. Rev.* 89: 1861–1873.
2. Henglein, A. 1993. Physicochemical properties of small metal particles in solution—Microelectrode reactions, chemisorption, composite metal particles, and the atom-to-metal transition. *J. Phys. Chem.* 97: 5457–5471.
3. Niemeyer, C. M. 2001. Nanoparticles, proteins, and nucleic acids: Biotechnology meets materials science. *Angew. Chem. Int. Ed. Engl.* 40: 4128–4158.
4. Roduner, E. 2006. Size matters: Why nanomaterials are different. *Chem. Soc. Rev.* 35: 583–592.
5. Schmid, G. 1992. Large clusters and colloids—Metals in the embryonic state. *Chem. Rev.* 92: 1709–1727.

6. Coronado, E., A. Ribera, J. Garcia-Martinez, N. Linares, and L. M. Liz-Marzán. 2008. Synthesis, characterization and magnetism of monodispersed water soluble palladium nanoparticles. *J. Mater. Chem.* 18: 5682–5688.

7. Eustis, S. et al. 2005. Growth and fragmentation of silver nanoparticles in their synthesis with a fs laser and CW light by photo-sensitization with benzophenone. *Photochem. Photobiol. Sci.* 4: 154–159.

8. Anderson, H. H. 1997. *Small Particles and Inorganic Clusters.* Springer-Verlag, Berlin, Germany.

9. Jain, P. K., K. S. Lee, I. H. El-Sayed, and M. A. El-Sayed. 2006. Calculated absorption and scattering properties of gold nanoparticles of different size, shape, and composition: Applications in biological imaging and biomedicine. *J. Phys. Chem. B.* 110: 7238–7248.

10. Milani, P. and W. A. Deheer. 1990. Improved pulsed laser vaporization source for production of intense beams of neutral and ionized clusters. *Rev. Sci. Instrum.* 61: 1835–1838.

11. Amendola, V. and M. Meneghetti. 2009. Laser ablation synthesis in solution and size manipulation of noble metal nanoparticles. *Phys. Chem. Chem. Phys.* 11: 3805–3821.

12. Fojtik, A. and A. Henglein. 1993. Laser ablation of films and suspended particles in a solvent–formation of cluster and colloid solutions. *Ber. Bunsen-Ges. Phys. Chem. Chem. Phys.* 97: 252–254.

13. Neddersen, J., G. Chumanov, and T. M. Cotton. 1993. Laser-ablation of metals–a new method for preparing SERS active colloids. *Appl. Spectrosc.* 47: 1959–1964.

14. Sibbald, M. S., G. Chumanov, and T. M. Cotton. 1996. Reduction of cytochrome c by halide-modified, laser-ablated silver colloids. *J. Phys. Chem.* 100: 4672–4678.

15. Rao, C. R. K., V. Lakshminarayanan, and D. C. Trivedi. 2006. Synthesis and characterization of lower size, laurylamine protected palladium nanoparticles. *Mater. Lett.* 60: 3165–3169.

16. Huang, N. M. et al. 2009. Gamma-ray assisted synthesis of silver nanoparticles in chitosan solution and the antibacterial properties. *Chem. Eng. J.* 155: 499–507.

17. Amendola, V., S. Polizzi, and M. Meneghetti. 2007. Free silver nanoparticles synthesized by laser ablation in organic solvents and their easy functionalization. *Langmuir* 23: 6766–6770.

18. Vogel, A. et al. 1999. Energy balance of optical breakdown in water at nanosecond to femtosecond time scales. *Appl. Phys. B.* 68: 271–280.

19. Momma, C. et al. 1996. Short-pulse laser ablation of solid targets. *Opt. Commun.* 129: 134–142.

20. Berthe, L., R. Fabbro, P. Peyre, L. Tollier, and E. Bartnicki. 1997. Shock waves from a water-confined laser-generated plasma. *J. Appl. Phys.* 82: 2826–2832.

21. Berthe, L., R. Fabbro, P. Peyre, and E. Bartnicki. 1999. Wavelength dependent of laser shock-wave generation in the water-confinement regime. *J. Appl. Phys.* 85: 7552–7555.

22. Sylvestre, J. P., A. V. Kabashin, E. Sacher, and M. Meunier. 2005. Femtosecond laser ablation of gold in water: Influence of the laser-produced plasma on the nanoparticle size distribution. *Appl. Phys. A.* 80: 753–758.

23. Zhu, S., Y. F. Lu, and M. H. Hong. 2001. Laser ablation of solid substrates in a water-confined environment. *Appl. Phys. Lett.* 79: 1396–1398.

24. Saito, K., K. Takatani, T. Sakka, and Y. H. Ogata. 2002. Observation of the light emitting region produced by pulsed laser irradiation to a solid–liquid interface. *Appl. Surf. Sci.* 197–198: 56–60.

25. Tsuji, T., Y. Okazaki, Y. Tsuboi, and M. Tsuji. 2007. Nanosecond time-resolved observations of laser ablation of silver in water. *Jpn. J. Appl. Phys.* 46: 1533–1535

26. Sasoh, A., K. Watanabe, Y. Sano, and N. Mukai. 2005. Behavior of bubbles induced by the interaction of a laser pulse with a metal plate in water. *Appl. Phys. A.* 80: 1497–1500.

27. Niu, K. Y. et al. 2010. Morphology control of nanostructures via surface reaction of metal nanodroplets. *J. Am. Chem. Soc.* 132: 9814–9819.

28. Tsuji, T., Y. Tsuboi, N. Kitamura, and M. Tsuji. 2004. Microsecond-resolved imaging of laser ablation at solid-liquid interface: Investigation of formation process of nano-size metal colloids. *Appl. Surf. Sci.* 229: 365–371.

29. Phuoc, T. X., B. H. Howard, D. V. Martello, Y. Soong, and M. K. Chyu. 2008. Synthesis of Mg(OH)2, MgO, and Mg nanoparticles using laser ablation of magnesium in water and solvents. *Opt. Lasers Eng.* 46: 829–834.

30. Yoo, J. H., S. H. Jeong, R. Greif, and R. E. Russo. 2000. Explosive change in crater properties during high power nanosecond laser ablation of silicon. *J. Appl. Phys.* 88: 1638–1649.
31. Nichols, W. T., T. Sasaki, and N. Koshizaki. 2006. Laser ablation of a platinum target in water. I. Ablation mechanisms. *J. Appl. Phys.* 100: 114911.
32. Nichols, W. T., T. Sasaki, and N. Koshizaki. 2006. Laser ablation of a platinum target in water. II. Ablation rate and nanoparticle size distributions. *J. Appl. Phys.* 100: 114912.
33. Nichols, W. T., T. Sasaki, and N. Koshizaki. 2006. Laser ablation of a platinum target in water. III. Laser-induced reactions. *J. Appl. Phys.* 100: 114913.
34. Mafuné, F. and T. Kondow. 2004. Selective laser fabrication of small nanoparticles and nano-networks in solution by irradiation of UV pulsed laser onto platinum nanoparticles. *Chem. Phys. Lett.* 383: 343–347.
35. Mafuné, F., J. Kohno, Y. Takeda, T. Kondow, and H. Sawabe. 2000. Formation and size control of silver nanoparticles by laser ablation in aqueous solution. *J. Phys. Chem. B.* 104: 9111–9117.
36. Mafuné, F., J. Kohno, Y. Takeda, T. Kondow, and H. Sawabe. 2001. Formation of gold nanoparticles by laser ablation in aqueous solution of surfactant. *J. Phys. Chem. B.* 105: 5114–5120.
37. Mafuné, F. and T. Kondow. 2003. Formation of small gold clusters in solution by laser excitation of interband transition. *Chem. Phys. Lett.* 372: 199–204.
38. Zamiri, R. et al. 2011. Preparation of starch stabilized silver nanoparticles with spatial self-phase modulation properties by laser ablation technique. *Appl. Phys. A.* 102: 189–194.
39. Zamiri, R. et al. 2012. Aqueous starch as a stabilizer in zinc oxide nanoparticle synthesis via laser ablation. *J. Alloy. Comp.* 516: 41–48.
40. Zamiri, R. et al. 2011. Preparation of silver nanoparticles in virgin coconut oil using laser ablation. *Int. J. Nanomed.* 6: 71–75.
41. Zamiri, R., A. Zakaria, H. A. Ahangar, A. R. Sadrolhosseini, and M. A. Mahdi. 2010. Fabrication of silver nanoparticles dispersed in palm oil using laser ablation. *Int. J. Mol. Sci.* 11: 4764–4770.
42. Besner, S., A. Kabashin, F. Winnik, and M. Meunier. 2008. Ultrafast laser based "green" synthesis of non-toxic nanoparticles in aqueous solutions. *Appl. Phys. A.* 93: 955–959.
43. Kabashin, A. V. and M. Meunier. 2003. Synthesis of colloidal nanoparticles during femtosecond laser ablation of gold in water. *J. Appl. Phys.* 94: 7941–7943.
44. Yang, G. W. 2007. Laser ablation in liquids: Applications in the synthesis of nanocrystals. *Prog. Mater. Sci.* 52: 648–698.
45. Kabashin, A. V., M. Meunier, C. Kingston, and J. H. T. Luong. 2003. Fabrication and characterization of gold nanoparticles by femtosecond laser ablation in an aqueous solution of cyclodextrins. *J. Phys. Chem. B.* 107: 4527–4531.
46. Pyatenko, A., K. Shimokawa, M. Yamaguchi, O. Nishimura, and M. Suzuki. 2004. Synthesis of silver nanoparticles by laser ablation in pure water. *Appl. Phys. A.* 79: 803–806.
47. Tsuji, T., T. Kakita, and M. Tsuji. 2003. Preparation of nano-size particles of silver with femtosecond laser ablation in water. *Appl. Surf. Sci.* 206: 314–320.
48. Tsuji, T., K. Iryo, N. Watanabe, and M. Tsuji. 2002. Preparation of silver nanoparticles by laser ablation in solution: Influence of laser wavelength on particle size. *Appl. Surf. Sci.* 202: 80–85.
49. Bogaerts, A. and Z. Chen. 2005. Effect of laser parameters on laser ablation and laser-induced plasma formation: A numerical modeling investigation. *Spectrochim. Acta B* 60: 1280–1307.
50. Shafeev, G. A., E. Freysz, and F. Bozon-Verduraz. 2004. Self-influence of a femtosecond laser beam upon ablation of Ag in liquids. *Appl. Phys. A.* 78: 307–309.
51. Procházka, M., P. Mojzeš, J. Štěpánek, B. Vlčková, and P.-Y. Turpin. 1997. Probing applications of laser-ablated Ag colloids in SERS spectroscopy: Improvement of ablation procedure and SERS spectral testing. *Anal. Chem.* 69: 5103–5108.
52. Takami, A., H. Kurita, and S. Koda. 1999. Laser-induced size reduction of noble metal particles. *J. Phys. Chem. B* 103: 1226–1232.
53. Menendez-Manjon, A. and S. Barcikowski. 2011. Hydrodynamic size distribution of gold nanoparticles controlled by repetition rate during pulsed laser ablation in water. *Appl. Surf. Sci.* 257: 4285–4290.

54. Sylvestre, J.-P. et al. 2004. Surface chemistry of gold nanoparticles produced by laser ablation in aqueous media. *J. Phys. Chem. B* 108: 16864–16869.
55. Petersen, S. and S. Barcikowski. 2009. In situ bioconjugation: Single step approach to tailored nanoparticle-bioconjugates by ultrashort pulsed laser ablation. *Adv. Funct. Mater.* 19: 1167–1172.
56. Petersen, S., J. Jakobi, and S. Barcikowski. 2009. In situ bioconjugation—Novel laser based approach to pure nanoparticle-conjugates. *Appl. Surf. Sci.* 255: 5435–5438.
57. Walter, J. G., S. Petersen, F. Stahl, T. Scheper, and S. Barcikowski. 2010. Laser ablation-based one-step generation and bio-functionalization of gold nanoparticles conjugated with aptamers. *J. Nanobiotechnol.* 8: 21.
58. Ellington, A. D. and J. W. Szostak. 1990. In vitro selection of RNA molecules that bind specific ligands. *Nature* 346: 818–822.
59. Robertson, D. L. and G. F. Joyce. 1990. Selection in vitro of an RNA enzyme that specifically cleaves single-stranded DNA. *Nature* 344: 467–468.
60. Tuerk, C. and L. Gold. 1990. Systematic evolution of ligands by exponential enrichment: RNA ligands to bacteriophage T4 DNA polymerase. *Science* 249: 505–510.
61. Petersen, S. and S. Barcikowski. 2009. Conjugation efficiency of laser-based bioconjugation of gold nanoparticles with nucleic acids. *J. Phys. Chem. C.* 113: 19830–19835.
62. Kinoshita, Y. et al. 2006. Expression of prostate-specific membrane antigen in normal and malignant human tissues. *World J. Surg.* 30: 628–636.
63. Huang, X. H., P. K. Jain, I. H. El-Sayed, and M. A. El-Sayed. 2007. Gold nanoparticles: Interesting optical properties and recent applications in cancer diagnostic and therapy. *Nanomedicine* 2: 681–693.
64. Jain, P. K., I. H. El-Sayed, and M. A. El-Sayed. 2007. Au nanoparticles target cancer. *Nano Today* 2: 18–29.
65. Loo, C., A. Lowery, N. J. Halas, J. West, and R. Drezek. 2005. Immunotargeted nanoshells for integrated cancer imaging and therapy. *Nano Lett.* 5: 709–711.
66. Kessentini, S. and D. Barchiesi. 2012. Quantitative comparison of optimized nanorods, nanoshells and hollow nanospheres for photothermal therapy. *Biomed. Opt. Express* 3: 590–604.
67. Doherty, G. J. and H. T. McMahon. 2009. Mechanisms of endocytosis. *Annu. Rev. Biochem.* 78: 857–902.
68. Sobhan, M. A., V. K. Sreenivasan, M. J. Withford, and E. M. Goldys. 2012. Non-specific internalization of laser ablated pure gold nanoparticles in pancreatic tumor cell. *Colloids Surf. B. Biointerfaces* 92: 190–195.
69. Conner, S. D. and S. L. Schmid. 2003. Regulated portals of entry into the cell. *Nature* 422: 37–44.
70. Sonnichsen, C. et al. 2002. Drastic reduction of plasmon damping in gold nanorods. *Phys. Rev. Lett.* 88: 077402.
71. Qian, X. M. and S. M. Nie. 2008. Single-molecule and single-nanoparticle SERS: From fundamental mechanisms to biomedical applications. *Chem. Soc. Rev.* 37: 912–920.
72. Stiles, P. L., J. A. Dieringer, N. C. Shah, and R. R. Van Duyne. 2008. Surface-enhanced Raman spectroscopy. *Annu. Rev. Anal. Chem.* 1: 601–626.
73. Wachsmann-Hogiu, S., T. Weeks, and T. Huser. 2009. Chemical analysis in vivo and in vitro by Raman spectroscopy-from single cells to humans. *Curr. Opin. Biotechnol.* 20: 63–73.
74. Le Ru, E. C., E. Blackie, M. Meyer, and P. G. Etchegoin. 2007. Surface enhanced Raman scattering enhancement factors: A comprehensive study. *J. Phys. Chem. C* 111: 13794–13803.
75. Nie, S. M. and S. R. Emery. 1997. Probing single molecules and single nanoparticles by surface-enhanced Raman scattering. *Science* 275: 1102–1106.
76. Allgeyer, E. S., A. Pongan, M. Browne, and M. D. Mason. 2009. Optical signal comparison of single fluorescent molecules and Raman active gold nanostars. *Nano Lett.* 9: 3816–3819.
77. Emory, S. R., R. A. Jensen, T. Wenda, M. Han, and S. Nie. 2006. Re-examining the origins of spectral blinking in single-molecule and single-nanoparticle SERS. *Faraday Discuss.* 132: 249–259.
78. Amendola, V. et al. 2011. SERS labels for quantitative assays: Application to the quantification of gold nanoparticles uptaken by macrophage cells. *Anal. Method.* 3: 849–856.

79. Ando, J., K. Fujita, N. I. Smith, and S. Kawata. 2011. Dynamic SERS imaging of cellular transport pathways with endocytosed gold nanoparticles. *Nano Lett.* 11: 5344–5348.
80. Kennedy, D. C., K. A. Hoop, L. L. Tay, and J. P. Pezacki. 2010. Development of nanoparticle probes for multiplex SERS imaging of cell surface proteins. *Nanoscale* 2: 1413–1416.
81. Petersen, S. et al. 2009. Co-transfection of plasmid DNA and laser-generated gold nanoparticles does not disturb the bioactivity of GFP-HMGB1 fusion protein. *J. Nanobiotechnol.* 7: 6.
82. Baumgart, J. et al. 2012. Off-resonance plasmonic enhanced femtosecond laser optoporation and transfection of cancer cells. *Biomaterials* 33: 2345–2350.
83. Salmaso, S. et al. 2009. Cell up-take control of gold nanoparticles functionalized with a thermoresponsive polymer. *J. Mater. Chem.* 19: 1608–1615.
84. Stefanadis, C. et al. 2001. Increased temperature of malignant urinary bladder tumors in vivo: The application of a new method based on a catheter technique. *J. Clin. Oncol.* 19: 676–681.
85. Amendola, V. et al. 2011. Top-down synthesis of multifunctional iron oxide nanoparticles for macrophage labelling and manipulation. *J. Mater. Chem.* 21: 3803–3813.
86. Chen, H., C. Khemtong, X. Yang, X. Chang, and J. Gao. 2011. Nanonization strategies for poorly water-soluble drugs. *Drug Discov. Today* 16: 354–360.
87. Nagare, S. and M. Senna. 2004. Reagglomeration mechanism of drug nanoparticles by pulsed laser deposition. *Solid State Ion.* 172: 243–247.
88. Nagare, S., J. Sagawa, and M. Senna. 2006. Chemical and structural properties of drug-protein nanocomposites prepared by pulsed laser deposition from conjugated targets. *J. Nanopart. Res.* 8: 37–42.
89. Kenth, S., J. P. Sylvestre, K. Fuhrmann, M. Meunier, and J. C. Leroux. 2011. Fabrication of paclitaxel nanocrystals by femtosecond laser ablation and fragmentation. *J. Pharm. Sci.* 100: 1022–1030.
90. Besner, S. and M. Meunier. 2010. Laser synthesis of nanomaterials. In *Laser Precision Microfabrication*, eds K. Sugioka, M. Meunier, and A. Piqué. pp. 163–187. Springer, Berlin, Germany.
91. Sasaki, K. and N. Takada. 2010. Liquid-phase laser ablation. *Pure Appl. Chem.* 82: 1317–1327.
92. Semaltianos, N. G. 2010. Nanoparticles by laser ablation. *Crit. Rev. Solid State Mat. Sci.* 35: 105–124.
93. Zeng, H. B. et al. 2012. Nanomaterials via laser ablation/irradiation in liquid: A review. *Adv. Funct. Mater.* 22: 1333–1353.

13

Nanomedicine in Brain Tumors

Gerardo Caruso, Maria Caffo, Lucia Merlo, Giuseppe La Fata,
Marcello Passalacqua, and Francesco Tomasello

CONTENTS

13.1 Introduction

Gliomas are the most common primary brain tumors in adults with a worldwide incidence of approximately 7 out of 100,000 individuals per year. The current treatment for glioma patients consists of surgical resection, followed by radiotherapy and chemotherapy [1]. Glioblastoma multiforme (GBM) is the most common type of brain tumor in adults, with a median survival of ~12–18 months post-diagnosis [2]. Despite recent advances in conventional therapeutic approaches including the gamma-knife (radiation) and temozolomide (chemotherapy) [3], mortality is still close to 100% and the average survival of patients with GBM is less than 1 year. Surgical treatment is invasive but represents the first approach for the vast majority of brain tumors due to difficulties arising in early stage detection. However, after surgical resection, the residual pool of invasive cells rises to recurrent tumor, which, in 96% of cases, arises adjacent to the resection margins [4]. Aggressive treatments have extended the median survival, but it is often associated with significant impairment in the quality of life. Radiation therapy and chemotherapy are noninvasive options often used as adjuvant therapy but may also be effective for curing early stage tumors.

Adjuvant radiotherapy gives limited benefits and causes debilitation side effects, which reduce its efficacy [5]. The effectiveness of systemic chemotherapy is limited by toxic effects on healthy cells, generally resulting in morbidity or mortality of the patient. In patients with recurrent GBM, the 6-month progression-free survival is only 21% after treatment with temozolomide [3]. Moreover, the presence of the blood–brain barrier (BBB) limits the passage of a wide variety of anticancer agents.

Gliomas are brain tumors with histological, immunohistochemical, and ultrastructural features of glial differentiation. Approximately 50% of primary brain tumors are gliomas, arising from astrocytes, oligodendrocytes or their precursors, and ependymal cells. Gliomas are classified from I to IV according to the World Health Association (WHO) malignancy scale. Grade I gliomas are benign with a slow proliferation rate and include pilocytic astrocytoma most common in pediatric age. Grade II gliomas are characterized by a high degree of cellular differentiation and grow diffusely into the normal brain parenchyma and are prone to malignant progression. They include astrocytoma, oligodendroglioma, and oligoastrocytoma. Grade III lesions include anaplastic astrocytoma, anaplastic oligoastrocytoma, and anaplastic oligodendroglioma. These tumors show a higher cellular density and a notable presence of atypia and mitotic cells. Grade IV tumors are the most malignant and the most frequent gliomas and include glioblastoma and gliosarcoma. These tumors presented microvascular proliferations and pseudopalisading necrosis. Recent advances in glioma molecular biology have evidenced various genes involved in cell growth, apoptosis, and angiogenesis. It is now generally understood that tumorigenesis occurs by either overexpression of oncogenes or inactivation of tumor suppressor genes.

Nanotechnology is an emerging field that deals with interactions between molecules, cells, and engineered substances such as molecular fragments, atoms, and molecules. The application of nanotechnology in medicine is now known as nanomedicine and the drug formulations using the nanomaterials are referred to as nano-formulations. The impact of nanotechnology in medicine can mainly be seen in diagnostic methods, drug-release techniques, and regenerative medicine. Nanomedicine could lead to overcoming the key-points related to malignant brain tumors treatment, such as 1) the lack of cancer-cells specificity in drug delivery and targeting; 2) the incomplete drug transit through the BBB and into the tumor cells; and 3) avoiding side effects in normal brain tissue. Reduction of toxicity to peripheral organs can also be achieved with these systems [6]. Nanoparticle (NP)-based drug-delivery systems, antisense approach to modify gene expression in cancer cell genome, and molecular-based cancer cell targeting represent important therapeutic strategies in gliomas treatment.

13.2 Glioma Biology

Genomic DNA aberrations are key genetic events in gliomagenesis. Recurrent genomic regions of alteration, including net gains and losses, have been found in gliomas. Whereas some of these regions contain known oncogenes and tumor suppressor genes, the biologically relevant genes within other regions remain to be identified. The phenotypic and genotypic heterogeneity indicates that no isolated genetic event accounts for gliomagenesis but rather the cumulative effects of a number of alterations that operate in a concerted manner. This pathological process includes various biological events, such as activation of growth factor receptor signaling pathways, downregulation of many apoptotic mechanisms, and unbalance of pro- and antiangiogenic factors. Several growth factor receptors, such as epidermal growth factor receptor (EGFR), platelet-derived growth factor receptor (PDGFR), C-Kit, and vascular endothelial growth factor receptor (VEGFR), are overexpressed, amplified, and/or mutated in gliomas.

13.2.1 Invasion, Angiogenesis, and Extracellular Matrix

Invasive behavior is a peculiar feature of malignant brain tumors. These tumors are almost invariably fatal recurring near the resection margin in almost all cases. Cerebral gliomas show a unique pattern of invasion and with rare exceptions do not metastasize outside of the brain. Invading glioma cells normally migrate to distinct anatomic structures. These structures include the basement membrane (BM) of blood vessels, the subependymal space, the glia limitans externa, and the parallel and intersecting nerve fiber tracts in the white matter. Glioma cells' adhesion to proteins of the surrounding extracellular matrix (ECM), degradation of ECM components, and migration of glioma cells are fundamental phases in glioma biology. ECM is composed of proteoglycans, glycoproteins, and collagens and also contains fibronectin, laminin, tenascin, hyaluronic acid, and vitronectin. Critical factors in glioma invasion include the synthesis and deposition of ECM components by glioma and mesenchymal cells, the release of ECM-degrading activities for remodeling interstitial spaces, the presence of adhesion molecules, and the effects of cell–matrix interactions on the behavior of glioma cells. ECM modification aids the loss of contact inhibition allowing tumor cells to freely migrate and invade the surrounding tissues. The proteolytic degradation of the BM is mediated by proteases, such as the matrix metalloproteinases (MMPs), secreted by tumor and stromal cells [7]. MMPs play an important role in human brain–tumor invasion, probably due to an imbalance between the production of MMPs and the tissue inhibitor of metalloproteinases-1 (TIMP-1) by the tumor cells [7]. MMP-1 is able to initiate breakdown of the interstitial collagens and to activate the other MMPs that allow glioma cell infiltration. Adhesion process is the binding of the cells to each other and to the ECM through adhesion molecules such as integrins, selectins, cadherins, the immunoglobulin superfamily, and lymphocyte homing receptors. The extracellular ligands that anchor these adhesions include laminin, fibronectin, vitronectin, and various collagens. Integrins are heterodimers of α- and β-subunits that regulate many aspects of the cell behavior including survival, proliferation, migration, and differentiation. Integrins are expressed on different cell types, including neurons, glial cells, and meningeal and endothelial cells. β2-integrins are specifically expressed by leukocytes and they are found on microglia (MG) and on infiltrating leukocytes within the central nervous system (CNS); downregulated β1-integrin protein levels in vivo probably affect interactions of glioma cells with ECM components, leading to reduced migration along vascular BMs [8].

Tumor angiogenesis is often activated during the early, preneoplastic stages of tumor development and is controlled by a number of positive or negative regulators produced by cancer cells and tumor-associated leukocytes. Angiogenesis, the formation of new blood vessels from existing microvessels, is a histological indicator of the degree of malignancy and prognosis. Angiogenesis also includes vessel penetration into avascular regions of the tissue and is critically dependent on the correct interactions among endothelial cells, pericytes, and surrounding cells and their association with the ECM and the vascular BM. The presence of endothelial glomeruloid-like proliferation and the presence of a positive immunoreaction at the level of BM of tumor vascular channel are predictive of active tumor invasiveness [9]. Macrophages (MPs) can exert a dual influence on blood vessel formation and function. On the one hand, MPs produce molecules that are proangiogenic; on the other hand, they can express antiangiogenic molecules and damage the integrity of blood vessels. In general, as for interaction with neoplastic cells, the proangiogenic functions of TAM prevail. In several studies, in human cancer, TAM accumulation has been associated with angiogenesis and with the production of angiogenic factors such as VEGF and PDGF [10]. MPs recruited in situ represent an indirect pathway of amplification of angiogenesis, in concert with angiogenic molecules directly produced by tumor cells.

13.3 Blood–Brain Barrier

The brain is one of the least accessible organs for the delivery of pharmacological compounds. Specific interfaces tightly regulate the exchange between the peripheral blood circulation and the cerebrospinal fluid (CSF) circulatory system. These barriers are the choroid plexus epithelium (blood–ventricular CSF), the arachnoid epithelium (blood–subarachnoid CSF), and the blood–brain interstitial fluid. Their function is to maintain a constant environment inside the brain, by strictly regulating the composition of the cerebral extracellular fluid, and to protect the brain against potentially toxic substances. The microvessel endothelial cells that form the BBB display important morphological characteristics such as the presence of tight and adherens junctions (AJs), the absence of fenestrations, and a low pinocytic activity. Endothelial cell tight junctions (TJs) limit the paracellular flux of hydrophilic molecules across the BBB. The endothelial cells secrete and are surrounded by a basal lamina (BL), with the end feet of astrocytic glial cells close to its opposite side. Astrocytes play many essential roles in the healthy CNS, including biochemical support of endothelial cells that form the BBB, regulation of blood flow, provision of nutrients to the nervous tissue, maintenance of extracellular ion balance, and in the repair and scarring process of the brain and spinal cord following traumatic injuries.

TJs are located on the apical region of endothelial cells and are structurally formed by a complex network made of a series of parallel, interconnected, transmembrane, and cytoplasmic strands of proteins [11]. TJs consist of three integral membrane proteins, namely, claudin, occludin, and junction adhesion molecules, and a number of cytoplasmic accessory proteins. The high level of integrity of TJs is reflected by the high electrical resistance of the BBB (1500–2000 Ω cm^2), which depends on a proper extracellular Ca^{2+} ion concentration. Cytoplasmic proteins link membrane proteins to actin, which is the primary cytoskeleton protein for the maintenance of structural and functional integrity of the endothelium. AJs are located below the TJs in the basal region of the lateral plasma membrane. They are composed of transmembrane glycoproteins (cadherins) linked to the cytoskeleton by cytoplasmic proteins, thus providing additional tightening structure between the adjacent endothelial cells at the BBB. In addition to supporting the barrier function, AJs mediate the adhesion of brain endothelial cells to each other, the initiation of cell polarity, and the regulation of paracellular permeability [12]. The transport of solutes and other compounds across the BBB is strictly constrained through both physical TJs and AJs. Generally speaking, a large number of medical compounds, such as antibiotics, chemotherapeutic agents, monoclonal antibodies (mAbs), and recombinant proteins for the CNS, are unable to pass through the barrier (Figure 13.1).

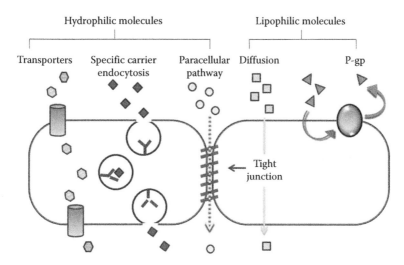

FIGURE 13.1
(See color insert.) Molecular transport across the BBB.

A wide range of CNS drugs including large-molecular-weight biological therapeutic peptides, proteins, and genes may gain entry into the brain with NPs being carriers.
Receptor-mediated transcytosis and adsorptive transcytosis are two major routes for NP
delivery across the BBB.

Receptor-mediated transcytosis across the BBB has been explored more actively
because of its high specificity. Large molecules, which are necessary for the normal function of the brain, are delivered to the brain by specific receptors. These receptors are
highly expressed on the endothelial cells forming the BBB. Therapeutic compounds are
able to cross the BBB after association/conjugation to these specific ligands. Receptor-
mediated transcytosis has been demonstrated for transferrin (Tf), insulin, insulin-like
growth factors (IGF-1 and IGF-2), leptin, and low-density lipoprotein receptor–related
protein (LRP). Transferrin receptor (TfR) is a transmembrane glycoprotein expressed
on hepatocytes, erythrocytes, intestinal cells, monocytes, choroid plexus epithelial cells,
neurons, and BBB endothelial cells. Drug targeting to the TfR can be achieved by using
the endogenous ligand Tf or by using antibodies directed against the TfR. Cationic liposomes coupled with Tf caused a significant enhancement of luciferase gene expression
activity in C6 glioma cells, primary hippocampal neurons, and primary cortical neurons
[13]. Ulbrich et al. developed human serum albumin (HSA) NPs coupled to Tf or TfR
mAbs (OX26) for the delivery of the loperamide [14]. The authors demonstrated an efficient drug transport to the brain using OX26-conjugated HSA NPs. OX26 mAb can avoid
competition with endogenous Tf in the circulation system because it binds to an extracellular domain of TfR (14). Recently, the OX26 mAb with liposomes was employed as
a targeting ligand to deliver therapeutic proteins and peptides to the brain [15]. Aktas
et al. recently designed OX26 mAb–bearing chitosan–PEG NPs and confirmed that
OX26 mAb is a critical functional moiety facilitating NPs to cross the BBB [16]. Similar
to OX26 mAb, 8D3 mAb and R17217 mAb are also able to bind to TfRs and can be trans-
cytosed across the BBB.

The insulin receptor (IR) is a large protein with a molecular weight of 300 kDa.
A novel study reports that a genetically engineered human/mouse chimeric form of the
human insulin receptor monoclonal antibody (HIRMAb) has been rapidly transported

into the brain of an adult Rhesus monkey after intravenous administration, suggesting its potential for delivering drugs across the BBB [17].

Low-density lipoprotein receptor–related proteins 1 and 2 (LRP1 and LRP2) are multifunctional, multi ligand scavenger and signaling receptors. They can interact with a diverse range of molecules and mediators including ApoE, tissue plasminogen activator (tPA), plasminogen activator inhibitor 1 (PAI-1), lactoferrin, melanotransferrin, α2-macroglobulin (α2-M), receptor-associated protein (RAP), HIV-1 TAT protein, heparin cofactor II, heat shock protein 96 (HSP-96), and engineered angiopeps. Several drugs that normally do not cross the BBB, including tubocurarine, loperamide, dalargin, 8-chloro-4-hydroxy-1-oxol, quinoline-5-oxide choline salt (MRZ 2/576), and doxorubicin, show higher concentrations in the brain when associated with polysorbate 80-coated NPs.

Another group of LRP ligands, known as angiopeps, has also been reported to be highly effective BBB targeting ligand. The most studied is angiopep 2, which has shown greater transcytosis capacity and parenchymal accumulation. Recently, Tosi et al. developed poly (D,L-lactide-co-glycolide) (PLGA) NPs with two surface-modified ligands: a BBB-penetrating peptide (similopioid peptide, g7) for transporting across the BBB and a sialic acid (SA) residue for the interaction with receptors in the brain tissue to prolong the NP residence time [18]. The researchers reported a remarkably high dose in the CNS over a prolonged period of time (24 h). These results were attributed to the ability of SA g7 NPs to cross the BBB and remain within the brain parenchyma.

Adsorptive-mediated transcytosis relies on the interaction of a ligand with moieties expressed at the luminal surface of cerebral endothelial cells. Several peptides allow the intracellular delivery of polar, biologically active compounds in vitro and in vivo. These peptides possess multiple positive charges, and some of them share common features, such as hydrophobicity and helical moment and ability to interact with a lipid membrane and to adopt a significant secondary structure upon binding to lipids. They have been successfully used as vectors for delivery of drugs that are P-gp substrates by effectively bypassing the P-gp in the BBB.

Adsorptive-mediated transcytosis (AMT) enables many poorly brain-penetrating drugs across the BBB and holds potential for promoting drug delivery into the brain. However, because it is a nonspecific process, the adsorptive process also occurs in the blood vessels and in other organs. This poses a challenge for both achieving therapeutic concentration in the brain and limiting the drug distribution in nontarget organ.

Cell-penetrating peptides (CPPs) and cationic proteins (e.g., albumin) are commonly used to enhance brain drug delivery via adsorptive-mediated transcytosis. A large variety of cargo molecules/materials have been effectively delivered into cells via CPPs, including small molecules, proteins, peptides, fragments of DNA, liposomes, and NPs. The transcription factor TAT, involved in the replication cycle of human immunodeficiency virus (HIV), was demonstrated to penetrate into cells [19]. One of the most interesting demonstrations of the effectiveness of TAT-shuttled nanocarriers across the BBB was accomplished by TAT-conjugated CdS:Mn/ZnS quantum dots (QDs) [20]. Histological data showed that TAT QDs migrated beyond endothelial cells and reached the brain parenchyma. Recently, Liu et al. produced compelling evidence that TAT facilitates human brain endothelial cell uptake of NPs self-assembled from TAT-PEG-b-cholesterol in vitro and that the NPs with TAT were able to cross the BBB and translocate around the cell nucleus of neurons [21]. Qin et al. prepared liposomes using cholesterol-PEG2000-TAT (TAT-LIP) and compared them to liposomes fabricated from cholesterol-PEG2000 polymer (LLIP) and conventional cholesterol formulation (LIP) in vitro and in vivo [22]. These data suggested that adsorptive transcytosis could be one of the mechanisms for TAT-LIP transport across the BBB and the positive charge of the TAT-LIP played an important role in enhancing this transport. SynB vectors are a new family of vectors derived from the antimicrobial peptide protegrin 1 (PG-1). These peptides are able to interact with the cell surface and cross the plasmatic membrane. Furthermore, the internalization of these peptide vectors does not depend on a chiral receptor, since the D-enantio form penetrates as efficiently as the parent peptide (L-form) and retro-inverso sequences exhibit identical penetrating activity. Adenot and colleagues studied brain uptake of a number of free and SynB3-vectorized chemotherapeutic agents using both in situ brain perfusion and in vitro BBB/cell model [23]. They reported that SynB3's conjugation with various poorly brain-penetrating drugs enhanced their brain penetration with no effect on TJ integrity.

13.4 Nanotechnology and Nanomedicine

Nanotechnology in biomedical research is emerging as an interdisciplinary science that has found its own value in clinical methodologies including therapeutics, imaging, and diagnostic. Nanotechnology refers to the understanding and control of matter at dimensions between approximately 1 and 100 nm. Nanotechnology takes up a large multidisciplinary field of science that adopts concepts from chemistry, biology, and physics and is expected to expand multidirectionally to generate innovations in the treatments to various human diseases, including cancer. The National Institutes of Health defines the nanomedicine as the application of nanotechnology to disease treatment, diagnosis, and monitoring and to the control of biological systems. The field of nanomedicine aims to use the properties and physical characteristics of nanomaterials, which have been extensively investigated for both diagnostic and therapeutic purposes. Nanosystems are characterized by submicron dimensions, which confer considerable advantages if compared to large-sized systems, including targeted delivery, higher and deeper tissue penetrability, greater cellular uptake, and greater ability to cross the BBB. NPs consist of molecules of different kinds and compositions and are capable to contain drugs and DNA–RNA fragments and to regulate their transport and intake into target tissues and cells. Nanomedicine is applied in many fields of biology and medicine, such as fluorescent biological labels; drug and gene delivery; detection of pathogens; detection of proteins; probing of DNA structure; tissue engineering; tumor destruction via heating, separation, and purification of biological molecules and cells; MRI contrast enhancement; and phagokinetic studies [24]. NP drug-delivery vehicles have shown the ability to encapsulate a variety of therapeutic agents such as small molecules (hydrophilic and/or hydrophobic), peptides, protein-based drugs, and nucleic acids. By encapsulating these molecules inside a nanocarrier, the solubility and stability of the drugs can be improved, providing an opportunity to reevaluate potential drugs previously ignored because of poor pharmacokinetics. Encapsulated molecules can be released from nanocarriers in a controlled manner over time to maintain a drug concentration within a therapeutic window, or the release can be triggered by some stimulus to the delivery site [25]. The surface of the nanocarrier can be engineered to increase the blood circulation and influence the biodistribution, while attachment of targeting ligands to the surface can result in enhanced uptake by target tissues. The net result of these properties is to lower the systemic toxicity of the therapeutic agent while increasing the concentration of the agent in the area of interest, resulting in a higher therapeutic index. In addition to therapeutic drugs, imaging agents can also be incorporated into nanocarriers to improve tumor detection and imaging [26]. Finally, NPs can be engineered to be multifunctional with the ability to target diseased tissue, carry imaging agents for detection, and deliver multiple therapeutic agents for combination therapy [27].

13.4.1 Nanoparticle Drug Delivery

Drug delivery can be defined as the process of releasing a bioactive agent at a specific rate and at a specific site. Most drugs have been formulated to accommodate the oral or injection delivery routes, which are not always the most efficient routes for a particular therapy. In addition, drugs can also be administered through other means, including transdermal, transmucosal, ocular, pulmonary, and implantation. New biological drugs such as proteins and nucleic acids require novel delivery technologies that will minimize side effects and lead to better patient compliance. As current advances in biotechnology and related areas are aiding the discovery of new classes of drugs, it is crucial to improve specific drug-delivery methods to turn these new advances into clinical effectiveness. Several drugs are limited by their poor solubility, high toxicity, and high dosage, aggregation due to poor solubility, nonspecific delivery, in vivo degradation, and short circulating half-lives. Innovative drug-delivery systems may make it possible to use certain chemical entities or biologics that were previously impractical because of toxicities or because they were impossible to administer. Nanostructure-mediated drug delivery has the potential to enhance drug bioavailability, improve the timed release of drug molecules, and enable precision drug targeting [28]. Anatomic features such as the BBB, the branching pathways of the pulmonary system, and the tight epithelial junctions of the skin make it difficult for drugs to reach many desired physiologic targets. Nanostructured drug carriers will help to penetrate or overcome these barriers to drug delivery. Courrier et al. have shown that the greatest efficiency for delivery into the pulmonary system is achieved for particle diameters of b100 nm [29]. Greater uptake efficiency has also been shown for gastrointestinal absorption [30] and transcutaneous permeation [31], with particles around 100 and 50 nm in size, respectively.

Drug-delivery carriers are macromolecular assemblies that can incorporate imaging and therapeutic compounds of distinct nature, such as small chemicals, fluorophores and biosensors, peptides and proteins, oligonucleotides, and genes. The use of targeted NPs to deliver chemotherapeutic agents in cancer therapy offers many advantages to improve drug delivery and to overcome many problems associated with conventional chemotherapy [32]. NPs via either passive targeting or active targeting have been shown to enhance the intracellular concentration of drugs in cancer cells while avoiding toxicity in normal cells. They can be designed to improve the solubility of these cargo molecules and their bioavailability and also to control their circulation, biodistribution in the body, and release rate, together enhancing their efficacy. Surface properties' modifications confer advantageous properties to the particles, such as increased solubility and biocompatibility useful in the crossing of biophysical barriers. The use of biodegradable materials in the NPs formulation permits the drug release for prolonged period. Advantages of nanostructure-mediated drug delivery include the ability to deliver drug molecules directly into cells and the capacity to target tumors within healthy tissue. DNA and RNA packaged within a nanoscale delivery system can be transported into the cell to fix genetic mutations or alter gene expression profiles. The mechanisms of cellular uptake of external particulates include clathrin- and caveola-mediated endocytosis, pinnocytosis, and phagocytosis. Nanoscale drug-delivery architectures are able to penetrate tumors due to the discontinuous nature of the tumor microvasculature, which typically contains pores ranging from 100 to 1000 nm in diameter. The efficiency of drug delivery to various parts of the body is directly affected by particle size and size distribution. Particle size and size distribution determine the in vivo distribution, biological fate, toxicity, and targeting ability of these delivery systems. In addition, they can influence drug loading, drug release, and stability of NPs [33]. Generally, NPs have relatively high cell uptake and are available to a wider range of cellular and intracellular targets due to their small size and mobility. Smaller particles have a larger surface area-to-volume ratio; therefore, most of the drug associated with small particles would be at or near the particle surface, leading to faster drug release. In contrast, larger particles have large cores, which allow more drugs to be encapsulated per particle and give slower release. Smaller particles also have a greater risk of aggregation during storage, transport, and dispersion.

Surface manipulation can control the extent of localization at interstitial sites and limit clearance. As nanomaterials are "stealthed" via hydrophilic PEGylation, their circulatory residence times increase. The NP zeta potential is commonly used to characterize the surface charge property of NPs [34]. It reflects the electrical potential of particles and is influenced by the composition of the particle and the medium in which it is dispersed. NPs with a zeta potential above ±30 mV have been shown to be stable in suspension, as the surface charge prevents aggregation of the particles. Endothelial damage or alteration may modify the distribution parameters of NPs. Inflammation, solid tumors, and deliberate disruption of endothelial contribute to an increased leakiness that provides vascular contents greater access to extravascular targets. The presence of disturbed, porous vascular beds at the tumor allows for selective targeting by this passive mechanism. Generally speaking, solubility, diffusion, and biodegradation of the particle matrix influence the drug-release process. Membrane coating acts as a drug-release barrier; therefore, drug solubility and diffusion in or across the polymer membrane become determining factors in drug release. Furthermore, the release rate also can be affected by ionic interactions between the drug and the auxiliary ingredients.

Drug loading and entrapment efficiency depend on drug solubility in the excipient matrix material (solid polymer or liquid dispersion agent), which is related to the matrix composition, molecular weight, drug–polymer interactions, and the presence of end-functional groups in either the drug or the matrix [35]. In addition, the drugs or proteins encapsulated in NPs show the greatest loading efficiency when they are loaded at or near their isoelectric point [36]. Through precise control of the drug carrier architecture, the release of the drug can be tuned to achieve a desired kinetic profile. In general, the drug-release rate depends on drug solubility, drug diffusion through the NP matrix, NP matrix erosion or degradation, and the combination of erosion and diffusion processes. Hence, solubility, diffusion, and biodegradation of the particle matrix govern the release process. It is evident that the method of incorporation has an effect on the release profile. If the drug is loaded by the incorporation method, then the system has a relatively small burst effect and sustained release characteristics. If the NP is coated by a polymer, the release is then controlled by the diffusion of the drug from the polymeric membrane.

13.4.2 Nanoparticle Functionalization and Targeting

Targeted therapy refers to the specific treatment of cancer cells while leaving healthy cells unharmed. Nanotechnology can be used to improve drug accumulation specifically to the tumor site using various mechanisms such as passive and active targeting. NPs functionalization represents the first step toward NP drug-delivery systems. Drug-delivery carriers can be functionalized to improve control of their circulation and biodistribution at the tissue, cellular, and subcellular level. This can be achieved by incorporating immune-evading moieties and/or affinity molecules that favor adhesion to either general or specific biological markers. When administered in vivo, therapeutic agents are recognized as foreign substances and rapidly cleared from the body. Clearance of foreign compounds in the body occurs mainly by the reticuloendothelial system (RES) and other elements of the immune system, as well as by renal filtration. For most applications, rapid clearance is detrimental as it minimizes the chances of the delivered agent to reach its targets in the body. An efficacious accumulation of the delivered therapeutic agent can be achieved by coating NPs with hydrophilic polymers/surfactants or formulating NPs with biodegradable copolymers with hydrophilic characteristics, for example, polyethylene glycol (PEG), polyethylene oxide, poloxamer, poloxamine, and polysorbate 80. The PEG minimizes interactions with plasma opsonins, complement, professional phagocytes, and lymphocytes. These interactions prolong the circulation in the bloodstream and favor lengthened medicinal effects and less frequent administrations [37]. Other strategy to minimize drug removal mimes the natural mechanism by which red blood cells avoid clearance by elements of the innate immune system. This is the case for CD47, a transmembrane protein that acts like a marker of the "self" by binding to its cognate receptor expressed on leukocytes, inhibitory receptor signal regulatory protein alpha (SIRPα). CD47 inhibits phagocytosis, in part via regulation of the cytoskeleton and inhibition of engulfing structures. Incorporation of CD47 on drug carrier surfaces reduces attachment to neutrophils and MPs prolonging circulation and inhibiting inflammation [38]. In addition, nanocarriers can also improve control of the drug efficacy upon release in the case of therapeutic interventions where administration is local. Localized implantation of bioactive agents embedded within porous matrices and/or hydrogels capable of responding to microenvironment properties can provide controlled release and effects. NP-based therapy can provide a significant advantage over standard chemotherapies by increasing the drug delivery specifically to the tumor site through either passive or active means.

Passive targeting refers to the accumulation of drugs to the required site due to physicochemical or pharmacological factors [39]. The increased vascular permeability coupled with the impaired lymphatic drainage in rapidly growing tumors allows an enhanced permeability and retention (EPR) effect of the nanosystems in the tumor. However, obstacles in this strategy are represented by mucosal barrier, nonspecific uptake of the particles, and nonspecific delivery of the drug. Appropriate size and functionalization with antibodies can provide means of enhanced delivery of drugs and reduced nonspecific toxicity. Doxil, a sterically stabilized PEGylated liposome that encapsulates doxorubicin, has shown good drug retention in the liposomal formulation. In experimental studies with EPR mechanism, some systems have shown significant improvements in tumor size reduction. Recently, Chytil et al. have exploited the EPR effect for targeting HPMA copolymer–based drug carriers with covalently bound hydrophobic substitutes for targeting solid tumors. Treatment of mice bearing EL-4 T-cell lymphoma with the aforementioned conjugates resulted in significant tumor regression. These nanoconjugates also enhanced tumor accumulation, indicating an important role of the EPR effect in anticancer activity [40].

Active targeting involves the attachment of a targeting ligand on the surface of the NP that recognizes receptors overexpressed on cancer cells. These targeting ligands can include antibodies, antibody fragments, peptides, aptamers, and small molecules such as folic acid or glucose that target the metabolism of cancer cells. The success of drug targeting depends on the selection of the targeting moiety, which should be abundant, have high affinity and specificity of binding to cell surface receptors, and should be well suited to chemical modification by conjugation. PEGylated gold NPs (AuNPs) are decorated with various amounts of human Tf by Choi et al. to enhance active targeting. Their results suggest that targeted NPs can provide greater intracellular delivery of therapeutic agents to the cancer cells within solid tumors than their nontargeted analogs [41].

Direct intratumor delivery of anticancer agents using NPs can be used in the treatment of local cancers such as prostate, head, and neck cancers. Recently, Sahoo et al. have demonstrated that Tf-conjugated paclitaxel (PTX) (Taxol (Tx))-loaded biodegradable NPs are more effective in demonstrating the antiproliferative effect of the drug than its solution or with unconjugated Tx-loaded NPs [42]. NPs are emerging as a promising tool for the intracellular delivery of practically insoluble drugs and sensitive drugs. Intracellular targeting refers to the delivery of therapeutic agents to specific compartments or organelles within the cell, and the delivered cargoes must gain access to intracellular compartments where their molecular targets are located. Interventions related to RNA interference (RNAi) or delivery of antisense oligonucleotides (AONs) require transport of these cargoes to the cytosol of the cell.

Gene therapy is a promising new approach for treating a variety of genetic and acquired diseases. These macromolecules are unstable and show a poor cellular uptake and are rapidly degraded by nucleases. To overcome these limitations, various chemical modifications of oligonucleotides have been tried. These modifications show some disadvantages such as decreased mRNA hybridization, elevated cytotoxicity, and increased nonspecific targeting. In order to overcome the disadvantages of viral carriers (high cytotoxicity, cost, small transgene size), nonviral carriers have been developed. The advantages associated with nonviral carriers include large-scale manufacture, low immunogenic response, versatile modifications, and the capacity to carry large inserts. Gene therapies require delivery to the cytosol, with subsequent transport to the cell nucleus. The drug can be delivered into target cell by simple diffusion or it may involve complex cellular machinery. The major route of intracellular therapeutic uptake is through endocytosis. This strategy is ideal in the case of delivery of therapeutic agents whose action is required at said subcellular compartments, such as in the case of carrier-assisted delivery of enzyme replacement for lysosomal storage disorders. Carriers themselves can also be designed to overcome endosomal membranes, such as in the case of pH-sensitive poly(acid) carriers and temperature-responsive poly(electrolyte) hydrogels [43]. Other strategies have been designed to directly overcome the plasmalemma; these include electroporation and ultrasound, where a local electrical or ultrasound pulse is exerted in the immediate post-administration period causing transient enhancement of the plasmalemma permeability [44], and biolistic particle delivery systems, where penetration into cells is gained by means of tungsten or gold particles that are propelled by a "gene gun" across the plasma membrane [45]. Amphiphilic and biodegradable cationic copolymers are efficient gene delivery systems, which can condense nucleic acid and form controlled nanosized complexes. Polyamidoamine (PAMAM) and poly[2-(dimethylamino) ethyl methacrylate] (PDMAEMA) are low-toxic polymers that have shown great potential as carriers. Poly(ε-caprolactone) (PCL) is another promising delivery system. PCL-g-PDMAEMA NP/DNA complexes could escape from the endosome and release their payloads effectively in the cytoplasm, which may be induced by the enhanced interaction between the complexes and cell membrane due to hydrophobic modification [46]. Small interfering RNA (siRNA) has attracted much attention because it enables sequence-specific manipulation of expression for multiple endogenous genes. The intracellular release of siRNA from pluronic/poly(ethylenimine) nanocapsules was achieved by changing the nanocapsules from a collapsed state to a swollen state using a brief cold shock treatment [47]. Weber et al. reported an amino-terminated carbosilane dendrimer–bound siRNA delivery system. These RNase-resistant carbosilane/siRNA dendriplexes have a high and prolonged gene-silencing effect and can be safely used in serum and antibiotics containing medium, without affecting cell viability and metabolic activity at relatively high dendrimer concentrations [48]. One of the most common methods used for the systemic delivery of siRNA involves their electrostatic interaction with cationic liposomes. Self-assembled liposome-protamine-hyaluronic acid NPs modified by DSPE-PEG with conjugated ligand have been used to overcome innate immune responses of siRNA-based therapy. The developed NP formulation has an siRNA encapsulation efficiency of 90% and showed a reduced systemic immunotoxicity [49].

13.4.3 Toxicity

Nanotoxicology evaluates the interactions of NPs with biological systems and the relationship between the physical and chemical properties of NPs with the induction of toxic biological responses. A number of factors can potentially influence NP toxicokinetics, such as size and surface properties of NPs, building blocks of NPs, and administration routes [50]. NPs exert some very special proprieties that are very relevant in the further design of toxicity testing of engineered nanomaterials. NPs own electronic, optical, and magnetic properties that are related to their physical dimensions and their breakdown could lead to a unique toxic effect that is difficult to predict. NP surfaces, also, are involved in many catalytic and oxidative processes potentially cytotoxic. Some NPs contain metals or compounds with known toxicity, and thus, the breakdown of these materials could elicit similar toxic responses to the components themselves. NPs could cause different types of effects such as mitochondrial damage, uptake through olfactory epithelium, platelet aggregation, and respiratory and cardiovascular effects.

High dose of single-walled carbon nanotubes causes reactive oxygen species (ROS) generation, mitochondrial dysfunction, oxidative stress, and change in cell morphology when incubated with keratinocytes and bronchial epithelial cell. Intratracheal instillation of carbon nanotubes in rodents causes chronic lung inflammation with foreign body granuloma formation and interstitial fibrosis. Fullerenes are recently evaluated as potential new antimicrobial agents in view of their potency for the induction of oxygen species after photoexcitation and cause lipid peroxidation and glutathione depletion. QDs, in vitro, induce lipid peroxidations and damage to plasma membranes, mitochondria, and nucleus. Metallic colloidal AuNPs, used as potential carriers for drug delivery, imaging molecules, and genes, as well as for the development of novel cancer therapy, present high cytotoxicity due to the stabilizer cetyl-trimethyl-ammonium bromide (CTAB) that remains as a residue even after washing. Only the PEG-modified AuNPs, where the excess of CTAB has been removed, did not show cytotoxicity.

Although targeted NPs have emerged as one strategy to overcome the lack of specificity of conventional chemotherapy, there are other potential risks and challenges associated with this novel strategy. Some cancer cell types would develop drug resistance rendering drugs released from the targeted NPs to be ineffective. Also the targeted NPs might change the stability, solubility, and pharmacokinetic properties of the carried drugs. The shelf life, aggregation, leakage, and toxicity of materials used to make NPs are other limitations for their use. Some materials used to make NPs show low toxicity but degrade quickly and do not circulate in tissues long enough for sustained drug/gene delivery. On the other hand, other materials such as carbon nanotubes and QDs are durable and can persist in the body making them potentially toxic and limiting their use for repeated treatments.

However, the toxicokinetics of NPs used in the CNS must be understood as it may affect the integrity of the BBB and cause acute or long-term toxic effects to the CNS. NP toxicity may induce TJ opening of the BBB. An in vitro study by Olivier and coworkers showed that poly(butylcyanoacrylate) (PBCA) NP induced a permeabilization of the BBB model, which was presumably attributed to the toxicity of the carrier [51]. PAMAM dendrimer toxicity is dose and generation dependent in vitro. QD toxicity is dependent on factors derived from both their inherent physicochemical properties and environmental conditions such as particle size, charge, concentration, outer coating bioactivity, and mechanical stability [52].

13.4.4 Therapeutic Nanoparticles

NPs serving as drug carriers play an essential role in brain drug delivery. They can be utilized to maintain drug levels in a therapeutically desirable range and increase half-lives, solubility, stability, and permeability of drugs. They can also be structurally adapted to deliver a variety of drugs, improve delivery efficiency, and reduce side effects by targeted delivery. Several major types of NP that have been widely used for construction of nanomedicines, such as polymeric NPs and micelles, liposomes, peptide NPs, and dendrimers, have also found applications in the CNS.

13.4.4.1 Polymeric and Polymer–Drug Conjugate Nanoparticles

Polymeric NPs are extensively used for the nanoencapsulation of various bioactive molecules and medicinal drugs. Polymeric NPs are structured in two different forms, nanospheres and nanocapsules. They are, respectively, characterized by a matrix system in which the drug is dispersed and a reservoir in which the drug is confined in a hydrophobic core surrounded by a single polymeric membrane (core–shell structure). These carriers show a higher stability in biological fluids and against the enzymatic metabolism. Polymers are being developed to create delivery systems with excellent drug and protein loading and release properties, a long shelf life, and little toxicity. The core matrix of these NPs can be composed by various biodegradable polymers, such as (PLGA), chitosan, poly(alkylcyanoacrylate) (PACA), PBCA, poly(lysine), PCL, and polyaspartate (PAsp). Polymer NPs have been used as transport vectors for various peptides CNS delivery after intravenous injection such as hexapeptide dalargin, loperamide, tubocurarine, and doxorubicin. PLGA is one of the most successfully used biodegradable nanosystem because it undergoes hydrolysis in the body to produce the biodegradable metabolites, lactic acid and glycolic acid. Surface modification of PLGA, drug encapsulation methods and particle size, additives added during formulation, molecular weight of drug, and ratio of lactide to glycolide moieties have strong influence on the release and effective response of formulated nanomedicines. PTX promotes the polymerization of tubulin causing cell death by disrupting the cell division process. It is one of the potent anticancer agent but less useful for clinical administration due to its poor solubility. PLGA intermingled with vitamin E and tocopheryl PEG succinate (TPGS) has been used to encapsulate this drug. This formulation has shown good activity and much faster administration in comparison to traditional formulation. Using some additive to the PLGA NPs, 100% drug encapsulation efficiency was achieved [53]. Polymers can also be used to coat other types of NPs. PEG is a hydrophilic polymer that has been used to coat the surface of NPs, which allows them to avoid clearance by the RES and cross the BBB [54]. In the 9L gliosarcoma model, PEG coating of an NP MRI contrast agent increased the amount of MRI signal intensity from the agent. Other hydrophilic polymers, including hydrogel (polyacrylamide), dextran, and polysorbate, have been used to coat the surface of NPs to prolong plasma circulation and improve delivery across the BBB.

Natural polymers have been extensively used in solid-oral dosage forms, where they have been used as binders, diluents, disintegrant, and matrixing agents. These substances occur widely in nature and are generally biocompatible, biodegradable, nontoxic, and non-immunogenic [55]. Starch is a common polysaccharide and occurs in plants where it acts as a storage material. Starch NPs have been employed to deliver insulin via noninvasive routes [56]. Cole et al. evaluated magnetized iron oxide NPs coated with starch as a means of targeting brain tumors. MR and histological data showed that surface modification with polyethylene oxide improved delivery to tumor cells resulting in a greater accumulation of particles in the neoplastic tissue [57]. Chitosan is composed of glucosamine and N-acetyl glucosamine linked by β-1–4 glucosidic bonds. NPs with chitosan were used to investigate the controlled release of the antiretroviral drug lamivudine [58]. Drug-release kinetics showed that the mechanism of drug release was by diffusion. Nano-complexes of chitosan and polyoxometalates were also tested as anticancer preparation [59].

Polymer–drug conjugates are formed through side-chain grafting of drugs to polymer chains, allowing them to deliver high doses of chemotherapeutic drugs. These agents bear numerous functional groups that are available for covalent binding to a variety of biochemically active groups, which direct them to malignant tumors where they can deliver functional drugs acting on several tumor targets [60]. Nanoconjugates that carry more than one functional group provide the capability to simultaneously inhibit several tumor pathways, deliver optimal drug concentrations to the site of treatment, and reduce adverse effects on healthy tissue. ProLindac (AP5346) is composed of a HPMA backbone copolymer with platinum grafted to the side chains through a pH-sensitive chelator designed for drug release in the tumor environment. Preclinical data demonstrate superior efficacy of the polymer–drug conjugates using multiple cancer models including a M5076 sarcoma platinum-resistant tumor xenograft mice model, multiple colon xenograft models, L1210 leukemia, and 0157 hybridoma models [61]. Nanoconjugates can overcome drawbacks of conventional chemotherapy such as drug resistance and toxicity by specifically targeting tumor cells, activating cancer cell uptake, and bypassing multidrug resistance transporters.

13.4.4.2 Micelle Nanoparticles

Micelle NPs (MNPs) are amphiphilic spherical structures composed of a hydrophobic core and a hydrophilic shell. Hydrophobic part is the inner core of the block copolymer, which encapsulates the poorly water-soluble drug, whereas the outer hydrophilic shell or corona of the block protects the drug from the aqueous environment and stabilizes the MNPs against recognition in vivo by the RES. The core can sometimes be made up of a water-soluble polymer that is rendered hydrophobic by the chemical conjugation of water-insoluble drug and by complexation of the two oppositely charged polyions. The polymer always contains a nonionic water-soluble segment and an ionic segment that can be neutralized by oppositely charged surfactant to form a hydrophobic core. The electrostatic interaction between the ionic segment of the block polymer and the surfactant group changes these segments from water soluble to water insoluble, leading to a hydrophobic core in the micelles [62]. MNPs can be engineered by means of ligand coupling or addition of pH-sensitive moieties according to the biological characteristics of the diseased site for active targeting. All these features related to MNPs make them ideal carriers for anticancer drugs and tumor targeting [63]. On reaching the target site, micelles are internalized into the cells via fluid-state endocytosis. Liu et al. synthesized cholesterol-terminated PEG and modified the other end of PEG with TAT peptide for aiding the transport of NPs to the brain [64]. TAT-PEG-cholesterol could self-assemble into micelles with a hydrophobic core of cholesterol for encapsulation of ciprofloxacin, an antibiotic against neuron inflammatory diseases. Their animal studies based on rats qualitatively confirmed that TAT-conjugated micelles were able to cross the BBB [64]. Surfactants are being incorporated into anticancer metal-based drugs. The surfactant dodecylamine reacts with selenious acid to produce a quaternary ammonium salt, which can be conjugated to copper or cobalt ions to form copper or cobalt cationic complexes. Initial studies demonstrated effectiveness in vitro against five human monolayer tumor cell lines: MCF7 (breast carcinoma), HEPG (2) (liver carcinoma), U-251 (glioma), HCT116 (colon carcinoma), and H-460 (lung carcinoma). Recently, the potential antitumor activity of NK012, a 7-ethyl-10-hydroxycamptothecin (SN-38) micellar formulation, and bevacizumab in human lung cancers has been evaluated [65].

Micelles with surface-attached specific antibodies, also called immunomicelles, provide an opportunity for targeting in terms of diversity of targets and specificity of interactions. Micelle-attached antibodies retain their ability to specifically interact with their antigens. Recently, it has been shown that certain nonpathogenic monoclonal antinuclear autoantibodies with the nucleosome-restricted specificity recognize the surface of numerous tumors, but not normal cells via tumor cells surface-bound nucleosomes [66]. EGF-coupled PMs are another approach for targeting to tumors. EGF is a ligand having nuclear translocation properties. Zeng et al. reported an EGF-conjugated PEG-b-poly(δ-valerolactone) micelle system that targets the EGF receptors overexpressed by the breast cancer cells [67]. These micelles were shown to localize in the nucleus of MDA-MB-468 breast cancer cells and perinuclear region. Thus, EGF conjugates are useful for nuclear targeting, which is critical for the delivery of anticancer drug whose site of action is located in the nucleus.

An interesting target for molecular cancer therapy is represented by heat shock protein 90 (HSP90), a molecular chaperone that under normal conditions is responsible for the prevention of protein aggregation [68]. HSP90 becomes overexpressed under conditions of stress, resulting in tumorigenesis and increased proliferation in a variety of cancers including lung, prostate, and breast. Tanespimycin, a derivative of the HSP90-inhibitor geldanamycin, has been explored clinically for chemotherapeutic purposes. The mechanism of action of tanespimycin involves the degradation of oncogenic signaling proteins, inducing cell death via apoptosis. In patients with multiple myeloma treated with tanespimycin, disease stabilization was observed [69].

13.4.4.3 Liposomes

Liposomes (Figure 13.2) are self-assembled phospholipid membranes with an inner core where the drugs can be entrapped. The lipids form a bilayer based on hydrophobic interactions in continuous parallel packing, with the hydrophilic head groups positioned toward the aqueous environment. Liposomes are relatively stable and biodegradable and do not elicit any immune response. Commonly, liposomes have such problems as short clearance time and low transport rate, which, however, can be overcome by incorporating PEG and targeting ligands. In addition, liposomes modified with PEG have improved structural stability against rapid release of drug molecules. Recently, a series of antibody-coupled PEGylated liposomes (immunoliposomes) for receptor-mediated delivery of various therapeutics including nucleic acids and drug molecules to the brain has been developed [70]. In a novel study, Feng et al. used anti-EGFR antibody-carrying immunoliposomes to deliver sodium borocaptate (BSH) for boron capture neuron therapy. This new delivery system selectively delivered a large amount of boron to glioma cells in vitro and in vivo. In immunoliposome-treated mice, BSH was detected in the tumor and surrounding regions at 24 h after injection and remained at a high level for another 24 h [71].

Gold nanoparticles   Liposome

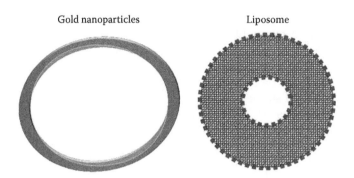

FIGURE 13.2
NPs utilized in brain drug delivery.

Lipid-based nanocapsules provide a novel approach for encapsulation of poorly soluble drugs like cisplatin or PTX. Recently, a novel strategy for the efficient encapsulation of cisplatin in a lipid formulation based on repeated freezing and thawing of a concentrated solution of cisplatin in the presence of negatively charged phospholipids (dioleoyl-phosphatidylserine and dioleoylphosphatidylcholine) has been developed. This strategy produced nanocapsules composed of small aggregates of cisplatin covered by a single lipid bilayer. These preparations showed an exceptional drug-to-lipid ratio and when tested on ovarian cancer cells evidenced higher cytotoxicity when compared to free drug. Kizelsztein et al. prepared sterically stabilized nanoliposomes by mixing the liposome-forming lipid, cholesterol, and the lipopolymer 2000PEG-DSPE and used it to deliver tempamine (TMN), a stable radical with antioxidant and proapoptotic activities. This system shows efficiency in inhibiting autoimmune encephalomyelitis (EAE) in mice. The penetration of TMN-loaded liposome dose into the brain was threefold to sixfold higher in EAE mice than in normal mice [72].

PEGylated liposomes were also evaluated to encapsulate vector/DNA plasmid complexes such as polyethylenimine/oligodeoxynucleotides (PEI/ODNs). In this way, the accumulation of ODN in the brain was significantly improved once TfR-specific antibody 8D3 was noncovalently bound to the PEGylated liposome through streptavidin–biotin binding. Grahn and coworkers employed non-PEGylated liposomes for simultaneous encapsulation of topotecan (topoCED) and gadodiamide (gadoCED), thus enabling therapy of GBM and real-time monitoring of NP distribution [73]. Desirable TPT pharmacokinetics and co-convection of gadodiamide have been demonstrated. A finding has reported successful curtailment of cellular resistance and effective cell death by constructing a cationic liposome incorporated with doxorubicin and siRNA targeted to MRP1 and BCL2 mRNA. In this study, a multifunctional cationic liposome has been fabricated that contains doxorubicin and siRNA to the multidrug resistance protein MRP1 and the antiapoptotic protein BCL2. Active targeting via novel stealth liposomes surface conjugated with fibronectin mimetics to specifically target tumor endothelial cells overexpressed with integrin alpha(5) beta(1) family has also been attempted [74]. Antisense therapy represents an innovative gene-silencing strategy. In a phase I study, patients with advanced solid tumors undergoing radiation therapy were administered a liposomal formulation, LErafAON, which encapsulates the raf AON, with the purpose of acting on c-raf, a protein that bestows cancer cells with resistance to radiation or chemotherapy. The results of this preliminary report showed the c-raf-1 mRNA was inhibited in three of five patients. Moreover, four of the twelve patients evaluable at the end of treatment exhibited a partial tumor response, four had stable disease, and the remaining four showed progressive disease [75].

13.4.4.4 Peptides Nanoparticles

Peptides that specifically interact with receptors overexpressed by cancer cells had successfully developed as targeting molecules for drug delivery and in vivo imaging. The interaction of peptides and proteins with the cell membrane results in their penetration into the cell or the formation of pores within the cell membrane. Because of their ability to target and enter cells, peptide and protein carriers hold great potential for the delivery of genes and AONs to cancer cells [76]. Bombesin (BBN) peptide and its analogs can be used to target gastrin-releasing peptide (GRP) receptors; in vivo GRP receptors are over-expressed in GBM, small cell lung, gastric, pancreatic, prostate, breast, cervical, and colon cancers. Recently, GNPs functionalized by a high load of thioctic acid–BBN peptide were used to target prostate tumor xenografts in SCID mice. Using normal and prostate tumor–bearing mice, they showed that this compound exhibits high binding affinity to the tumor and confirmed that these constructs are GRP receptor specific and accumulate with high selectivity in GRP receptor-rich pancreatic acini [77]. Fibroblast growth factor analogs can be used to target cells expressing fibroblast growth factor receptors (FGFRs). This receptor family is often expressed on both tumor cells and neo-vasculature. Truncated human basic fibroblast growth factor peptide (tbFGF) was recently used to achieve target-ing of liposomes carrying chemotherapeutic drugs [78]. This peptide contains both the bFGF receptor–binding site and a part of the heparin-binding site, which allows it to bind FGFRs on cell surface, without stimulating cellular proliferation. The albumin-bound PTX (Abraxane) is currently tested as a first-line therapy or in combination with other drugs for metastatic breast cancer and other cancers that have been shown to be sensitive to taxane drugs, such as ovarian and prostate. Preclinical studies have shown that the con-centration of PTX bound to albumin in endothelial cells and in the extravascular space was significantly increased [79]. Peptides can, also, act as therapeutic agents conjugated to NPs. Melittin is a cytolytic peptide that represents a potential candidate for cancer chemo-therapy. Melittin is a nonspecific cytolytic peptide that attacks all lipid membranes lead-ing to significant toxicity when injected intravenously. Recently, it has been demonstrated that synthetic nanoscale vehicles like PFC NPs can deliver melittin by passive and active molecular targeting to kill both established solid tumors and precancerous lesion cells [80].

13.4.4.5 Dendrimers

A dendrimer (Figure 13.3) is generally defined as a macromolecule, which is characterized by its highly branched 3D structure that provides a high degree of surface functionality. The generation number and the chemical composition of the core, branches, and surface functional groups determine the size, shape, and reactivity of dendrimers. Dendrimers can function as drug carriers either by encapsulating drugs within the dendritic structure or by interacting with drugs at their terminal functional groups via electrostatic or covalent bonds (prodrug). Dendrimers have been studied extensively for targeting and delivery of therapeutic agents and of contrast agents for magnetic resonance imaging. The avidimers are dendrimers targeted to tumor vasculature using a methotrexate PAMAM bioconjugate platform functionalized with small targeting ligands [81]. The authors demonstrated in vitro that drug-free dendrimer conjugates were not cytotoxic and that drug-loaded den-drimer conjugates had no effect on folate receptor–negative cells. Polyanionic PAMAM dendrimers showed rapid serosal transfer rates in crossing adult rat intestine in vitro and had low tissue deposition. The transport of PAMAM and surface-modified PAMAM across cell monolayer follows endocytosis-mediated cellular internalization. Various studies

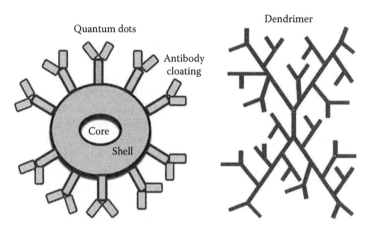

FIGURE 13.3
NPs utilized in brain tumor diagnosis and brain drug delivery.

report that PEG-modified dendrimers show reduction of cytotoxicity and immunogenicity, high exocytosis rate and low accumulation in endothelial cells, and excellent solubility and favorable pharmacokinetic [82]. Multimodal dendrimer–conjugated magnetofluorescent nanoworms called dendriworms were developed recently for siRNA delivery. The magnetic core in dendriworms enables in vivo imaging of dendriworms with MR, while PAMAM dendrimers conjugated to the magnetic core allow nucleic acid delivery and targeting. Dendriworms accumulate in the lungs and the reticuloendothelial filtration organs following systemic delivery. Dendriworms administered with CED efficiently delivered EGFR siRNA to suppress the expression of EGFR in GBM in a mouse model [83]. Yang et al. prepared epidermal growth factor (EGF)–carrying boronated PAMAM dendrimer G4.0 for neutron capture therapy of brain tumors [84]. Doxorubicin was conjugated to RGD-coupled PEGylated PAMAM dendrimer via a degradable disulfide spacer for controlled release in the treatment of glioma tumors [85]. PAMAM dendrimers have been tested as genetic material carriers. SuperFect–DNA complexes, a transfection reagent consisting of activated dendrimers, are characterized by high stability and provide more efficient transport of DNA into the nucleus. The high transfection efficiency of dendrimers may be due to their well-defined shape but also due to the low pK of the amines [86].

13.5 Nanomedicine Applications in Brain Tumors

In the last decade, various studies have demonstrated the value of nanotechnology in brain tumor treatment. Nanotechnology and nanomedicine have been utilized to perform new therapeutic intracerebral drug-delivery systems and to develop treatments for various diseases and disorders. Nanomedicine can give many ideal devices for the delivery of specific compounds to brain tumors, loading them into NP-based carriers via a variety of chemical methods including encapsulation, adsorption, and covalent linkage. By using nanotechnology in drug design and delivery, it will be possible to deliver the drug to the targeted tissue and cells across the BBB, to release the drug at the controlled rate, and to be able to escape from degradation processes.

Diagnostic methods are essential for the early detection of diseases to enable their prompt treatment minimizing possible damage to the rest of the organism. Conventional imaging technologies represent static images of tumors rather than a continuous visualization of tumor proliferation. Nanodiagnostics is defined as the use of nanotechnology for clinical diagnostic purposes, and it was developed to meet the demand for increased sensitivity in clinical diagnoses and earlier disease detection. NP-based system imaging allows an early detection of tumor as well as opportunities for real-time monitoring, thereby increasing both sensitivity and accuracy of anticancer therapies. Nanotechnology has produced advances in imaging diagnosis, developing novel methods and increasing the resolution and sensitivity of existing techniques. These systems include positron-emission tomography (PET), single-photon-emission CT (SPECT), fluorescence reflectance imaging, fluorescence-mediated tomography (FMT), fiber-optic microscopy, optical frequency-domain imaging, bioluminescence imaging, laser-scanning confocal microscopy, and multiphoton microscopy. Molecular imaging can identify tumor cell location within the body and aims to provide information such as metabolism, expression profile, and stage of the disease. Furthermore, molecular imaging can reveal early tumor response to therapy that will aid in improving treatment regimens. These new cellular targeting–based imaging detection methods can reach the specific and selective molecular recognition only for tumor cells, through the recognition of tumor-specific molecules into ligand receptor, antibody–antigen interaction, or other interaction processes between NP drug-loaded systems and cancer cells, leading to a diffuse and complete delivering of drug into cancer cells. QDs (Figure 13.3) are colloidal semiconductor nanocrystals and present photophysical properties: when excited, they can vary their size emitting fluorescence at different wavelengths. QDs can be used as probes for high-resolution molecular imaging of cellular components, for tracking cell activities and movement inside the body, and for specific targeting interaction through antibodies linked onto NP surfaces with specific tumor-associated antigens expressed on cancer cell surface. In a recent a study, the utility of the QD–aptamer(Apt)–doxorubicin (Dox) conjugate [QD-Apt(Dox)] as novel targeted cancer imaging, therapy, and sensing system has been demonstrated. The targeted QD imaging system (QD-Apt) was capable of differential uptake and imaging of prostate cancer cells that express the prostate-specific membrane antigen (PSMA) [87]. Arndt-Jovin et al. applied QDs coupled to EGF or anti-EGFR (Her1) mAbs (mouse MAb 528, H-11, and H199.12) to map human GBM with fluorescence microscopy. QDs coupled to EGF could be taken up and the level of their uptake varied, consistent with the wild-type and mutant EGFR expression in the cell lines [88]. Recently, the capacity of QDs to stimulate the EGFR activation promoting specific detection of intracellular EGFR subpopulation [89] has been demonstrated. Superparamagnetic iron oxide NPs have been explored for various biomedical applications such as magnetic drug targeting and enhanced magnetic resonance imaging. Iron oxide NPs, in particular, have received significant attention owing to their proven biocompatibility and biodegradability. These NPs develop superparamagnetism at the nanometer scale, as each particle becomes a single magnetic domain that is free to rotate at room temperature. The relaxivity of iron oxide NPs, or their ability to provide contrast in MRI, can be improved by tuning the size, shape, and defect type of the NP core. NP targeting methods, iron oxide NP-based MRI contrast agents, targeted NP-based MRI contrast agents, and intraoperative NP-enabled brain tumor delineation may produce contrast enhancement at an earlier stage of gliomas. Veiseh et al. designed a nanoprobe (NPCP-CTX-Cy5.5) comprised of an iron oxide NP coated with a PEGylated chitosan, to which a targeting ligand (chlorotoxin [CTX]) and a near-infrared fluorophore (NIRF, Cy.5.5) were conjugated. CTX was used as a tumor-targeting ligand because it selectively binds to a variety of cancers including glioma, prostate cancer, sarcoma, and intestinal cancer [90]. In a novel research, it has been demonstrated that Tf-conjugated SPIO is suitable for MR imaging; the results suggested that Tf-SPIONs could act as potential and specific MR targeting contrast agent for brain glioma possessing the characteristics of long retention and higher T2 relaxation rate [91]. In vivo application of NP-based platforms in brain tumor is limited by insufficient accumulation and retention within tumors owing to limited specificity for the target. Recently, a nanoprobe composed of an iron oxide NP coated with biocompatible PEG-grafted chitosan copolymer, to which a tumor-targeting agent, CTX, and an NIRF are conjugated, has been formulated. This nanoprobe shows a reduced toxicity and a sustained retention in tumors. This NP, for the versatile affinity of the targeting ligand and the flexible conjugation chemistry, represents a potential tool for the diagnosis and treatment of brain tumors [92].

Recent advances in molecular studies have evaluated new glioma-associated biomarkers and their implications in cancer progression. The possibility to inhibit contemporary more pathways into glioma progression by molecular-based targeted approaches, using a nanocarrier loaded with anticancer agent, represents a promising therapeutic strategy. The future challenges of this approach are represented by the possibility to modify the cell genome and induce it to a reversion to the wild-type conditions, the enhancing of immune system antitumor capacity, and the targeted drug delivery into brain tumor cells. It is known that NPs may trigger an inflammatory process resulting in the release of different proinflammatory cytokines, chemokines, ROS, and transcription factors that could explain some known side effects. These mechanisms may involve MG and interact in different ways with normal brain tissue and glioma cells [10]. All these aspects are crucial to modulate important molecular pathway, such as neovascularization, invasiveness, and interaction between tumor and perilesional tissue. We think that, using the intrinsic capacity of NPs, it should be very interesting to structure a new NP-based molecular approach against two or more molecular targets, contemporary. In our laboratories, we are trying to create a carrier loaded with an antisense molecule against hypoxic ischemic factor-1α (HIF-1α) and interleukin-8 (IL-8). HIF-1α appears to be a highly involved factor in the tumor development influencing growth rate, invasiveness, and metastasis. Local hypoxia, due to increased proliferation or insufficient oxygen supply and inactivation of tumor suppressors, oncogenes, and growth factors, along with other cell types, such as MPs, contributes to form a tumor microenviroment capable to modulate the HIF response itself. Antisense inhibition of HIF may be a strong target for antiangiogenic therapy. Our study group has recently showed high expression levels of PGES-1 (prostaglandin E-1 synthase) and IL-8 in high-grade glioma cells and microglial cells, strongly correlated with grading tumor [24,93]. The efficiency of nanoplatform systems has been also enhanced by various molecular targeting, such as IL-13 expression and TfR and LDL surface receptor pathways. Glioma cells show an upregulation of expression of IL-13 receptor-α2 on their surface cells. In a recent study, the improvement of internalization of doxorubicin-loaded nanoliposomes, targeted with conjugated IL-13, and cytotoxicity in U251 glioma cells has been shown. In an in vivo animal model, the authors demonstrated the inhibition of the growth of subcutaneously implanted gliomas [94].

Gene therapy involves the delivery of DNA molecules to cancer cells to modify a gene in an effort to treat the disease. The delivery of DNA can be accomplished using a variety of vectors including viruses, cell-based systems, and synthetic vectors. For glioma gene therapy, viral vectors have been used to deliver suicide genes, proapoptotic genes, p53, cytokines, and caspases. These studies have shown promising preclinical results, but clinical trials have been limited by the fact that transduced cells were found only within a very short distance of the delivery site. To overcome these limitations, synthetic vectors have been developed to more safely deliver DNA. In a recent study, the authors investigate targeted gene delivery to C6 glioma cells in a xenograft mouse model using CTX-labeled NPs [95]. The developed nanovector consists of an iron oxide NP core, coated with a copolymer of chitosan, PEG, and PEI. The CTX promotes specific uptake of nanovectors into glioma cells, exposing a higher proportion of target cells to the delivered payload. Another gene therapy approach is the injection in cancer cells with genes that can destroy the cells. A prodrug or an inactive form of a toxic drug is administered to the patients, and this drug will kill off any cancer cells with the suicide genes in them. The use of specific NPs can represent a valid alternative to overcome possible toxic and infective effects of classic gene therapy.

One of the most promising new strategies is the RNAi-based approach, consisting of small double-stranded RNA molecules that can sequence and specifically inhibit the expression of targeted oncogenes. To harness the full potential of this approach, the prime requirements are to deliver the siRNA molecules with high selectivity and efficiency into tumor cells and to monitor both siRNA delivery and the resulting knockdown effects at the single cell level. The use of NP nonviral gene delivery systems such as carbon nanotubes blocks a selective genic function without toxic effect on cellular phenotype. These systems have the ability to carry short single genes and also short DNA fragments or siRNA molecules that exert RNAi on target gene expression after their internalization [96]. Recently, the synthesis and target-specific delivery of multifunctional siRNA-QD constructs for selectively inhibiting the expression of EGFR variant III (EGFRvIII) in target human U87 GBM cells has been evaluated. This study demonstrates the multifunctional siRNA-QD strategy focusing on targeted delivery, high transfection efficiency, and multimodal imaging/tracking. This strategy could also provide highly useful information regarding biosurface chemistry of nanomaterials. In addition, the application of multifunctional siRNA-QDs to modulate the key cancer signaling pathways is important not only for selective chemotherapeutic strategy but also for dissecting signaling cascades triggered by inhibiting specific proteins [97].

Antiangiogenic approaches in glioma therapy have been strongly directed against VEGF pathway. In an in vivo murine model, created by implantation of U-87 MG malignant glioma cells in mouse, the suppression of capacity of glioma cells to form tumors in mice has been demonstrated. This result was obtained after transfection of antisense VEGF cDNA in an antisense orientation through the recombinant adenoviral vector Ad5CMV-alphaVEGF. Infection of U-87 MG malignant glioma cells resulted in the reduction of the level of the endogenous VEGF mRNA and in reduced production of the VEGF-targeted secretory form [98]. Agemy et al. have proposed a multifunctional theranostic NP in which the CGKRK peptide provides the targeting function that takes the NPs to tumor vascular cells and into their mitochondria. The NP uses the mitochondria-targeted $_D$[KLAKLAK]$_2$ peptide as the drug and iron oxide as a diagnostic component for MRI. In addition, the NP was combined with the tumor-penetrating peptide iRGD, which enhances the NP penetration into the extravascular tumor tissue. Systemic treatment of GBM-bearing mice eradicated most tumors in one GBM mouse model and significantly delayed tumor development in another [99]. An important molecular target used to selectively detect glioma cells is IL-13, based on upregulated expression of IL-13α2 on the surface of GBM cells. The efficiency of liposomes has been increased through surface ligand targeting via mAbs to specific receptor upregulated in glioma cell surfaces such as TfRs, LDL receptors, and IL-13 receptors. A biopolymeric gene delivery NP has recently been shown to be effective in vivo in delaying tumor growth. This polymeric NP-based nonviral gene delivery vector is a cationic albumin-conjugated PEGylated NP in which a plasmid encoding proapoptotic

Apo2 ligand/tumor necrosis factor–related apoptosis-inducing ligand (Apo2L/TRAIL) is incorporated. The incorporation of plasmid DNA into the host cell genome caused inhibition of tumor growth and prolonged survival in mice bearing implanted C6 gliomas [100]. Schneider et al. have evaluated a "double-punched" approach to overcome the escape of GBM cells to the immune surveillance, through an active-specific immunization with Newcastle-disease-virus-infected tumor cells and blocked the TGF-β production by delivery of TGF-β AONs using polybutylcyanoacrylate NPs. This approach induced a significant decrease in plasma TGF-β2 level as well as an increase in rate of high-affinity IL-2 receptor (CD25) on lymphocytes and consequently of antitumoral cytotoxicity [101]. By using anti-FAK phosphorothioate AONs packaging into liposomes in U251 MG cells, the downregulation of expression levels of FAK, and the activation of apoptosis, through increase in caspase-3 activity, has been also demonstrated [102].

Although immunotherapy is being investigated as an adjunct treatment, the ability of gliomas to escape immune response represents a significant obstacle to this strategy. A possible strategy is represented by the activation of the innate immune system by toll-like receptor (TLR) agonists such as CpG oligonucleotides. Because TLR9, CpG receptor, is located intracellularly, the authors have hypothesized that enhance CpG internalization may also potentiate its immunostimulatory response. In this study, it has been reported that carbon nanotubes enhanced CpG uptake by tumor-associated phagocytic cells resulting in an overall proinflammatory response against glioma cells both in vitro and in vivo. Furthermore, a single injection of low-dose CNTCpG complexes eradicated intracranial gliomas through activation of NK and CD8 cells. These findings demonstrate that CNTs are nontoxic vehicles that can improve CpG uptake into tumor-associated inflammatory cells, leading to a more robust antitumor response [103]. Alizadeh et al. evaluated the mechanism of cyclodextrin-based NP (CDP-NP) uptake into a murine glioma model. Using mixed in vitro culture systems, the authors demonstrated that CDP-NP was preferentially taken up by BV2 and N9 MG cells as compared to GL261 glioma cells. Fluorescent microscopy and flow cytometry analysis of intracranial GL261 gliomas confirmed these findings and demonstrated a predominant CDP-NP uptake by MPs and MG within and around the tumor site. These studies demonstrate that MP and MG could potentially be used as NP drug carriers into malignant brain tumors [104].

Bernardi et al. evaluated the efficacy of immunonanoshells in vitro against medulloblastoma and malignant glioma cell lines. The authors used an antibody against human EGFR (HER2) to target gold–silica nanoshells. In glioma cell lines, they showed the capacity of these immunonanoshells of leading cell death in U373 and U87 malignant glioma cell lines. The same authors conjugated gold–silica nanoshells to an antibody specific to IL-13 receptor alpha 2 (IL-13Rα2), strongly expressed in gliomas, demonstrating that these immunonanoshells are capable of leading to cell death in U373 and U87 malignant glioma cell lines [105].

Convection-enhanced delivery (CED) was developed to address the diffusion of agents directly delivered to the brain. This strategy has been used to deliver proteins and small particles, including liposomes and polymeric NPs. Combining polymeric controlled release with CED could improve the drug distribution limitations of implantable wafers while also offering spatiotemporal distribution control that is lacking from CED. PLGA is capable of encapsulating and releasing a variety of agents, including chemotherapy drugs, for long periods of time. The efficacy of CED of surface-modified, drug-loaded, PLGA NPs to treat intracranial glioma using the topoisomerase I inhibitor camptothecin (CPT) was evaluated. CPT is an attractive drug for delivery by controlled release because it has known anticancer activity but is limited by low solubility and serious systemic toxicity. The NPs were shown effective both in culture and in vivo, with a statistically significant survival benefit observed in all animals treated [106]. A significant therapeutic effect was found after CED of both IONPs and EGFRvIIIAb-IONPs in mice. Dispersion of the NPs over days may potentially target infiltrating tumor cells outside the tumor that are potentially responsible for tumor recurrence. Use of bioconjugated magnetic NPs may permit the advancement of CED in the treatment of malignant gliomas due to their sensitive imaging qualities on standard T2-weighted MRI and therapeutic effects [107].

A polymeric nanobioconjugate drug based on biodegradable, nontoxic, and nonimmunogenic polymalic acid was used for design and synthesis of nanomedicine drug for i.v. treatment of brain tumors. The polymeric drug crosses the brain tumor barrier (BTB) and tumor cell membrane using tandem mAbs targeting the BTB and tumor cells. The inhibition of tumor angiogenesis was obtained by specifically blocking the synthesis of a tumor neovascular trimer protein, laminin-411, by attached AONs. The AONs were released into the target cell cytoplasm via pH-activated trileucine, an endosomal escape moiety. Introduction of a trileucine endosome escape unit resulted in significantly increased AON delivery to tumor cells, inhibition of laminin-411 synthesis in vitro and in vivo, specific accumulation in brain tumors, and suppression of intracranial glioma growth [108].

An alternative therapeutic strategy is the photodynamic therapy (PDT) with targeted delivery systems. PDT involves the intratumoral release of photosensitizers combined with local excitation by an appropriate wavelength of light, resulting in the production of singlet oxygen and other ROS, which initiate apoptosis, and cytotoxic activity within treated neoplastic tissues. PDT is an interesting approach resulting in a very selective locoregional therapeutic approach with an important improvement in local control of tumor and a significantly improved survival [109]. Wang et al. in a recent study have demonstrated, using a molecular targeting of glioma cells through CD133 antigen overexpressed on the surface of GBM cells, a prominent photothermal selective damage of targeted glioma cells. Molecular targeting in this case has been performed using carbon nanotubes conjugated with anti-CD133 mAbs [110].

PTX, one of the most successful anticancer drugs, is the first of a class of microtubule-stabilizing agents and has demonstrable antitumor activity in glioma cell lines. However, because of the poor aqueous solubility and low therapeutic index of PTX, the clinical application is extremely limited. The efficiency of liposomal and other NP-based drug-delivery systems such as colloidal nonlipidic coated polymeric NPs, microspheres, and micelles is enhanced by targeting various molecules. A recent study demonstrated that drug-loaded MPEGylated PCL long-circulating NPs provided a sustained release of the embedded drug and higher or at least comparable in vitro cytotoxicity to that of Tx injection against C6 glioblastoma cells. The therapeutic improvement of MPEG-NP/PTX in vivo against intracranial C6 glioblastoma was obtained based on the effect of passive tumor targeting [111]. Recently, a novel sequential-targeting nanomedicine by encapsulating a drug conjugate that can target glioma cells into a BBB-penetrating micelle was developed. Tf was grafted onto the micelle and cyclo-[Arg-Gly-Asp-D-Phe-Lys] (c[RGDfK]), a peptide that specifically binds to the integrin overexpressed glioma cells, was conjugated to PTX to obtain c[RGDfK]-PTX conjugate (RP). Tf conjugation significantly enhanced the ability of a micelle to transport RP into primary BMCEs in vitro and brain in vivo, while c[RGDfK] modification improved the drug selectivity to integrin overexpressed. Retention of drug in glioma and peritumoral tissue was, also, obtained [112]. In an ongoing phase I clinical trial, PTX albumin-stabilized NP formulation is being used in treating advanced cancers such as bladder cancer and brain and CNS tumors. The authors demonstrated in a subset of patients a decrease in tumor vascular permeability [113]. In a novel study, in vivo murine U87MG GBM model, liposomal topotecan increased survival more than 20-fold [114].

Poly(ethylene glycol)-co-poly(ε-caprolactone) (PEGePCL) NPs were conjugated to Angiopep for enhanced delivery across the BBB as well as for targeting the tumor via LRP-mediated endocytosis. Angiopep-conjugated PEGePCL NPs were internalized by U87 MG glioma cells and displayed higher cell uptake and stronger inhibition and apoptosis toward glioma cells due to LRP-mediated endocytosis. Besides, angiopep-conjugated PEGePCL NPs constructed increase the transport of the NPs across the BBB and target the brain glioma by in vitro co-culture model and in vivo imaging of brain fluorescence [115]. Majoros et al. evaluated a multifunctional dendrimer conjugated with fluorescein isothiocyanate (for imaging studies), folic acid (for targeting cancer cells overexpressing folate receptors), and PTX. The authors demonstrated in vitro that drug-free dendrimer conjugates were not cytotoxic and that drug-loaded dendrimer conjugates had no effect on folate receptor–negative cells [81].

Cyclooxygenase (COX)-2 is the key enzyme in arachidonic acid metabolism resulting in prostaglandin production and is induced by several factors, such as growth factors, cytokines, and tumor promoters. In particular, COX-2 expression and prostaglandin production are associated with tumorigenesis and tumor progression. Celecoxib, a selective COX-2 inhibitor, has been reported to mediate growth-inhibitory effects and to induce apoptosis in various cancer cell lines. PLGA NPs incorporating celecoxib showed the same cytotoxicity against U87MG tumor cells as celecoxib itself. Furthermore, celecoxib did not affect the degree of migration of U87MG cells. When C6 rat glioma cells were used, PLGA NPs incorporating celecoxib showed dose-dependent cytotoxicity similar to that of celecoxib itself. Neither celecoxib nor PLGA NPs incorporating celecoxib affected COX-2 expression in C6 cells on Western blot assay [116]. Curcumin is a polyphenolic compound derived from the Indian spice turmeric. NanoCurc™, a recently described polymeric NP formulation of curcumin, was used to treat medulloblastoma and GBM cells. This formulation caused a dose-dependent decrease in growth of multiple brain tumor cell cultures, including the embryonal tumor–derived lines DAOY and D283Med and the glioblastoma neurosphere lines HSR-GBM1 and JHH -GBM14. The reductions in viable cell mass observed were associated with a combination of G2/M arrest and apoptotic induction. Curcumin also significantly decreased anchorage-independent clonogenic growth and reduced the CD133-positive stemlike population. These data suggest that curcumin NPs can inhibit malignant brain tumor growth through the modulation of cell proliferation, survival, and stem cell phenotype [117]. Recently, Tian et al. showed the feasibility of encapsulating the alkylating agent, temozolomide (3,4-dihydro-3-methyl-4-oxoimidazo [5,1-d]-as-tetrazine-8-carboxamide [TMZ]), into PBCA NPs by polymerization. Compared with TMZ solution, TMZ-PBCA NPs exhibited sustained release in vitro. Furthermore, based on the pattern of distribution in body organs, higher concentrations of TMZ can be detected in the brain after binding to PBCA NPs coated with polysorbate 80, which may be more useful for treating brain tumors. The prepared formulation may also reduce the toxicity of chemotherapy [118]. Steiniger et al. demonstrated in a murine glioblastoma model a statistically significant increase in survival time of glioblastoma-bearing rats treated with doxorubicin bound to polysorbate-coated NPs compared to the control groups treated with other doxorubicin formulations. More than 20% of the animals showed a long-term remission and no drug–NP complex neurotoxicity was observed [119].

An interesting approach consists of coating an NP with polysorbate 80, which adsorbs apolipoproteins B and E and allows receptor-mediated endocytosis by brain capillary endothelial cells. In this study, 40% of the rats treated with doxorubicin-loaded NPs survived for the duration of the study (6 months) with no evidence of residual tumor. Similarly, PEGylated doxorubicin-loaded solid lipid NPs can enhance delivery across the BBB after intravenous administration in rabbits [120]. Doxorubicin was present in the brain only after administration of the NP formulation, and the extent of doxorubicin transport was dependent on the extent of PEG modification. Recently, a promising chemotherapeutic drug (SN-38) incorporated in micelles was compared with CPT-11, a prodrug of SN-38, for the GBM treatment in mice. The growth-inhibitory effects of the drug-loaded micelles were 34- to 444-fold more potent than those of CPT-11. In addition, when the drug was incorporated in the nanovectors, a significantly potent antitumor activity against an orthotopic GBM xenograft and significantly longer survival rate than CPT-11 were observed [121].

A new strategy to achieve selective drug delivery to tumor tissue is magnetic targeting. This approach has the advantage of enhancing the attraction of drug-loaded magnetic NPs in cancer cells by using an externally applied magnetic field [122]. The nanoconjugate Polycefin (based on polymalic acid) has been studied in animal models of human glioma, using intracranial injections of human cancer cells. Antiangiogenic results have been obtained in rats by injection, in vivo, of human glioma U87MG xenografts [123]. A novel study aimed to examine the applicability of PEI-modified magnetic NPs (GPEI) as a potential vascular drug/gene carrier to brain tumors. The obtained data show that cationic magnetic NP GPEI exhibits high cell penetration ability and low cell toxicity. In addition, GPEI could be magnetically captured in glioma lesions following clinically viable intracarotid administration. Furthermore, the extent of GPEI accumulation was 5.2-fold higher than that of G100 in the tumor lesions, but not in the contralateral normal brain, revealing higher target selectivity of cationic NPs [124]. Recently, Etame et al. have demonstrated, in a rat model, using MRgFUS, a focal enhanced delivery of AuNPs with therapeutic potential into the cerebral hemisphere [125].

The use of stem cells, as cellular carriers, is a promising therapeutic strategy to deliver specific drug-loaded NPs. Some studies indicate that MSCs promote tumor development either by providing a niche for cancer stem cells, through impairing immune surveillance, or by differentiation into cancer-associated fibroblast-like cells. The applications of NP systems with stem cells include long-term labeling of stem cells with superparamagnetic iron oxide NPs, fluorochrome-loaded NPs, or QDs to monitor their fate and regenerative potential [126].

13.6 Conclusions

The brain cancer treatment represents a hard challenge in neurosurgery and oncology. Malignant gliomas involve, in their progression, multiple aberrant signaling pathways, and the BBB restricts the delivery of many chemotherapeutic agents. Targeted therapies have successfully been applied in cancers, but their efficacy remains low in malignant brain tumors. There are several factors underlying the disappointing results in brain cancer therapeutics including limited tumor cell drug uptake, intracellular drug metabolism, inherent tumor sensitivity to chemotherapy, and cellular mechanisms of resistance. An optimal realization of a system that overcomes the problems associated with novel strategies in brain tumor treatments requires the identification of specific neoplastic markers, the development of technology for the biomarker-targeted delivery of therapeutic agents, and the simultaneous capability of avoiding biological and biophysical barriers. Nanotechnology provides a unique opportunity to combat cancer on the molecular scale through careful engineering of nanomedicines to specifically interact with cancer cells and inhibit cancer cell function. It is also possible to take into neoplastic tissue novel selective contrast enhancement molecules to visualize brain tumor and to study in vivo all of its characteristics, such as cellular proliferation, angiogenesis, necrosis, tumor-safe tissue interface, and edema. NP-based delivery systems could increase the overcoming of BBB by the drug with a targeted-cell specificity modality. This approach permits the use of a lower dose of drug and a selective drug delivery to target tumor cells, both into the central core of tumor and into the distal foci of tumor cells within areas often characterized from integrity of BBB [6]. This aspect is very important in early diagnosis, in recurrences, in preoperative histological and grade diagnosis, and in preoperative treatment planning. The ability to monitor biodistribution of treatments, migration of cells throughout the body, and tumor development and evolution in real time can elucidate new pathways cancer cells rely on. Despite the significant advancements that have been made, nanotechnology is still a relatively young field, and little is known about the long-term effects of exposure to nanomaterials, especially in clearance organs such as the liver, spleen, and kidneys. Furthermore, the potential toxicity associated with the wide variety of nanomaterials available ranges from completely inert to highly toxic, which could slow their advancement into the clinic. Moreover, the efficiency of targeting NPs to the tumor is not very high and the targeting is always not perfect. The object of debate is the results about the long-term effects of interactions between NPs, coating molecules, and target cells. In order to develop new and specific materials for definite treatment, other factors need to be selected in order to design better-targeted NPs. These factors include the particle size, shape, sedimentation, drug encapsulation efficacy, desired drug-release profiles, distribution in the body, circulation, and cost. A major account for the slow development of effective targeted NPs has been due to the lack of knowledge about the distribution and location of targeted NPs after either oral administration or injection. These important steps in the nanoplatform-based drug delivery should be investigated to improve the knowledge about systemic ways of administration and their advantages and limits, acute and chronic, and local and systemic toxic effects. Most studies have not examined the targeting efficiency of NPs real time in vivo; thus, precise biodistribution and subsequently therapeutic effects are not well known. Therefore, detecting malignant cells in the body and monitoring treatment effects on these cells in real time is another challenge needed to be overcome to develop efficient-targeted NPs. This will rely on the development of better characterization tools and methodologies and more reproducible synthesis strategies so that accurate and broadly applicable conclusions can be drawn. Basic knowledge of cell biology, tumor biology, immunology, and cancer biology is necessary to the rational design of NPs for brain tumor therapy.

References

1. Wen, P.Y. and Kesari, S. 2008. Malignant gliomas in adults. *N. Engl. J. Med.* 359:492–507.
2. Grossman, S.A., Ye, X., Piantadosi, S. et al. 2010. Survival of patients with newly diagnosed glioblastoma treated with radiation and temozolomide in research studies in the United States. *Clin. Cancer Res.* 16:2443–2449.
3. Stupp, R., Hegi, M.E., van den Bent, M.J. et al. 2006. Changing paradigms—An update on the multidisciplinary management of malignant glioma. *Oncologist* 11:165–180.

4. Burger, P.C., Dubois, P.J., Schold Jr., S.C. et al. 1983. Computerized tomographic and pathological studies of the untreated, quiescent, and recurrent glioblastoma multiforme. *J. Neurosurg.* 58:159–169.
5. Ruben, J.D., Dally, M., Bailey, M., Smith, R., Mclean, C.A., and Fedele, P. 2006. Cerebral radiation necrosis: Incidence, outcomes, and risk factors with emphasis on radiation parameters and chemotherapy. *Int. J. Radiat. Oncol. Biol. Phys.* 65:499–508.
6. Caruso, G., Caffo, M., Alafaci, C. et al. 2011. Could nanoparticles systems have a role in the treatment of cerebral gliomas? *Nanomedicine* 7:744–752.
7. Uhm, J.H., Dooley, N.P., Villemure, J.-G. and Yong, V.W. 1997. Mechanisms of glioma invasion: Role of matrix metalloproteinases. *Can. J. Neurol. Sci.* 24:3–15.
8. Gladson, C.L. 1996. Expression of integrin alpha v beta 3 in small blood vessels of glioblastoma tumors. *J. Neuropathol. Exp. Neurol.* 55:1143–1149.
9. Caffo, M., Germanò, A., Caruso, G. et al. 2004. An immunohistochemical study of extracellular matrix proteins laminin, fibronectin and type IV collagen in paediatric glioblastoma multiforme. *Acta Neurochir.* 146:1113–1118.
10. Caffo, M., Caruso, G., Barresi, V. et al. 2012. Immunohistochemical study of CD68 and CR3/43 in astrocytic gliomas. *J. Anal. Oncol.* 1:42–49.
11. Cardoso, F.L., Brites, D., and Brito, M.A. 2010. Looking at the blood–brain barrier: Molecular anatomy and possible investigation approaches. *Brain Res. Rev.* 64:328–363.
12. Hawkins, B.T. and Davis, T.P. 2005. The blood-brain barrier/neurovascular unit in health and disease. *Pharmacol. Rev.* 57:173–185.
13. da Cruz, M.T., Simoes, S., and de Lima, M.C. 2004. Improving lipoplex-mediated gene transfer into C6 glioma cells and primary neurons. *Exp. Neurol* 187:65–75.
14. Ulbrich, K., Hekmatara, T., Herbert, E., and Kreuter, J. 2009. Transferrin- and transferrin-receptor-antibody modified nanoparticles enable drug delivery across the blood-brain barrier (BBB). *Eur. J. Pharm. Biopharm.* 71:251–256.
15. Zhang, Y. and Pardridge, W.M. 2001. Conjugation of brain-derived neurotrophic factor to a blood-brain barrier drug targeting system enables neuroprotection in regional brain ischemia following intravenous injection of the neurotrophin. *Brain Res.* 889:49–56.
16. Aktas, Y., Yemisci, M., Andrieux, K. et al. 2005. Development and brain delivery of chitosan-PEG nanoparticles functionalized with the monoclonal antibody OX26. *Bioconjug. Chem.* 16:1503–1511.
17. Boado, R.J., Zhang, Y., and Pardridge, W.M. 2007. Humanization of anti-human insulin receptor antibody for drug targeting across the human blood–brain barrier. *Biotechnol. Bioeng.* 96:381–391.
18. Tosi, G., Vergoni, A.V., Ruozi, B.Z. et al. 2010. Sialic acid and glycopeptides conjugated PLGA nanoparticles for central nervous system targeting: In vivo pharmacological evidence and bio-distribution. *J. Control. Release* 145:49–57.
19. Rapoport, N., Marin, A.P., and Timoshin, A.A. 2000. Effect of a polymeric surfactant on electron transport in HL-60 cells. *Arch. Biochem. Biophys.* 384:100–108.
20. Santra, S., Yang, H., Holloway, P.H., Stanley, J.T., and Mericle, R.A. 2005. Synthesis of water dispersible fluorescent, radio-opaque, and paramagnetic CdS:Mn/ZnS quantum dots: A multi-functional probe for bioimaging. *J. Am. Chem. Soc.* 127:1656–1657.
21. Liu, L., Guo, K., Lu, J. et al. 2008a. Biologically active core/shell nanoparticles self-assembled from cholesterol-terminated PEG-TAT for drug delivery across the blood-brain barrier. *Biomaterials* 29:1509–1517.
22. Qin, Y., Chen, H., Yuan, W. et al. 2011. Liposome formulated with TAT-modified cholesterol for enhancing the brain delivery. *Int. J. Pharm.* 420:304–312.
23. Adenot, M., Merida, P., and Lahana, R. 2007. Applications of a blood-brain barrier technology platform to predict CNS penetration of various chemotherapeutic agents. 2. Cationic peptide vectors for brain delivery. *Chemotherapy* 53:73–76.
24. Caruso, G., Caffo, M., Raudino, G., Alafaci, C., Salpietro, F.M., and Tomasello, F. 2010a. Antisense oligonucleotides as an innovative therapeutic strategy in the treatment of high-grade gliomas. *Rec. Pat. CNS Drug Discov.* 5:53–69.

25. Moghimi, S.M. 2006. Recent developments in polymeric nanoparticle engineering and their applications in experimental and clinical oncology. *Anticancer Agents Med. Chem.* 6:553–561.
26. Kim, K., Lee, M., Park, H. et al. 2006. Cell-permeable and biocompatible polymeric nanoparticles for apoptosis imaging. *J. Am. Chem. Soc.* 128:3490–3491.
27. Nasongkla, N., Bey, E., Ren, J. et al. 2006. Multifunctional polymeric micelles as cancer-targeted, MRI-ultrasensitive drug delivery systems. *Nano Lett.* 6:2427–2430.
28. Dubin, C.H. 2004. Special delivery: Pharmaceutical companies aim to target their drugs with nano precision. *Mech. Eng. Nanotechnol.* 126:10–12.
29. Courrier, H.M., Butz, N., and Vandamme, T.F. 2002. Pulmonary drug delivery systems: Recent developments and prospects. *Crit. Rev. Ther. Drug Carrier Syst.* 19:425–498.
30. Hussain, M., Jaitley, V., and Florence, A.T. 2001. Recent advances in the understanding of uptake of microparticulates across the gastrointestinal lymphatics. *Adv. Drug Delivery Rev.* 50:107–142.
31. Kohli, A.K. and Alpar, H.O. 2004. Potential use of nanoparticles for transcutaneous vaccine delivery: Effect of particle size and charge. *Int. J. Pharm.* 275:13–17.
32. Heidel, J.D. and Davis, M.E. 2011. Clinical developments in nanotechnology for cancer therapy. *Pharm. Res.* 28:187–99.
33. Singh, R. and Lillard Jr., J.W. 2009. Nanoparticle-based targeted drug delivery. *Exp. Mol. Pathol.* 86:215–223.
34. Couvreur, P., Barratt, G., Fattal, E., Legrand, P., and Vauthier, C. 2002. Nanocapsule technology: A review. *Crit. Rev. Ther. Drug Carrier Syst.* 19:99–134.
35. Panyam, J., Williams, D., Dash, A., Leslie-Pelecky, D., and Labhasetwar, V. 2004. Solid-state solubility influences encapsulation and release of hydrophobic drugs from PLGA/PLA nanoparticles. *J. Pharm. Sci.* 93:1804–1814.
36. Calvo, P., Remunan-Lopez, C., Vila-Jato, J.L., and Alonso, M.J. Chitosan and chitosan/ethylene oxide-propylene oxide block copolymer nanoparticles as novel carriers for proteins and vaccines. *Pharm. Res.* 14:1431–1436.
37. Musacchio, T. and Torchilin, V.P. Recent developments in lipid-based pharmaceutical nanocarriers. *Front. Biosci.* 16:1388–1412.
38. Stachelek, S.J., Finley, M.J., Alferiev, I.S. et al. 2011. The effect of CD47 modified polymer surfaces on inflammatory cell attachment and activation. *Biomaterials* 32:4317–4326.
39. Kim, J.H., Kim, Y.S., Park, K. et al. 2008. Antitumor efficacy of cisplatin-loaded glycol chitosan nanoparticles in tumor-bearing mice *J. Control. Release.* 127:41–49.
40. Chytil, P., Etrych, T., Konak, C. et al. 2008. New HPMA copolymer-based drug carriers with covalently bound hydrophobic substituents for solid tumour targeting. *J. Control. Release* 127:121–130.
41. Choi, C.H., Alabi, C.A., Webster, P., and Davis, M.E. 2010. Mechanism of active targeting in solid tumors with transferrin-containing gold nanoparticles. *Proc. Natl. Acad. Sci. USA* 107:1235–1240.
42. Sahoo, S.K. and Labhasetwar, V. 2005. Enhanced antiproliferative activity of transferrin-conjugated paclitaxel-loaded nanoparticles is mediated via sustained intracellular drug retention. *Mol. Pharm.* 2:373–383.
43. Oishi, M., Kataoka, K., and Nagasaki, Y. 2006. pH-responsive three-layered PEGylated polyplex micelle based on a lactosylated ABC triblock copolymer as a targetable and endosome-disruptive nonviral gene vector. *Bioconjug. Chem.* 17:677–688.
44. Trollet, C., Scherman, D., and Bigey, P. 2008. Delivery of DNA into muscle for treating systemic diseases: Advantages and challenges. *Methods Mol. Biol.* 423:199–214.
45. O'Brien, J.A., Holt, M., Whiteside, G., Lummis, S.C., and Hastings, M.H. 2001. Modifications to the hand-held Gene Gun: Improvements for in vitro biolistic transfection of organotypic neuronal tissue. *J. Neurosci. Methods* 112:57–64.
46. Guo, S., Huang, Y., Zhang, W. et al. 2011. Ternary complexes of amphiphilic polycaprolactone-graft-poly(N,N-dimethylaminoethylmethacrylate), DNA and polyglutamic acid-graft-poly(ethylene glycol) for gene delivery. *Biomaterials* 32:4283–4292.
47. Lee, S.H., Choi, S.H., Kim, S.H., and Park, T.G. 2008. Thermally sensitive cationic polymer nanocapsules for specific cytosolic delivery and efficient gene silencing of siRNA: Swelling induced physical disruption of endosome by cold shock. *J. Control. Release* 125:25–32.

48. Weber, N., Ortega, P., Clemente, M.I. et al. 2008. Characterization of carbosilane dendrimers as effective carriers of siRNA to HIV-infected lymphocytes. *J. Control. Release* 132:55–64.
49. Chono, S., Li, S.D., Conwell, C.C., and Huang, L. 2008. An efficient and low immunostimulatory nanoparticle formulation for systemic siRNA delivery to the tumor. *J. Control. Release* 131:64–69.
50. Powers, K.W., Palazuelos, M., Moudgil, B.M., and Roberts, S.M. 2007. Characterization of the size, shape, and state of dispersion of nanoparticles for toxicological studies. *Nanotoxicology* 1:42–51.
51. Olivier, J.C., Fenart, L., Chauvet, R., Pariat, C., Cecchelli, R., and Couet, W. 1999. Indirect evidence that drug brain targeting using polysorbate 80-coated polybutylcyanoacrylate nanoparticles is related to toxicity. *Pharm. Res.* 16:1836–1842.
52. Hardman, R. 2006. A toxicologic review of quantum dots: Toxicity depends on physicochemical and environmental factors. *Environ. Health Perspect.* 114:165–172.
53. Fonseca, C., Simoes, S., and Gaspar, R. 2002. Paclitaxel-loaded PLGA nanoparticles: Preparation, physiochemical characterization and in vitro anti-tumoral activity. *J. Control. Release* 83:273–286.
54. Koo, Y.L., Reddy, G.R., Bhojani, M. et al. 2006. Brain cancer diagnosis and therapy with nanoplatforms. *Adv. Drug Deliv. Rev.* 58:1556–1557.
55. Anwunobi, A.P. and Emeje, M.O. 2011. Recent application of natural polymers in nanodrug delivery. *J. Nanomed. Nanotechnol.* S4:002.
56. Mahkam, M. 2010. Modified chitosan cross-linked starch polymers for oral insulin delivery. *J. Bioact. Compat. Polym.* 25:406–418.
57. Cole, A.J., David, A.E., Galbian, C.J., and Yang, V.C. 2011. Magnetic brain tumor targeting and biodistribution of long-circulating PEG-modified, cross-linked starch-coated iron oxide nanoparticles. *Biomaterials* 32:6291–6301.
58. Dev, A., Binulal, N.S., Anitha, A. et al. 2010. Preparation of poly (lactic acid)/chitosan nanoparticles for anti-HIV drug delivery applications. *Carbohydr. Polym* 80:833–838.
59. Menon, D., Thomas, R.T., Narayanan, S. et al. 2011. A novel chitosan/polyoxometalate nanocomplex for anti-cancer applications. *Carbohydr. Polym* 84:887–893.
60. Pinhassi, R.I., Assaraf, Y.G., Farber, S. et al. 2010. Arabinogalactan-folic acid-drug conjugate for targeted delivery and target-activated release of anticancer drugs to folate receptor-overexpressing cells. *Biomacromolecules* 11:294–303.
61. Rice, J.R., Gerberich, J.L., Nowotnik, D.P., and Howell, S.B. 2006. Preclinical efficacy and pharmacokinetics of AP5346, a novel diaminocyclohexane-platinum tumor targeting drug delivery system. *Clin. Cancer. Res.* 1:2248–2254.
62. Savic, R., Eisenberg, A., and Maysinger, D. 2006. Block copolymer micelles as delivery vehicles of hydrophobic drugs: micelle-cell interactions. *J. Drug Target* 14:343–355.
63. Shen, Y. and Jiasheng, T. 2009. Synthesis and characterization of low molecular weight hyaluronic acid-based cationic micelles for efficient siRNA delivery. *Carbohydr. Polym.* 77:95–104.
64. Liu, L., Venkatraman, S.S., Yang, Y.Y. et al. 2008b. Polymeric micelles anchored with TAT for delivery of antibiotics across the blood–brain barrier. *Biopolymers* 90:617–623.
65. Kenmotsu, H., Yasunaga, M., Goto, K. et al. 2010. The antitumor activity of NK012, an SN-38-incorporating micelle, in combination with bevacizumab against lung cancer xenografts. *Cancer* 116:4597–4604.
66. Torchillin, V.P., Iakaubov, L.Z., and Estrov, Z. 2003. Therapeutic potential of antinuclear autoantibodies in cancer. *Cancer Ther.* 1:179–190.
67. Zeng, F., Lee, H., and Allen, C. 2006. Epidermal growth factor conjugated poly(ethylene glycol)-block-poly(δ-valerolactone) copolymer micelles for targeted delivery of chemotherapeutics. *Bioconjug. Chem.* 17:399–409.
68. Khong, T. and Spencer, A. 2011. Targeting heat shock protein 90 induces apoptosis and inhibits critical survival and proliferation pathways in multiple myeloma. *Mol. Cancer Ther.* 10:1909–1917.
69. Richardson, P.G., Ghana-Khan, A.A., Alsina, M. et al. 2010. Tanespimycin monotherapy in relapsed multiple myeloma: Results of a phase 1 dose-escalation study. *Br. J. Haematol.* 150:438–445.

70. Olivier, J.C., Huertas, R., Lee, H.J., Calon, F., and Pardridge, W.M. 2002. Synthesis of pegylated immunonanoparticles. *Pharm. Res.* 19:1137–1143.
71. Feng, B., Tomizawa, K., Michiue, H. et al. 2009. Delivery of sodium borocaptate to glioma cells using immunoliposome conjugated with anti-EGFR antibodies by ZZ-His. *Biomaterials* 30:1746–1755.
72. Kizelsztein, P., Ovadia, H., Garbuzenko, O., Sigal, A., and Barenholz, Y. 2009. Pegylated nano-liposomes remote-loaded with the antioxidant tempamine ameliorate experimental autoimmune encephalomyelitis. *J. Neuroimmunol.* 213:20–25.
73. Grahn, A.Y., Bankiewicz, K.S., Dugich-Djordjevic, M. et al. 2009. Non-PEGylated liposomes for convection-enhanced delivery of topotecan and gadodiamide in malignant glioma: Initial experience. *J. Neurooncol.* 95:185–197.
74. Saad, M., Garbuzenko, O.B., and Minko, T. 2008. Co-delivery of siRNA and an anticancer drug for treatment of multidrug-resistant cancer. *Nanomedicine* 3:761–776.
75. Dritschilo, A., Huang, C.H., Rudin, C.M. et al. 2006. Phase I study of liposome-encapsulated c-raf antisense oligodeoxyribonucleotide infusion in combination with radiation therapy in patients with advanced malignancies. *Clin. Cancer Res.* 12:1251–1259.
76. Abes, S., Ivanova, G.D., Abes, R. et al. 2009. Peptide-based delivery of steric-block PNA oligo-nucleotides. *Methods Mol. Biol.* 480:85–99.
77. Chanda, N., Kattamuri, V., Shukla, V. et al. 2010. Bombesin functionalized gold nanoparticles show in vitro and in vivo cancer receptor specificity. *Proc. Natl. Acad. Sci. USA* 107:8760–8765.
78. Chen, X., Wang, X., Wang, Y. et al. 2010. Improved tumor-targeting drug delivery and therapeutic efficacy by cationic liposome modified with truncated bFGF peptide. *J. Control. Release* 145:17–25.
79. Desai, A., Vyas, T., and Amiji, M. 2008. Cytotoxicity and apoptosis enhancement in brain tumor cells upon coadministration of paclitaxel and ceramide in nanoemulsion formulations. *J. Pharm. Sci.* 97:2745–2756.
80. Pan, H., Soman, N.R., Schlesinger, P.H., Lanza, G.M., and Wickline, S.A. 2009. Cytolytic peptides nanoparticles ('NanoBees') for cancer therapy. *Wires Nanomed. Nanobiotechnol.* 3:318–327.
81. Majoros, I.J., Myc, A., Thomas, T., Mehta, C.B., and Baker, Jr. J.R. 2006. PAMAM dendrimer-based multifunctional conjugate for cancer therapy: Synthesis, characterization, and functionality. *Biomacromolecules* 7:572–579.
82. Desai, P.N., Yuan, Q., and Yang, H. 2010. Synthesis and characterization of photocurable polyamidoamine dendrimer hydrogels as a versatile platform for tissue engineering and drug delivery. *Biomacromolecules* 11:666–673.
83. Agrawal, A., Min, D.H., Singh, N. et al. 2009. Functional delivery of siRNA in mice using dendriworms. *ACS Nano* 3:2495–2504.
84. Yang, W., Barth, R.F., Wu, G. et al. 2011. Convection enhanced delivery of boronated EGF as a molecular targeting agent for neutron capture therapy of brain tumors. *J. Neurooncol.* 95:355–365.
85. Zhang, L., Zhu, S., Qian, L., Pei, Y., Qiu, Y., and Jiang, Y. 2011. RGD-modified PEG-PANAM-DOX conjugates: In vitro and in vivo studies for glioma. *Eur. J. Pharm. Biopharm.* 79:232–240.
86. Klajnert, B. and Bryszewska, M. 2001. Dendrimers: properties and applications. *Acta Biochim. Pol.* 48:199–208.
87. Bakalgot, V., Zhang, L., Levy-Nissembaum, E. et al. 2007. Quantum dot-aptamer conjugates for synchronous cancer imaging, therapy and sensing of drug delivery based on bi-fluorescence resonance energy transfer. *Nano Lett.* 7:3065–3070.
88. Arndt-Jovin, D.J., Kantelhardt, S.R., Caarls, W., de Vries, A.H., Giese, A., and Jovin Ast, T.M. 2009. Tumor-targeted quantum dots can help surgeons find tumor boundaries. *IEEE Trans. Nanobiosci.* 8:65–71.
89. Dudu, V., Rotari, V., Eng, B., and Vazquez, M. 2011. Targeted extracellular nanoparticles enable intracellular detection of activated epidermal growth factor receptor in living brain cancer cells. *Nanomedicine* 7:896–903.

90. Veiseh, O., Sun, C., Fang, C. et al. 2009. Specific targeting of brain tumors with an optical/magnetic resonance imaging nanoprobe across the blood-brain barrier. *Cancer Res.* 69:6200–6207.

91. Jiang, W., Xie, H., Ghoorah, D. et al. 2012. Conjugation of functionalized SPIONs with transferrin for targeting and imaging brain glial tumors in rat model. *PloS ONE* 7:e37376.

92. Jain, K.K. 2011. Role of nanobiotechnology in the personalized management of glioblastoma multiforme. *Nanomedicine* 6:411–414.

93. Caruso, G., Caffo, M., Raudino, G., Alafaci, C., and Tomasello, F. 2010b. New therapeutic strategies in gliomas treatment. In *Brain Tumors—Current and Emerging Therapeutic Strategies*, ed. A.L. Abujamra, pp. 281–306. Rijeka, Croatia: InTech.

94. Madhankumar, A.B., Slage-Webb, B., Mintz, A., Sheehan, J.M., and Connor, J.R. 2006. Interleukin-13 receptor-targeted nanovesicles are a potential therapy for glioblastoma multiforme. *Mol. Cancer Ther.* 5:3162–3169.

95. Kievit, F.M., Veiseh, O., Fang, C. et al. 2010. Chlorotoxin labeled magnetic nanovectors for targeted gene delivery to glioma. *ACS Nano* 4:4587–4594.

96. Zhang, Z., Yang, X., Zhang, Y. et al. 2006. Delivery of telomerase reverse transcriptase small interfering RNA in complex with positively charged single walled carbon nanotubes suppresses tumor growth. *Clin. Cancer Res.* 12:4933–4939.

97. Jung, J., Solanki, A., Memoli, K.A. et al. 2010. Selective inhibition of human brain tumor cell proliferation via multifunctional quantum dot-based siRNA delivery. *Angew. Chem. Int. Ed. Engl.* 49:103–107.

98. Im, S.A., Gomez-Manzano, C., Fueyo, J. et al. 1999. Antiangiogenesis treatment for gliomas: Transfer of antisense-vascular endothelial growth factor inhibits tumor growth in vivo. *Cancer Res.* 59:895–900.

99. Agemy, L., Friedmann-Morvinski, D., Kotamraju, V.R. et al. 2011. Targeted nanoparticle enhanced proapoptotic peptide as potential therapy for glioblastoma. *PNAS* 42:17450–17455.

100. Lu, W., Sun, Q., Wan, J., She, Z., and Jiang, X. 2006. Cationic albumin-conjugated pegylated Nanoparticles allow gene delivery into brain tumors via intravenous administration. *Cancer Res.* 66:11878–11886.

101. Schneider, T., Becker, A., Ringe, K. et al. 2008. Brain tumor therapy by combined vaccination and antisense oligonucleotide delivery with nanoparticles. *J. Neuroimmunol.* 195:21–27.

102. Wu, Z.M., Yuan, X.H., Jiang, P.C., Li, Z.Q., and Wu, T. 2006. Antisense oligonucleodes targeting the focal adhesion kinase inhibit proliferation, induce apoptosis and cooperate with cytotoxic drugs in human glioma cells. *J. Neuro-Oncol.* 77:117–123.

103. Zhao, D., Alizadeh, D., Zhang, L. et al. 2011. Carbon nanotubes enhance CpG uptake and potentiate antiglioma immunity. *Clin. Cancer Res.* 17:771–782.

104. Alizadeh, D., Zhang, L., Hwang, J., Schluep, T., and Badie, B. 2010. Tumor-associated macrophages are predominant carriers of cyclodextrin-based nanoparticles into gliomas. *Nanomedicine* 6:382–390.

105. Bernardi, R.J., Lowery, A.R., Thompson, P.A., Blaney, S.M., and West, J.L. 2008. Immunonanoshells for targeted photothermal ablation in medulloblastoma and glioma: An in vitro evaluation using human cell lines. *J. Neurooncol.* 86:165–172.

106. Sawyer, A.J., Saucier-Sawyer, J.K., Booth, C.J. et al. 2011. Convection-enhanced delivery of camptothecin-loaded polymer nanoparticles for treatment of intracranial tumors. *Drug Deliv. Transl. Res.* 1:34–42.

107. Hadjipanayis, C.G., Machaidze, R., Kaluzova, M. et al. 2010. EGFRvIII antibody conjugated iron oxide nanoparticles for MRI guided convection-enhanced delivery and targeted therapy of glioblastoma. *Cancer Res.* 70:6303–6312.

108. Ding, H., Inoue, S., Ljubimov, A.V. et al. 2010. Inhibition of brain tumor growth by intravenous poly(β-L-malic acid) nanobioconjugate with pH-dependent drug release. *PNAS* 107:18143–18148.

109. Reddy, G.R., Bhojani, M.S., McConville, P. et al. 2006. Vascular targeted nanoparticles for imaging and treatment of brain tumors. *Clin. Cancer Res.* 12:6677–6686.

110. Wang, C., Huang, Y., and Peng, C. 2009. Photothermal ablation of stem-cell like glioblastoma using carbon nanotubes functionalized with anti-CD133. *ICBME Proc.* 23:888–891.

111. Xin, X., Chen, L., Gu, J. et al. 2010. Enhanced anti-glioblastoma efficacy by PTX-loaded PEGylated poly(-caprolactone) nanoparticles: In vitro and in vivo evaluation. *Int. J. Pharmacol.* 402:238–247.

112. Zhang, P., Hu, L., Yin, Q., Feng, L., and Li, Y. 2012. Transferrin-modified c[RGDfK]-paclitaxel loaded hybrid micelle for sequential blood-brain barrier penetration and glioma targeting therapy. *Mol. Pharmaceut.* 9:1590–1598.

113. Chien, A.J., Illi, J.A., Ko, A.H. et al. 2009. A phase I study of a 2-day lapatinib chemosensitization pulse preceding nanoparticle albumin-bound Paclitaxel for advanced solid malignancies. *Clin. Cancer Res.* 15:5569–5575.

114. Orringer, D.A., Koo, Y., Kopelman, R., and Sagher, O. 2009. Small solutions for big problems: The application of nanoparticles to brain tumor diagnosis and therapy. *Clin. Pharmacol. Ther.* 85:531–534.

115. Xin, H., Jiang, X., Gu, J. et al. 2011. Angiopep-conjugated poly(ethylene glycol)-co-poly (e-caprolactone) nanoparticles as dual-targeting drug delivery system for brain glioma. *Biomaterials* 32:4293–4305.

116. Kim, T., Jeong, Y., Jin, S. et al. 2011. Preparation of polylactide-co-glycolide nanoparticles incorporating celecoxib and their antitumor activity against brain tumor cells. *Int. J. Nanomed.* 6:2621–2631.

117. Lim, K.J., Bisht, S., Bar, E.E., Maitra, A., and Eberhart, C.G. 2011. A polymeric nanoparticle formulation of curcumin inhibits growth, clonogenicity and stem-like fraction in malignant brain tumors. *Cancer Biol. Ther.* 11:464–473.

118. Tian, X., Lin, X., Wei, F. et al. 2011. Enhanced brain targeting of temozolomide in polysorbate-80 coated polybutylcyanoacrylate nanoparticles. *Int. J. Nanomed.* 6:445–452.

119. Steiniger, S.C.J., Kreuter, J., Khalansky, A.S. et al. 2004. Chemotherapy of glioblastoma in rats using doxorubicin-loaded nanoparticles. *Int. J. Cancer* 109:759–767.

120. Zara, G.P., Cavalli, R., Bargoni, A., Fundarò, A., and Vighetto, D. 2002. Intravenous administration to rabbits of non-stealth and stealth doxorubicin-loaded solid lipid nanoparticles at increasing concentrations of stealth agent: Pharmacokinetics and distribution of doxorubicin in brain and other tissues. *J. Drug Target* 10:327–335.

121. Kuroda, J., Kuratsu, J., and Yasunaga, M. 2009. Potent antitumor effect of SN-38-incorporating polymeric micelle, NK012, against malignant glioma. *Int. J. Cancer* 124:2505–2511.

122. Chertok, B., David, A.E., Huang, Y., and Yang, V.C. 2007. Glioma selectivity of magnetically targeted nanoparticles: A role of abnormal tumor hydrodynamics. *J. Control. Release* 122:315–323.

123. Ljubimova, J.Y., Fujita, M., Khazenzon, N.M. et al. 2008. Nanoconjugate based on polymalic acid for tumor targeting. *Chem. Biol. Interact.* 171:195–203.

124. Chertok, B., David, A.E., and Yang, V.C. 2010. Polyethyleneimine-modified iron oxide nanoparticles for brain tumor drug delivery using magnetic targeting and intra-carotid administration. *Biomaterials* 31:6317–6324.

125. Etame, A.B., Diaz, R.J., O'Reilly, M.A. et al. 2012. Enhanced delivery of gold nanoparticles with therapeutic potential into the brain using MRI-guided focused ultrasound. *Nanomedicine* 8:1133–1142.

126. Yukawa, H., Mizufune, S., Mamori, C. et al. 2009. Quantum dots for labeling adipose tissue-derived stem cells. *Cell Transplant.* 18:591–599.

14

Inorganic Nanoparticle Materials for Controlled Drug Delivery Systems

R.M. Freire, E. Longhinotti, E.H.S. Sousa, and P.B.A. Fechine

CONTENTS

14.1 Nanotechnology: Development of Drug Delivery

The history of drug delivery began with a German scientist called Paul Ehrlich (1854–1915) [1]. The reasoning for the design of active agents should be selective and only attack the disease-causing organism. However, to reach its target, the drug could be delivered along with the agent of selectivity forming a kind of *magic bullet*, created to only kill the organism targeted [2]. This idea was used as a script for the movie *Dr. Ehrlich's Magic Bullet* in 1940. Hence, in the 1950s, the first related studies using microencapsulated drug particles appeared [3]. In the 1960s, leading research provided an initial understanding of pharmacokinetics [4] and therapeutic drug monitoring. The utilization of polymers for drug delivery began in the 1960s when scientists had developed systems combining previous understanding about pharmacokinetics, biological interface, and biocompatibility. However, only a few studies had succeeded in developing new drug delivery systems due to the slow release of large molecular weight compounds. The work of Davis [5,6] and Gimbrone et al. [7] was one of these that achieved partial success. However, the polymers used in these studies proved to be unsuitable, since they caused inflammation in animal tissue. In 1976, the first drug delivery system based on polymers for slow release of macromolecules without inflammatory issues was described [8,9]. After this, several other systems were developed, but only in 1980 novel proposals emerged reporting a pH-sensitive releasing system [10] and a cell-specific targeting strategy using liposomes [11,12]. Later in 1987, it was described long-circulating liposome providing new grounds for effective drug delivery [13]. Since the 1990s, the use of polyethylene glycol (PEG) has gained full attention and intense usage due to its ability to increase circulation time of liposome [14] and polymer nanoparticles [15].

The nanotechnology has been always present in certain way in the drug delivery systems developed such as in the process of microencapsulation of drugs and liposome preparation. The development of the nanotechnology combined with controlled-release systems has brought a series of advantages [3]:

- Improved delivery of poorly water-soluble drugs
- Targeted delivery of drugs in a cell- or tissue-specific manner
- Transcytosis of drugs across tight epithelial and endothelial barriers
- Delivery of large macromolecule drugs to intracellular sites of action
- Codelivery of two or more drugs or therapeutic modality for combination therapy
- Visualization of sites of drug delivery by combining therapeutic agents with imaging modalities
- Real-time read on the *in vivo* efficacy of a therapeutic agent

All these advantages are possible only because of the manipulation of objects at the nanometer scale (10^{-9} m), at the level of atoms, molecules, and supramolecular structures [16]. The relative small size of the polymeric nanoparticles (1–100 nm) or liposomes makes these species excellent vehicles with the possibility to deliver drugs in the action site leading to high selectivity, an essential feature in the treatment of cancer, for example [17]. New drug carriers are needed for enhancing the pharmacokinetics and biodistribution of drugs that suffer from poor solubility, stability, and unwanted toxicity [18], issues that, without solution, could lead to drug disapproval by regulatory agencies. However, various parameters such as composition, size, surface chemistry, shape, and mechanical flexibility should be observed when designing such materials [19]. These characteristics may influence their therapeutic function as well as impact the circulating half-life of the particles and their biodistribution [9]. Several types of nanotechnology platform that include polymeric particles, liposomes, dendrimer, solid lipid nanoparticle, nanotubes, nanocrystal, and inorganic nanoparticles that can be used as drug carriers have been developed [17]. Despite the variety of nanomaterials developed, their introduction into the body is still considered one of the main problems; to achieve the goal of delivering the drug at the target site, it is necessary to escape the body's protective machinery [20]. In this context, interdisciplinary research has been carried out focused on exploiting inorganic nanoparticles. This has been motivated due to the possibility of preparing efficient formulation through synthetic strategies, leading to nanomaterials with appropriate characteristics (size, shape, surface, etc.) for biomedical applications such as drug delivery.

14.2 Drug Delivery and Inorganic Nanoparticles

During the last years, a number of inorganic nanoparticles have drawn considerable attention for biomedical applications, since their small size on the nanometer scale allows their application as drug carrier in cancer therapy and treatment of other ailments [2]. However, their therapeutic function is influenced by physicochemical properties such as composition, surface chemistry, shape, and mechanical flexibility modulating how it behaves into the human body [18,19]. Regarding this, inorganic nanoparticles have many advantages with respect to other materials (liposomes, dendrimers, nanotubes, etc.) due to numerous synthetic strategies successfully developed leading to nanomaterials with improved properties according to the desired application [21]. Thus, among novel inorganic materials for drug delivery, it can be listed gold nanoparticles, silica, quantum dots (QDs), and magnetic nanoparticles (MNPs) [20]. These are versatile tools with key features allowing their use for therapeutic and/or diagnostics.

14.2.1 Magnetic Nanoparticles

The use of magnetic particles for biomedical applications started a long time ago, in the seventeenth century, with the use of bar magnets for the removal of iron from the body [22]. At the end of the nineteenth century, the electromagnet solved the problem of the deep penetrating energy transfer in clinical use [23]. The use of MNPs was suggested in 1960 by Freeman et al. [22]. In their work, the authors studied the influence of iron concentration in the bloodstream based on the assumption that the MNP could be transported through the vascular system to a specific part of the body from an external magnetic field. These studies were performed with six patients that adsorb up to 120 mg of alpha iron under the tongue. No toxic effects in the blood were observed. In 1963, Meyers et al. [24] propose the use of the metallic iron particles as contrast agents for localized radiation therapy.

The use of MNP as the drug vehicle gained strength in the late 1970s with several publications reporting the synthesis of magnetic microspheres to deliver doxorubicin (anticancer drug) in a specific site applying an external magnetic field [25–27]. During the first half of the 1980s, several studies were conducted using animal models such as Holtzman rats [28,29] and new approaches involving magnetic microcapsules [30]. In the latter, the authors observed that the microcapsules synthesized could be magnetically controlled in the artery and urinary bladder of the animals investigated. In 1989, Gallo et al. [31] report the use of a physiological pharmacokinetic model to describe the disposition of doxorubicin following its administration. Despite several reports on the potential biological application of MNP at the nanometer length, it was just in the middle of the 1990s with the assessment of magnetic fluids in preclinical experiments in laboratory scale that this was described [2,32]. In 1996, Lubbe and coworkers reported the development of a magnetic fluid capable of binding epirubicin (anticancer drug) and releasing it later on via the desorption process [33]. Thus, these authors had created a type of drug–ferrofluid complex, designated magnetic drug, which was investigated under the influence of a magnetic field. Tests conducted *in vivo* revealed that the combination of epirubicin bound to a ferrofluid made possible to accumulate and deliver into the tumor a sufficient dosage to prevent tumor remission after 20 min of treatment. Thus, given the enormous potential of magnetic systems in the last decade, MNPs have been widely investigated for biomedical applications including cellular therapy and cell labeling [34], magnetohyperthermia treatment [35], and magnetic resonance imaging (MRI) [36].

Iron oxide nanoparticles (IONPs) represent a class of inorganic nanomaterial extremely important due to their unique properties such as superparamagnetism (depending on size), for example [37]. These properties avoid nanoparticle agglomeration into human body, since in the absence of the magnetic field, the magnetization disappears [2]. Thus, a possible embolization of capillary vessels is prevented making the use of IONPs sustainable for biomedical applications. The most widely studied of them is magnetite (Fe_3O_4) due to its low cytotoxicity and strong magnetic properties [38]. Furthermore, these IONPs are easily synthesized by coprecipitation [39], a chemical method that promotes high homogeneity of the material obtained and a low particle size distribution [40]. Fe_3O_4 belongs to a class of materials called ferrites, and their interesting magnetic properties can be explained by structural composition of Fe^{2+} and Fe^{3+}. Given this, it is imperative to know the crystallographic arrangement of ferrites to understand the origin of their magnetic properties.

14.2.2 Ferrites

14.2.2.1 Structure

The ferrites are ceramic materials based on iron oxide (Fe_2O_3) in the composition together with other metal oxides. Their formula can be represented by $MO \cdot Fe_2O_3$ or MFe_2O_4, where M is typically a divalent metal cation, that is, a metal with valence 2+, for example, Fe^{2+}, Ni^{2+}, Mn^{2+}, and Zn^{2+} [40]. The crystalline arrangement of ferrite is characteristic of a spinel-type structure (space group *Fd3m*). This may be represented by the formula (A) $[B]_2O_4$, where A and B represent divalent and trivalent cations, respectively. It is worth noting the nomenclature used in this formula for the spinel structure in which the presence of parentheses indicates the metal located in the tetrahedral site and brackets assigned to the metal present in the octahedral sites [41]. The spinel unit cell is composed of 32 oxygen atoms (O^{2-}) arranged in a cubic close-packed (fcc) structure forming 64 tetrahedral sites and 32 octahedral sites. However, only 24 are occupied, 16 octahedral and 8 tetrahedral, because of considerations relating to charge balancing [42]. With full occupancy of all sites, the positive charge would be greater than the negative charge and the structure would not be electrically neutral. Thus, only 24 sites in the two sublattices (tetrahedral and octahedral) are occupied to maintain neutrality. So, the metal ions present in the spinel structure can occupy tetrahedral (A) or octahedral (B) sites.

According to the distribution of cations at each site, the spinel ferrites can be classified as normal or inverse. In normal structure, divalent ions only occupy site A, while the trivalent ions only occupy site B. $ZnFe_2O_4$ is a clear example of ferrite in this configuration and their structure may be represented by $(Zn^{2+})\left[Fe_2^{3+}\right]O_4$ [43]. For an inverse structure, the opposite is true. The trivalent cations prefer to occupy site A and half of the site B. The other half of the latter site is occupied by divalent cations. It is well established that Fe_3O_4 shows an inverse spinel structure where the occupations of the cations may be designated as $(Fe^{3+})[Fe^{2+}Fe^{3+}]O_4$. Furthermore, there are ferrites that show a random distribution of the cations leading to a state intermediate between the normal and inverse structures. Thus, for these MNPs, the general formula more complete may be represented by $\left(M_{1-x}^{2+}Fe_x^{3+}\right)\left[Fe_{2-x}^{3+}M_x^{2+}\right]O_4$, where x represents the inversion parameter and values close at 0 and 1 stand for the normal and inverse cases, respectively. $MnFe_2O_4$ shows structure predominantly normal. However, precise cation distributions include the notation $\left(Mn_{0.8}^{2+}Fe_{0.2}^{3+}\right)\left[Fe_{1.8}^{3+}Mn_{0.2}^{2+}\right]O_4$ in which $x = 0.2$ [44]. In addition, mixed ferrites can also be synthesized. In these structures, certain amount of the divalent cation is replaced by another of the same valency. Generally, Zn^{2+} is chosen as a replacement because of its preference for tetrahedral sites [43]. This is known and used as an advantage to generate a large magnetic moment. Several studies in the literature report the use of Zn^{2+} as a substituent in mixed ferrites [45–48].

14.2.2.2 Magnetic Properties

The magnetic properties of the materials with spinel structure, for example, ferrites, originate from the interaction between the magnetic moments of the cations present on the two sites tetrahedral (A) and octahedral (B) of the sublattices. However, before trying to understand how magnetism occurs in nanoparticles, it is important to have an overview of the different magnetic behaviors. Magnetic materials can be classified according to their interaction with an external magnetic field. A characteristic property used for this purpose is the magnetic susceptibility (χ), which mathematically can be express by the following equation:

$$\chi = \frac{M}{H} \tag{14.1}$$

where
  M is the magnetization
  H the external field applied

The values found for χ serve as a parameter to evaluate the magnetic behavior of the material and classify it in two different groups: "weak" magnetic (diamagnetic and paramagnetic) and "strong" magnetic (ferromagnetic, antiferromagnetic, and ferrimagnetic). Table 14.1 shows the different orientations of the dipole moments and the range of values of χ for the various magnetic behaviors in each group. So, the interactions between moments on the various sites of the spinel structure occur according to a behavior reported leading to a resulting magnetic moment. However, one should consider the distances between the cations that occupy sites A and B, the oxygen ion that links them, and also the angle between the three ions to evaluate the strength of this interaction [42]. Taking Fe_3O_4 as an example, first, you must know the distribution of the cations Fe^{3+} and Fe^{2+} in sites A and B. The literature [49] reports that Fe_3O_4 has a completely inverse spinel structure with their magnetic moment exclusively due to the Fe^{2+} cations present at site B. Since the Fe^{3+} moments in sites A and B neutralize each other due to antiparallel orientation (antiferromagnetic) leaving only Fe^{2+} moments. Thus, the magnetic properties of these materials with spinel structure are derived from the superposition of moments of cations present in the two sublattices (A and B).

TABLE 14.1

Different Orientations of the Dipole Moments for Each Magnetic Behavior and Their Range of Values Found for χ

Groups	Magnetic Behavior	Dipole Orientation	χ
"Weak"	Diamagnetic	—	-10^{-6} to -10^{-3}
	Paramagnetic	← ↑ ↗ ↘ ↓	10^{-6} to 10^{-3}
"Strong"	Ferromagnetic	↑ ↑ ↑ ↑ ↑	10^{-2} to 10^{6}
	Antiferromagnetic	↑ ↓ ↑ ↓ ↑	0 to 10^{-2}
	Ferrimagnetic	↑ ↓ ↑ ↓ ↑	10^{-2} to 10^{6}

Source: Ribeiro, T.G., Síntese e Caracterização de nanopartículas magnéticas de óxidos mistos de $MnFe_2O_4$ recobetas com quitosana. Estudo da influência da dopagem com Gd^{3+} nas propriedades estruturais e magnéticas thesis, Universidade de São Paulo, São Paulo, Brazil, 2008; Varandan, V. et al., *Nanomedicine: Design and Applications of Magnetic Nanomaterials, Nanosensors and Nanosystems*, 1st edn., Wiley, Chichester, U.K., 2008.

Besides all the magnetic behavior reported earlier, there is also the superparamagnetism, an essential magnetic behavior for biomedical applications [2]. This phenomenon is closely related to the size of magnetic particle. For understanding this phenomenon, it is required to know that in a magnetic particle, their spins are organized in several regions called magnetic domains. When the particle diameter decreases to a point where it is comparable to the size of the domain, it reaches the critical diameter (d_{cr}). Upon reaching this limit, the formation of the domain wall becomes energetically unfavorable so that each particle comprises a single domain. Theoretically, this can be predicted using the following equation [52]:

$$d_{cr} = \frac{9}{4\pi} \frac{\sigma}{M_s^2}$$ (14.2)

where
σ is the domain wall energy
M_s is the saturation magnetization of the particle

Another important point to be discussed to evaluate the existence of the superparamagnetic behavior is the relaxation time (τ), that is, the time to spin out from an equilibrium state to another. This can be estimated using the following equation [53]:

$$\tau = \tau_0 e^{\Delta E/k_B T}$$ (14.3)

where
ΔE is the energy barrier to moment reversal
T is the temperature
k_B is the Boltzmann constant
τ_0 is inversely proportional to the jump attempt frequency of the particle magnetic moment between the opposite directions of the magnetization easy axis

The height of ΔE, that is, the energy required to perform the moment reversal, is proportional to the KV, where K and V represent anisotropy constant and particle volume, respectively. Thus, there is a decline in the height of ΔE toward $k_B T$ with decreasing particle size, and the moment reversal can be thermally activated preventing the existence of a stable magnetization in the absence of a magnetic field.

Experimentally, the superparamagnetism of very small particles can be observed through the hysteresis curve and ^{57}Fe Mössbauer spectroscopy. A parameter of great importance to be discussed is the measured time (τ_m) for these two techniques, since it might influence the experimental results. In typical experiments, τ_m for magnetization measurements may vary in the range of 10^2–10^{-5} s, while for ^{57}Fe Mössbauer spectroscopy, τ_m has values that can be found from 10^{-7} to 10^{-9} s [51]. If $\tau \ll \tau_m$, the relaxation proceeds very fast relatively to the experimental time window of the technique, and particles can be considered as in the superparamagnetic regime. However, if $\tau \gg \tau_m$, the relaxation time is very slow and the particles are in a state so-called blocked regime, once quasi-static properties are observed for the material studied [53]. Figure 14.1 shows a typical graph of MNP in superparamagnetic regime investigated by the hysteresis curve and ^{57}Fe Mössbauer spectroscopy. In these two techniques, the absence of the area within the hysteresis loop (Figure 14.1a) and the doublet in the center of the spectrum (Figure 14.1b) are experimental evidences of the MNP as in the superparamagnetic regime [48,54].

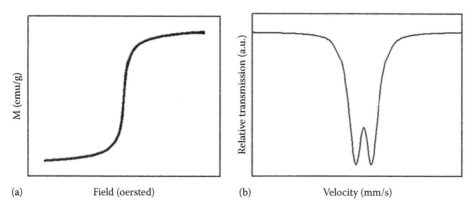

(a)　　　　Field (oersted)　　　　(b)　　　　Velocity (mm/s)

FIGURE 14.1
Typical graph of MNP that exhibits the phenomenon of relaxation superparamagnetic investigated by: (a) hysteresis curve; (b) ^{57}Fe Mössbauer spectroscopy.

14.2.3 Magnetic Nanoparticles in Drug Delivery: Properties and Applications

Before making any comment about the use of MNP in therapeutic applications such as drug delivery, it is important to clarify a serious problem: the biocompatibility. Some nanoparticles do not exhibit this characteristic, an essential parameter for biomedical applications [55]. Then, initially these materials could not be used as drug carriers in the human body. However, MNP exhibits excellent surface reactivity enabling conjugation to various ligands that can be used to anchor biocompatible polymers exploiting electrostatic, hydrophobic, chelating, and covalent interactions [56]. The most common functional groups used for surface modification are alkoxysilane compounds and carboxylic acid groups [37]. This last one is often used together with MNP due to the ease and fast process of surface modification [57–59]. Figure 14.2 shows several examples of the functional groups that can be used to anchor biocompatible polymers on the surface of MNP.

The choice of the surface coating is an important aspect to be evaluated, since this process can provide chemical stabilization under physiological conditions [55]. In addition, the surface chemistry affects a variety of factors including circulation time in bloodstream, target site accumulation, and endocytosis [18]. The literature reports the development of a number of magnetic systems coated with PEG [60,61], dextran [62], chitosan [63], poly (d,l-lactide-co-glycolide) (PLGA) [64], and Pluronics P123 [65] and F127 [66], among others. Furthermore, the advances in nanotechnology and the increasing amount of studies in this area have led to the development of a variety of new coating methods such as lipid film hydration (LFH) [67], micro- [68] and miniemulsion [69], and layer by layer (LbL) [70], among others. Using the LFH method, Nappini et al. [71] synthesized magnetoliposomes (MLs) containing cobalt ferrite (CoFe$_2$O$_4$) as magnetic core and carboxyfluorescein (CF), an important dye with the ability to label a population of lymphocytes. These authors found that magnetic properties of the CoFe$_2$O$_4$ nanoparticles may influence the CF release. The movement of the MNP within the MLs, when exposed to a low-frequency magnetic field (0.2, 2.84, and 5.82 kHz), affects the permeability of the lipid bilayer leading to a fast CF release. This caused a change in the releasing mechanism from Fickian diffusion (unexposed) to zero order (exposed).

Another important parameter to be discussed is the average size of the MNP. This factor can directly influence MNO elimination from the human body [55]. According to the literature, nanoparticulate systems smaller than 5.5 nm are readily removed by renal clearance [72], while larger particles >200 nm are sequestered by phagocytic cells

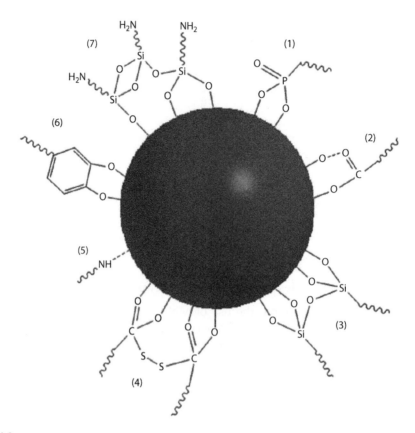

FIGURE 14.2
Some functional groups that can be used to anchor polymer onto MNP: (1) Phosphonic acid; (2) Carboxylic acid; (3) Trimethoxy silane; (4) Cysteine; (5) Amine; (6) Dopamine; (7) Aminopropyltriethoxysilane (APTES).

of the spleen [73]. Based on these data, the ideal range for biological application would be from 5.5 up to 200 nm. However, for MNP, this range becomes even smaller due to the requirement of superparamagnetic regime. Table 14.2 summarizes estimated values of d_{cr} for some MNPs with spherical morphology. In this context, there are many reports showing great control of the nanoparticle size based on the changes in the synthetic conditions such as buffers [74], stirring velocity [39], and temperature [75]. Currently, there is quite a significant ability to control the MNP size by manipulating synthetic parameter.

In addition, the intensification of research around the properties of nanoparticles has led to control also another important parameter for drug delivery that is the shape.

TABLE 14.2

Estimated Values of d_{cr} for Some MNPs with Spherical Morphology

MNP	d_{cr} (nm)	Reference
Fe_3O_4	128	[76]
γ-Fe_2O_3	166	[77]
$CoFe_2O_4$	28	[78]
$NiZnFe_2O_4$	21.6	[79]

Several synthetic methodologies have been developed for the production of different shapes of MNP [80–82]. Yoo et al. [18] reported that nonspherical particles experience different hydrodynamic forces during its passage through the bloodstream. This affects their biodistribution and elimination from human body. The motion of spherical particles through a vessel is only affected by the presence of external forces such as gravitational, magnetic field, van der Waals, and electrostatic interactions [83]. Nevertheless, Geng et al. [84] showed in their study that filamentous particles experimented persistent circulation (up to a week) in the bloodstream, while spherical particles were removed within 2 days. Thus, a particle with nonspherical shape has presented enormous potential for therapeutics.

For asymmetrical shapes (discs, rods, etc.), the tumbling and rolling motion led to lateral deviation from one side to the other of the capillary toward the vessel wall [85]. This will certainly influence the passage of particles through the organism filters or bifurcations in the vessels. Furthermore, the asymmetry characteristic of nonspherical particles may appear as an advantage. Some filtration units of the body such as the spleen, for example, are described as slits (asymmetric) [86]. Thus, clearly the symmetry or asymmetry of the particle can influence the filtration rate and consequently affect the blood circulation time and biodistribution [87]. In addition, studies report that the degree of flexibility of the particle should also be considered independently of its shape [18,88]. A clear example of the impact of this parameter is what happens in the passage of red blood cells through spleen. Spherical particles smaller than 200 nm in diameter are able to pass, while flexible red blood cells of about 10 μm can also pass without any problem [89]. So, the degree of flexibility contributes to the increase in the blood circulation time [84].

14.3 Silica Nanoparticles

In 1992, mesoporous silica was synthesized by Kresge and coworkers [90] and since then has gained much attention because of several key advantages such as high surface area (>700 m^2/g), large pore volume (>0.9 cm^3/g), and adjustable structure and pore size. In addition to these textural properties, silica is chemically inert and thermally stable and has abundant silanol groups (Si–OH) on the surface, which facilitates its conjugation with many different functional groups.

Monodisperse silica spheres were first synthesized in 1968 by Stöber [91] through hydrolysis and condensation of a silica precursor, tetraethyl orthosilicate (TEOS), in ethanolic solution of ammonia. Figure 14.3 illustrates a general reaction scheme for hydrolysis and condensation of a silica precursor through basic catalysis.

Stober's method has been modified by using different precursors, solvents, and surfactants as structure-directing reagent. Figure 14.4 illustrates how mesoporous silica was prepared by co-condensation reaction using a neutral surfactant as a template. In a typical procedure, template particles are coated by condensation polymerization of inorganic silica precursors such as TEOS [92]. Then the template can be removed by heating or extraction in Soxhlet system.

The choice of these variables and the control of experimental conditions such as pH, temperature, and stirring speed may define the morphological properties of the silica materials. Some reviews can be found in the literature [93–96] that present several synthetic strategies developed for the production of different shapes, porosities, and morphologies of mesoporous silica nanoparticles (MSNs) aiming specific applications [97–103].

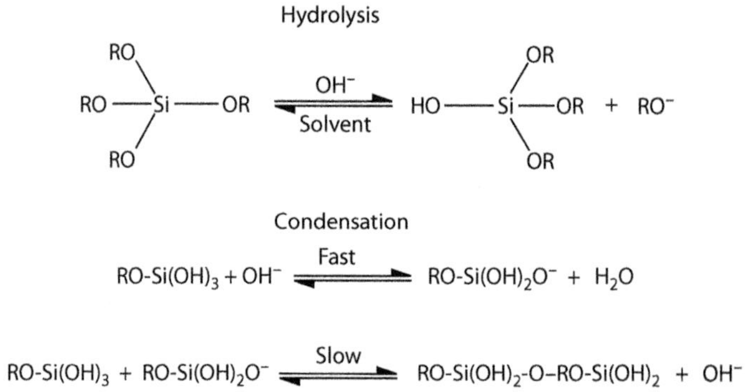

Hydrolysis

RO—Si(OR)—OR $\xrightleftharpoons[\text{Solvent}]{\text{OH}^-}$ HO—Si(OR)—OR + RO⁻

Condensation

RO-Si(OH)₃ + OH⁻ $\xrightleftharpoons{\text{Fast}}$ RO-Si(OH)₂O⁻ + H₂O

RO-Si(OH)₃ + RO-Si(OH)₂O⁻ $\xrightleftharpoons{\text{Slow}}$ RO-Si(OH)₂-O–RO-Si(OH)₂ + OH⁻

FIGURE 14.3
Base catalyzed hydrolysis and condensation of a general silica precursor by conventional methods.

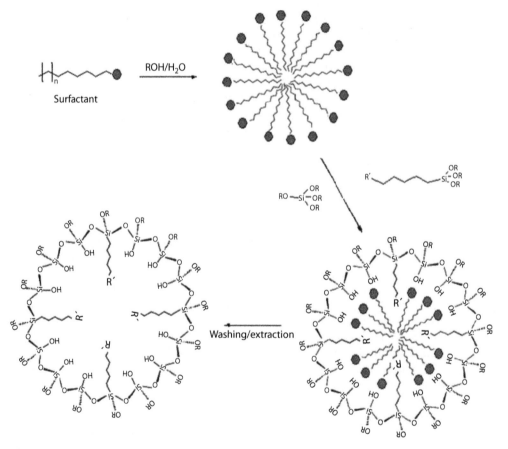

FIGURE 14.4
General procedures for mesoporous silica production by co-condensation method.

Huh at al. [104] were able to prepare mesoporous silica with different morphologies (spheres, tubes, rods) using a co-condensation method, base catalyzed with low surfactant concentration. The morphology and particle size were controlled using different organo-alkoxysilanes as structure-directing reactants. In addition, the presence of the precursors during the base-catalyzed condensation greatly influenced the final particle shape.

MSN obtained by polymerization in the presence of surfactants has showed interesting properties for drug delivery such as larger surface area, tunable pore size, and volume. In addition, therapeutic drugs or tracing molecules could be incorporated into the nanochannels of MSN to be delivered into cells. To increase the efficiency of MSN internalization into cells, the external surface may be modified with receptor-specific ligands such as proteins, small molecules, and DNA [76]. Additionally, several studies have been conducted to evaluate biodegradability and biocompatibility of MSN since these are important properties to support their biomedical applications.

He et al. [105] have synthesized MSN by condensation using diluted TEOS and surfactant along with ammonia solution as a catalyst. In this study, the effect of the MSN shape was investigated regarding to cellular uptake and cellular function. These authors demonstrated a relationship between MSN shape and cellular responses such as particle uptake, cell viability, early apoptosis, adhesion, migration, and cytoskeleton formation.

Hudson and coworkers [106] investigated the biocompatibility and cytotoxicity of mobil composition of material number 41 (MCM – 41), Santa Barbara amorfa number 15 (SBA – 15), and mesostructured cellular foam (MCF) mesoporous silicates with different particle sizes. *In vitro* studies using mesoporous silicates showed a significant degree of toxicity at high concentrations with mesothelial cells. Similar results were obtained by Tao et al. [107] that examined the cytotoxicity and endocytosis of MCM-41 and SBA-15 mesoporous silica and their ammonium-functionalized counterparts. They observed that the cytotoxicity of mesoporous silicate is particle dependent and concentration dependent. No significant cell death was detected when the ammonium-functionalized counterparts were used, suggesting a toxicity reduction by functionalization; however, this effect is dependent of the cell type.

Functionalized MSNs have been shown to be a much better biocompatible material, which can be listed as a great candidate for application in drug delivery systems.

14.4 Mathematical Models for Drug Delivery

Fifty years ago, Takeru Higuchi started to develop mathematical models to evaluate the release of the active agent (usually a drug) from a matrix system as a function of time [108]. Based on the complex concepts of mass transport, he was able to describe a simple model incorporating a series of physical parameters such as the drug diffusion coefficient and the drug solubility, since both modify the release rates [109]. The equation developed by Higuchi can be written as [110,111]

$$\frac{M_t}{M_\infty} = K_H \sqrt{t} + b \tag{14.4}$$

where

M_t is the amount of drug released at the time (t)

M_∞ denotes the total amount of drug

K_H and b are regression parameters that represent the Higuchi release constant and initial amount of drug in solution ($t = 0$)

According to the literature [112], this initial amount (b) can be attributed to a kind of *burst effect* explained by the release of existing drug on the surface of the matrix system or changes in its structure leading to immediate release of the drug. The Higuchi equation is an expression essentially based on the diffusion process of drugs through the matrix system obeying Fick's laws. Thus, the use of this model for a system that presents case-II transport, that is, relaxation of the polymer chains to transport the drug, is not recommendable. Nevertheless, the Higuchi model became the standard of analysis for pharmaceutical systems. Its success and applicability have led to comparison by Paul [113] to other expressions such as $E = mc^2$, developed by Albert Einstein.

However, the Higuchi model has limitations and required the development of other mathematical models. In order to better understand and identify the release mechanism, Korsmeyer et al. [114,115] developed a simple expression relating exponentially the drug release with t. Mathematically, this equation can be written as

$$\frac{M_t}{M_\infty} = Kt^n + b \qquad (14.5)$$

where
  K represents a kinetic constant that considers structural and geometric features of the matrix system
  n denotes the diffusional exponent

This last parameter was improved by Peppas [116] and then can be used to characterize the different mechanisms that lead to drug release. Later, Ritger and Peppas [117,118] related n to the geometric conformation of the matrix systems swellable and nonswellable. Table 14.3 shows n values and the mechanism of diffusional release from various swellable and nonswellable controlled-release systems. Thus, for determination of the exponent n, it is necessary to have full knowledge of the characteristics of the matrix system. Furthermore, Costa et al. [119] report that only 60% ($M_t/M_\infty < 0.6$) of the experimental release data should be used for the calculation of n. The Korsmeyer–Peppas model is generally used for release experiments where the mechanism is not well known or when there is a possibility of a combination between diffusion and case-II transport mechanisms.

However, in all the available models reported earlier, there was no relationship that could evaluate the contribution of each mechanism (diffusion and case-II transport)

TABLE 14.3

Diffusional Exponent and Mechanism of Diffusional Release from Various Swellable and Nonswellable Controlled-Release Systems

Geometric Shape	Diffusional Exponent		Drug Release Mechanism
	Swellable	Nonswellable	
Thin film	0.5	0.5	Diffusion
	$0.5 < n < 1.0$	$0.5 < n < 1.0$	Anomalous transport[a]
	1.0	1.0	Case-II transport
Cylindrical	0.45	0.45	Diffusion
	$0.45 < n < 0.89$	$0.45 < n < 1.0$	Anomalous transport[a]
	0.89	1.0	Case-II transport
Spherical	0.43	0.43	Diffusion
	$0.43 < n < 0.85$	$0.43 < n < 1.0$	Anomalous transport[a]
	0.85	1.0	Case-II transport

[a] The diffusion and case-II transport mechanisms combined.

that leads to drug release. In 1989, Jennifer J. Sahlin and again Nikolaos A. Peppas elaborated a methodology following a heuristic approach where two phenomena can be considered as additive [120]. So, the expression formulated by them brought together terms that reflect the relative contributions of both mechanisms of drug release:

$$\frac{M_t}{M_\infty} = K_1 t^m + K_2 t^{2m} \tag{14.6}$$

where K_1 and K_2 are constants that based on their magnitude it is possible to infer their relative contributions on the diffusion and case-II transport mechanisms, respectively. The coefficient m is the diffusional exponent for the matrix system, which is independent of geometric shape. The Sahlin–Peppas model is extremely useful, since there are several information that can be extracted. Besides the relative contribution of each mechanism, it is possible to calculate the percentage of drug released due to diffusional (F) and case-II transport (R) mechanism [120]. To determine F, the following equation is employed:

$$F = \frac{1}{1 + (K_2/K_1)t^m} \tag{14.7}$$

R can be calculated as the ratio between the contributions of the mechanisms. So, the equation used is

$$\frac{R}{F} = \frac{K_2}{K_1} t^m \tag{14.8}$$

In addition to all models reported in this section, various others have been developed over time: order zero, first order [121], Baker–Lonsdale [122], and Hopfenberg [123], among others. This shows that during the past five decades, the fields of pharmaceutical sciences have had a significant advance in the description and investigation of drug release. In the literature, there are interesting reviews that describe a number of mathematical models for drug delivery as well [119,124,125].

14.5 Quantum Dots

Nanotechnology has provided new exciting tools to monitor biological events and deliver cargos in a selective way as it has been discussed. One exciting type of nanoparticle called QDs was among the first ones to show promising biological applications. This type of nanoparticle made of semiconductor compounds, for example, CdS and CdSe, in a nanometric size has presented outstanding fluorescent behavior. These QDs show many desired key features for an "ideal" fluorophore such as high quantum yield, high molar extinction coefficients (10–100-fold higher than organic dyes), narrow and symmetrical emission bands along with broad absorption band, large Stokes shifts, and high resistance to photobleaching. These properties are due to a quantum confinement effect promoted by the nanometric dimensions of this material. This allows us to prepare a wide range of QD emitters by manipulating their size to reach specific emission wavelengths (Figure 14.5).

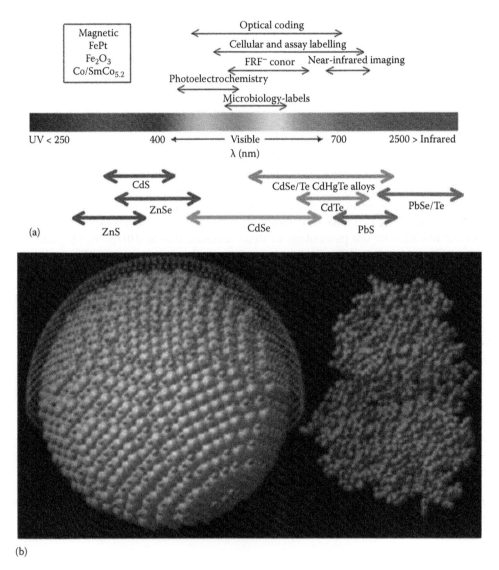

FIGURE 14.5
(See color insert.) Quantum-dots general features. (a) Representation of the type of applications of semiconductor quantum-dot nanomaterials, their common formulation along with their emission wavelength range (from Reference 126). (b) It shows in molecular scale the size of a quantum-dot made of CdSe/ZnS with 6 nm of diameter in comparison to a protein called MBP (maltose-binding protein, MW 44 KD) with an overall dimensions of $3 \times 4 \times 6.5$ nm.

Currently, these properties have been widely explored and applied in biology and medicine for fluorescence imaging and also for delivery. The first use of QD was replacing organic fluorophores in labeling isolated biomolecules, which has moved forward by incorporating them in cellular structure and whole-cell labeling [126,127]. One useful feature of these QDs is their application for simultaneous protein labeling and investigation; since their emissions are narrow and can be taken apart for analytical purposes, it has been used in multiplex assays enabling simultaneous investigations *in vivo* [128]. Besides imaging, these nanoparticles have also been used in delivering drugs and biomolecules.

Since QDs are usually made of toxic heavy metal (Cd, Te, Pb, Hg), there is still a concern in their preparation and broad biological application of these materials. Some QDs are covered by a shell, for example, CdSe can be recovered by a layer of ZnS that also improves its emitting properties. However, a quite diverse number of coating and bioconjugation strategies have been conducted to improve QD solubility and performance and also to prevent metal leaching (e.g., cadmium), thus making it less toxic and biocompatible [126,129]. Nonetheless, there are also novel QDs made of less toxic semiconductor core, for example, silicon and InP [130,131], which are interesting alternatives. Among the strategies for coating and bioconjugation that emerged during the last years, sulfur-based incorporation has been adopted where disulfide bonds are made with ZnS shells, for example, allowing incorporating bifunctional groups. Polymers, silica, and gold have also been used to coat QDs [126,129]. These technologies have led to the large popularity of QD in biology and expanding its application.

QDs for bioimaging have increasingly required more selectivity and have exhibited interesting properties to work as delivery platform, once they could also report their cargo release by imaging. Selective delivery or homing of QD has been achieved by using several exciting strategies incorporating many small and large molecules onto nanoparticle surfaces such as

1. Small peptides: Several short peptide sequences, also known as penetrating peptides, like polyarginine residues, that can internalize large cargos [132], others even more selective such as tumor homing peptide F3 [133]), peptide T18 targeting injured intestinal cells [134], or TAT-HA2 sequence targeting cardiac myocytes to deliver troponin C [135]. This last report has combined two interesting strategies, first a small peptide sequence (TAT) that internalizes cargos in a receptor-independent way but lacks a route to scape from a vesicle where it is internalized. However, by incorporating also an HA2 pH-sensitive sequence, it has provided a route to scape vesicle so transporting QD to cellular cytosol where it can work delivering its cargo.

2. Proteins: Some proteins have been anchored onto QDs providing great internalization and cell selectivity using, for example, specific routes for endocytosis by employing Shiga toxin, ricin, or transferrin taking advantage of cell surface receptor binding [136]. Antibodies have also been used coupled to QD conferring high cell specificity; one example was reported for cancer treatment using an epidermal growth factor antibody (EGF) that binds strongly and selectively to EGF receptor overexpressed in cancer cells; using this strategy, it is possible to image and also target these cells in a very selective manner delivering anticancer drugs [137,138].

3. Small molecules or polymers: Folate has been explored for imaging and delivery of cargo to lymphoma cells *in vivo* based on the overexpression of receptors in cancer cells [139]. Another exciting data comes from the use of beta-cyclodextrin modified with arginine and histidine that delivers simultaneously siRNA and doxorubicin, an anticancer drug [140]. Jiang and Park have used hyaluronic acid conjugated with polyethyleneimine (PEI) as a promising vehicle for imaging and treatment of liver and kidney tumors delivering siRNA, taking advantage of cell surface receptors for hyaluronic specific on these tissues [141]. By delivering antivascular endothelial growth factor siRNA using QD strategy, it was possible to suppress tumor growth and also cause reduction of vascular endothelial growth factor (VEGF) production.

4. Aptamers: QDs have also been modified with small sequences of DNA or RNA providing selectivity toward specific targets enabling imaging and delivery to cells [142,143]. One exciting example has been reported that takes advantage of a DNA aptamer that binds with high affinity to mucin 1, a cell surface–associated mucin protein. This protein is mutated and strongly overexpressed in 90% of late stage of epithelial ovarian cancers and associated metastatic lesions. Based on this, a QD also containing such DNA sequence to provide selectivity was incorporated a cancer drug (doxorubicin) aiming to treat. Additionally, they were able to image ovarian cancer cells during this treatment [143].

These exciting strategies have not just provided a right and sometimes unique address to QD but also provided alternatives for internalization of these nanoparticles, otherwise not easily achieved. Due to these successful strategies, many opportunities have emerged to use QD for imaging, delivery, and therapy, a multifunctional approach.

Based on what we have described here, it is clear that inorganic nanoparticle research is an extremely hot field that has offered many outstanding opportunities to advance our way to treat and monitor disease.

References

1. Strebhardt, K. and A. Ullrich. 2008. Paul Ehrlich's magic bullet concept: 100 years of progress. *Nature Reviews Cancer* 8: 473–480.
2. Arruebo, M., R. Fernandez-Pacheco, M. R. Ibarra, and J. Santamaria. 2007. Magnetic nanoparticles for drug delivery. *Nano Today* 2: 22–32.
3. Rosen, H. and T. Abribat. 2005. The rise and rise of drug delivery. *Nature Reviews Drug Discovery* 4: 381–385.
4. Levy, G. 1965. Pharmacokinetics of salicylate elimination in man. *Journal of Pharmaceutical Sciences* 54: 959–967.
5. Davis, B. K. 1972. Control of diabetes with polyacrylamide implants containing insulin. *Experientia* 28: 348–348.
6. Davis, B. K. 1974. Diffusion in polymer gel implants. *Proceedings of the National Academy of Sciences USA* 71: 3120–3123.
7. Gimbrone, M. A., R. S. Cotran, S. B. Leapman, and J. Folkman. 1974. Tumor growth and neovascularization: An experimental model using rabbit cornea. *Journal of the National Cancer Institute* 42: 413–427.
8. Langer, R. and J. Folkman. 1976. Polymers for the sustained release of proteins and other macromolecules. *Nature* 263: 797.
9. Farokhzad, O. C. and R. Langer. 2009. Impact of nanotechnology on drug delivery. *ACS Nano* 3: 16–20.
10. Yatvin, M. B., W. Kreutz, B. A. Horwitz, and M. Shinitzky. 1980. pH-sensitive liposomes: Possible clinical implications. *Science* 210: 1253.
11. Leserman, L. D., J. Barbet, F. Kourilsky, and J. N. Weinstein. 1980. Targeting to cells of fluorescent liposomes covalently coupled with monoclonal antibody or protein A. *Nature* 288: 602.
12. Heath, T. D., R. T. Fraley, and D. Papahdjopoulos. 1980. Antibody targeting of liposomes: Cell specificity obtained by conjugation of F(ab′)2 to vesicle surface. *Science* 210: 539.
13. Allen, T. M. and A. Chonn. 1987. Large unilamellar liposomes with low uptake into the reticuloendothelial system. *FEBS Letters* 223: 42.

14. Klibanov, A. L., K. Maruyama, V. P. Torchilin, and L. Huang. 1990. Amphipathic polyethylene-glycols effectively prolong the circulation time of liposomes. *FEBS Letters* 268: 235.
15. Gref, R. et al. 1994. Biodegradable long-circulating polymeric nanospheres. *Science* 263: 1600.
16. Parveen, S., R. Misra, and S. K. Sahoo. 2012. Nanoparticles: A boon to drug delivery, therapeutics, diagnostics and imaging. *Nanomedicine: Nanotechnology, Biology and Medicine* 8: 147–166.
17. Faraji, A. H. and P. Wipf. 2009. Nanoparticles in cellular drug delivery. *Bioorganic & Medicinal Chemistry* 17: 2950–2962.
18. Yoo, J.-W., N. Doshi, and S. Mitragotri. 2011. Adaptive micro and nanoparticles: temporal control over carrier properties to facilitate drug delivery. *Advanced Drug Delivery Reviews* 63: 1247–1256.
19. Doshi, N. and S. Mitragotri. 2009. Designer biomaterials for nanomedicine. *Advanced Functional Materials* 24: 3843–3854.
20. Papasani, M. R., G. Wang, and R. A. Hill. 2012. Gold nanoparticles: The importance of physiological principles to devise strategies for targeted drug delivery. *Nanomedicine: Nanotechnology, Biology, and Medicine* xx: 1–11, doi:10.1016/j.nano.2012.1001.1008.
21. Cushing, B. L., V. L. Kolesnichenko, and C. J. O'Connor. 2004. Recent advances in the liquid-phase syntheses of inorganic nanoparticles. *Chemical Reviews* 35: 3893–3946.
22. Freeman, M. W., A. Arrott, and J. H. L. Watson. 1960. Magnetism in medicine. *Journal of Applied Physics* 31: S404.
23. Fiorentini, G. and A. Szasz. 2006. Hyperthermia today: Electric energy, a new opportunity in cancer treatment. *Journal of Cancer Research and Therapeutics* 2: 41–46.
24. Meyers, P. H., F. Cronic, and C. M. Nice. 1963. Experimental approach in the use and magnetic control of metallic iron particles in the lymphatic and vascular system of dogs as a contrast and isotopic agent. *American Journal of Roentgenology, Radium Therapy, and Nuclear Medicine* 90: 1068–1077.
25. Widder, K. J., A. E. Senyei, and D. F. Ranney. 1979. Magnetically responsive microspheres and other carriers for the biophysical targeting of antitumor agents. In *Advances in Pharmacology*, eds. S. Garattini, A. Goldin, F. Hawking, I. J. Kopin, and R. J. Schnitzer, pp. 213–271. Academic Press, San Diego, CA.
26. Senyei, A., K. Widder, and G. Czerlinski. 1978. Magnetic guidance of drug-carrying microspheres. *Journal of Applied Physics* 49: 3578–3583.
27. Widder, K., A. Senyei, and D. G. Scarpelli. 1978. Magnetic microspheres: A model system for site specific drug delivery in vivo. *Proceedings of the Society for Experimental Biology and Medicine* 58: 141–146.
28. Widder, K. J. et al. 1983. Selective targeting of magnetic albumin microspheres to the Yoshida sarcoma: Ultrastructural evaluation of microsphere disposition. *European Journal of Cancer and Clinical Oncology* 19: 141–147.
29. Widder, K., R. M. Morris, G. A. Poore, D. P. Howard, and A. Senyei. 1983. Selective targeting of magnetic albumin microspheres containing low-dose doxorubicin: Total remission in Yoshida sarcoma-bearing rats. *European Journal of Cancer & Clinical Oncology* 19: 135–139.
30. Kato, T. et al. 1984. Magnetic microcapsules for targeted delivery of anticancer drugs. *Applied Biochemistry and Biotechnology* 10: 199–211.
31. Gallo, J. M., C. T. Hung, P. K. Gupta, and D. G. Perrier. 1989. Physiological pharmacokinetic model of adriamycin delivered via magnetic albumin microspheres in the rat. *Journal of Pharmacokinetics and Pharmacodynamics* 17: 305–326.
32. Lubbe, A. S. et al. 1996. Preclinical experiences with magnetic drug targeting: Tolerance and efficacy. *Cancer Research* 56: 4694–4701.
33. Lubbe, A. S. et al. 1996. Clinical experiences with magnetic drug targeting: A phase I study with 4′-epidoxorubicin in 14 patients with advanced solid tumors. *Cancer Research* 56: 4686–4693.
34. Feng, B. et al. 2008. Synthesis of Fe3O4/APTES/PEG diacid functionalized magnetic nanoparticles for MR imaging. *Colloids and Surfaces A: Physicochemical and Engineering Aspects* 328: 52–59.
35. Guedes, M. H. A. et al. 2004. Proposal of a magnetohyperthermia system: Preliminary biological tests. *Journal of Magnetism and Magnetic Materials* 272–276, Part 3: 2406–2407.

36. Lee, J.-H. et al. 2007. Artificially engineered magnetic nanoparticles for ultra-sensitive molecular imaging. *Nature Medicine* 13: 95–99.
37. Boyer, C., M. R. Whittaker, V. Bulmus, J. Liu, and T. P. Davis. 2010. The design and utility of polymer-stabilized iron-oxide nanoparticles for nanomedicine applications. *NPG Asia Materials* 2: 23–30.
38. Zhao, Y., Z. Qiu, and J. Huang. 2008. Preparation and analysis of Fe3O4 magnetic nanoparticles used as targeted-drug carriers. *Chinese Journal of Chemical Engineering* 16: 451–455.
39. Valenzuela, R. et al. 2009. Influence of stirring velocity on the synthesis of magnetite nanoparticles (Fe3O4) by the co-precipitation method. *Journal of Alloys and Compounds* 488: 227–231.
40. Valenzuela, R. 1994. *Magnetic Ceramics.* Cambridge University Press, Cambridge, New York.
41. Mozaffari, M., M. Eghbali Arani, and J. Amighian. 2010. The effect of cation distribution on magnetization of ZnFe2O4 nanoparticles. *Journal of Magnetism and Magnetic Materials* 322: 3240–3244.
42. Goldman, A. 2006. *Modem Ferrite Technology.* Springer, New York.
43. Choi, E. J., Y. Ahn, and K.-C. Song. 2006. Mossbauer study in zinc ferrite nanoparticles. *Journal of Magnetism and Magnetic Materials* 301: 171–174.
44. Mahmoud, M. H., H. H. Hamdeh, J. C. Ho, M. J. O'Shea, and J. C. Walker. 2000. Mossbauer studies of manganese ferrite fine particles processed by ball-milling. *Journal of Magnetism and Magnetic Materials* 220: 139–146.
45. Akther Hossain, A. K. M., H. Tabata, and T. Kawai. 2008. Magnetoresistive properties of Zn1-xCoxFe2O4 ferrites. *Journal of Magnetism and Magnetic Materials* 320: 1157–1162.
46. Vaidyanathan, G., S. Sendhilnathan, and R. Arulmurugan. 2007. Structural and magnetic properties of Co1-xZnxFe2O4 nanoparticles by co-precipitation method. *Journal of Magnetism and Magnetic Materials* 313: 293–299.
47. Arulmurugan, R., B. Jeyadevan, G. Vaidyanathan, and S. Sendhilnathan. 2005. Effect of zinc substitution on Co-Zn and Mn-Zn ferrite nanoparticles prepared by co-precipitation. *Journal of Magnetism and Magnetic Materials* 288: 470–477.
48. Singhal, S., S. K. Barthwal, and K. Chandra. 2006. Structural, magnetic and Mossbauer spectral studies of nanosize aluminum substituted nickel zinc ferrites. *Journal of Magnetism and Magnetic Materials* 296: 94–103.
49. Sun, S. and H. Zeng. 2002. Size-controlled synthesis of magnetite nanoparticles. *Journal of the American Chemical Society* 124: 8204–8205.
50. Ribeiro, T. G. 2008. Síntese e Caracterização de nanopartículas magnéticas de óxidos mistos de MnFe2O4 recobetas com quitosana. Estudo da influência da dopagem com Gd3+ nas propriedades estruturais e magnéticas thesis, Universidade de São Paulo, Brazil.
51. Varandan, V., L. Chen, and X. Xie. 2008. *Nanomedicine: Design and Applications of Magnetic Nanomaterials, Nanosensors and Nanosystems,* 1st edn., Wiley, Chichester, U.K.
52. Aliev, S. M. et al. Study of domain structure of magnetic powder particles by Mossbauer spectroscopy. *Journal of Magnetism and Magnetic Materials* 322: 424–426.
53. Knobel, M. et al. 2008. Superparamagnetism and other magnetic features in granular materials: a review on ideal and real systems. *Journal of Nanoscience and Nanotechnology* 8: 2836–2857.
54. Malik, R. et al. 2010. Mossbauer and magnetic studies in nickel ferrite nanoparticles: Effect of size distribution. *Journal of Magnetism and Magnetic Materials* 322: 3742–3747.
55. Sun, C., J. S. H. Lee, and M. Zhang. 2008. Magnetic nanoparticles in MR imaging and drug delivery. *Advanced Drug Delivery Reviews* 60: 1252–1265.
56. Latham, A. H. and M. E. Williams. 2008. Controlling transport and chemical functionality of magnetic nanoparticles. *Accounts of Chemical Research* 41: 411–420.
57. Yin, H., H. P. Too, and G. M. Chow. 2005. The effects of particle size and surface coating on the cytotoxicity of nickel ferrite. *Biomaterials* 26: 5818–5826.
58. Jain, T. K. et al. 2008. Magnetic nanoparticles with dual functional properties: Drug delivery and magnetic resonance imaging. *Biomaterials* 29: 4012–4021.
59. Zhang, L., R. He, and H.-C. Gu. 2006. Oleic acid coating on the monodisperse magnetite nanoparticles. *Applied Surface Science* 253: 2611–2617.

60. Yamaoka, T., Y. Tabata, and Y. Ikada. 1994. Distribution and tissue uptake of poly(ethylene glycol) with different molecular weights after intravenous administration to mice. *Journal of Pharmaceutical Sciences* 83: 601–606.
61. Ferrari, M. 2005. Nanovector therapeutics. *Current Opinion in Chemical Biology* 9: 343–346.
62. Hong, R. Y. et al. 2008. Synthesis, characterization and MRI application of dextran-coated Fe3O4 magnetic nanoparticles. *Biochemical Engineering Journal* 42: 290–300.
63. Chang, Y.-C., S.-W. Chang, and D.-H. Chen. 2006. Magnetic chitosan nanoparticles: Studies on chitosan binding and adsorption of Co(II) ions. *Reactive and Functional Polymers* 66: 335–341.
64. Lee, S.-J. et al. 2005. Magnetic enhancement of iron oxide nanoparticles encapsulated with poly(d,l-lactide-co-glycolide). *Colloids and Surfaces A: Physicochemical and Engineering Aspects* 255: 19–25.
65. Chen, S. et al. 2007. Temperature-responsive magnetite/PEO-PPO-PEO block copolymer nanoparticles for controlled drug targeting delivery. *Langmuir* 23: 12669–12676.
66. Park, J. et al. 2009. Antibiofouling amphiphilic polymer-coated superparamagnetic iron oxide nanoparticles: Synthesis, characterization, and use in cancer imaging in vivo. *Journal of Materials Chemistry* 19: 6412–6417.
67. Sabaté, R., R. Barnadas-Rodriguez, J. Callejas-Fernandez, R. Hidalgo-Álvarez, and J. Estelrich. 2008. Preparation and characterization of extruded magnetoliposomes. *International Journal of Pharmaceutics* 347: 156–162.
68. Dresco, P. A., V. S. Zaitsev, R. J. Gambino, and B. Chu. 1999. Preparation and properties of magnetite and polymer magnetite nanoparticles. *Langmuir* 15: 1945–1951.
69. Landfester, K. and L. P. Ramírez. 2003. Encapsulated magnetite particles for biomedical application. *Journal of Physics: Condensed Matter* 15: S1345–S1361.
70. Kim, H. S. et al. 2002. Multifunctional layer-by-layer self-assembly of conducting polymers and magnetic nanoparticles. *Thin Solid Films* 419: 173–177.
71. Nappini, S., F. B. Bombelli, M. Bonini, B. Norden, and P. Baglioni. 2009. Magnetoliposomes for controlled drug release in the presence of low-frequency magnetic field. *Soft Matter* 6: 154–162.
72. Choi, H. S. et al. 2007. Renal clearance of quantum dots. *Nature Biotechnology* 25: 1165–1170.
73. Chen, L. T. and L. Weiss. 1973. The role of the sinus wall in the passage of erythrocytes through the spleen. *Blood* 41: 529–537.
74. Aquino, R., F. A. Tourinho, R. Itri, M. C. F. L. e Lara, and J. Depeyrot. 2002. Size control of MnFe2O4 nanoparticles in electric double layered magnetic fluid synthesis. *Journal of Magnetism and Magnetic Materials* 252: 23–25.
75. Mozaffari, M., S. Manouchehri, M. H. Yousefi, and J. Amighian. 2010. The effect of solution temperature on crystallite size and magnetic properties of Zn substituted Co ferrite nanoparticles. *Journal of Magnetism and Magnetic Materials* 322: 383–388.
76. Lu, A.-H., E. L. Salabas, and F. Schüth. 2007. Magnetic nanoparticles: synthesis, protection, functionalization, and application. *Angewandte Chemie International Edition* 46: 1222–1244.
77. Leslie-Pelecky, D. L. and R. D. Rieke. 1996. Magnetic properties of nanostructured materials. *Chemistry of Materials* 8: 1770–1783.
78. Maaz, K., A. Mumtaz, S. K. Hasanain, and A. Ceylan. 2007. Synthesis and magnetic properties of cobalt ferrite (CoFe2O4) nanoparticles prepared by wet chemical route. *Journal of Magnetism and Magnetic Materials* 308: 289–295.
79. Caizer, C. and M. Stefanescu. 2002. Magnetic characterization of nanocrystalline Ni-Zn ferrite powder prepared by the glyoxylate precursor method. *Journal of Physics D: Applied Physics* 35: 3035.
80. Shukla, N. and M. M. Nigra. 2007. Synthesis and self-assembly of magnetic nanoparticles. *Surface Science* 601: 2615–2617.
81. Garcia, C. B. W., Y. Zhang, S. Mahajan, F. DiSalvo, and U. Wiesner. 2003. Self-Assembly approach toward magnetic silica-type nanoparticles of different shapes from reverse block copolymer mesophases. *Journal of the American Chemical Society* 125: 13310–13311.
82. Song, Q. and Z. J. Zhang. 2004. Shape control and associated magnetic properties of spinel cobalt ferrite nanocrystals. *Journal of the American Chemical Society* 126: 6164–6168.

83. Decuzzi, P., S. Lee, B. Bhushan, and M. Ferrari. 2005. A theoretical model for the margination of particles within blood vessels. *Annals of Biomedical Engineering* 33: 179–190.
84. Geng, Y. et al. 2007. Shape effects of filaments versus spherical particles in flow and drug delivery. *Nature Nanotechnology* 2: 249–255.
85. Decuzzi, P., R. Pasqualini, W. Arap, and M. Ferrari. 2009. Intravascular delivery of particulate systems: Does geometry really matter? *Pharmaceutical Research* 26: 235–243.
86. Moghimi, S. M., A. C. Hunter, and J. C. Murray. 2001. Long-circulating and target-specific nanoparticles: theory to practice. *Pharmacological Reviews* 53: 283–318.
87. Venkataraman, S. et al. 2011. The effects of polymeric nanostructure shape on drug delivery. *Advanced Drug Delivery Reviews* 63: 1228–1246.
88. Veiseh, O., J. W. Gunn, and M. Zhang. 2009. Design and fabrication of magnetic nanoparticles for targeted drug delivery and imaging. *Advanced Drug Delivery Reviews* 62: 284–304.
89. Champion, J. A., Y. K. Katare, and S. Mitragotri. 2007. Particle shape: A new design parameter for micro- and nanoscale drug delivery carriers. *Journal of Controlled Release* 121: 3–9.
90. Kresge, C. T., M. E. Leonowicz, W. J. Roth, J. C. Vartuli, and J. S. Beck. 1992. Ordered mesoporous molecular sieves synthesized by a liquid-crystal template mechanism. *Nature* 359: 710–712.
91. Stober, W., A. Fink, and E. Bohn. 1968. Controlled growth of monodisperse silica spheres in the micron size range. *Journal of Colloid and Interface Science* 26: 62–69.
92. Prado, A. G. S., A. O. Moura, and A. R. Nunes. 2011. Nanosized silica modified with carboxylic acid as support for controlled release of herbicides. *Journal of Agricultural and Food Chemistry* 59: 8847–8852.
93. Slowing, I. I., B. G. Trewyn, S. Giri, and V. S. Y. Lin. 2007. Mesoporous silica nanoparticles for drug delivery and biosensing applications. *Advanced Functional Materials* 17: 1225–1236.
94. Slowing, I. I., J. L. Vivero-Escoto, C.-W. Wu, and V. S. Y. Lin. 2008. Mesoporous silica nanoparticles as controlled release drug delivery and gene transfection carriers. *Advanced Drug Delivery Reviews* 60: 1278–1288.
95. Jin, Y. et al. 2009. Amorphous silica nanohybrids: Synthesis, properties and applications. *Coordination Chemistry Reviews* 253: 2998–3014.
96. Slowing, I. I., J. L. Vivero-Escoto, B. G. Trewyn, and V. S. Y. Lin. 2010. Mesoporous silica nanoparticles: structural design and applications. *Journal of Materials Chemistry* 20: 7924–7937.
97. Schacht, S., Q. Huo, I. G. Voigt-Martin, G. D. Stucky, and F. Schuth. 1996. Oil-water interface templating of mesoporous macroscale structures. *Science* 273: 768–771, doi:10.1126/science.273.5276.768.
98. Grün, M., I. Lauer, and K. K. Unger. 1997. The synthesis of micrometer- and submicrometer-size spheres of ordered mesoporous oxide MCM-41. *Advanced Materials* 9: 254–257.
99. Huo, Q., J. Feng, F. Schuth, and G. D. Stucky. 1997. Preparation of hard mesoporous silica spheres. *Chemistry of Materials* 9: 14–17.
100. Yang, H., G. Vovk, N. Coombs, I. Sokolov, and G. A. Ozin. 1998. Synthesis of mesoporous silica spheres under quiescent aqueous acidic conditions. *Journal of Materials Chemistry* 8: 743–750.
101. Lu, F., S.-H. Wu, Y. Hung, and C.-Y. Mou. 2009. Size effect on cell uptake in well-suspended, uniform mesoporous silica nanoparticles. *Small* 5: 1408–1413.
102. Fowler, C. E., D. Khushalani, B. Lebeau, and S. Mann. 2001. Nanoscale materials with meso-structured interiors. *Advanced Materials* 13: 649–652.
103. Radu, D. R., C.-Y. Lai, J. Huang, X. Shu, and V. S. Y. Lin. 2005. Fine-tuning the degree of organic functionalization of mesoporous silica nanosphere materials via an interfacially designed co-condensation method. *Chemical Communications*: 1264–1266.
104. Huh, S., J. W. Wiench, J.-C. Yoo, M. Pruski, and V. S. Y. Lin. 2003. Organic functionalization and morphology control of mesoporous silicas via a co-condensation synthesis method. *Chemistry of Materials* 15: 4247–4256.
105. Huang, X., X. Teng, D. Chen, F. Tang, and J. He. 2010. The effect of the shape of mesoporous silica nanoparticles on cellular uptake and cell function. *Biomaterials* 31: 438–448.
106. Hudson, S. P., R. F. Padera, R. Langer, and D. S. Kohane. 2008. The biocompatibility of mesoporous silicates. *Biomaterials* 29: 4045–4055.

107. Tao, Z., B. B. Toms, J. Goodisman, and T. Asefa. 2009. Mesoporosity and functional group dependent endocytosis and cytotoxicity of silica nanomaterials. *Chemical Research in Toxicology* 22: 1869–1880.
108. Siepmann, J. and N. A. Peppas. 2011. Higuchi equation: Derivation, applications, use and misuse. *International Journal of Pharmaceutics* 418: 6–12.
109. Siepmann, J. and N. A. Peppas. 2011. In honor of Takeru Higuchi. *International Journal of Pharmaceutics* 418: 1–2.
110. Higuchi, T. 1961. Rate of release of medicaments from ointment bases containing drugs in suspension. *Journal of Pharmaceutical Sciences* 50: 874–875.
111. Higuchi, T. 1963. Mechanism of sustained-action medication. Theoretical analysis of rate of release of solid drugs dispersed in solid matrices. *Journal of Pharmaceutical Sciences* 52: 1145–1149.
112. Lopes, C. M., J. M. S. Lobo, and P. Costa. 2005. Formas farmacêuticas de liberação modificada: polímeros hidrifílicos. *Revista Brasileira de Ciências Farmacêuticas* 41: 143–154.
113. Paul, D. R. 2011. Elaborations on the Higuchi model for drug delivery. *International Journal of Pharmaceutics* 418: 13–17.
114. Korsmeyer, R. W., R. Gurny, E. Doelker, P. Buri, and N. A. Peppas. 1983. Mechanisms of solute release from porous hydrophilic polymers. *International Journal of Pharmaceutics* 15: 25–35.
115. Korsmeyer, R. W. and N. A. Peppas. 1981. Macromolecular and modeling aspects of swelling-controlled systems. In *Controlled Release Delivery Systems*, eds. T. J. Rosemam and S. Z. Mansdorf, pp. 77–90. Marcel Dekker Inc, New York.
116. Peppas, N. A. 1985. Analysis of Fickian and non-Fickian drug release from polymers. *Pharmaceutica Acta Helvetiae* 60: 110–111.
117. Ritger, P. L. and N. A. Peppas. 1987. A simple equation for description of solute release II. Fickian and anomalous release from swellable devices. *Journal of Controlled Release* 5: 37–42.
118. Ritger, P. L. and N. A. Peppas. 1987. A simple equation for description of solute release I. Fickian and non-fickian release from non-swellable devices in the form of slabs, spheres, cylinders or discs. *Journal of Controlled Release* 5: 23–36.
119. Costa, P. and J. M. Sousa Lobo. 2001. Modeling and comparison of dissolution profiles. *European Journal of Pharmaceutical Sciences* 13: 123–133.
120. Peppas, N. A. and J. J. Sahlin. 1989. A simple equation for the description of solute release. III. Coupling of diffusion and relaxation. *International Journal of Pharmaceutics* 57: 169–172.
121. Gibaldi, M. and S. Feldman. 1967. Establishment of sink conditions in dissolution rate determinations. Theoretical considerations and application to nondisintegrating dosage forms. *Journal of Pharmaceutical Sciences* 56: 1238–1242.
122. Baker, R. W. and H. S. Lonsdale. 1974. Controlled release: mechanisms and rates. In *Controlled Release of Biologically Active Agents*, eds. A.C. Taquary and R.E. Lacey, pp. 15–71. Plenum Press, New York.
123. Hopfenberg, H. B. 1976. *Controlled release polymeric formulations*, pp. 7–23. American Chemical Society, Washington, DC.
124. Manadas, R., M. E. Pina, and F. Veiga. 2002. A dissolução in vitro na previsão da absorção oral de fármacos em formas farmacêuticas de liberação modificada. *Brazilian Journal of Pharmaceutical Sciences* 38: 375–399.
125. Lao, L. L., N. A. Peppas, F. Y. C. Boey, and S. S. Venkatraman. 2011. Modeling of drug release from bulk-degrading polymers. *International Journal of Pharmaceutics* 418: 28–41.
126. Medintz, I. L., H. T. Uyeda, E. R. Goldman, and H. Mattoussi. 2005. Quantum dot bioconjugates for imaging, labelling and sensing. *Nature Materials* 4: 435–446, doi:Doi 10.1038/Nmat1390.
127. Michalet, X. et al. 2005. Quantum dots for live cells, in vivo imaging, and diagnostics. *Science* 307: 538–544, doi:Doi 10.1126/Science.1104274.
128. Medintz, I. L. et al. 2004. A fluorescence resonance energy transfer-derived structure of a quantum dot-protein bioconjugate nanoassembly. *Proceedings of the National Academy of Science USA* 101: 9612–9617, doi:Doi 10.1073/Pnas.0403343101.
129. Algar, W. R. et al. 2011. The controlled display of biomolecules on nanoparticles: A challenge suited to bioorthogonal chemistry. *Bioconjugate Chemistry* 22: 825–858, doi:Doi 10.1021/Bc200065z.

130. Bharali, D. J., D. W. Lucey, H. Jayakumar, H. E. Pudavar, and P. N. Prasad. 2005. Folate-receptor-mediated delivery of InP quantum dots for bioimaging using confocal and two-photon microscopy. *Journal of the American Chemical Society* 127: 11364–11371, doi:Doi 10.1021/Ja051455x.

131. He, Y. et al. 2009. Photo and pH stable, highly-luminescent silicon nanospheres and their bioconjugates for immunofluorescent cell imaging. *Journal of the American Chemical Society* 131: 4434–4438, doi:Doi 10.1021/Ja808827g.

132. Medintz, I. L. et al. 2008. Intracellular delivery of quantum dot-protein cargos mediated by cell penetrating peptides. *Bioconjugate Chemistry* 19: 1785–1795, doi:10.1021/bc800089r.

133. Derfus, A. M., A. A. Chen, D. H. Min, E. Ruoslahti, and S. N. Bhatia. 2007. Targeted quantum dot conjugates for siRNA delivery. *Bioconjugate Chemistry* 18: 1391–1396, doi:10.1021/bc060367e.

134. Costantini, T. W. et al. 2009. Targeting the gut barrier: Identification of a homing peptide sequence for delivery into the injured intestinal epithelial cell. *Surgery* 146: 206–212, doi:Doi 10.1016/J.Surg.2009.05.007.

135. Koshman, Y. E. et al. 2008. Delivery and visualization of proteins conjugated to quantum dots in cardiac myocytes. *Journal of Molecular and Cellular Cardiology* 45: 853–856, doi:Doi 10.1016/J.Yjmcc.2008.08.006.

136. Tekle, C., B. van Deurs, K. Sandvig, and T. G. Iversen. 2008. Cellular trafficking of quantum dot-ligand bioconjugates and their induction of changes in normal routing of unconjugated ligands. *Nano Letters* 8: 1858–1865, doi:Doi 10.1021/Nl0803848.

137. Diagaradjane, P. et al. 2008. Imaging epidermal growth factor receptor expression in vivo: Pharmacokinetic and biodistribution characterization of a bioconjugated quantum dot nanoprobe. *Clinical Cancer Research* 14: 731–741, doi:Doi 10.1158/1078-0432.Ccr-07-1958.

138. Deepagan, V. G. et al. 2012. In vitro targeted imaging and delivery of camptothecin using cetuximab-conjugated multifunctional PLGA-ZnS nanoparticles. *Nanomedicine-Uk* 7: 507–519, doi:Doi 10.2217/Nnm.11.139.

139. Schroeder, J. E., I. Shweky, H. Shmeeda, U. Banin, and A. Gabizon. 2007. Folate-mediated tumor cell uptake of quantum dots entrapped in lipid nanoparticles. *J Controlled Release* 124: 28–34, doi:Doi 10.1016/J.Jconrel.2007.08.028.

140. Freire, R. M., A. C. H. Barreto, and P. B. A. Fechine. 2012. Nanopartículas Magnéticas: aplicações terapêuticas no combate ao cancêr. *Revista ABM - Metalurgia, Materiais e Mineração* 68: 54–56.

141. Jiang, G., K. Park, J. Kim, K. S. Kim, and S. K. Hahn. 2009. Target specific intracellular delivery of siRNA/PEI-HA complex by receptor mediated endocytosis. *Molecular Pharmaceutics* 6: 727–737, doi:Doi 10.1021/Mp800176t.

142. Ko, M. H. et al. 2009. In vitro derby imaging of cancer biomarkers using quantum dots. *Small* 5: 1207–1212, doi:Doi 10.1002/Smll.200801580.

143. Savla, R., O. Taratula, O. Garbuzenko, and T. Minko. 2011. Tumor targeted quantum dot-mucin 1 aptamer-doxorubicin conjugate for imaging and treatment of cancer. *Journal of Controlled Release* 153: 16–22, doi:Doi 10.1016/J.Jconrel.2011.02.015.

15

Gold Nanoshells and Carbon Nanotubes as a Therapeutic Tool for Cancer Diagnosis and Treatment

José M. Lanao, Laura Martínez-Marcos, and Cristina Maderuelo

CONTENTS

15.1 Introduction to Gold Nanoshells

In recent decades, there have been numerous advances in the field of synthetic polymer-based nanoparticles and other materials as well as in the field of biological cellular carriers with potential applications in diagnosis and treatment of cancer [1–4]. More recently, the use of metal nanoparticles has been explored with different biomedical applications [5].

The metal nanoshells are nanoparticles formed by a core of silicon and a metal cover usually gold. Other metals such as copper, silver, platinum, palladium, nickel, or combinations of these metals can also be used [6–8]. Nanomaterials based on the use of gold particles and gold nanoshells were initially proposed by West and Halas in 2003 as materials for biomedical applications [9].

Possibly one of the most interesting applications of metallic nanoshells is as contrast agents in imaging for diagnosis of various diseases and mainly for the diagnosis of cancer and also for thermal ablation therapies in the treatment of cancer [7,10–12].

The gold nanoshells are a kind of metallic nanoparticles alternative to gold nanospheres or microspheres, formed by a silicon core and a thin covering of gold with a size that typically ranges between 1 and 100 nm [13,14]. Alternatively to gold nanoshells, the gold nanorods have been developed showing differences in the shape and a smaller size than gold nanoshells.

These kinds of nanoparticles show interesting optical properties with a specific surface plasmon resonance (SPR) wavelength and extent of the plasmon dependent on the size, shape, and composition of the gold nanoparticles. These differences in their optical properties allow optical tunability with implications in light-scattering and absorption-based applications in biomedicine [15].

The gold nanoshells have some advantages and disadvantages for their use in biomedical applications. The main advantages of the gold nanoshells can be summarized as follows [7,16]:

1. Biocompatibility and low systemic toxicity.
2. The gold nanoshells can respond to different stimuli such as physical, chemical, and biological with numerous potential applications.
3. Gold shells permit the conjugation of peptides onto its surface or the association with macromolecular nucleic acids.
4. Gold nanoshells may be used as near-infrared (NIR) contrast agents and conjugated with antibodies or other targeting moieties, which may be used for molecular imaging and diagnosis of different diseases.
5. Gold nanoshells may be used for the treatment of different pathologies and especially for cancer treatment.

The main disadvantage of the gold nanoshells is the possibility of aggregation in a physiological environment although the aggregation phenomena can be minimized using gold nanoshells coated with polyethylene glycol (PEG) [17].

Extended reviews about the potential toxicity of gold nanoparticles conclude that the reported data are contradictory and that the critical size of gold nanoparticles to irreversibly bind to biomolecules and to produce functional changes in the cells is about 1–2 nm [18].

The pharmacokinetics and biodistribution of gold nanoshells show that they selectively accumulate in organs and tissues of the reticuloendothelial system. Gold nanoparticles coated with PEG minimize the aggregation and show good biodistribution properties [17]. Studies with PEG-coated gold nanoshells in mice after intravenous (IV) administration showed selective accumulation of gold in organs such as the spleen, the liver, and the kidney [19,20].

15.2 Methods of Obtaining Gold Nanoshells

Metal nanoshells may be produced by different methods and usually using layer-by-layer (LbL) self-assembly [6,21–23]. LbL self-assembly is based on electrostatic interactions between positively and negatively charged film monolayers to produce a thin film and is an alternative to other surface modification methods [22].

The production of polyelectrolyte multilayer self-assembly involves different steps such as the preparation of the first monolayer by immersing the loaded substrate in a solution of an oppositely charged polymer, washing with an appropriate solvent to remove unbound polyelectrolyte and prevent contamination, and finally forming successive layers [22,24].

Drugs or other therapeutic substances may be incorporated in the gold nanoparticles in three different ways: (1) loaded inside the hollow core of the gold nanoshell, (2) conjugated with the surface of the gold nanoshell, and (3) loaded in nanoliposomes or polymeric matrices and incorporated in the gold microsphere [25]. Gold nanoshells may be characterized using different methods such us transmission electron microscopy (TEM), UV–VIS spectroscopy, Fourier transform infrared (FTIR) spectroscopy, Raman spectroscopy, and x-ray photoelectron spectroscopy [26].

15.3 Gold Nanoshells for Cancer Diagnosis

Gold nanoshells may be used for cancer diagnosis using different strategies such as NIR imaging and conjugation with antibodies.

15.3.1 Imaging Methods

NIR spectroscopy is a powerful imaging noninvasive tool with biomedical applications especially in the field of diagnosis. This technique is based on the fact that the light in the NIR region has slightly higher wavelengths than red light. These wavelengths can penetrate the skin and bones getting information about tissue changes that can be used for diagnostic purposes.

Recently, the use of nanoparticle-based NIR contrast agents using gold nanoshells has been proposed [10,27]. The use of gold nanoshells with NIR is based on the phenomenon known as SPR. Plasmons are quasiparticles formed by photons that inciding on a metal are trapped by the free electrons allowing the transport of light through the metal and forming the basis of optical biosensors. The manipulation of the structure and size of metal nanoshells allows tuning the optical SPR wavelengths in a wide range including the NIR region as shown in Figure 15.1 [7,28–31]. Moreover, gold nanoshells show a higher sensitivity of SPR in comparison with gold solid colloids of a similar size [31].

The use of gold nanoshells combining different nanoshell radius and shell thicknesses may scatter light throughout the visible and the NIR regions, and therefore, the same nanoshells may be used in light-based microscopy studies or in NIR tissue imaging studies [29].

Gold nanoshells may be used as luminescent contrast agents when they are excited with NIR lasers [32]. NIR light-absorbing gold nanoshells are introduced into tumors and may be used for diagnosis and treatment of cancer [25]. For imaging purposes, NIR gold nanoshells exhibit good signal intensities and do not demonstrate photo or chemical instability associated with other contrast agents [10]. Gold nanoshells tuned to absorb light in the NIR region have been used by several authors as contrast agents to visualize or diagnose different kinds of tumors such as colon [27].

Other authors have demonstrated that gold nanoparticles may be used as a contrast agent for *in vivo* tumor imaging using photoacoustic imaging [33]. Photoacoustic computed tomography is a kind of 3D imaging system based on the delivery of electromagnetic

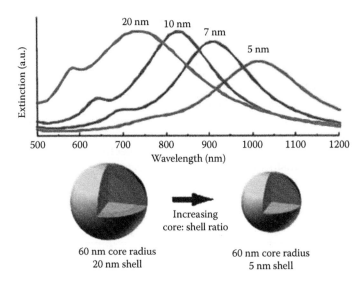

FIGURE 15.1
Optical resonance of gold nanoshells depending on the core/shell ratio. (Reproduced with permission from Springer Science+Business Media: *Ann. Biomed. Eng.*, Metal nanoshells, 34, 2006, 15, Hirsch, L.R. et al.)

induction of ultrasonic pulses in a target tissue. Part of the delivered energy is absorbed and converted into heat, and the thermoacoustic effect results in the production of an ultrasonic emission that can be measured by use of ultrasonic transducers. This technique has potential applications for cancer diagnosis and treatment [34,35]. As an example, PEGylated gold nanoparticles between 20 and 50 nm administered by IV route as a contrast agent have been used successfully for imaging of tumors in mice using photoacoustic tomography [33].

Optical coherence tomography (OCT) is a noninvasive imaging technique based on low-coherence interferometry that permits cross-sectional imaging of biological systems [36]. Three-dimensional imaging of tissues in real time based on OCT may be enhanced using different kinds of nanoparticles. As an example, OCT, using photothermal excitation gold nanoshells that selectively detect regions containing nanoshells, has been applied for tumor imaging in *ex vivo* human breast tissues [37].

Recently, plasmonic vesicle nanostructures assembled from amphiphilic gold nanoparticles with a hollow cavity coated with two types of different polymers (hydrophobic and hydrophilic) analogous to block copolymers have been developed with potential use for imaging or drug delivery in cancer therapeutics. The disassembly of these plasmonic vesicles is a pH-dependent process mediated by the acidic intracellular compartment that induces changes in scattering properties and surface-enhanced Raman scattering (SERS) signals with interesting applications for cell imaging and traceable drug delivery [38].

15.3.2 Antibodies

The inert surface of gold nanoshells permits conjugation to monoclonal antibodies and their application for tumor targeting and diagnosis [19,39,40].

In the scientific literature, we can find some examples of bioconjugation of gold nanoshells with monoclonal antibodies for cancer imaging. The human epidermal growth factor receptor-2 (HER-2) is an oncogene that has been used as a tumor marker in breast cancer. Anti-HER2-conjugated gold nanoshells have been used for imaging of breast adenocarcinoma cells *in vitro* [40,41]. More recently, gold nanoshells targeted to anti-HER2 antibodies have been used to achieve optical contrast in an HER2-overexpressing breast cancer cell line [42].

Also, synthesized bioconjugates of gold nanospheres with monoclonal antibodies have been tested in epithelial cancer cells to recognize epidermal growth factor receptor, and antibody-conjugated gold–gold sulfide nanoparticles have been used to visualize cancerous cells *in vitro* [43,44].

15.4 Gold Nanoparticles and Gold Nanoshells for Cancer Treatment

Currently the gold nanoparticles especially the gold nanoshells are a promising alternative to other delivery systems based on nanoparticles for cancer treatment. Therapeutic alternatives with this kind of metal nanoparticles are based on the use of chemotherapy, photothermal therapy, and gene therapy.

15.4.1 Chemotherapy

Chemotherapy applied for the treatment of cancer is considered to be an effective option though usually it depends on the introduction of a drug carrier to achieve the drug target like cancerous cells, and therefore, chemotherapy drugs can be an effective treatment when they are directly targeted to cancer tissues.

The combined therapy using cytostatic drugs and nanosized drug delivery systems appears to be a good strategy to increase the amount of the drug in the tumor increasing the efficacy of anticancer drugs [45].

Prostate cancer is one of the most frequent kinds of cancer in the male population in developed countries. Functionalized gold nanorods based on thiolated PEG and polyacrylic acid (PAA) attached with doxorubicin have been developed and constitute a promising drug delivery system for the delivery of drugs to prostate cancer cells [46]. In addition, antiandrogen gold nanoparticles selectively accumulate in hormone-insensitive and chemotherapy-resistant prostate cancer cells and engage two different kinds of receptors [47].

Drug conjugates of anticancer drugs with gold nanoshells show a better anticancer efficacy *in vitro* in comparison with other nanodelivey systems. In comparison to doxorubicin InP quantum dots, doxorubicin-conjugated gold nanoparticles demonstrate a higher efficacy against a doxorubicin-resistant B16 melanoma cell line [48]. Also, gold nanoparticles stabilized with a monolayer of L-aspartate as the attachment agent containing cytostatic drugs such as doxorubicin, cisplatin, and capecitabine successfully accumulate in hepatic cancer cells *in vitro* [49].

15.4.2 Nanoshell Phototherapy

Photothermal ablation of cancer cells or tissues is a method in which light is absorbed by the tissue and converted to heat, through light-absorbing dyes or metallic nanoparticles, resulting in the killing of cancer cells. This method is known as photothermal ablation therapy (PTA). The success of this kind of therapy is based on a strong optical absorption

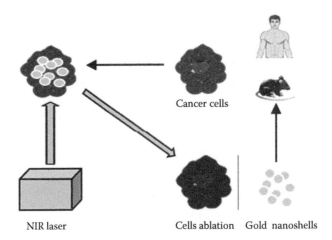

FIGURE 15.2
PTA based on the use of gold nanoshells for cancer treatment.

and a high efficiency of photothermal conversion at the cancer cells [11,12,25,50]. The PTA may be applied in three different ways: using light only, using light associated with molecules for photothermal conversion, and using light and plasmonic metallic nanoparticles for photothermal conversion [50].

The gold nanoshells may be used, among other metal nanoparticles, for phototherapy treatment of cancer [50]. The gold nanoshells can accumulate selectively in tumors. Once this occurs, they can be activated via an external stimulus such as NIR laser light that is converted into heat that can actually kill cancer cells. This method has been tested for cancer therapy in both *in vitro* and *in vivo* experiments. Figure 15.2 shows a scheme of the application of PTA for cancer treatment.

In vivo studies in mice using a murine colon carcinoma model have demonstrated that PTA using NIR-absorbing gold nanoshells was successful for tumor ablation [51]. Similar results using NIR gold nanoshells were obtained for prostate cancer photothermal ablation both *in vitro* using human prostate cancer cell lines and *in vivo* using an ectopic murine tumor model with over 90% of tumor necrosis [52,53].

The spatiotemporal temperature distribution using MR temperature imaging demonstrates enhanced tumor-specific heating for laser ablation via gold nanoshells in a xenograft model of prostate cancer [54]. Moreover, a higher temperature elevation at the tumor surface using a combination of laser heating and gold nanoshells in comparison with laser heating only was observed in mice previously inoculated with prostate cancer cells. In addition, the heat shock protein (HSP) expression associated with tumor survival was diminished when nanoshells were used in combination with laser therapy [55].

Recently, an *in vitro* model using a glioblastoma cell line and gold nanorods for photothermal therapy was used. For the experiment, brain astrocytoma cells were used. The study was based on irradiation of gold nanorods with a continuous wave laser inducing a decrease in the cancer cell viability [56].

NIR gold nanoparticles conjugated with antibodies have also been proposed for cancer theranostics. Gold–gold sulfide nanoparticles exposed to NIR laser can be used for imaging cancerous cells *in vitro*. The same nanoparticles were conjugated with anti-HER2 antibodies and specifically bind to SK-BR-3 breast carcinoma cells that overexpress the HER2 receptor allowing imaging of cancer cells using either multiphoton microscopy when a lower-power laser excitation was used or thermal ablation of cancer cell if a higher-power laser excitation was used. Also, anti-HER2 gold nanoshells have been successfully tested for fluorescent optical imaging and photothermal therapy of drug-resistant ovarian cancer cells *in vitro* [44,57].

In the same way, targeting and photothermal ablation of trastuzumab-resistant breast cancer cells *in vitro* using anti-HER2-conjugated silica–gold nanoshells and NIR lasers were successfully tested and suggest a new therapeutic alternative for the treatment of advanced HER2+ breast cancer [58]. In addition in mouse and human xenograft models of breast cancer, residual breast cancer stem cells related with accelerated repopulation and disease recurrence can be sensitized to radiation treatment using local hyperthermia after IV administration of gold nanoshells [59].

Recently, studies using PEGylated gold nanoshells and gold nanorods injected intravenously in mice with a subcutaneous tumor xenograft model have demonstrated that the efficiency of the tumor accumulation depends on the size of the gold nanoparticles and the dosage regimen used. The multiple dosing administration of gold nanoparticles provides a higher particle accumulation in tumor compared with a single dose. Moreover, the smaller nanorods show a higher accumulation than the larger nanoshells [60]. In another similar research using gamma radiation therapy and PEG-coated gold nanoparticles with different sizes, the authors conclude that the particles with intermediate size (12.1 and 27.3 nm) produce a higher accumulation and higher radiosensitivity than other sizes, both *in vitro* and *in vivo* [61].

15.4.3 Combined Photothermal Ablation and Chemotherapy

One inconvenience of the PTA is the heterogeneous heat distribution into the tumor that can lead to sublethal thermal dose in some tumor regions. For this reason, different studies point to the use of different kinds of gold nanoparticles containing anticancer drugs for a synergistic effect combining simultaneously PTA with chemotherapy for cancer treatment [46,62,63]. EphB4 is a kind of tyrosine kinase receptor that is overexpressed in different kinds of tumors such as breast or prostate, among others. A recent study has demonstrated that doxorubicin-loaded gold nanospheres targeted to EphB4 receptor combined with PTA were very effective in the treatment of experimental tumors in mice [63].

A novel cancer therapy for drug-resistant cancer cells using gold nanoparticles called plasmonic nanobubble-enhanced endosomal escape (PNBEE) has been proposed. Plasmonic nanobubbles are vapor bubbles generated in the cytoplasm of cancer cells when a laser light strikes a plasmon through a gold nanoparticle and is converted into heat. If the gold nanoparticle incorporates a drug or specific antibody, a drug plasmonic nanobubble is optically generated releasing the drug inside the cancer cell. This method has been tested in animals and increases the therapeutic efficacy against resistant cells and reduces the dose, the treatment time, and the nonspecific toxicity [64].

15.4.4 Gene Therapy

Gene therapy for cancer treatment is based on introducing genetic material into cancer cells so that the gene expression stops cell growth or leads to cell death. Usually this type of gene therapy is based mainly on the use of antisense oligonucleotides or small interfering RNA (siRNA). The siRNA is a double-stranded RNA that can decrease the expression of the chimeric protein and induce apoptosis [65]. Antisense oligonucleotides can prevent the production of proteins involved in cancer cell growth and are constituted by artificial sequences of DNA complementary to the messenger RNA of the gene that it is intended to inhibit [66].

The usual strategies for gene therapy commonly used carriers such as nanoparticulate systems or viral vectors, among others. Gold nanoshells can be used as carriers of therapeutic DNA or RNA for cancer treatment as an alternative to other delivery systems [67,68]. The NIR light-absorbing properties of gold nanoshells may be used for the light-triggered release of the genetic material from gold nanoparticles allowing temporal control of the antisense DNA release [67].

Gold nanoparticles formed by charge reversal polymers have been proposed for the delivery of siRNA and plasmid DNA in cancer cells. The charge reversal polymers may produce complexes with nucleic acids and can be used for the transfection of genetic material [69,70]. Gold nanoparticles coated with chitosan and a pH-responsive charge-reversible polymer (PAH-Cit) were developed for the delivery of siRNA. This delivery system protects siRNA against enzymatic degradation and facilitates the siRNA release at acidic pH by a pH-induced mechanism of PAH-Cit both *in vitro* and *in vivo* [70].

15.5 Introduction to Carbon Nanotubes

Carbon nanotubes (CNTs) as gold nanoshells are one of the possible cancer treatments studied nowadays and also used as diagnostic systems with application in chemotherapy and gene therapy. Moreover, recent studies have demonstrated that CNTs are capable of acting as potential drug delivery systems.

These small structures are formed by sp2 carbon benzene ring (graphite) sheets rolled into a characteristic cylindrical shape that were discovered in the 1990s and proposed some years ago with biomedical applications [71]. Depending on the number of sheets composing CNT, these can be classified as single-walled carbon nanotubes (SWCNTs) if there is only one carbon graphite sheet or as multiwalled carbon nanotubes (MWCNTs) if there are two or more carbon graphite sheets being part of these novel structures [72]. Generally, synthesized SWCNT diameter varies from 0.4 to 2 nm, whereas MWCNTs are characterized by a diameter of 1–3 nm of the inner tubes and 2–100 nm diameter of the outer tubes [73]. These carbon nanostructures can be applied in the fields of nanoelectronics, composite materials, and biomedicine, among others [74,75]. There has been evidence that suggests that these nanostructures can be applied within the areas of cancer imaging and chemotherapy, immunotherapy, or gene therapy [74–77].

CNTs are able to act as carriers of chemotherapy drugs that can be injected in order to alter the replication process of cancerous cells. Regarding chemotherapy, CNTs have shown effectiveness when injected into live cells to determine the damage level caused by chemotherapy drugs not only to cancerous cells but also to live cells and tissues. Therefore, it is possible to have certain knowledge concerning chemotherapy drug activity within a patient suffering from cancer.

Another application of CNTs consists of drug delivery due to its morphology features, which allow drug attachment in the inside and outside of CNTs. In addition, the surface of CNTs can be modified in order to perform diverse activities, once they have reached the affected cells or tissues [78].

The main disadvantage of CNTs concerning their application as drug delivery carriers is their poor degradation property. However, several studies promote the theory that functionalized CNTs can be used as adjuvant drug delivery systems, in combination with other novel techniques such as photothermal ablation, photoacoustic ablation, and radio-frequency ablation of tumors [79].

The potential toxicity of CNTs is one of their main disadvantages in comparison with their potential benefits as nanodelivery systems [80]. Pulmonary toxicity of CNTs has been identified after different incubation periods of time in mice population. The graphite sheets that compound these structures can lead to fine particle residues that, if inhaled, can cause pulmonary diseases and, therefore, toxicity [81]. Other studies also demonstrate an inflammatory response at pulmonary level at higher doses after the inhaled adminis-tration either SWCNT or MWCNT [82,83]. However, another toxicity study *in vitro* and in a mouse model using MWCNTs and PEGylated MWCNTs at lower doses demonstrates good biocompatibility and the absence of short- and long-term toxicities [84]. In addition, a recent study suggests potential teratogenicity of MWCNTs in mice [85]. The toxicity prob-lems constitute one of the main disadvantages of CNTs, apparently related to the level of exposure and need to be deeply investigated.

15.6 Methods of Obtaining Carbon Nanotubes

Different techniques have been developed to obtain CNTs. The synthesis of these struc-tures depends on the method applied, considering therefore the carbon arc-discharge technique, the laser-ablation technique, and the chemical vapor deposition (CVD) tech-nique [86].

The arc-discharge method is the most simple and common way to produce CNTs. The main disadvantage of this method is the generation of multiple components; thus, it is neces-sary to perform a purification step. This method is based on the application of two electrodes into a vacuum chamber or container in order to produce an arc discharge that will form afterward, and carbon deposits in one of the electrodes involved (the negative electrode). In addition, it is necessary to introduce an inert gas such as helium or argon at the beginning of the process, to accelerate the transition from the gas to the solid phase [72,86,87].

Another important issue is the energy applied in the process. This originates through a power supply. At this moment, the two electrodes are brought closer to produce the elec-tric arc. The period of time used to apply the power is the factor that determines the CNT length. The range of CNT length that can be obtained with this method goes from microm-eters to a few centimeters, and its diameter varies from 5 to 20 µm. Regarding SWCNTs and MWCNTs, the elements presented during this process differs. In the case of SWCNT, a cobalt catalyst is required [72,86,87].

The laser-ablation technique consists of laser pulses carried out into a flow tube heated to a very high temperature (1200°C) whose target is graphite mixed with a catalyst as well as the use of an inert gas to form SWCNTs. The nanotubes obtained by this method can be from 5 to 20 nm of diameter [88–90].

In both methods, the arc-discharge and the laser-ablation techniques have some disad-vantages regarding the obtention of tangled bundles and the consequent step of purifica-tion. Therefore, it increases the cost of the process. However, the high yield production generated by these methods is an advantage.

The CVD technique also called the catalytic growth method constitutes another way of production of CNTs. This method uses hydrocarbon gas and an alternative energy to origi-nate the decomposition of those hydrocarbons inside a CVD reactor. Consequently, carbon filaments are produced and transformed into CNTs. It is possible to obtain SWCNTs and MWCNTs with this method. Nevertheless, it is necessary to use a catalyst for the generation of SWCNTs. The main advantage of this method is the absence of a very high temperature, as it is not required to carry out this process. In addition, it permits an excellent control of the nanotube characteristics such as diameter, length, alignment, and growth rate [72,87].

15.7 Carbon Nanotubes for Cancer Diagnosis

SWCNTs have shown a high absorbance of above 1,100 nm in which cancerous tissues cannot be well depicted through imaging techniques on their own, but when SWCNTs have been administered and systematically irradiated at the tumor zone with NIR light, the affected zone becomes reachable by imaging approaches [91,92]. Another application of CNTs in chemotherapy concerns the diagnosis and treatment of pancreatic cancer, which can detect positive lymph nodes. In this study, the use of magnetic multiwalled carbon nanotubes (mMWCNTs) showed promise for application of nanotubes for tumor diagnosis [93].

Circulating tumor cells (CTCs) are defined as the tumor cells that circulate in the blood stream and often indicate metastasis usually of endothelial origin like lung, breast, or prostate cancer. It is still necessary to provide an efficient detection method of these cells in patients who suffer from cancer due to the small number of them that arise with a very high number of blood cells. Therefore, the isolation process of CTCs constitutes a challenge that would allow the early detection and possible prevention of this disease if a new end capable technique is discovered.

One of the potential detection strategies being studied nowadays is the use of photoacoustic flowmetry in combination with nanoparticles that detect antigens located in the cell surface with application to melanoma cell detection [94]. Gold-plated CNTs conjugated with folic acid have been used as a contrast agent for photoacoustic imaging for the detection of CTCs magnetically captured in mice allowing the prevention of metastasis [95].

15.8 Carbon Nanotubes for Cancer Treatment

15.8.1 Chemotherapy

The use of CNTs as an alternative to other delivery systems to selectively direct the drugs to target tumors has been proposed and experimentally tested in recent years [74,96–99].

Regarding the appropriate CNTs to be used for drug delivery, it has been suggested that SWCNTs have ideal properties for this purpose as opposed to MWCNTs. The main favorable property is the high surface area of SWCNTs that promotes efficient drug loading into the CNT. Also the functionalization of SWCNTs increases their hydrophilicity and the binding to cancer cells [73,100]. Other authors using human lungs and pharynx carcinoma cell lines have shown that the physicochemical properties such as the dispersion status of the SWCNTs influence the cell viability [101].

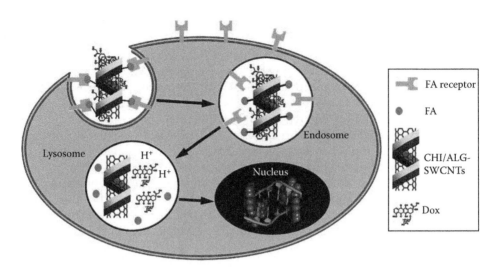

FIGURE 15.3
(See color insert.) Scheme of the uptake by endocytosis of SWCNT loading doxorubicin by cancer cells. (Reproduced from *Biomaterials*, 33, Meng, L., Zhang, X., Lu, Q., Fei, Z., and Dyson, P.J., Single walled carbon nanotubes as drug delivery vehicles: targeting doxorubicin to tumors, 1689, Copyright 2012, with permission from Elsevier.)

Figure 15.3 shows a scheme of the uptake by endocytosis of SWCNT loading doxorubicin by cancer cells, accumulating into lysosomes where the drug is released [92].

The requirements that CNTs have to present in order to be considered as drug carriers focus on an aqueous solubility necessary for the adequate absorption and transport of drugs and the optimal functionalization of CNTs that can be achieved by adding a surfactant, a solvent, or any biomolecular support. Thus, it is also acceptable to obtain functionalized CNTs through the modification of the side walls that constitute the tube structure. It has been proved that functionalizing CNTs leads to a reduction of their cytotoxicity and an increased biocompatibility.

There are two types of functionalization called noncovalent and covalent. Regarding the noncovalent functionalization, there are some critical aspects of CNTs that are involved in this process. Due to the hydrophobic property of drugs, the loading process of drugs into CNTs requires some ionic modifications of its surface, and consequently, the CNT will be able to incorporate drugs by adsorption [98].

The covalent functionalization of CNTs is obtained by oxidation processes that confer them a hydrophilic part with, for example, OH and COOH groups. The use of acid reactions often leads to shorter nanotubes that are usually preferred to act as drug delivery carriers although it can damage the CNT structure if the acid reaction is very strong [98].

One of the main properties of CNTs refers to their capability to enter cells and, therefore, be considered as drug delivery carriers for different kinds of drugs. This feature is mainly presented in functionalized water-soluble CNTs. Another way to obtain functionalized CNTs is based on the addition of PEG or addition of amphiphilic polymers to the surface of SWCNTs [102,103].

One of the most commonly used chemotherapy drugs is doxorubicin, widely used for breast cancer, among others. Despite the effectiveness shown with doxorubicin treatment, it is necessary to establish a carrier for this drug regarding the low solubility and nonspecific distribution properties that are usually common in this kind of drugs [92].

Diverse studies have demonstrated the potential efficacy of different kinds of CNTs loaded with doxorubicin as delivery systems for cancer treatment involving either SWCNTs or MWCNTs.

SWCNTs coated with doxorubicin using hydrazone bonds provide a pH-dependent drug release that facilitates the doxorubicin release in the acidic microenvironment of the tumor reducing the systemic toxicity of the drug. This system was successfully tested *in vitro* using a human hepatocellular liver carcinoma cell line [104].

In addition, modified SWCNTs with chitosan and folic acid loaded with doxorubicin were effective in the killing of hepatocellular carcinoma human cell lines and in the inhibition of liver cancer in mice [105]. Similar results *in vitro* were obtained with functionalized SWCNTs loaded with etoposide against human alveolar carcinoma epithelial cell lines [106], and efficacy *in vitro* and *in vivo* was demonstrated using functionalized SWCNTs bearing pirarubicin for the treatment of bladder cancer [107]. Functionalized SWCNTs have also been tested with other antineoplastic drugs such as daunorubicin, gemcitabine, and cisplatin [108–110].

Diverse studies have demonstrated the efficacy achievable when doxorubicin was delivered by PEGylated SWCNTs for the treatment of cancer in mice population [111], and in another research, noncovalent SWCNTs modified with tocopheryl PEG succinate and carrying doxorubicin showed a longest retention in tumor and greater efficacy in a sarcoma-bearing mouse model [112].

Oxidized multiwalled CNTs (O-MWCNTs) have demonstrated a high adsorption capacity for doxorubicin, and PEGylated O-MWCNTs modified with angiopep-2 demonstrate efficacy *in vitro* and *in vivo* using an animal model in the treatment of brain glioma [113]. Functionalized MWCNTs have been developed with other anticancer drugs such as epirubicin and daunorubicin. In this case, gold nanoparticles were introduced in MWCNTs and loaded with daunorubicin obtaining a new nanosystem increasing drug intracellular concentrations in human hepatoma cells *in vitro* [114,115].

CNTs may also be developed as magnetic nanoparticles and manipulated under a magnetic field. Recently, a magnetic dual targeted nanocarrier based on the combination of MWCNTs and iron oxide magnetic nanoparticles carrying doxorubicin have been proposed. This nanodelivery system showed intracellular drug delivery and enhanced cytotoxicity against human glioblastoma cells [116]. A similar strategy was evaluated using magnetic functionalized MWCNTs bearing gemcitabine. *In vivo* experimental administration of gemcitabine magnetic CNTs inhibits the lymph node metastasis under a magnetic field [117].

Other chemotherapy drugs such as paclitaxel attached to PEGylated SWCNTs by an ester union have provided evidence of higher effectiveness in reducing tumor growth than Taxol in breast cancer and the low toxicity of these nanostructures for chemotherapy drug delivery. Its main attraction is the possibility they offer to deliver low drug doses and the stability they present as a drug delivery system [102,118]. Also functionalized MWCNTs with poly(citric acid) to increase their hydrophilicity and loaded with paclitaxel showed a pH-dependent release and a cytotoxic effect against human alveolar epithelial and ovarian carcinoma cell lines [119].

15.8.2 Photothermal Therapy

Like gold nanoshells, the CNTs show a strong optical absorption in the NIR region and can be used for photothermal ablation in cancer therapy [87]. When CNTs are undergoing laser irradiation, they absorb the NIR electromagnetic radiation energy and an increase of the temperature is produced depending on the amount of irradiated energy [120,121]. Previous studies demonstrate the potential of this method in photothermal ablation of cancer cells *in vitro* [122].

Functionalized SWCNTs and MWCNTs have been evaluated with different kinds of experimental cancer. The moderate laser irradiation of SCCVII carcinoma in mice after intratumoral injection of SWCNTs produced the eradication of carcinoma cells [123]. In addition, PEGylated SWCNTs facilitate the tumor uptake of CNTs in mice after IV administration and can be used for photothermal ablation of tumors by NIR laser irradiation [111].

Also, functionalized MWCNTs with human serum albumin (HSA) selectively target the Gp60 receptor located on the membrane of malignant cancer cells and have been successfully tested for photothermal ablation of liver cancer cells *in vitro*. This kind of HSA nanotubes was also successfully tested for nanophotothermolysis of cancer tissue by external laser irradiation after intra-arterial administration of HSA-MWCNTs using an *ex vivo* human pancreatic adenocarcinoma model [124,125]. Also, photothermal ablation of human colorectal cancer cell line with NIR laser using amphiphilic functionalized MWCTs with a novel biocompatible polyhedral oligomeric silsesquioxane polyurethane (POSS-PCU) has been developed [97].

A new anticancer synergistic therapy combining chemotherapy and photothermal ablation induced by CNTs has been tried. In this study, noncovalent functionalized SWCNTs conjugated with docetaxel combined with NIR radiation showed efficacy in the suppression of tumor growth using an *in vitro* human prostate cancer cell line and a murine sarcoma cancer model [126].

15.8.3 Gene Therapy

The use of CNTs and especially SWCNTs in the field of gene therapy for cancer treatment is an alternative to other known strategies such as the use of viral vectors.

One of these studies highlights that SWCNTs have the ability to act as siRNA transfer systems whose aim is to deliver the siRNA to be introduced into dendritic cells and, therefore, modulate their immune response against cancerous cell growth [77].

With applications in breast cancer therapy, functionalized siRNA-loaded SWCNTs have been proposed for selective targeting to MDM2 oncogene and successfully tested by the inhibition *in vitro* of breast carcinoma cells [127], and functionalized SWCNTs bearing p53, a gene that recognizes damaged DNA and induces apoptosis in cancer cells, produced an increase in the cell killing of breast cancer cells *in vitro* [128].

In addition, an increase in the transfection efficiency has been shown when SWCNTs loaded with RNA were combined with the protein protamine and the drug chloroquine [129].

Also, solubilized SWCNTs have been tested *in vitro* for the delivery of antigen peptides into antigen-presenting cells such as dendritic cells and macrophages to induce IgG responses against tumor antigens [130].

As in the case of cancer chemotherapy, synergistic combined therapeutic effects using gene therapy and photothermal therapy mediated by CNTs have been recently proposed. Functionalized SWCNTs with polyethylenimine and loaded with siRNA have been used as a delivery system suppressing the proliferation of human prostate cancer cell lines *in vitro* and with a high anticancer activity in an experimental cancer mice model. With the same *in vivo* model, the antitumor therapeutic efficiency was enhanced by combining the use of photothermal ablation with gene therapy [126].

Also magnetic CNTs give rise to a viable alternative in the case of cancer gene therapy, and recently the use of water-soluble fluorescein isothiocyanate (FITC)-labeled magnetic carbon nanotubes (FITC-mCNT) has been proposed that allow under a magnetic field to target cytoplasma and the nucleus of the cell for monocyte-based gene therapies using human monocytic leukemia cell lines [131].

15.9 Conclusion

Gold nanoshells and functionalized CNTs are types of novel biocompatible nanosystems with important therapeutic implications for diagnosis and imaging of cancer and the treatment of this disease based on the use of chemotherapy, PTA, or gene therapy.

Although in recent years there have been promising advances in the research with this type of nanosystems, most of these studies have been conducted using *in vitro* models based on the use of different types of cancer cell lines or *in vivo* using animal models of experimental cancer; however, there is an evident default of clinical studies that confirm the clinical application of this kind of novel therapies.

Currently, the research activity in this field is very intense, and in the coming years, an introduction of this type of nanotechnology into clinical practice for both the diagnosis and treatment of cancer is expected.

References

1. Brigger, I., C. Dubernet, and P. Couvreur. 2012. Nanoparticles in cancer therapy and diagnosis. *Adv Drug Deliv Rev* 54(5): 631–651.
2. Mazzaferro, S., K. Bouchemal, and G. Ponchel. 2012. Oral delivery of anticancer drugs III: Formulation using drug delivery systems. *Drug Discov Today* 18(1–2): 99–104.
3. Jabir, N. R. et al. 2012. Nanotechnology-based approaches in anticancer research. *Int J Nanomed* 7: 4391–4408.
4. Gutierrez Millan, C., C. I. Colino Gandarillas, M. L. Sayalero Marinero, and J. M. Lanao. 2012. Cell-based drug-delivery platforms. *Ther Deliv* 3: 25–41.
5. J. Blackman (ed.). 2009. *Metallic Nanoparticles*, 1st edn., Elsevier, Amsterdam, the Netherlands.
6. Jankiewicz, B. J., D. Jamiola, J. Choma, and M. Jaroniec. 2012. Silica-metal core-shell nanostructures. *Adv Colloid Interface Sci* 170: 28–47.
7. Hirsch, L. R. et al. 2006. Metal nanoshells. *Ann Biomed Eng* 34: 15–22.
8. Su, C. H. et al. 2007. Nanoshell magnetic resonance imaging contrast agents. *J Am Chem Soc* 129: 2139–2146.
9. West, J. L. and N. J. Halas. 2003. Engineered nanomaterials for biophotonics applications: Improving sensing, imaging, and therapeutics. *Annu Rev Biomed Eng* 5: 285–292.
10. Altinoglu, E. I. and J. H. Adair. 2010. Near infrared imaging with nanoparticles. *Wiley Interdiscip Rev Nanomed Nanobiotechnol* 2: 461–477.
11. Young, J. K., E. R. Figueroa, and R. A. Drezek. 2012. Tunable nanostructures as photothermal theranostic agents. *Ann Biomed Eng* 40: 438–459.
12. Singh, A. K. et al. 2012. Multi-dye theranostic nanoparticle platform for bioimaging and cancer therapy. *Int J Nanomed* 7: 2739–2750.
13. Chithrani, B. D., A. A. Ghazani, and W. C. Chan. 2006. Determining the size and shape dependence of gold nanoparticle uptake into mammalian cells. *Nano Lett* 6: 662–668.

14. Mody, V. V., R. Siwale, A. Singh, and H. R. Mody. 2010. Introduction to metallic nanoparticles. *J Pharm Bioallied Sci* 2: 282–289.

15. Jain, P. K., K. S. Lee, I. H. El-Sayed, and M. A. El-Sayed. 2006. Calculated absorption and scattering properties of gold nanoparticles of different size, shape, and composition: Applications in biological imaging and biomedicine. *J Phys Chem B* 110: 7238–7248.

16. Delcea, M., H. Mohwald, and A. G. Skirtach. 2011. Stimuli-responsive LbL capsules and nanoshells for drug delivery. *Adv Drug Deliv Rev* 63: 730–747.

17. Hirsch, L. R. et al. 2003. Nanoshell-mediated near-infrared thermal therapy of tumors under magnetic resonance guidance. *Proc Natl Acad Sci USA* 100: 13549–13554.

18. Khlebtsov, B. and N. Khlebtsov. 2008. Enhanced solid-phase immunoassay using gold nanoshells: Effect of nanoparticle optical properties. *Nanotechnology* 19: 435703.

19. Kaur, S. et al. 2012. Recent trends in antibody-based oncologic imaging. *Cancer Lett* 315: 97–111.

20. James, W., Hirsch, L. R., West, J. L., O'Neal, P. D., and Payne, J. D. 2007. Application of inaa to the build-up and clearance of gold nanoshells in clinical studies in mice. *J Radioanal Nucl Chem* 271: 455–459.

21. Ariga, K., Y. M. Lvov, K. Kawakami, Q. Ji, and J. P. Hill. 2011. Layer-by-layer self-assembled shells for drug delivery. *Adv Drug Deliv Rev* 63: 762–771.

22. de Villiers, M. M., D. P. Otto, S. J. Strydom, and Y. M. Lvov. 2011. Introduction to nanocoatings produced by layer-by-layer (LbL) self-assembly. *Adv Drug Deliv Rev* 63: 701–715.

23. Spuch-Calvar, M., J. Perez-Juste, and L. M. Liz-Marzan. 2007. Hematite spindles with optical functionalities: growth of gold nanoshells and assembly of gold nanorods. *J Colloid Interface Sci* 310: 297–301.

24. Lvov, Y., K. Ariga, M. Onda, I. Ichinose, and T. Kunitake. 1999. A careful examination of the adsorption step in the alternate layer-by-layer assembly of linear polyanion and polycation. *Colloid Surfaces Physicochem Eng Aspect* 146: 337–346.

25. Melancon, M. P., M. Zhou, and C. Li. 2011. Cancer theranostics with near-infrared light-activatable multimodal nanoparticles. *Acc Chem Res* 44: 947–956.

26. T. Pham, J. B. J., N.J. Halas, and T. Randall. 2002. Preparation and characterization of gold nanoshells coated with self-assembled monolayers. *Langmuir* 18(12): 4915–4920.

27. Puvanakrishnan, P. et al. 2009. Near-infrared narrow-band imaging of gold/silica nanoshells in tumors. *J Biomed Opt* 14: 024044.

28. Wang, Y. et al. 2004. Photoacoustic tomography of a nanoshell contrast agent in the *in vivo* rat brain. *Nano Lett* 4: 1689–1692.

29. Loo, C. et al. 2004. Nanoshell-enabled photonics-based imaging and therapy of cancer. *Technol Cancer Res Treat* 3: 33–40.

30. Rasch, M. R., K. V. Sokolov, and B. A. Korgel. 2009. Limitations on the optical tunability of small diameter gold nanoshells. *Langmuir* 25: 11777–11785.

31. Sun, Y. and Y. Xia. 2002. Increased sensitivity of surface plasmon resonance of gold nanoshells compared to that of gold solid colloids in response to environmental changes. *Anal Chem* 74: 5297–5305.

32. Gao, L., T. J. Vadakkan, and V. Nammalvar. 2011. Nanoshells for *in vivo* imaging using two-photon excitation microscopy. *Nanotechnology* 22: 365102.

33. Zhang, Q. et al. 2009. Gold nanoparticles as a contrast agent for *in vivo* tumor imaging with photoacoustic tomography. *Nanotechnology* 20: 395102.

34. Yang, X., E. W. Stein, S. Ashkenazi, and L. V. Wang. 2009. Nanoparticles for photoacoustic imaging. *Wiley Interdiscip Rev Nanomed Nanobiotechnol* 1: 360–368.

35. Jung, Y. et al. 2012. Multifunctional nanoprobe to enhance the utility of optical based imaging techniques. *J Biomed Opt* 17: 016015.

36. Huang, D. et al. 1991. Optical coherence tomography. *Science* 254: 1178–1181.

37. Zhou, C. et al. 2010. Photothermal optical coherence tomography in *ex vivo* human breast tissues using gold nanoshells. *Opt Lett* 35: 700–702.

38. Song, J., J. Zhou, and H. Duan. 2012. Self-assembled plasmonic vesicles of SERS-encoded amphiphilic gold nanoparticles for cancer cell targeting and traceable intracellular drug delivery. *J Am Chem Soc* 134: 13458–13469.

39. T. A. Erickson and J. W. Tunnel. 2009. *Gold Nanoshells in Biomedical Applications. Nanomaterials for the Life Sciences.* Vol. 3: Mixed Metal Nanomaterials Wiley-VCH Verlag GmbH & Co, Weinheim, Germany.

40. Loo, C. et al. 2005. Gold nanoshell bioconjugates for molecular imaging in living cells. *Opt Lett* 30: 1012–1014.

41. Loo, C., A. Lowery, N. Halas, J. West, and R. Drezek. 2005. Immunotargeted nanoshells for integrated cancer imaging and therapy. *Nano Lett* 5: 709–711.

42. Bickford, L. R., G. Agollah, R. Drezek, and T. K. Yu. 2010. Silica-gold nanoshells as potential intraoperative molecular probes for HER2-overexpression in *ex vivo* breast tissue using near-infrared reflectance confocal microscopy. *Breast Cancer Res Treat* 120: 547–555.

43. Sokolov, K. et al. 2003. Real-time vital optical imaging of precancer using anti-epidermal growth factor receptor antibodies conjugated to gold nanoparticles. *Cancer Res* 63: 1999–2004.

44. Day, E. S. et al. 2010. Antibody-conjugated gold-gold sulfide nanoparticles as multifunctional agents for imaging and therapy of breast cancer. *Int J Nanomed* 5: 445–454.

45. Kratz, F. and A. Warnecke. 2012. Finding the optimal balance: Challenges of improving conventional cancer chemotherapy using suitable combinations with nano-sized drug delivery systems. *J Control Release* 164(2): 221–235.

46. Wang, T. et al. 2011. Fabrication of doxorubicin functionalized gold nanorod probes for combined cancer imaging and drug delivery. *Dalton Trans* 40: 9789–9794.

47. Dreaden, E. C. et al. 2012. Antiandrogen gold nanoparticles dual-target and overcome treatment resistance in hormone-insensitive prostate cancer cells. *Bioconjug Chem* 23: 1507–1512.

48. Zhang, X., H. Chibli, D. Kong, and J. Nadeau. 2012. Comparative cytotoxicity of gold-doxorubicin and InP-doxorubicin conjugates. *Nanotechnology* 23: 275103.

49. Tomuleasa, C. et al. 2012. Gold nanoparticles conjugated with cisplatin/doxorubicin/capecitabine lower the chemoresistance of hepatocellular carcinoma-derived cancer cells. *J Gastrointestin Liver Dis* 21: 187–196.

50. Rozanova, N. and J. Zhang. 2009. Photothermal ablation therapy for cancer based on metal nanostructures. *Sci China Ser B-Chem* 52: 1559–1575.

51. O'Neal, D. P., L. R. Hirsch, N. J. Halas, J. D. Payne, and J. L. West. 2004. Photo-thermal tumor ablation in mice using near infrared-absorbing nanoparticles. *Cancer Lett* 209: 171–176.

52. Stern, J. M. et al. 2007. Efficacy of laser-activated gold nanoshells in ablating prostate cancer cells in vitro. *J Endourol* 21: 939–943.

53. Stern, J. M., J. Stanfield, W. Kabbani, J. T. Hsieh, and J. A. Cadeddu. 2008. Selective prostate cancer thermal ablation with laser activated gold nanoshells. *J Urol* 179: 748–753.

54. Stafford, R. J. et al. 2011. MR temperature imaging of nanoshell mediated laser ablation. *Int J Hyperthermia* 27: 782–790.

55. Rylander, M. N., R. J. Stafford, J. Hazle, J. Whitney, and K. R. Diller. 2011. Heat shock protein expression and temperature distribution in prostate tumours treated with laser irradiation and nanoshells. *Int J Hyperthermia* 27: 791–801.

56. Smith, M. A. et al. 2012. Wolbachia and DNA barcoding insects: patterns, potential, and problems. *PLoS One* 7: e36514.

57. Chen, W. et al. 2010. A molecularly targeted theranostic probe for ovarian cancer. *Mol Cancer Ther* 9: 1028–1038.

58. Carpin, L. B. et al. 2011. Immunoconjugated gold nanoshell-mediated photothermal ablation of trastuzumab-resistant breast cancer cells. *Breast Cancer Res Treat* 125: 27–34.

59. Atkinson, R. L. et al. 2010. Thermal enhancement with optically activated gold nanoshells sensitizes breast cancer stem cells to radiation therapy. *Sci Transl Med* 2: 55ra79.

60. Puvanakrishnan, P., J. Park, D. Chatterjee, S. Krishnan, and J. W. Tunnell. 2012. *In vivo* tumor targeting of gold nanoparticles: Effect of particle type and dosing strategy. *Int J Nanomed* 7: 1251–1258.

61. Zhang, X. D. et al. 2012. Size-dependent radiosensitization of PEG-coated gold nanoparticles for cancer radiation therapy. *Biomaterials* 33: 6408–6419.

62. Shi, P. et al. 2012. pH-responsive NIR enhanced drug release from gold nanocages possesses high potency against cancer cells. *Chem Commun (Camb)* 48: 7640–7642.

63. You, J. et al. 2012. Effective photothermal chemotherapy using doxorubicin-loaded gold nanospheres that target EphB4 receptors in tumors. *Cancer Res* 72: 4777–4786.
64. Lukianova-Hleb, E. Y., A. Belyanin, S. Kashinath, X. Wu, and D. O. Lapotko. 2012. Plasmonic nanobubble-enhanced endosomal escape processes for selective and guided intracellular delivery of chemotherapy to drug-resistant cancer cells. *Biomaterials* 33: 1821–1826.
65. Wohlbold, L. et al. 2003. Inhibition of bcr-abl gene expression by small interfering RNA sensitizes for imatinib mesylate (STI571). *Blood* 102: 2236–2239.
66. Dean, N. M. and C. F. Bennett. 2003. Antisense oligonucleotide-based therapeutics for cancer. *Oncogene* 22: 9087–9096.
67. Barhoumi, A., R. Huschka, R. Bardhan, M. W. Knight, and N. J. Halas. 2009. Light-induced release of DNA from plasmon-resonant nanoparticles: Towards light-controlled gene therapy. *Chem Phys Lett* 482: 171–179.
68. Bardhan, R., S. Lal, A. Joshi, and N. J. Halas. 2011. Theranostic nanoshells: From probe design to imaging and treatment of cancer. *Acc Chem Res* 44: 936–946.
69. Guo, S. et al. 2010. Enhanced gene delivery and siRNA silencing by gold nanoparticles coated with charge-reversal polyelectrolyte. *ACS Nano* 4: 5505–5511.
70. Han, L. et al. 2012. Enhanced siRNA delivery and silencing gold-chitosan nanosystem with surface charge-reversal polymer assembly and good biocompatibility. *ACS Nano* 6: 7340–7351.
71. Bekyarova, E. et al. 2005. Applications of carbon nanotubes in biotechnology and biomedicine. *J Biomed Nanotechnol* 1: 3–17.
72. Tamburro, J. and K. Canadiate. 2009. Carbon Nanotubes: Cancer Treatment Of The Future. *Ninth Annual Freshman Conference*, University of Pittsburgh, PA, Vol. B4, pp. 1–6.
73. Madani, S. Y., N. Naderi, O. Dissanayake, A. Tan, and A. M. Seifalian. 2011. A new era of cancer treatment: carbon nanotubes as drug delivery tools. *Int J Nanomed* 6: 2963–2979.
74. Elhissi, A. M., W. Ahmed, I. U. Hassan, V. R. Dhanak, and A. D'Emanuele. 2012. Carbon nanotubes in cancer therapy and drug delivery. *J Drug Deliv* 2012: 837327.
75. Beg, S. et al. 2011. Advancement in carbon nanotubes: Basics, biomedical applications and toxicity. *J Pharm Pharmacol* 63: 141–163.
76. Khazaei, A., M. N. Rad, and M. K. Borazjani. 2010. Organic functionalization of single-walled carbon nanotubes (SWCNTs) with some chemotherapeutic agents as a potential method for drug delivery. *Int J Nanomed* 5: 639–645.
77. Yang, R. et al. 2006. Single-walled carbon nanotubes-mediated *in vivo* and in vitro delivery of siRNA into antigen-presenting cells. *Gene Ther* 13: 1714–1723.
78. Tang, M. F., L. Lei, S. R. Guo, and W. L. Huang. 2010. Recent progress in nanotechnology for cancer therapy. *Chin J Cancer* 29: 775–780.
79. Liu, Z., J. T. Robinson, S. M. Tabakman, K. Yang, and H. Dai. 2011. Carbon materials for drug delivery and cancer therapy. *Materials Today* 14: 316–323.
80. Gulati, N. and H. Gupta. 2012. Two faces of carbon nanotube: Toxicities and pharmaceutical applications. *Crit Rev Ther Drug Carrier Syst* 29: 65–88.
81. Lam, C. W., J. T. James, R. McCluskey, and R. L. Hunter. 2004. Pulmonary toxicity of single-wall carbon nanotubes in mice 7 and 90 days after intratracheal instillation. *Toxicol Sci* 77: 126–134.
82. Teeguarden, J. G. et al. 2011. Comparative proteomics and pulmonary toxicity of instilled single-walled carbon nanotubes, crocidolite asbestos, and ultrafine carbon black in mice. *Toxicol Sci* 120: 123–135.
83. Morimoto, Y. et al. 2012. Pulmonary toxicity of well-dispersed single-wall carbon nanotubes after inhalation. *Nanotoxicology* 6: 766–775.
84. Tang, S. et al. 2012. Short- and long-term toxicities of multi-walled carbon nanotubes *in vivo* and in vitro. *J Appl Toxicol* 32: 900–912.
85. Fujitani, T. et al. 2012. Teratogenicity of multi-wall carbon nanotube (MWCNT) in ICR mice. *J Toxicol Sci* 37: 81–89.

86. Harris, A. T., C. H. See, J. Liu, O. Dunens, and K. MacKenzie. 2008. Towards the large-scale synthesis of carbon nanotubes in fluidised beds. *J Nanosci Nanotechnol* 8: 2450–2457.

87. Yakobson, B. and R. Smalley. 1997. Fullerene nanotubes: $C_{1,000,000}$ and beyond. *American Scientist* 85: 324–337.

88. Guo, T. et al. 1995. Self-assembly of tubular fullerenes. *J Phys Chem* 99: 10694–10697.

89. Guo, T., P. Nikolaev, A. Thess, D. T. Colbert, and R. E. Smalley. 1995. Catalytic growth of single-walled manotubes by laser vaporization. *Chem Phys Lett* 243: 49–54.

90. Scott, C. D., S. Arepalli, P. Nikolaev, and R. E. Smalley. 2001. Growth mechanisms for single-wall carbon nanotubes in a laser-ablation process. *Appl Phys A* 72: 573–580.

91. Pastorin, G. 2009. Crucial functionalizations of carbon nanotubes for improved drug delivery: A valuable option? *Pharm Res* 26: 746–769.

92. Meng, L., X. Zhang, Q. Lu, Z. Fei, and P. J. Dyson. 2012. Single walled carbon nanotubes as drug delivery vehicles: targeting doxorubicin to tumors. *Biomaterials* 33: 1689–1698.

93. Yu, X., Y. Zhang, C. Chen, Q. Yao, and M. Li. 2010. Targeted drug delivery in pancreatic cancer. *Biochim Biophys Acta* 1805: 97–104.

94. Yu, M., S. Stott, M. Toner, S. Maheswaran, and D. A. Haber. 2011. Circulating tumor cells: approaches to isolation and characterization. *J Cell Biol* 192: 373–382.

95. Galanzha, E. I. et al. 2009. *In vivo* magnetic enrichment and multiplex photoacoustic detection of circulating tumour cells. *Nat Nanotechnol* 4: 855–860.

96. Fabbro, C. et al. 2012. Targeting carbon nanotubes against cancer. *Chem Commun (Camb)* 48: 3911–3926.

97. Tan, A., S. Y. Madani, J. Rajadas, G. Pastorin, and A. M. Seifalian. 2012. Synergistic photothermal ablative effects of functionalizing carbon nanotubes with a POSS-PCU nanocomposite polymer. *J Nanobiotechnol* 10: 34.

98. Zhang, W., Z. Zhang, and Y. Zhang. 2011. The application of carbon nanotubes in target drug delivery systems for cancer therapies. *Nanoscale Res Lett* 6: 1–22.

99. Lay, C. L., J. Liu, and Y. Liu. 2011. Functionalized carbon nanotubes for anticancer drug delivery. *Expert Rev Med Devices* 8: 561–566.

100. Madani, S. Y., A. Tan, M. Dwek, and A. M. Seifalian. 2012. Functionalization of single-walled carbon nanotubes and their binding to cancer cells. *Int J Nanomed* 7: 905–914.

101. Hitoshi, K., M. Katoh, T. Suzuki, Y. Ando, and M. Nadai. 2011. Differential effects of single-walled carbon nanotubes on cell viability of human lung and pharynx carcinoma cell lines. *J Toxicol Sci* 36: 379–387.

102. Prakash, S., M. Malhotra, W. Shao, C. Tomaro-Duchesneau, and S. Abbasi. 2011. Polymeric nanohybrids and functionalized carbon nanotubes as drug delivery carriers for cancer therapy. *Adv Drug Deliv Rev* 63: 1340–1351.

103. Bottini, M., N. Rosato, and N. Bottini. 2011. PEG-modified carbon nanotubes in biomedicine: Current status and challenges ahead. *Biomacromolecules* 12: 3381–3393.

104. Gu, Y. J., J. Cheng, J. Jin, S. H. Cheng, and W. T. Wong. 2011. Development and evaluation of pH-responsive single-walled carbon nanotube-doxorubicin complexes in cancer cells. *Int J Nanomed* 6: 2889–2898.

105. Ji, Z. et al. 2012. Targeted therapy of SMMC-7721 liver cancer in vitro and *in vivo* with carbon nanotubes based drug delivery system. *J Colloid Interface Sci* 365: 143–149.

106. Chen, C. et al. 2012. EGF-functionalized single-walled carbon nanotubes for targeting delivery of etoposide. *Nanotechnology* 23: 045104.

107. Chen, G. et al. 2012. In vitro and *in vivo* studies of pirarubicin-loaded SWNT for the treatment of bladder cancer. *Braz J Med Biol Res* 45: 771–776.

108. Taghdisi, S. M., P. Lavaee, M. Ramezani, and K. Abnous. 2011. Reversible targeting and controlled release delivery of daunorubicin to cancer cells by aptamer-wrapped carbon nanotubes. *Eur J Pharm Biopharm* 77: 200–206.

109. Arsawang, U. et al. 2011. How do carbon nanotubes serve as carriers for gemcitabine transport in a drug delivery system? *J Mol Graph Model* 29: 591–596.

110. Guven, A., I. A. Rusakova, M. T. Lewis, and L. J. Wilson. 2012. Cisplatin@US-tube carbon nano-capsules for enhanced chemotherapeutic delivery. *Biomaterials* 33: 1455–1461.
111. Liu, X. et al. 2011. Optimization of surface chemistry on single-walled carbon nanotubes for *in vivo* photothermal ablation of tumors. *Biomaterials* 32: 144–151.
112. Wang, L. et al. 2011. Synergistic enhancement of cancer therapy using a combination of docetaxel and photothermal ablation induced by single-walled carbon nanotubes. *Int J Nanomed* 6: 2641–2652.
113. Ren, J. et al. 2012. The targeted delivery of anticancer drugs to brain glioma by PEGylated oxidized multi-walled carbon nanotubes modified with angiopep-2. *Biomaterials* 33: 3324–3333.
114. Chen, Z. et al. 2011. Adsorption behavior of epirubicin hydrochloride on carboxylated carbon nanotubes. *Int J Pharm* 405: 153–161.
115. Li, Q., D. Guo, R. Zhang, and X. Wang. 2012. Increasing anticancer drug internalization induced by new Au-MWCNTs nanocomposite. *J Nanosci Nanotechnol* 12: 2192–2198.
116. Lu, Y. J., K. C. Wei, C. C. Ma, S. Y. Yang, and J. P. Chen. 2012. Dual targeted delivery of doxorubicin to cancer cells using folate-conjugated magnetic multi-walled carbon nanotubes. *Colloids Surf B Biointerfaces* 89: 1–9.
117. Yang, F. et al. 2011. Magnetic functionalised carbon nanotubes as drug vehicles for cancer lymph node metastasis treatment. *Eur J Cancer* 47: 1873–1882.
118. Liu, Z. et al. 2008. Drug delivery with carbon nanotubes for *in vivo* cancer treatment. *Cancer Res* 68: 6652–6660.
119. Sobhani, Z., R. Dinarvand, F. Atyabi, M. Ghahremani, and M. Adeli. 2011. Increased paclitaxel cytotoxicity against cancer cell lines using a novel functionalized carbon nanotube. *Int J Nanomed* 6: 705–719.
120. Ghosh, S. et al. 2009. Increased heating efficiency and selective thermal ablation of malignant tissue with DNA-encased multiwalled carbon nanotubes. *ACS Nano* 3: 2667–2673.
121. Moon, H. K., S. H. Lee, and H. C. Choi. 2009. In vivo near-infrared mediated tumor destruction by photothermal effect of carbon nanotubes. *ACS Nano* 3: 3707–3713.
122. Chakravarty, P. et al. 2008. Thermal ablation of tumor cells with antibody-functionalized single-walled carbon nanotubes. *Proc Natl Acad Sci USA* 105: 8697–8702.
123. Huang, N. et al. 2010. Single-wall carbon nanotubes assisted photothermal cancer therapy: animal study with a murine model of squamous cell carcinoma. *Laser Surg Med* 42: 638–648.
124. Iancu, C. et al. 2011. Enhanced laser thermal ablation for the in vitro treatment of liver cancer by specific delivery of multiwalled carbon nanotubes functionalized with human serum albumin. *Int J Nanomed* 6: 129–141.
125. Mocan, L. et al. 2011. Selective *ex-vivo* photothermal ablation of human pancreatic cancer with albumin functionalized multiwalled carbon nanotubes. *Int J Nanomed* 6: 915–928.
126. Wang, L. et al. 2013. Synergistic anticancer effect of RNAi and photothermal therapy mediated by functionalized single-walled carbon nanotubes. *Biomaterials* 34: 262–274.
127. Chen, H. et al. 2012. Functionalization of single-walled carbon nanotubes enables efficient intracellular delivery of siRNA targeting MDM2 to inhibit breast cancer cells growth. *Biomed Pharmacother* 66: 334–338.
128. Karmakar, A. et al. 2011. Ethylenediamine functionalized-single-walled nanotube (f-SWNT)-assisted in vitro delivery of the oncogene suppressor p53 gene to breast cancer MCF-7 cells. *Int J Nanomed* 6: 1045–1055.
129. Sanz, V., H. M. Coley, S. R. Silva, and J. McFadden. 2012. Protamine and chloroquine enhance gene delivery and expression mediated by RNA-wrapped single walled carbon nanotubes. *J Nanosci Nanotechnol* 12: 1739–1747.
130. Villa, C. H. et al. 2011. Single-walled carbon nanotubes deliver peptide antigen into dendritic cells and enhance IgG responses to tumor-associated antigens. *ACS Nano* 5: 5300–5311.
131. Gul-Uludag, H., W. Lu, P. Xu, J. Xing, and J. Chen. 2012. Efficient and rapid uptake of magnetic carbon nanotubes into human monocytic cells: Implications for cell-based cancer gene therapy. *Biotechnol Lett* 34: 989–993.

16

Bioinspired Nanomaterials for Bone Tissue Engineering

Eilis Ahern, Timothy Doody, and Katie B. Ryan

CONTENTS

16.1 Introduction

The loss or failure of an organ or tissue is one of the most frequent, devastating, and costly problems in human health care [1,2]. Existing approaches to repair or replace damaged bone have focused on bone grafting techniques; however, they are limited by supply problems, disease transmission, immune responses, and cost [3]. The need for high-quality bone graft substitutes has resulted in research focusing on tissue engineering using biomaterials to provide alternatives to support bone regeneration [4,5]. Tissue engineering typically utilizes suitable biomaterials to prepare a scaffold support that is implanted and subsequently facilitates remodeling of the defect site. However, the regeneration of natural bone tissue is a complex, coordinated temporal process involving molecular, cellular, biochemical, and mechanical cues [6]. Most biomaterials alone cannot match the efficacy of bone grafts because they lack both the osteogenic cells and osteoinductive proteins that make autograft bone so advantageous [7,8]. In the pursuit of a clinically effective alternative to bone grafting, current state of the art envisages the use of tissue-engineered constructs that play a more active role in the regenerative process [9]. Different approaches are being explored including using biomaterials that act as mimetics of the extracellular matrix (ECM), to design multifunctional constructs that combine stem or progenitor cells to support osteogenesis, facilitate the spatial and temporal presentation of signaling molecules including growth factors (GFs) to promote osteoinduction, respond to environmental stimuli, and support revascularization of the de novo tissue [3,10–12]. Alternatively, another promising approach–focusing on the implantation of acellular scaffolds to recruit cells into the substrate and stimulate tissue development, circumventing the need to culture and expand cell-seeded scaffolds ex vivo prior to implantation [13].

To enhance regeneration, tissue engineering approaches have taken their inspiration from nature and in particular from an enhanced understanding of the structure and function of the ECM to design biomimetic materials that recreate the complexity and functionality of tissue-specific microenvironments, to maintain and regulate cell behavior, and direct functional tissue regeneration. Materials are being engineered to replicate the structural hierarchy of native bone tissue on multiple length scales from the nano- to the macroscale [14] and to recapitulate the compositional and surface features to elicit the appropriate biochemical and physical microenvironmental cues intrinsic to functional tissue development [9,15,16]. The considerations for scaffold design are complex and a range of materials including natural and synthetic polymers [17] have been widely investigated for their potential application in tissue engineering [18–20]. Bioengineering approaches underpinned by advances in nanotechnology are being used to prepare sophisticated biomaterials that bring together the more favorable aspects of natural and synthetic polymers in a single material as well as to produce complex macromolecular self-assemblies that can act as ECM substitutes in tissue regeneration. Generic substrates can be customized for tissue-specific applications through functionalization of the physical properties using patterning techniques to control the nanotopography or by controlling the spatial and temporal presentation of biochemical signaling cues to direct cellular behavior [21]. Interfacial bioactivity may be controlled through the inclusion of bioactive glasses and calcium phosphates (CaPs) [22,23], which have been studied extensively in hard tissue regeneration and other biomedical applications due to their biocompatibility and bioactivity [24–26].

Traditionally, scaffold fabrication methods have focused on controlling structural features relating to porosity and internal pore architecture so as to maximize the diffusion of nutrients as well as conferring appropriate mechanical properties [27]. However, advances in nanotechnology are increasingly facilitating the design and fabrication of sophisticated constructs that tailor the presentation of bioactive signals, in addition to recapitulating the nanoscale dimensions encountered by cells in native tissues [28]. Nanomaterials are described by having at least one of their dimensions less than 100 nm in size and may be distinguished from bulk materials of the same composition by virtue of the dramatic increase in surface area and surface area-to-volume ratio [29,30]. In the context of bone regeneration, the significant increase in surface area offers the prospect for improved cell attachment due to the increased availability of binding sites [31,32]. Another compelling justification for nanoscale engineering of constructs centers on the fact that cells are equipped to interact with nanometric features of the surrounding environment [33]. They contain nanoscale features such as integrins and focal contacts as well as fine processes (e.g., cilia and filopodia) that are compatible in size with ECM structures and many biologically functional signaling molecules, and which play an integral role in interaction with their environment [34].

This chapter will illustrate how an increased knowledge of the biological composition, structural complexity, and functionality of native bone tissue, in addition to a better understanding of the molecular processes at play in bone regeneration, have provided an important foundation for the design of biomimetic materials to support functional tissue regeneration. The chapter highlights how advances in nanotechnology have impacted the fundamental elements of tissue engineering, namely, the design and fabrication of biomaterials and scaffolds and spatiotemporal presentation of cell signaling molecules including GFs, to promote bone repair.

16.2 Bone Tissue Regeneration

Following slight loss or damage to the bone tissue, a cascade of events is initiated resulting in recruitment of mesenchymal cells to the wound site, their differentiation into osteoblasts, synthesis of osteoid, matrix production, and, ultimately, calcification of the ECM leading to self-repair and regeneration of the tissue [35]. However, regeneration varies depending on a variety of factors including the cause and the size of the defect as well as patient factors (e.g., age) [36]. In cases of impaired healing where large or critical-size defects are caused by trauma or disease, strategies to aid in the recruitment of cells into the injury site to support skeletal reconstruction are required [13]. Popular approaches to facilitate the renewal of bone tissue include the use of bone graft techniques, which involve transplanting regeneration-competent cells to a particular tissue defect [15]. The cells may be sourced from the patient (autologous cells) or from a human donor (allogeneic) [37] and may be typically harvested from the iliac crest due to the plentiful supply of progenitor cells and GFs, quantity of bone available, and relative ease of harvest [38]. Other sources include the distal femur, proximal tibia, fibula, and distal radius [39]. Although the graft does not provide any mechanical stability [40], it is particularly advantageous as it allows for transplantation of bone containing live cells, bioactive molecules, and in some cases, a vascular supply that enhances cell survival, tissue regeneration, and healing [41,42]. The cells present in the graft are capable of responding to local stimuli and releasing GFs of their own, which accelerates angiogenesis and bone formation [39].

Worldwide, 2.2 million bone graft procedures are performed annually, representing about 10% of all orthopedic operations [43,44]. The drawbacks with this approach are well documented and include the limited supply of tissue that can be harvested in the case of autografts as well as pain, donor site morbidity, and prolonged recovery times [45]. In the case of allografts, the risk of pathogen transmission and the generation of immune and inflammatory responses in the recipient are of particular concern [42]. Demineralized bone matrix (DBM) has been investigated to overcome the problems of foreign body immunogenic reactions associated with allografts. An acid extraction demineralization process removes the antigenic surface structure from the allograft resulting in a material with variable osteoconductive and osteoinductive potential compared to normal bone. Its merit as a graft substitute is attributed to the presence of collagen and other proteins as well as bone morphogenetic proteins (BMPs) that are retained after the demineralization process [46]. Bone xenografts have also been investigated but the potential risk of virus infection and host rejection is considered to outweigh any potential benefit [13].

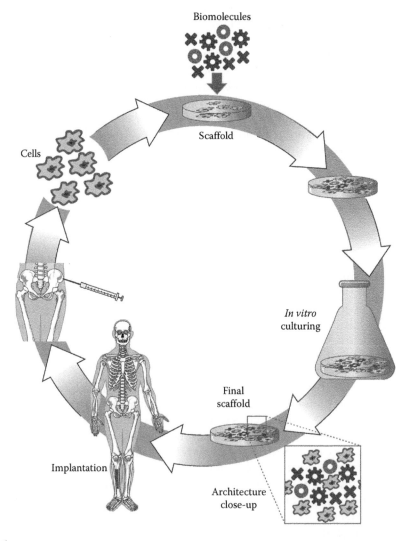

FIGURE 16.1
(See color insert.) Bone tissue engineering utilizing 3-D scaffold structures. Cells obtained from healthy bone tissue followed by isolation and expansion in culture. Cells are seeded onto a suitable scaffold with appropriate biomolecules. The construct is cultured *in vitro*. Transplantation of the final scaffold incorporating cells and biomolecules into the defect area.

The evolving field of tissue engineering has sought to develop a clinically effective solution to support the regeneration of bone defects in an effort to address the limitations associated with bone grafting [47]. Tissue engineering approaches can be divided into three categories. The first entails direct injection of a cell bolus into the target or systemic circulation; however, cell survival postinjection is a concern. The second category focuses on guided tissue approaches to deliver low-molecular-weight drugs, proteins, or oligonucleotides. The incorporation of signaling molecules can stimulate regeneration *in vivo* by attracting regeneration-competent cells and stimulating their proliferation and differentiation into tissue-specific cells, or by blocking regeneration-suppressing signals [2,15,19,45]. A number of osteoinductive products are employed clinically to help promote bone growth, including OP-1 (Stryker, USA) that combines collagen type I of bovine origin with recombinant, human-derived BMP-7 (rhBMP-7). Another example includes INFUSE (Medtronic, MN), which contains a collagen mesh combined with genetically engineered BMP (rhBMP-2), used in the treatment of certain spinal conditions [48]. In 2007, the market for this bone device alone accounted for more than $760 million in sales [49]. However, despite their commercial successes, these products have poor control over protein release and fall short of ideal requirements for functional tissue regeneration. The third strategy depicted in Figure 16.1 involves combining stem cells, such as MSCs, or mature cells (for osteogenesis) with biocompatible materials or scaffolds (for osteoconduction) and appropriate GFs (for osteoinduction), in order to generate and maintain bone [10,50–53]. In most approaches, some form of cell culturing *in vitro* is undertaken to ensure sufficient cell numbers, stem cell differentiation, and tissue maturation prior to implantation [54]. Bioreactor systems have been investigated to enhance this *in vitro* step because they can mechanically stimulate cells and dynamic culture conditions can support nutrient exchange [55]. During the initial stages as the cells proliferate, they deposit a mostly collagenous, bioactive ECM that becomes increasingly mineralized in prolonged cultures. The entire construct (both cells and scaffold) is then transplanted into the host [42].

16.3 Biological Considerations

16.3.1 Bone Composition

Bone is the highly vascular, hard connective substance that forms the basis of the skeletal system and is primarily composed of bone cells, inorganic minerals, and organic material. It provides a variety of important functions such as protection, support, structure, movement, homeostasis, cell production, and storage [56,57]. The majority of the bone organic phase is composed of type I collagen, with the remainder composed of noncollagenous proteins including GFs, cytokines, proteoglycans, and glycoproteins [58]. Inorganic mineral deposits consisting mainly of calcium (Ca) and phosphate are dispersed throughout the gaps in the collagen framework [59]. These deposits have an apatitic structure, the most common being carbonated hydroxyapatite [60]. There are a number of biological apatites due to the presence of impurities (hydrogen phosphate, sodium, magnesium, and citrate) in the nanostructure as well as variations in the CaP ratio and stoichiometry, giving rise to diverse morphological forms and biological properties [61,62].

16.3.2 Bone Remodeling

Bone has a dynamic ability to remodel itself under changing conditions and regenerate after damage [63], although an upper fracture size limit exists for this self-repair process [64]. Bone remodeling is a continuous process, regulated by the actions of osteoclasts and osteoblasts that control bone resorption and formation [65]. It is influenced by a number of factors, including mechanical requirements, Ca homeostasis, hormones, GFs, and cytokines. The term osteoblast describes a diverse population of cells including immature osteoblast lineage cells, differentiating, and mature matrix-producing osteoblasts that are important in bone formation in a process called osteogenesis [66]. Working in groups along the layer of new bone being produced [63], osteoblasts lay down premineralized bone matrix, termed osteoid, and subsequently facilitate its mineralization through the deposition of Ca salts [67]. As the ECM develops, matures, and undergoes mineralization, matrix proteins are synthesized and deposited [68,69]. During bone formation, a subpopulation of osteoblasts undergo terminal differentiation, become trapped in the osteoid [66], and form a network extending throughout the mineralized bone [66]. They are subsequently termed osteocytes and exhibit decreased metabolic activity compared to active osteoblasts. Although, they are believed to have a role in mediating osteoblastic and osteoclastic activity [63,68]. Osteoclasts are multinucleate, terminally differentiated myeloid cells that are responsible for bone resorption [66] and are influenced by systemic hormones and cytokines [63]. During the process of bone resorption, the bone matrix undergoes acidification and proteolysis, hydroxyapatite (HA) crystals are mobilized, and residual collagen fibers are digested. The balance that exists between the production and function of osteoblasts and osteoclasts is an essential determinant in bone homeostasis [70].

16.3.3 Hierarchical Structure of Bone

Bone has a complex, composite structure that may be described at different lengths from the macro- to the nanoscale (depicted in Figure 16.2) [57,71]. At the macroscopic level, the outer part of bone is composed of hard cortical tissue with a compact pattern and is much stronger than the inner cancellous (trabecular) bone [72,73] that is a highly porous structure and contains bone marrow and blood vessels. The differences in structure and composition give rise to variability in the mechanical properties, as shown in Table 16.1 [73,74]. Cortical bone is seen to have a greater compressive strength, flexural strength, fracture toughness, and a higher Young's modulus. However, it has a lower strain to fracture percentage than cancellous bone [73,75]. The microstructure consists of osteons (Haversian systems) that arise when sheets of lamellae form layers around a central Haversian canal that contains blood vessels and nerve fibers. At the submicrostructure level, lamellae of mineralized collagen fibers are present [72,76,77]. At the nanoscale, ECM constituents including collagenous and noncollagenous proteins as well as inorganic minerals (biological apatites such as carbonated hydroxyapatite) are present. Triple-helical type I collagen molecules self-assemble into fibrils and are arranged such that individual triple helices are staggered in the axial direction by 40 nm relative to its neighboring molecule and overlap the adjacent molecule by 27 nm giving rise to a characteristic axial periodicity, D of 67 nm [78–80]. Crystals of apatite occur in a discontinuous manner and occupy the gaps between the collagen molecules. The presence of noncollagenous organic proteins such as osteopontin, osteonectin, and osteocalcin is thought to influence mineral deposition and has an effect on the orientation, size, and crystal habit [72,77].

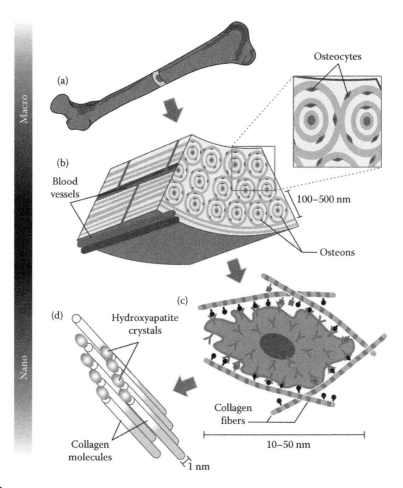

FIGURE 16.2
(See color insert.) Hierarchical organization of bone tissue from the macro-(centimeter) scale to the nanoscale structure: (a) at the macrostructural level, bone may be described on the exterior by a hard, calcified, cortical (compact) layer; (b) at the submicron level, osteons (Haversian systems) are evident; (c) cells interact with ligand sites within the ECM through their membrane receptors; and (d) the distinctive nanostructure that comprises the ECM is composed of an inorganic CaP mineral phase and an organic component chiefly composed of type I collagen. (From Stevens, M.M. and George, J.H., *Science,* 310, 1135–1138. Copyright 2005. Redrawn and reproduced with permission from AAAS.)

TABLE 16.1

Biomechanical Properties of Cortical and Cancellous Bone

Property	Cortical Bone	Cancellous Bone
Compressive strength (MPa)	100–230 [81]	2–12 [82]
Flexural, tensile strength (MPa)	50–150 [81]	10–20 [82]
Strain to failure (%)	1–3 [83]	5–7 [83]
Fracture toughness (MPa.m$^{1/2}$)	2–12 [81]	0.1 [82]
Young's modulus (GPa)	7–30 [81,83]	0.5–0.05 [82,83]

Source: Adapted from Hench, L.L. and Wilson, J., eds., Introduction, in *An Introduction to Bioceramics*, pp. 1–24, World Scientific, River Edge, NJ, 1993.

16.3.4 The Cellular Environment

The cell is immersed in a dynamic landscape and is sensitive to a diverse array of signals from its environment including those from neighboring cells, physical signals, soluble bioactive factors (e.g., GFs), and insoluble macromolecules of the ECM [50]. Normal tissue cells are anchorage dependent [84] and in the body, it is the ECM, a multicomponent structure within or on which cells adhere. The primary constituents include fibrous structural proteins (e.g., collagens, laminins, fibronectin, vitronectin, and elastin), specialized proteins (e.g., GFs, small matricellular proteins, and small integrin-binding glycoproteins [SIBLINGs]), and proteoglycans [85,86]. ECM composition and organization changes from tissue to tissue, throughout development, and during aging and disease [50]. Collagens, the most abundant proteins, play an important role in the structural framework [87], while other noncellular ECM components including fibronectin, laminin, and glycosaminoglycans (GAGs) have important roles in cell adhesion, protein adhesion, and GF binding [88].

The significance of the ECM has generally been described in terms of its structural support for cell adhesion and role in the physical integrity of tissue. However, advances in the understanding surrounding ECM composition, architecture, and function highlight its importance in regulating diverse cellular processes including cell growth, survival, differentiation, and in morphogenesis and the maintenance of functional homeostasis [48,89–91]. The ECM has been shown to play an instructive role by providing structural information to integrins and other cell surface molecules [92], and it is believed that ECM proteins and inorganic minerals represent an abundance of nanoscale structures that can play a role in this cell-matrix signaling [93]. The ECM can bind soluble/secreted factors, thereby acting as a reservoir for GFs and cytokines. In addition to the efficient temporal and spatial presentation of these signaling molecules to cell surface receptors [90,94], it protects against their degradation and modulates their synthesis [88].

Cellular interaction with the ECM and its subsequent behavior is mediated by integrin-dependent adhesions called focal adhesions [95], which play an important role in the bidirectional transfer between extracellular molecules and the cytoskeleton [96]. Fundamental to this are integrins, the cell adhesion receptors, which are a family of heterodimeric transmembrane receptors on the cell surface [97]. In many cases, recognition and binding of integrins with ligands is mediated by short peptide sequences within the ECM protein molecules [98]. The most notable example discussed in the literature is the arginine–glycine–aspartic acid (RGD) recognition sequence present in the hydrophilic loop of several ECM proteins including fibronectin, vitronectin, and tenascin. Ligand binding leads to a complex array of events including ligand clustering, cytoskeletal rearrangements, gene expression, and protein synthesis [15,52]. Consequently, cellular interaction with ECM effector molecules embedded in the ECM can have implications for the regulation of cellular adhesion, differentiation, proliferation, migration, and ultimately, tissue remodeling [15,90]. In addition to the ECM composition, cells can sense and respond to geometric cues at the micro- and nanoscale. The local topography including roughness and fiber diameter, as well as the type of ordered topography, for example, ridges, steps, grooves, pillars, and pits, in addition to molecular conformation, can influence cell behavior [32,99]. It is proposed that the diversity in these features can alter cell activities by affecting ligand clustering, the distribution of focal adhesions, and the modulation of intracellular signaling pathways that control transcription and gene expression [99]. A greater understanding regarding how cells interpret and process these nanoscale features to promote greater cellular interaction and tissue regeneration will have a key influence on the design of novel materials in tissue engineering [100].

The ECM is responsive to cellular activity and is constantly undergoing remodeling as cells build and reshape the ECM by degrading and reassembling it. ECM remodeling can induce and be affected by extracellular and cellular tension [85]. It is governed by the proteolytic activity of specific matrix metalloproteinases (MMPs) and serine proteases [101] and is associated with a change in the chemical cues present in the microenvironment as well as physical changes in matrix elasticity [50]. The breakdown process is important in modulating extracellular signaling events as MMPs degrade tissue matrix releasing small bioactive peptides and GFs stored within the ECM [85]. In certain cases (e.g., collagens and some laminin isoforms), the RGD sequences only become exposed after degradation or proteolytic cleavage [97]. MMPs also play a role in the initiation of cell migration by generating chemotactic signals [102,103]. While aberrations in MMP functioning have been observed to be characteristic in many disease states (e.g., metastatic tumors).

16.4 Scaffold Requirements

An important facet of tissue engineering is controlling the interplay between materials (scaffolds), cells, and signaling molecules including GFs to create environments that promote the regeneration of functional tissues [104]. Many studies employ artificial scaffolds either based on solid polymeric materials or hydrogel structures to replicate the function of the ECM, although microparticles, nanoscale inorganic mineral phases, and polymer nanoparticles have also been investigated [105]. However, particulate systems lack the dimensional stability of scaffold structures when first surgically implanted, and fabrication of 3-D porous, composite, or hybrid scaffolds that act as structural templates for bone regeneration represent a viable solution [106]. The scaffold provides mechanical support, shape, and cell-scale architecture for *de novo* tissue construction both *in vitro* and *in vivo* as cells expand, organize, and develop into the desired tissue [85,107,108]. It may also play a role in the localization and delivery of cells and/or GFs to specific sites in the body [109]. During tissue formation, the starter scaffold matrix is degraded, resorbed, or metabolized and eventually replaced by the functional tissue [2]. The degree of remodeling depends on the tissue itself (cancellous bone, 3–6 months; cortical bone, 6–12 months), host anatomy, and physiology [75]. It is known that cells are responsive to the structure, mechanical properties, and the chemical features of the material surface [110]. Therefore, the scaffold plays a fundamental role in the regenerative process, and rational scaffold design is important to provide appropriate physical, chemical, and biological cues to promote cell attachment, migration, growth, and tissue regeneration. Despite the lack of consensus on the precise features, a number of key requirements outlined in Table 16.2 have been described to guide the design and development of scaffolds suitable for bone tissue engineering.

TABLE 16.2

Considerations Related to the Design and Production of Scaffolds Employed
in Bone Tissue Engineering

Property	Features
Biocompatible	The biomaterials used to construct the scaffold as well as the breakdown products should be biocompatible to avoid foreign body responses that might otherwise lead to rejection or interfere with the healing cascade [111].
Degradable	Scaffold breakdown should be contemporaneous with tissue regeneration and ensure that tissue volume along with its physical and mechanical properties is preserved as the scaffold is replaced by native tissue [20,112].
Scaffold architecture	A porous 3-D structure is required to fill the size and shape of the tissue void, accommodate cells, and guide their growth and subsequent matrix deposition while supporting tissue vascularization [113].
	Pore size is an important determinant in the success and fate of tissue regeneration. It is dependent on cell and tissue type. In the case of bone, pore sizes in the range $\geq$100–400 μm have been suggested to facilitate cell migration, tissue ingrowth, and vascularization [31,35,75,111,114,115].
	An interconnected porous network is central to the vascularization of the tissue and nutrient/waste transfer [75,111].
Mechanical features	Sufficient to facilitate handling and surgical implantation [111].
	Replicate those of the target tissue to provide important cues that instruct tissue development. Cells convert these signals into biochemical signals by "mechanotransduction" [110,116,117].
	Maintain the requisite functional strength and stiffness until sufficient tissue regeneration has occurred to support itself and replace the scaffold [118,119].
Nanotopography	Important in directing cellular interaction with the substrate and guiding tissue development [120–123]. Can influence cell proliferation and differentiation [124–126].
Biomaterial properties	Biomaterials should ideally be osteoconductive and osteoinductive and support osseointegration [127].
	Exert spatiotemporal control over presentation of bioactive signals including GFs, e.g., BMPs, adhesive ligands, and proteases that have roles in self-renewal, migration, differentiation, as well as the cell fate of uncommitted stem or progenitor cells [15,52].
Production	Control *in vitro* microenvironment prior to implantation, as cells respond to the biomechanical and biochemical cues present in the culture, which may alter subsequent cell behavior [128,129].
	Control over the processing of natural materials and proteins to minimize structural breakdown to avoid/minimize variability between batches and immune responses [19].
	Technology should facilitate translation from bench scale to the clinic, production according to good manufacturing practice standards, and be cost-effective [20,111].

16.5 Biomaterials for Bone Regeneration

16.5.1 Introduction

In order to recreate physiologically relevant cell behavior in the scaffold environment, biomaterial substrates are increasingly being designed to mimic the natural cellular environment to promote cell attachment, migration, and guide the 3-D development of functional tissue [130]. In particular, the success of grafting procedures, which has been attributed to the physical and biological similarity of the transplanted tissue with that of the host tissue, has provided inspiration for designing the ideal bone substitute. To this end, many biologically derived and synthetic materials are being engineered to recreate the osteoinductive, osteoconductive, and osseointegration features intrinsic to bone regeneration. From a structural perspective, solid polymer scaffolds, hydrogels, and micro- and nanoparticles are being investigated to design scaffolds on multiple length scales to better emulate biological matrices compared to those with single-scale features [14]. For example, scaffolds with nanofibrous architecture are being designed to recapitulate the structural architecture of the nanofibrous collagen ECM [131]. From a compositional perspective, polymers are being combined with inorganic nanocrystalline mineral phases that resemble those in the native bone to better replicate the structural complexity and consequently provide improved cell signaling and regenerative success [64,132]. Biomaterials are also being engineered to direct cell behavior at the biomolecular level [15], through the inclusion of chemotactic factors and binding proteins (or peptide motifs), to enhance cell recruitment and attachment [35]. In many cases, more sophisticated biomaterials are being designed to modulate cellular activity and orchestrate the complex regenerative process in a manner analogous to that in native ECM by incorporating multiple features including bioresponsive and bioinstructive properties [111].

16.5.2 Polymers

16.5.2.1 Natural Polymers

A number of naturally derived polymers have been investigated in the manufacture of synthetic ECM substitutes for bone regeneration including proteins such as collagen [133] and silk [134] as well as polysaccharides including alginate [135], chitosan [136], and hyaluronan [137]. They possess a number of intrinsic advantages including biological recognition, presentation of receptor-binding ligands, and the tendency toward cellular degradation [15,85]. Drawbacks associated with biologically derived materials include purification problems, batch to batch reproducibility, pathogen transmission, and the potential for host immune responses. Collagen is the primary protein component of skin, cartilage, and bone and to date, more than 20 distinct types have been characterized. Apart from differences in physical structure and composition, variability in hierarchical spatial arrangement helps explain the diversity in properties across various tissues [78,138]. Collagen type I, the most abundant in the ECM, plays an important role in bone tissue architecture and its tensile strength properties [78]. It possesses a number of other favorable properties that explain its popularity in tissue engineering applications (particularly as a matrix for cartilage repair) including its biodegradability, biocompatibility, ease of availability, and versatility [139,140]; however, its mechanical properties limit its use in orthopedic applications, particularly in load-bearing circumstances [111].

Combining collagen with other materials (polymers, HA) has been investigated as an approach to improve its mechanical properties and tissue functioning. Cross-linking with other biopolymers has been used to tailor its tensile strength, stability, and biodegradation [139]. Scaffolds containing collagen and GAG were prepared using a freeze-drying method. The polypeptide chains of the collagen were subsequently covalently cross-linked via dehydrothermal process to enhance their mechanical properties [31]. Studies on collagen–GAG scaffold systems were conducted to assess the impact of mean pore sizes ranging from 85 to 325 μm on MC3T3-E1 osteoblast cell attachment and proliferation. The study suggested that the specific surface area of the scaffold played an important role in initial cell adhesion, but the enhanced migration associated with larger pore sizes explained the highest cell numbers in scaffolds with a mean pore size of 325 μm [141]. Early studies were also conducted to assess the influence of the physicochemical and mechanical environment on stem cell behavior. The impact of altering the stiffness values using 1-ethyl-3-3-dimethylaminopropyl carbodiimide (EDAC) as an additional cross-linking treatment highlighted that stiffer scaffold gels (1.5 kPa) directed MSC differentiation toward an osteogenic lineage as determined using levels of RUNX2 expression, whereas SOX9 expression, which was used as a measure of chondrogenic lineage, was unregulated in gels when the stiffness decreased to 0.5 kPa. Altering the GAG type in the scaffold was also shown to be important in directing MSC differentiation, with chondroitin sulfate favoring osteogenic differentiation and hyaluronic acid favoring chondrogenic differentiation [142].

Apart from collagen, a number of silk biomaterials including silk fibroin, spider silk, and recombinant silk proteins have been used and have been investigated as scaffold substrates in bone regeneration applications [100]. Silk fibroin possesses a number of desirable characteristics including tunable material properties, cell compatibility, and robust mechanical properties supporting its use; however, its slow degradation rate may be a limiting factor [143]. In vitro studies have shown that silk-based scaffolds support human mesenchymal stem cell (hMSC) attachment and spreading with variability in the bone tissue responses as a consequence of processing conditions employed, the material's surface properties, and degradation rate [144].

16.5.2.2 Synthetic Polymers

Many of the new biomaterials used to develop ECM scaffolds are based on synthetic polymers. The driving force for their development stems from drawbacks with biologically derived polymers and the fact that synthetic materials have reproducible material and mechanical properties [118]. While these materials facilitate scale-up and permit further tailoring of the material properties (e.g., molecular weight, functional groups, polymer chain conformation) [105], they often lack the nano- and microscale biological motifs that direct host tissue response in a manner analogous to the native ECM. In addition, they also require systematic evaluation to ensure their biocompatibility [19]. A wide variety of synthetic polymers including poly(ethylene glycol) (PEG), poly(methyl methacrylate), poly(urethane), poly(lactic-co-glycolic acid) (PLGA), poly(propylene fumarate) (PPF), and poly(ε-caprolactone) (PCL) have been investigated as ECM substitutes and scaffolds for tissue regeneration, with varying degrees of success [130]. Studies have predominantly centered on degradable polymers that are eliminated from the body to create nonpermanent scaffolds that can be completely replaced by the generation of new tissue [20]. Physicochemical properties such as molecular weight, crystal property, copolymer ratio, solubility, degree of cross-linking, biodegradation, and biocompatibility influence the selection of polymers [145]. From a biological perspective, the ability of the biomaterial to support cellular adhesion and proliferation is of considerable importance but is often limited by the polymer's hydrophobic characteristics. Poor surface wetting can limit protein adsorption, cell attachment, and spreading [146]. Consequently, combinations of polymers have been engineered to overcome this problem.

Poly(α-hydroxy acids) in particular PLGA and poly(lactic acid) (PLA) alone or in combination with other polymers (e.g., PEG) are among the most extensively investigated synthetic polymers for drug delivery and tissue engineering applications [119,147–149] because of their biocompatibility, both of the starting material and breakdown products, and the potential to control payload release. The properties of PLA vary depending on the stereochemical form (L-PLA [P_LLA], D-PLA [P_DLA], or racemic mixture D, L-PLA [P_{DL}LA]) used. Degradation occurs via bulk hydrolysis at physiological pH, and the breakdown rate may be tailored from months to years by altering the polymer composition and molecular weight [150] or by introducing peptide segments that may be cleaved by cell proteases such as the MMPs [151]. This has the advantage of tailoring resorption in line with tissue development. However, changes in polymer composition should be considered in light of the potential effect on cell adhesion and ECM production [86]. Limitations to PLA/ PLGA usage in tissue engineering include its hydrophobicity and the generation of acidic species as it degrades, which further exacerbates polymer breakdown through an auto-catalytic process. Yang et al. observed poor cell growth on PLA films and PLGA (75:25) scaffolds, which they attributed to polymer hydrophobicity. However, peptide and protein surface modifications demonstrated enhanced cell attachment and spreading [152]. One approach to promote cellular interaction involved preparation of PLA scaffolds with nano-fibrous pore walls. This appeared to address the hydrophobicity problem and resulted in enhanced protein adsorption. These scaffolds exhibited selective take-up of vitronec-tin and fibronectin and greater osteoblastic cell attachment compared to scaffolds with solid pore walls [153]. These nanofibrous scaffolds designed to mimic the morphologi-cal function of collagen fibrils also substantially improved biomineralization compared to solid-walled scaffolds. Indices of osteoblastic differentiation (alkaline phosphatase [ALP] activity, RUNX2 protein, and sialoprotein mRNA) were also enhanced in the case of nano-fibrous scaffolds [154].

PCL is a biocompatible, semicrystalline, slowly degrading polymer that is widely investi-gated for bone regeneration due to its relative lack of toxicity and low cost, particularly where electrospinning is the method of fabrication [112]. A number of studies have demonstrated its potential in bone tissue engineering. It is considered to possess more favorable mechanical properties than PLGA with respect to trabecular bone [105] and it is comparatively cheaper than other polymers [155]. However, its use is constrained by its hydrophobicity, but this may be alleviated by combining it with other polymers and mineral apatites [156]. A nano-fibrous PCL scaffold fabricated using electrospinning and seeded with MSCs in osteogenic differentiation medium resulted in differentiation along an osteogenic lineage. The scaffolds supported cellular penetration and multilayer deposition, in addition to mineralization and the presence of type I collagen [157]. Li et al. also showed that nanofibrous PCL 3-D scaffolds designed to morphologically replicate collagen fibrils could support multilineage differentia-tion, depending on differentiation media. The authors suggested the potential value of this in multiphase tissue-engineered constructs [158].

16.5.2.3 Hydrogels

Hydrogels are cross-linked polymer networks that swell to form gels in the presence of water and may be described as physical, chemical, or biological depending on the cross-linking mechanism employed [20,159]. These versatile structures are composed of hydrated hydro-philic polymers and a number of biologically derived polymers including alginate, gelatin, and hyaluronic acid as well synthetic polymers including, PEG, poly(vinyl)alcohol (PVA), and poly(acrylic acid) (PAA), and copolymer derivatives PLA–PEG, have been investigated to form hydrogel networks for bone regeneration [160–164]. These substrates can be engineered to create tissue-specific microenvironments to increase the viability and efficacy of trans-planted cells [15]. This is a particularly promising strategy in the case of materials that lack the necessary biological activity due to the nonadhesive properties of the polymer [164,165]. A multitude of physical and chemical interactions between the cell and the hydrogel net-work are possible and controlling the complexity at the cell interface is an important element in enhancing tissue specificity [166]. Modification of the basic surface chemistry has been suggested to control cellular adhesion and behavior [167], while functionalization of the polymer structure with protein or peptide sequences to create cell adhesive surfaces is a common biomimetic strategy [151,168]. Osteogenic differentiation of hMSC cells was shown to occur in the case of phosphate-functionalized PEG hydrogels, whereas *t*-butyl-functionalized hydrogels induced adipogenesis [169]. Increasingly, more sophisticated hydrogel networks are being engineered with greater biofunctional specificity to mimic the chemical and biophysical structure and function of the ECM [170]. Complex multimembrane hydrogels have been suggested as a platform to encapsulate multiple cell types and achieve cellular organization to mimic the complexity of native tissue within a single construct. These layered structures also offer the potential to incorporate multiple GFs and achieve complex release profiles or establish GF gradients to instruct cell behavior [170,171].

Empirical evidence has suggested that the performance of tissue engineering scaffolds can be enhanced by using materials that incorporate bioresponsive functionalities and react to biological stimuli [172] to control the dynamic interplay with cells in a manner analogous to the native ECM. Nanoscale engineering of the polymer through the inclusion of protease-sensitive degradation sites has been employed to tailor polymer breakdown, GF release, and cell migration [151,166]. Hubbell and Lutolf employed the principles of biological recognition to engineer PEG hydrogel networks that were substrates for MMP-mediated matrix remodeling. Vinylsulfone-functionalized PEG was cross-linked with thiol-containing peptides through a Michael-type addition reaction that rendered them susceptible to cell-mediated proteolysis, while pendant oligopeptides were incorporated to enhance cell adhesion [151,173]. Alternatively, remodeling can be accomplished by cross-linking with substrates for the cell-associated protease, plasmin. Evaluation of plasmin-sensitive PEG hydrogels functionalized with integrin-binding adhesion ligands and BMP-2 demonstrated bone healing in a rat model [174]. These approaches based on cell-mediated scaffold degradation have been purported to more closely align with the temporal regeneration of tissue overcoming problems associated with the sudden loss of mechanical strength and tissue construct integrity that can be a feature of constructs that degrade by hydrolysis mechanisms [175].

16.5.3 Inorganic Phases

Bioactive ceramics including CaP as well as bioactive glasses (e.g., 45S5 Bioglass®) have been employed in bone tissue engineering to recapitulate the structural and compositional properties of native bone. They can react with physiologic fluids to form strong bonds with bone [119] and it is increasingly recognized that nanoscale inorganic components are more likely to be bioactive than microscaled compositions [71]. They are limited by an inherent brittleness that undermines their capacity to match the fracture toughness of bone and they are difficult to form into complex shapes such as highly porous structures [71,176].

16.5.3.1 Ceramics

CaPs, primarily of apatitic structure, are native constituents of bone tissue [177]. Commonly utilized CaP includes tricalcium phosphate (TCP) and HA that are biocompatible and osteoconductive and have been shown to have good affinity and integration with bone on implantation [23,178,179]. The different CaPs can be distinguished based on their biological response at the host site, which arises due to differences in their crystallinity, with HA undergoing a slower resorption compared to TCP [38]. Hydroxyapatite ($Ca_{10}(PO_4)_6(OH)_2$) is attractive owing to its structural similarity to the natural minerals found in bone [180,181].

However, biological apatites contain carbonate in their structure, which increases their solubility and bioresorbability with respect to HA [61,182]. The structure of HA has been modified to achieve remodeling efficiencies comparable to normal bone and a silicon (Si)-containing HA, ACTIFUSE™ (Baxter, USA), is available clinically for spinal fusion procedures [183]. CaPs are implicated in hMSC differentiation by sequestering charged proteins, such as the acidic sialoprotein, osteopontin, thereby promoting adhesion and osteogenesis of hMSCs [169].

16.5.3.2 Bioactive Glasses

Bioactive glasses have been extensively studied for bone regeneration and a number of bioglass-containing products (e.g., NovaBone) have been used clinically [183,184]. Bioglass usage in tissue engineering scaffolds conveys a number of advantages including the potential for osteoinduction, the capacity to support enzyme activity [185], and vascularization [186,187], in addition to a positive impact on osteoblast activity [188]. The use of bioglass implants in living tissue has shown cell differentiation, colonization, and the formation of new bone [189]. The primary constituents of most bioactive glasses include Si, Ca, and phosphate, a classic example of which is 45S5 Bioglass, developed by Hench, that consists of 45% SiO_2, 24.5% Na_2O, 24.5% CaO, and 6% P_2O_5 [183,190]. The incorporation of Si is attributed to its significant role in bone mineralization and gene activation [23]. Bioglasses form a strong bond with bone following immersion in biological fluids through the formation of a carbonated HA layer at the interface by surface mechanisms [26] and through the release of ionic dissolution products as the glass resorbs, which can stimulate osteoblasts [191] and upregulate families of genes found in osteoblasts [192]. Other examples of bioglass compostions include those based on cobalt, borate, and borosilicate [193]. Substitution of strontium (Sr) for Ca in bioactive glasses has been investigated as a means to combine the intrinsic regenerative properties of bioglass with those of Sr. Sr, which is present in trace quantities in bone and marketed for the treatment of osteoporosis [194–196], has been shown to work by stimulating osteoblasts in bone formation and preventing osteoclasts from resorbing bone [197,198]. *In vitro* evaluation of Sr bioglasses cultured with Saos-2 cells demonstrated that proliferation and ALP activity were promoted with increasing Sr substitution [199].

Other recent research has focused on engineering structures with superior biomedical performance. Bioglasses are being modified to incorporate and present GFs such as transforming growth factor beta (TGF-β) to promote angiogenesis and vascularization [186], while nanoscale fabrication of bioglasses using techniques such as solgel processing, gas phase synthesis, and laser spinning is being investigated to enhance regenerative capacity [190,200,201]. Despite these advantages, bioglasses, in a similar manner to the CaP materials, are brittle and cannot match the fracture toughness of bone; thereby limiting their use in load-bearing applications [71,202].

16.5.4 Composite and Hybrid Systems

Interactions at the cell–material interface may be affected by many properties of the biomaterial [203], such as surface chemistry, surface energy, availability of binding sites, surface area, and topography. Simple polymers do not adequately replicate the complex interactions between cells and the ECM, which has consequences for cell attachment, differentiation, proliferation, and functional tissue regeneration [107,204,205]. Anatomical and physiological inspiration has spurred the development of composite systems that bring together the bioactivity of inorganic mineral phases with polymer versatility and toughness [20,190,206,207]. Addition of inorganic phases to polymer substrates, particularly hydrophobic materials, can enhance the initial spread of serum proteins creating a more favorable environment for cell interaction [119]. In addition, the basic resorption products of HA or TCP may help to mitigate against the acidic by-products that occur with some polymers (e.g., PLGA). These breakdown products can lead to an unfavorable environment for cells, which can be exacerbated in cases of poor vascularization [119]. Generally, bioactive glass composites exhibit greater bioactivity than HA composites as 45S5 Bioglass™ is a class A bioactive material with osteoconductive and osteoinductive properties, whereas HA is a class B bioactive material, which exhibits only osteoconductive behavior [208]. However, despite this, CaP materials are more widely used in the clinic due to processing challenges with bioactive glasses [106].

There are a number of suitable techniques for fabricating these nanocomposites including thermally induced phase separation (TIPS) and electrospinning [209]. One of the major challenges associated with composite systems is ensuring maximum presentation of the bioactive particles at the biological interface to promote cell interaction [202]. Approaches to circumvent the occlusion of the mineral phase by the polymer component have used nanoscale inorganic phases to enable interaction at the molecular level as well as choosing techniques to selectively increase exposure of the inorganic phase. Gas foaming–particulate leaching (GF/PL) was employed to maximize the exposure of HA nanoparticles in PLGA/HA composite scaffolds. Compared to solvent casting and particulate leaching (SC/PL), the GF/PL method increased the exposure of the HA and in vitro cell culture evaluation of the composite scaffolds with rat calvarial osteoblasts demonstrated higher cell growth, ALP activity, and mineralization compared to SC/PL [210].

The preparation of inorganic/organic hybrids is viewed as a viable solution to maintain the biological properties of the bioactive phase. Solgel processes have been used to synthesize these structures, in which the distinct components interact at the molecular level either by molecular entanglements, for example, hydrogen bonding and/or van der Waals forces (class I hybrids) or by covalent bonding (class II hybrids) [202,211]. These hybrid structures have been suggested to better align the degradation rate of the constituent components and to create a material that acts as a single-phase bioactive material [115,211]. The expectation is that larger cells (20–50 μm in length) would interact with a more uniform bioactive composite [183].

Calcium phosphates such as HA or TCP are commonly employed to coat the surface of polymers [212,213] to enhance interfacial properties, in addition to producing hybrid [214] and composite [215] biomaterials. In one case, a nanocrystalline, carbonated apatite was deposited on electrospun PCL scaffolds by immersion in simulated body fluid for 7 days to increase wettability [216]. In another case, the addition of a CaP mineral layer to a PLGA 75:25/CaP composite scaffold was observed to mediate the foreign body response commonly seen at the polymeric interface [215]. This was also reported to increase mechanical and biological performance [217]. The deposition of a CaP coating (containing HA and brushite) on collagen scaffolds using a biphasic immersion technique was shown to considerably improve the mechanical properties of highly porous scaffolds to 90 kPa compared to just 0.3 kPa for a pure collagen scaffold, thus enhancing its potential application in bone repair [218]. Mineralization of collagen-GAG scaffolds by employing the same immersion technique was shown to enhance effective healing in a critical-size cranial defect rat model compared to a nonmineralized collagen–GAG scaffold. The benefit was attributed not only to the mineral properties, but also to the increased stiffness of these scaffolds as a result of CaP inclusion [219]. The addition of nano amorphous CaP particles to PCL scaffolds demonstrated that increasing mineral content promoted osteoblast adhesion and proliferation [220]. However, the benefit of employing nanoscale inorganic phases not only offers the prospect to emulate the structural complexity of native bone but to increase the mechanical properties of composite structures. The inclusion of nanoscale HA phases to poly(ethylene-co-acrylic acid) composites resulted in more uniform particle dispersion compared to those containing micron-sized particles [221], which is important in the overall context of structural integrity.

Combining bioglasses with polymers to create composite or hybrid materials is seen as a way to combine the advantages of both while addressing the limitations of the individual substituents. Composites of BG 45S5 and PLGA demonstrated greater osteoconductive, osteoinductive, mechanical and structural properties [222], and an earlier peak in ALP activity as well as the presence of osteogenic markers on composites seeded with hMSCs without osteogenic media stimulation [223]. There are also many examples of composites incorporating nanoscale bioglasses to increase the regenerative capability arising from the increased surface area, improved solubility, and scaffold stiffness [190] associated with the reduced particle size. One study that compared the size of bioglasses combined with PCL showed greater bioactivity and mechanical stability in the case of the nanofiber composites [224]. Hydrogel scaffolds composed of a dense nanofibrillar collagen gel framework hybridized with nanoscale bioactive glass particles demonstrated accelerated mineralization that increased scaffold stiffness. This was attributed to the high surface area and reactivity of the nanoscale bioglass. Furthermore, cell culturing in the absence of osteogenic factors showed that ALP activity was greater in cells cultured in hybrid scaffolds compared to dense collagen controls [225]. Another interesting example of biomimicry is based on the covalent coupling of bioactive silica to the polypeptide, poly(γ-glutamic acid) (γ-PGA). This hybrid structure is inspired by the composition and structure of native bone [226,227] and includes the biodegradable, biocompatible, and naturally occurring γ-PGA polymer [228]. This polymer is composed of constituent glutamic monomers reminiscent of the glutamic sequences in collagen that act as nucleation sites for HA during bone formation [229].

16.6 Bioactive Signals

16.6.1 Introduction

Despite the fact that the complex interactions and processes that underlie the coordination of bone regeneration have not been fully elucidated [230], next-generation biomaterials are being designed to combine bioactive and bioresorbable properties to direct *in vivo* mechanisms of tissue regeneration, stimulating the body to heal itself and leading to replacement of the scaffold by the regenerating tissue [176]. Functionalization of biomaterials either through the incorporation of GF signals, which may be released in an appropriate spatiotemporal manner to control cell differentiation and morphogenesis, or by engineering materials to become biomimetic through the incorporation of bioinstructive ligands has dominated recent bone regeneration research [48]. Advances in nanotechnology offer the opportunity to design innovative, multifunctional structures that more precisely control the presentation of the necessary physical and biochemical cues to regulate and direct cell activities inherent in the regenerative process.

16.6.2 Growth Factors

Natural ECM impacts on tissue development through its ability to bind, store, and release bioactive ECM effectors and direct them to the right place at the right time. Signaling molecules can be broadly grouped into the overlapping categories of mitogens (stimulate cell division), GFs (multiple functions including proliferation-inducing effects), and morphogens (control generation of tissue form) [52]. In the context of bone regeneration, a number of GFs have been identified to play key roles in directing chemotaxis, cell proliferation and differentiation, and ECM ossification. Of particular note is the TGF-β super family, including BMPs, as well as platelet-derived growth factors (PDGF), fibroblast growth factors (FGFs), and insulin-like growth factors (IGFs) [230,231]. In particular, the BMPs form a unique group of proteins within the TGF-β superfamily of genes and have pivotal roles in the regulation of bone induction, maintenance, and repair (reviewed in [53]). BMPs and recombinant BMPs (rBMPs) have been the focus of much research, specifically BMP-2 and BMP-7 [232–235]. These proteins induce osteogenic differentiation *in vitro* as well as bone defect healing *in vivo* [236,237]. A review of the biological functions and roles of GFs used in tissue regeneration and fracture healing can be found in Lee et al. [52] and Dimitriou et al. [10].

Biomolecules such as GFs are routinely incorporated into biomaterials in an effort to recapitulate ECM presentation of biological cues and control the complex cellular processes and biological responses fundamental to tissue development and regeneration [238]. It is thought that GF binding to the ECM raises their local concentration to levels appropriate for signaling, localizes their morphogenetic activity, and protects them from enzymatic degradation [15]. The INFUSE® bone graft/LT cage device (Medtronic, TN) approved by the FDA for anterior lumbar interbody fusion procedures [239], highlights the challenges associated with designing successful tissue engineering approaches. This device consists of a collagen sponge, which is combined with rhBMP-2 at a 1.5 mg/cc concentration [239] prior to implantation. Endogenously, nanogram quantities of BMP (per gram of bone matrix) are sufficient to initiate bone repair; however, the dose in this treatment is necessitated, in part, by challenges associated with controlling release kinetics [13]. Other important considerations associated with formulation and delivery of GF-based therapeutics include (1) the short protein half-life *in vivo*, (2) potential side effects associated with administration of multiple or supraphysiological doses, (3) undefined key GFs for a specific tissue, and (4) potential denaturation of the protein during formulation and handling [240–242]. Additional factors, including ECM degradation and cell target location, are important considerations [52]. The design of new delivery strategies to refine the spatial and temporal delivery of GFs may take inspiration from advances in nanotechnology to achieve more precise tailoring of release profiles and from the body, which has sophisticated regulatory mechanisms to ensure spatiotemporal control. These mechanisms include the production of precursors that become activated at their particular sites of action (e.g., TGF-β) or the requirement for cofactors such as heparan sulfate for receptor activation [92,243]. Alternatively, the mechanism by which ECM proteins bind GFs and regulate their distribution, activation, and presentation to cells presents a myriad of options that may be explored [244]. To date, a number of strategies for incorporating bioactive molecules into biomaterials to mimic endogenous GF presentation and to preserve the bioactivity of protein cargoes have been pursued including adsorption on the biomaterial surface and encapsulation within the scaffold matrix. An alternative approach includes incorporation of gene delivery mechanisms into the scaffold matrix; however, this approach is less well developed.

16.6.2.1 Surface Adsorption

The bioactive molecule may be immobilized onto the surface of a scaffold or fiber through transient physical bonds or by covalent tethering; the latter can be engineered to be bioresponsive by using chemical bonds that are cleaved to trigger GF release in response to cell-secreted proteases [19]. GF immobilization on scaffolds can allow for control over the density of presentation; however, it is essential to ensure the active domains of the signaling factor are accessible to direct cell activity [245]. This approach may also be utilized to pattern gradients of molecular signals on biomaterials to augment cellular activities intrinsic to morphogenetic processes in tissue remodeling and regeneration including cell migration [15]. One approach to tether BMP-2 to scaffolds involved conjugation of heparin to PLG scaffolds that subsequently facilitated BMP binding. This is analogous to the physiological context where heparin and heparan sulfate proteoglycans bind GFs (e.g., VEGF, FGF) and are thought to act as reservoirs [244]. In one study, heparin binding of BMP-2 to PLGA prolonged release of the protein for at least 14 days *in vitro* and induced bone formation to a greater extent compared to scaffolds where heparin or BMP-2 was absent. The benefits were attributed to the sustained and controlled presentation of the GF as a result of heparin binding. It was noted in this case that polymer functionalization with heparin was optimized using a star-shaped rather than the linear form of PLGA [246].

Alternatively, rather than tethering the entire GF to the biomaterial, short sequences of BMP proteins have been investigated to functionalize materials. The rationale in this case is to ensure a better match between GF size and the nanostructured features of the substrate onto which it is functionalized, especially when trying to replicate geometrical features in native bone, for example, nanosized HA (nHA) particles. The use of the shorter sequences has the advantage of simplifying the process and facilitating functionalization of nanoscaled substrates with bioactive sequences, which otherwise might not be possible with the larger protein structure. One study investigated three short peptides of BMP-7 (SNVILKKYRN, KPSSAPTQLN, and KAISVLYFDDS) and noted that the sequence KPSSAPTQLN on its own and in combination with SNVILKKYRN enhanced osteoblast proliferation *in vitro* [247]. Sequences of BMP-2 have also been investigated to functionalize alginate gels [248] and α-TCP scaffolds [249] in bone regeneration studies.

16.6.2.2 Encapsulation Strategies

Growth factors may be physically encapsulated in the bulk of a biodegradable material, such as a polymer system using techniques such as gelation or coacervation. The release of the GF cargo from the matrix may be characterized by an initial burst component associated with surface-adsorbed GF followed by a more controlled, slower release of the cargo corresponding to matrix diffusion or polymer degradation [52,238]. Release can be sustained or delayed depending on polymer composition and degree of cross-linking [250]. This delivery approach can achieve more sustained release profiles compared to the burst profiles commonly associated with bolus injections and also has the advantage of localizing the exogenous factors to the target site [251]. Alternatively, GF encapsulation in polymer microspheres or nanoparticles embedded in the bulk polymer may be used to delay release [13]. Preparation of nanofibers with core–sheath configurations using techniques such as coaxial electrospinning has been investigated to overcome the problem of burst release; however, their mechanical properties limit their use in hard tissue regeneration [252,253].

The incorporation of BMP-2 in P_LLA/collagen nanofibers prepared by electrospinning resulted in enhanced hMSC adherence, although proliferation was reduced. BMP-2 inclusion also corresponded to an increase in gene expression of ALP, osteocalcin, and type I collagen. Furthermore, the presence of BMP-2 resulted in an upregulation of focal adhesion kinase [254], a mediator of GF signaling [255]. Other materials including silk fibroin [256,257] and self-assembled structures have also been investigated to control the presentation of BMP-2. Delivery of BMP-2 from a nanostructured 3-D hydrogel formulation showed that the *in vivo* osteoinductive activity of BMP-2 was greatly influenced by its incorporation into self-assembled peptide amphiphile (PA) nanofibers. The aqueous formulation composed of the PA and GF self-assembled and gelled on injection in the back subcutis of rats and controlled protein release for 20 days. Histological examination demonstrated induction of ectopic bone formation, in contrast to sites injected with an aqueous solution of BMP-2 or PA alone [258]. BMP delivery from hybrid systems has also been investigated. In one study, an electrospun nanofiber mesh tube used to guide bone regeneration was filled with a peptide-modified alginate hydrogel that acted to sustain rhBMP-2 release [259]. Further studies demonstrated that bone regeneration and biomechanical function was significantly increased in a segmental defect model using the hybrid delivery system compared to collagen sponge delivery. The authors also noted that the nanofibrous mesh tube was crucial in promoting mineralization and guiding an integrated pattern of bone formation. They concluded that spatiotemporal strategies to localize GF delivery presentation were important in enhancing clinical outcomes [260]. The benefits of using hybrid PCL mesh/alginate systems have also been demonstrated by other researchers. In this case, the system yielded greater connectivity by week 4 and a 2.5-fold increase in bone volume by week 12 compared to a collagen delivery system. This was attributed to the enhanced protein retention in the hybrid system [261].

It is recognized that many GFs with diverse functionalities are involved in regeneration and they are expressed during different phases of the restoration process. Studies to mimic events *in vivo* have investigated delivery of multiple GFs and various controlled release strategies to replicate the intricacies of cooperative regenerative processes. One of the most direct ways to customize the temporal presentation of multiple GFs involves employing a variety of materials with distinct features (e.g., degradation profiles, affinity for the cargo) and exploiting these differences to tailor delivery. In one example, BMP-2 was encapsulated in PLGA microspheres prior to loading in a PPF scaffold, which was then subsequently coated with a VEGF loaded gelatin hydrogel. *In vivo* evaluation of release using ^{125}I-labeled GF showed a large initial burst of VEGF within the first 3 days, while BMP-2 release was maintained over the entire implantation period (56 days). VEGF enhanced development of a supportive vascular network and ectopic bone formation compared to BMP-2 alone, but similar effects were not apparent in the case of orthotopic bone regeneration [235]. In another example, the sequential delivery of GFs from a chitosan-based fibrous scaffold was achieved by first encapsulating the proteins, BMP-2 and BMP-7, in PLGA and poly(3-hydroxybutyrate-co-3-hydroxyvalerate) (PHBV) nanoparticles, respectively. Consequently, the release of BMP-7 was slowed compared to BMP-2 release [262,263].

16.6.2.3 Gene Delivery

As an alternative means of GF presentation, cells are being genetically modified to express the desired gene for encoding the bioactive agent of interest. Although this approach is less well developed, it offers many advantages including the potential for directed, sustained, and regulated expression [264]. A range of viral and nonviral gene vectors have been investigated with some reports of bone regeneration using viral vectors such as adenovirus [265] or retroviruses [266]. However, owing to safety concerns with viral vectors, a number of nonviral vectors including polymers and inorganic minerals have been explored. Polyamidoamine dendrimers investigated as a vector for the BMP-2 gene resulted in a low transfection efficiency; however, despite this, qualitative and quantitative indicators of osteogenesis were higher in transfected compared to nontransfected cells [267]. Nanosized HA particles have been used as a nonviral vector for plasmid DNA (pDNA) encoding BMP-2 and have been added to porous collagen scaffolds to prepare gene-activated matrices to promote stem cell–mediated bone formation [268]. Condensing pDNA with poly(ethyleneimine) (PEI) has been investigated to overcome limitations of poor transfection efficiency. Polyplexes of pDNA–PEI have been incorporated into PLGA [269] and nHA–collagen [270] scaffolds to prolong gene expression and to provide structural support for new tissue development [270]. The complexation of pDNA with PEI within PLGA scaffolds was shown to have an important bearing on quantity and distribution of bone formation in a critical-size cranial defect model compared to scaffolds containing uncondensed DNA where bone formation was significantly less and confined to the edges of the defect [269].

16.6.3 Extracellular Matrix Components

ECM components are valuable building blocks for the preparation of biomaterials involved in tissue engineering [271]. Tissue-specific ECMs contain a diverse array of specific protein sequences, which are likely to have a fundamental influence on cell behavior. A number of approaches have focused on inclusion of purified ECM proteins or short peptide sequences, by immobilization directly onto a material as an effective means of promoting adhesion and directing cell behavior [272]. This has been accomplished through the use of native long-chain ECM proteins as well as short peptide sequences derived from intact ECM proteins that are recognized as the minimal recognition sequence necessary to control material–cell interactions [273]. There has been some debate whether such simple peptide sequences can really replicate the complexity of cell ECM interactions and direct the coordinated processes necessary for functional tissue regeneration [274,275]. They have, however, shown promise in promoting cell adhesion, and these short sequences may be preferable due to their cost-effectiveness, decreased immunogenicity, the ability for reproducible manufacture, and control of functionalization parameters (e.g., pattern and density) [204,276].

The most widely studied adhesive peptide in the biomaterials field is the triamino acid sequence RGD, which is the principal integrin-binding domain present within ECM proteins such as fibronectin [277]. This inexpensive, simple sequence can be easily modified to enable functionalization and its role is implicated in a number of tissue engineering applications to trigger specific cell responses such as cell adhesion [274,278]. Biomaterial functionalization to control cell interaction can be achieved by surface and bulk modification [273], while the use of nanofibers and nanoparticles that replicate the organization of native tissues has the added potential to present these peptide sequences at high density [279].

There are many studies demonstrating the potential of substrate functionalization with the peptide sequence RGD. Modification of PLA films with RGD enhanced attachment and growth of human osteoprogenitor cells, while modification of 3-D porous PLGA scaffolds with the peptide sequence resulted in cell migration, expansion, and differentiation [152]. One of the seminal studies highlighting the impact of RGD modification of alginate hydrogels demonstrated that covalent coupling of the peptide promoted osteoblast cell adhesion and spreading in comparison to unmodified hydrogels where minimal cell adhesion was observed. A minimum ligand density was identified as necessary to initiate proliferation, which increased with increasing ligand density. *In vivo* evaluation by subcutaneous injection into the back of a SCID mouse model showed that modified alginates containing primary rat calvarial cells significantly increased the rate of new bone formation when compared with unmodified alginate [161].

Many other examples of peptide sequences have been examined in a diverse array of tissue engineering applications to promote integrin and cell binding including the recognition sequences isoleucine–lysine–valine–alanine–valine (IKVAV) and the pentapeptide tyrosine–isoleucine–glycine–serine–arginine (YIGSR) present in laminin and the binding sequence glycine–phenylalanine–hydroxyproline–glycine–glutamate–arginine (GFOGER) present in collagens I, II, and III [98]. There are also other peptides with additional distinctive roles including specific sequences that are susceptible to cleavage by MMPs or plasmin, those that impact on surface, or GF binding, as well as sequences that have the potential to self-assemble [280]. Other peptide sequences apart from RGD have also shown promise with regard to mimicking the ECM. Peptide fragments isolated from fibronectin (FN III9–10) were shown to be important in integrin-dependent cellular interactions with the ECM and MSC osteogenic differentiation [281]. Engineering of nanofibrous M13 phage tissue matrices with collagen I derived aspartic acid–glycine–glutamate–alanine (DGEA) peptides, which have been implicated in osteoblast differentiation [282], were shown to induce early osteogenic differentiation in preosteoblast cells of murine origin [283]. Other peptide sequences such as P-15, a 15-amino acid residue identical to the cell binding domain of type I collagen, have been investigated for their biomimetic potential [284]. Microporous HA particulates functionalized with P-15 and cultured with human bone marrow stromal cells demonstrated increased osteogenic markers including ALP activity and BMP-2 gene expression, in addition to promoting cell attachment and spreading [285].

There is a wealth of possibility contained within the ECM, in addition to those sequences that are exposed, fragments of many ECM proteins possess bioactivity with effects ranging from cell migration, differentiation, and proliferation, which are only revealed to cells when the ECM is modified by proteolytic cleavage or conformational change [175,286]. It is likely that as knowledge increases with respect to their functions and additional sequences are discovered, this will enable materials to be designed with increasing complexity that achieve greater selectivity in directing cell behavior [107]. In addition to ligand type, it is increasingly evident that factors such as organization at the nanoscale within biomaterials including their density and patterning parameters have implications for modulating integrin binding and the direction of cell fate [287,288].

16.7 Cells to Promote Bone Repair

The success of grafting techniques has highlighted the potential of cells to promote bone repair and prompted considerable research around the use of transplanted cells in combination with constructs to promote tissue regeneration, either directly or by regulating the process through the secretion of trophic factors [289]. A number of tissue-specific cells including primary adult osteoblasts as well as stem cells have been investigated for their regenerative capability. The latter have attracted a lot of interest because of their differentiation potential along multiple lineages including bone, muscle, and fat and their capacity for self-renewal and proliferation. Additionally, they can secrete large quantities of cytokines and trophic factors [290,291]. Considerable research efforts have focused on exploiting the reparative potential to regenerate bone and restore tissue function, especially in cases where intervention is necessary to support healing [292]. The human body has been used as a source of these reparative cells, which can be isolated from bone marrow and expanded in culture prior to injection *in vivo* either directly or contained in some construct [293,294]. As an alternative to adult stem cells, embryonic stem cells (ESCs) have the primary advantage of differentiating into any adult cell type, thus representing an immense potential to engineer multiple tissue types, especially compared to adult stem cells that are restricted to certain lineages [130]. Furthermore, they have the potential to be maintained in culture for long periods, generating cell quantities far in excess of that directly derived from tissue sources [295]; however, despite their clear advantages, ethical and safety concerns exist. These limitations prompted research focused on reprogramming of somatic cells into embryonic stem cell-like pluripotent cells, but this approach is also problematical.

Controlling stem cell differentiation toward the desired lineage is an important step in functional tissue development. Central to this process is the need to provide the necessary physicochemical and biological cues to regulate cell fate and form [130,296]. *In vivo*, stem cells reside in a dynamic environment termed the "stem cell niche" and are influenced by a diverse range of interactions with other cells, components of the ECM, and soluble factors that define the local biochemical and mechanical environment that governs their behavior [297–299]. It is accepted that recreating or at least simulating this complex environment may be critical in directing cell behavior *in vitro* and *in vivo* and promoting engraftment especially in the case of stem cell–based therapies [298]. However, this may be further complicated if the quality of engineered bone is dependent on cell source. Stevens' group highlighted the need to elucidate the distinct mechanisms and cues that drive the regenerative process in the case of discrete cell sources. They investigated the consequences of using cells from three different sources including neonatal osteoblasts, adult bone marrow–derived MSCs, and ESCs to form a bone-like mineralized material *in vitro*. Using a combination of analytical techniques including biochemical evaluations and micro-Raman, they demonstrated that osteoblasts and MSCs formed a material with many features intrinsic to native bone, in contrast to that created from ESCs that differed in its composition, stiffness, and nanolevel architecture [300].

From a functional perspective *in vivo*, scaffold mechanical properties provide cues important in the regulation of stem cell fate and tissue development [110,116,301]. Anchored cells probe their surroundings by pulling on it and exerting traction forces [84]. Cells respond to the mechanical properties of their environment including the ECM using the mechanosensitive nature of their adhesions and by adjusting the cytoskeleton [84,116,117]. Studies characterizing the mechanosensitivity of MSCs to matrix elasticity have shown that changing the stiffness of a polyacrylamide substrate on which MSCs were cultured was observed to influence their differentiation. Soft matrices favored MSC differentiation into neuronal-like cells, medium stiffness promoted myogenic differentiation, while a rigid matrix stimulated osteogenic differentiation [302]. However, recapitulating the properties of native tissue and integrating the complexity of features known to be important including compositional, biological, physical, and mechanical in a controlled manner are exceedingly challenging [125,130,303]. This is made even more difficult given that the precise mechanisms involved in signaling and differentiation [304], as well as the dynamic interplay of stem cells with biomaterials used in tissue-engineered constructs, are not fully understood.

A number of novel approaches are being pursued to simplify the process associated with exploiting the therapeutic potential of cells in bone regeneration. One concept involves implanting biomaterials that recruit stem cells already in the body to a targeted site as an alternative to seeding cells on biomaterial constructs *ex vivo* [299]. This may be augmented by incorporating chemoattractants in the biomaterial. For example, a gelatin hydrogel was used to deliver stromal cell–derived factor-1, which plays an important role in targeting stem cells to specific sites in the body [305]. These approaches could dispense with the need to expand cells and constructs *in vitro* (e.g., in bioreactors) prior to implantation and all the associated challenges. A particularly novel approach to bone engineering has focused on using the body as a bioreactor. In this technique, an artificial space is created between the tibia and the periosteum membrane that covers it, followed by injection of a gel to the pluripotent rich mesenchymal layer. It was proposed that the healing response that occurs can be manipulated to create large volumes of bone tissue, which is biomechanically identical to native bone [306].

16.8 Scaffold Fabrication Techniques

16.8.1 Introduction

Advances in nanotechnology have enabled the engineering of nanostructured materials in an effort to better address the current challenges in bone regeneration. Designing nanofibrous scaffolds that possess the properties and hierarchical organization of ECM is potentially useful for bone tissue engineering [189,307]. A variety of techniques to prepare scaffolds with nanoscale architecture and characteristics conducive to bone formation including phase separation, electrospinning, and macromolecular self-assembly have been suggested. For example, techniques such as electrospinning have been used to prepare nanofibrous scaffolds that have an enhanced capacity to absorb proteins compared to microstructured scaffolds due to the increased surface area, and by implication can present more binding sites for cell attachment [32]. There are many challenges associated with each of these fabrication techniques, not least the reproducible manufacture of scaffolds on a large scale. From this perspective, machine-made scaffolds offer several advantages over manually fabricated scaffolds including more cost-effective production, they are easier to standardize, and the potential to custom make the scaffold to fit 3-D defects [308]. This has spurred interest in rapid prototyping techniques including laser sintering, 3-D printing, and fusion deposition modeling but require further development.

16.8.2 Thermally Induced Phase Separation

TIPS is a simple and versatile multistage technique used to produce ECM-like fibers with nano-sized diameters. A number of fabrication steps are involved including dissolution of polymer, separation of polymer-rich and polymer-poor phases, gelation, extraction of solvent, freezing, and freeze-drying [131,309]. Phase separation is initiated by subjecting the system to some transition in the processing conditions, for example, lowering of temperature. This causes thermodynamic instability, leading to the separation of polymer-rich and polymer-poor phases [209]. The latter is removed, leaving behind the polymer rich phase to produce the nanofibrous matrix. The advantages and disadvantages of this technique are outlined in Table 16.3.

TABLE 16.3

Features of Processing Techniques Used to Prepare Nanoscaled Matrices

Technique	Advantages	Disadvantages	Fabricated Substrates
Thermally induced phase separation	Flexible process enables tailoring of scaffold properties [119,209]	Use of solvents [209]	P$_L$LA incorporating nHA [310]
	Possible to combine with other methods to control pore size and interconnectivity [131]	Poor control over fiber orientation and diameter [131]	PLGA/nHA [311]
		User sensitive [119]	
Electrospinning	Versatile and flexible process [131,312]	Small pore size [131]	PCL [157]
	Can produce thin mats or 3-D structures [312]	Pore size decreases as fiber thickness increases [313]	PCL/collagen/ nHA scaffolds [319]
	Control of scaffold characteristics, e.g., porosity, fiber orientation, and thickness [131,313–315]	Mechanical properties [313]	Silk fibroin/nHA fiber scaffolds containing BMP-2 [256]
	Produces long nanofibers, with a high surface area to volume ratio [315]	Residual solvents [318]	PLGA/amorphous TCP nanocomposite [320]
	Used to process a wide range of polymers, composites [209,189,316], and proteins [256,317]		PLGA/HA composite scaffolds incorporating DNA/ chitosan nanoparticles [321]
Self-assembly	Possible to tailor features of nanofibrous network [322,323]	Complex design	Inclusion of the BMP receptor– binding peptides termed osteopromotive domains in PA [327]
	Functionalization of the PA possible [324]	Used for peptide sequences rather than proteins [131]	Co-assembly of an RGDS functionalized PA with phosphorylated serine PA at a ratio of 95:5 [47]
	Produces fine fiber diameters [325] Inclusion of biomolecules [313] and cell entrapment possible [131,326]	Mechanical properties limit use in weight bearing bone applications [71,131]	Self-assembled gels to control BMP-2 release [258]

16.8.3 Electrospinning

Electrospinning has generated substantial interest as a simple and versatile technique to fabricate scaffolds for bone regeneration. The resulting scaffolds are highly porous and are composed of nonwoven fibers with nanoscale diameters [313,328] and have been investigated as ECM mimetics due to their structural similarity to fibrous collagen proteins in native bone [329]. The process uses an electric field to control the formation and deposition of polymer fibers [328]. The solution containing the polymer is placed into a syringe and

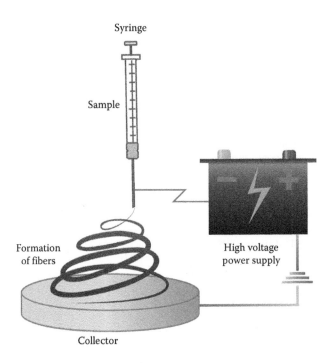

FIGURE 16.3
Schematic depicting the basic components of the electrospinning setup that may be used to produce nanofibrous substrates for bone tissue engineering. These fibers may be produced by applying a high voltage to the polymer solution to induce liquid jet formation. Fiber capture on a flat surface can be used to produce nonwoven mats. Alterations in the process variables including the applied voltage and distance between the sample and collector plate, and formulation variables including polymer molecular weight and viscosity of the starting polymer solution can affect the properties of the electrospun fiber.

the syringe needle is then directed to a collector, and a high voltage is applied creating opposite polarities between the needle and the collector [330]. The charge separation overcomes the surface tension of the polymer solution, and a thin stream is drawn from the needle toward the collector, Figure 16.3. The solvent evaporates, resulting in the formation of a polymer fiber [331] that is laminated together by heating the mesh slightly above the polymer melting point. The choice of solvent has an important impact on the process and must take factors such as biocompatibility, dissolution, and spinnability into account [209].

16.8.4 Peptide Self-Assembly

Self-assembly of natural or synthetic macromolecules including peptides has been used to produce ordered 3-D structures and fibers in the nanoscale domain [332]. Examples of self-assembly in nature and an increased understanding of how proteins and peptides fold and assemble have spurred the development of biocompatible nanofibrous scaffolds to replicate the hierarchical organization inherent in native tissue [32,258,333]. Some of the foremost research in this area has been undertaken by Zhang and coworkers [323], and by Stupp's group. Stupp and coworkers investigated the production of composite materials through the self-assembly and mineralization of peptide amphiphiles (PA) that contained distinct structural regions in an effort to recapitulate the characteristic structural arrangement of collagen and HA in native bone. Central to this are the key structural features of the amphiphilic molecule (Figure 16.4a), which

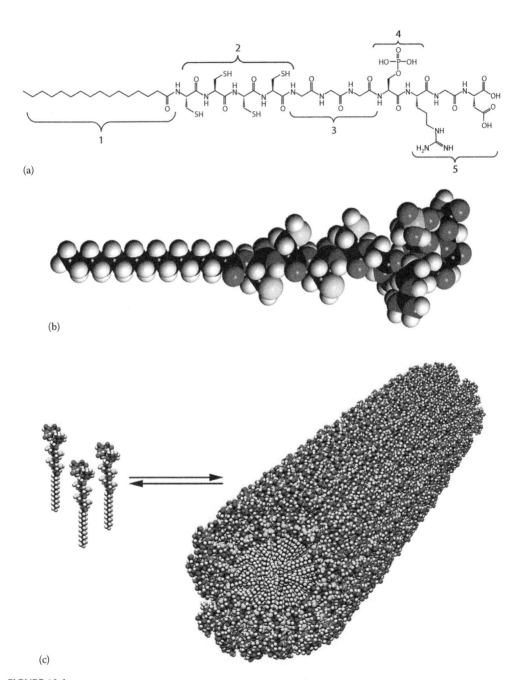

FIGURE 16.4
(See color insert.) (a) Peptide amphiphile structure highlighting the distinctive features intrinsic to the self-assembly process. Region 1 represents the alkyl chain that conveys the hydrophobic properties on the molecule. Region 2 highlights four cysteine residues that oxidize and can form disulfide bonds. Region 3 contains three glycine residues that impart flexibility to the hydrophilic head group. Region 4 depicts a phosphorylated serine residue that can interact with Ca ions and help direct mineralization of HA. Region 5 represents the cell adhesion ligand RGD. (b) Molecular model of the PA showing the overall conical shape of the molecule going from the narrow hydrophobic tail to the bulkier peptide region. (c) Schematic showing the cylindrical nanostructure of the self-assembled PA. (From Hartgerink, J.D. et al., *Science*, 294, 1684–1688. Copyright 2001. Reproduced with permission from AAAS.)

includes an alkyl chain that conveys the hydrophobic properties coupled to a bulkier peptide region (Figure 16.4b) that becomes repeatedly exposed on the surface after assembly, while the hydrophobic moieties orientate in the core (Figure 16.4c). The peptide segment contains amino acids that have a strong tendency to produce β-sheets that facilitate the formation of high aspect ratio cylindrical nanofibers as opposed to spherical structures [325,334]. Inherent in its functionality is the inclusion of residues to nucleate HA such as phosphorylated groups considered important for CaP formation and promoting biomineralization. The suitability of these self-assembled nanostructures for bone formation can be tailored further to include the peptide sequences of cell adhesion ligands (e.g., RGD) to enhance bioactivity by promoting cell adhesion on the nanofiber structure [325]. A series of cysteine residues were included to form disulfide bonds between adjacent molecules after oxidation and thus ensure the structural integrity of the supramolecular assembly. Alternatively, combination of the biomineralization and integrin-mediated functionality has been achieved by coassembly of two PAs, with one containing phosphorylated serine residues to promote HA nucleation and the other containing the biomimetic RGDS peptide sequence [47].

16.9 Conclusions and Future Perspectives

Given the limitations with current bone grafting strategies, there is an increasing need for clinical interventions to promote functional bone regeneration in cases where bone loss due to trauma, disease, or deformity precludes natural restoration of the tissue. In concert with advances in nanotechnology and molecular biology, considerable progress is being made toward developing biomimetic materials that emulate the sophisticated biological recognition and signaling functions inherent in native tissues. Nanostuctured biomaterials offer the opportunity to recreate the composition, structural organization, and topography inherent in native bone tissues. This has the potential to provide the appropriate cues to instruct cell behavior and actively support tissue regeneration, thus moving the field beyond the provision of passive constructs. Together with advancements in fabrication techniques, these bioinspired materials offer the potential to extend the repertoire of tissue-engineered constructs that will enable functional tissue regeneration.

However, despite the exciting prospect of engineering biomaterials to resemble native bone architecture and act as functional replacements, translation of clinically effective alternatives to bone grafts has been slow and hampered by a lack of understanding surrounding the dynamic interplay that occurs at the interface between cells and the substrates, as well as the precise mechanisms that regulate cell fate and form. Ongoing challenges remain in developing innovative biomaterials with the requisite level of complexity and appropriate mechanical features. Another inherent challenge includes the need to control and sequence the dynamics and spatial organization associated with the presentation of multiple biological signals. Despite the drive to design constructs that recapitulate native bone tissue at the nanoscale, the successful development of novel tissue engineering constructs will hinge on the ease with which complex and often sophisticated technologies may be translated to the clinic and reproducibly manufactured on a large scale, in a cost-effective manner.

References

1. Langer, R. and J. P. Vacanti. 1993. Tissue engineering. *Science* 260: 920–926.
2. Stock, U. A. and J. P. Vacanti. 2001. Tissue engineering: Current state and prospects. *Annual Review of Medicine* 52: 443–451, doi:10.1146/annurev.med.52.1.443.
3. Kanczler, J. M. et al. 2010. The effect of the delivery of vascular endothelial growth factor and bone morphogenic protein-2 to osteoprogenitor cell populations on bone formation. *Biomaterials* 31: 1242–1250, doi:10.1016/j.biomaterials.2009.10.059.
4. Seebach, C., J. Schultheiss, K. Wilhelm, J. Frank, and D. Henrich. 2010. Comparison of six bone-graft substitutes regarding to cell seeding efficiency, metabolism and growth behaviour of human mesenchymal stem cells (MSC) in vitro. *Injury* 41: 731–738, doi:10.1016/j.injury.2010.02.017.
5. Kirby, G. T. S. et al. 2011. PLGA-based microparticles for the sustained release of BMP-2. *Polymers* 3: 571–586.
6. Kneser, U., D. J. Schaefer, E. Polykandriotis, and R. E. Horch. 2006. Tissue engineering of bone: The reconstructive surgeon's point of view. *Journal of Cellular and Molecular Medicine* 10: 7–19.
7. Nelson, E. R. et al. 2008. New bone formation by murine osteoprogenitor cells cultured on corti-cocancellous allograft bone. *Journal of Orthopaedic Research* 26: 1660–1664, doi:10.1002/jor.20676.
8. Vinatier, C., D. Mrugala, C. Jorgensen, J. Guicheux, and D. Noël. 2009. Cartilage engineering: A crucial combination of cells, biomaterials and biofactors. *Trends in Biotechnology* 27: 307–314, doi:10.1016/j.tibtech.2009.02.005.
9. Shi, J., A. R. Votruba, O. C. Farokhzad, and R. Langer. 2010. Nanotechnology in drug delivery and tissue engineering: From discovery to applications. *Nano Letters* 10: 3223–3230, doi:10.1021/nl102184c.
10. Dimitriou, R., E. Tsiridis, and P. V. Giannoudis. 2005. Current concepts of molecular aspects of bone healing. *Injury* 36: 1392–1404, doi:10.1016/j.injury.2005.07.019.
11. Hollister, S. J. 2005. Porous scaffold design for tissue engineering. *Nature Materials* 4: 518–524.
12. Kanczler, J. M. et al. 2008. The effect of mesenchymal populations and vascular endothelial growth factor delivered from biodegradable polymer scaffolds on bone formation. *Biomaterials* 29: 1892–1900.
13. Mehta, M., K. Schmidt-Bleek, G. N. Duda, and D. J. Mooney. 2012. Biomaterial delivery of morphogens to mimic the natural healing cascade in bone. *Advanced Drug Delivery Reviews* 64: 1257–1276, doi:10.1016/j.addr.2012.05.006.
14. Tan, J. and W. M. Saltzman. 2004. Biomaterials with hierarchically defined micro- and nanoscale structure. *Biomaterials* 25: 3593–3601, doi:10.1016/j.biomaterials.2003.10.034.
15. Lutolf, M. P. and J. A. Hubbell. 2005. Synthetic biomaterials as instructive extracellular microenvironments for morphogenesis in tissue engineering. *Nature Biotechnology* 23: 47–55.
16. Lutolf, M. P., P. M. Gilbert, and H. M. Blau. 2009. Designing materials to direct stem-cell fate. *Nature* 462: 433–441.
17. Galvin, P. et al. Nanoparticle-based drug delivery: Case studies for cancer and cardiovascular applications. *Cellular and Molecular Life Sciences*: 1–16, doi:10.1007/s00018-011-0856-6.
18. Langer, R. and N. A. Peppas. 2003. Advances in biomaterials, drug delivery, and bionanotechnology. *AIChE Journal* 49: 2990–3006.
19. Romano, N. H., D. Sengupta, C. Chung, and S. C. Heilshorn. 2011. Protein-engineered biomaterials: Nanoscale mimics of the extracellular matrix. *Biochimica et Biophysica Acta: General Subjects* 1810: 339–349, doi:10.1016/j.bbagen.2010.07.005.
20. Place, E. S., J. H. George, C. K. Williams, and M. M. Stevens. 2009. Synthetic polymer scaffolds for tissue engineering. *Chemical Society Reviews* 38: 1139–1151, doi:10.1039/b811392k.
21. Sanchez, C., H. Arribart, and M. M. Giraud Guille. 2005. Biomimetism and bioinspiration as tools for the design of innovative materials and systems. *Nature Materials* 4: 277–288.

22. Zhang, R. and P. X. Ma. 1999. Poly(α-hydroxyl acids)/hydroxyapatite porous composites for bone-tissue engineering. I. Preparation and morphology. *Journal of Biomedical Materials Research* 44: 446–455, doi:10.1002/(sici)1097–4636(19990315)44:4<446::aid-jbm11>3.0.co;2-f.

23. Rezwan, K., Q. Z. Chen, J. J. Blaker, and A. R. Boccaccini. 2006. Biodegradable and bioactive porous polymer/inorganic composite scaffolds for bone tissue engineering. *Biomaterials* 27: 3413–3431, doi:10.1016/j.biomaterials.2006.01.039.

24. O'Sullivan, C. et al. 2011. A modified surface on titanium deposited by a blasting process. *Coatings* 1: 53–71.

25. Ji, L., G. Jell, Y. Dong, J. R. Jones, and M. M. Stevens. 2011. Template synthesis of ordered macroporous hydroxyapatite bioceramics. *Chemical Communications* 47: 9048–9050.

26. Hench, L. L., R. J. Splinter, W. C. Allen, and T. K. Greenlee. 1971. Bonding mechanisms at the interface of ceramic prosthetic materials. *Journal of Biomedical Materials Research* 5: 117–141, doi:10.1002/jbm.820050611.

27. Taboas, J. M., R. D. Maddox, P. H. Krebsbach, and S. J. Hollister. 2003. Indirect solid free form fabrication of local and global porous, biomimetic and composite 3D polymer-ceramic scaffolds. *Biomaterials* 24: 181–194, doi:10.1016/s0142–9612(02)00276–4.

28. Kriparamanan, R., P. Aswath, A. Zhou, L. Tang, and K. T. Nguyen. 2006. Nanotopography: Cellular responses to nanostructured materials. *Journal of Nanoscience and Nanotechnology* 6: 1905–1919.

29. Zhang, L. et al. 2007. Nanoparticles in medicine: Therapeutic applications and developments. *Journal of Clinical Pharmacology and Therapeutics* 83: 761–769.

30. Zhang, L. and T. J. Webster. 2009. Nanotechnology and nanomaterials: Promises for improved tissue regeneration. *Nano Today* 4: 66–80, doi:http://dx.doi.org/10.1016/j.nantod.2008.10.014.

31. O'Brien, F. J., B. A. Harley, I. V. Yannas, and L. J. Gibson. 2005. The effect of pore size on cell adhesion in collagen-GAG scaffolds. *Biomaterials* 26: 433–441, doi:10.1016/j.biomaterials.2004.02.052.

32. Stevens, M. M. and J. H. George. 2005. Exploring and engineering the cell surface interface. *Science* 310: 1135–1138, doi:10.1126/science.1106587.

33. Curtis, A. and M. Dalby. 2009. Cell response to nanofeatures in biomaterials. In *Cellular Response to Biomaterials*, ed. L. Di Silvio, pp. 429–461. Woodhead Publishing Limited, Cambridge, U.K.

34. Kim, D.-H., P. P. Provenzano, C. L. Smith, and A. Levchenko. 2012. Matrix nanotopography as a regulator of cell function. *The Journal of Cell Biology* 197: 351–360, doi:10.1083/jcb.201108062.

35. Boyan, B. D., T. W. Hummert, D. D. Dean, and Z. Schwartz. 1996. Role of material surfaces in regulating bone and cartilage cell response. *Biomaterials* 17: 137–146, doi:10.1016/0142–9612(96)85758–9.

36. Sokolsky-Papkov, M., K. Agashi, A. Olaye, K. Shakesheff, and A. J. Domb. 2007. Polymer carriers for drug delivery in tissue engineering. *Advanced Drug Delivery Reviews* 59: 187–206, doi:10.1016/j.addr.2007.04.001.

37. Griffith, L. G. and G. Naughton. 2002. Tissue engineering—Current challenges and expanding opportunities. *Science* 295: 1009–1014, doi:10.1126/science.1069210.

38. Giannoudis, P. V., H. Dinopoulos, and E. Tsiridis. 2005. Bone substitutes: An update. *Injury* 36: S20–S27, doi:10.1016/j.injury.2005.07.029.

39. Sen, M. K. and T. Miclau. 2007. Autologous iliac crest bone graft: Should it still be the gold standard for treating nonunions? *Injury* 38: S75–S80, doi:10.1016/j.injury.2007.02.012.

40. Khan, S. N. et al. 2005. The biology of bone grafting. *Journal of the American Academy of Orthopaedic Surgeons* 13: 77–86.

41. de Boer, H. and M. Wood. 1989. Bone changes in the vascularised fibular graft. *Journal of Bone and Joint Surgery, British Volume* 71(3)-B: 374–378.

42. Ishaug, S. L. et al. 1997. Bone formation by three-dimensional stromal osteoblast culture in biodegradable polymer scaffolds. *Journal of Biomedical Materials Research* 36: 17–28, doi:10.1002/(sici)1097–4636(199707)36:1<17::aid-jbm3>3.0.co;2-o.

43. Lohmann, H., G. Grass, C. Rangger, and G. Mathiak. 2007. Economic impact of cancellous bone grafting in trauma surgery. *Archives of Orthopaedic and Trauma Surgery* 127: 345–348.

44. Planell, J. A. 2009. *Bone Repair Biomaterials*. Woodhead; CRC, Cambridge, U.K.

45. Silva, E. A. and D. J. Mooney. 2004. Synthetic extracellular matrices for tissue engineering and regeneration. In *Current Topics in Developmental Biology*, ed. Gerald Schatten, Vol. 64, pp. 181–205. Academic Press, Cambridge, MA.

46. Zimmermann, G. and A. Moghaddam. 2011. Allograft bone matrix versus synthetic bone graft substitutes. *Injury* 42, Supplement 2: S16–S21, doi:10.1016/j.injury.2011.06.199.

47. Mata, A. et al. 2010. Bone regeneration mediated by biomimetic mineralization of a nanofiber matrix. *Biomaterials* 31: 6004–6012, doi:10.1016/j.biomaterials.2010.04.013.

48. Ratcliffe, A. 2011. Difficulties in the translation of functionalized biomaterials into regenerative medicine clinical products. *Biomaterials* 32: 4215–4217, doi:10.1016/j.biomaterials.2011.02.028.

49. Huebsch, N. and D. J. Mooney. 2009. Inspiration and application in the evolution of biomaterials. *Nature* 462: 426–432.

50. Sands, R. W. and D. J. Mooney. 2007. Polymers to direct cell fate by controlling the microenvironment. *Current Opinion in Biotechnology* 18: 448–453, doi:10.1016/j.copbio.2007.10.004.

51. Badylak, S. F. and R. M. Nerem. 2010. Progress in tissue engineering and regenerative medicine. *Proceedings of the National Academy of Sciences of the United States of America* 107: 3285–3286, doi:10.1073/pnas.1000256107.

52. Lee, K., E. A. Silva, and D. J. Mooney. 2011. Growth factor delivery-based tissue engineering: general approaches and a review of recent developments. *Journal of the Royal Society Interface* 8: 153–170, doi:10.1098/rsif.2010.0223.

53. Bessa, P. C., M. Casal, and R. L. Reis. 2008. Bone morphogenetic proteins in tissue engineering: The road from the laboratory to the clinic, part I (basic concepts). *Journal of Tissue Engineering and Regenerative Medicine* 2: 1–13, doi:10.1002/term.63.

54. Yeatts, A. B. and J. P. Fisher. 2011. Bone tissue engineering bioreactors: Dynamic culture and the influence of shear stress. *Bone* 48: 171–181, doi:10.1016/j.bone.2010.09.138.

55. Zhang, Z.-Y. et al. 2010. A comparison of bioreactors for culture of fetal mesenchymal stem cells for bone tissue engineering. *Biomaterials* 31: 8684–8695, doi:http://dx.doi.org/10.1016/j.biomaterials.2010.07.097.

56. Sommerfeldt, D. and C. Rubin. 2001. Biology of bone and how it orchestrates the form and function of the skeleton. *European Spine Journal* 10: 86–95.

57. Weiner, S. and H. D. Wagner. 1998. The material bone: structure-mechanical function relations. *Annual Review of Materials Science* 28: 271–298.

58. Sivasubramanyam, K. S. 2011. Osteoinduction by biomaterials; towards unravelling the underlying process. PhD thesis submitted Queen Mary University of London, London, U.K., http://qmro.qmul.ac.uk/jspui/handle/123456789/1279.

59. Glimcher, M. 1990. The possible role of collagen fibrils and collagen-phosphoprotein complexes in the calcification of bone in vitro and in vivo. *Biomaterials* 11: 7.

60. Wenk, H. R. and F. Heidelbach. 1999. Crystal alignment of carbonated apatite in bone and calcified tendon: Results from quantitative texture analysis. *Bone* 24: 361–369.

61. Kumta, P. N., C. Sfeir, D.-H. Lee, D. Olton, and D. Choi. 2005. Nanostructured calcium phosphates for biomedical applications: Novel synthesis and characterization. *Acta Biomaterialia* 1: 65–83, doi:10.1016/j.actbio.2004.09.008.

62. Kuhn-Spearing, L., C. Rey, H. M. Kim, and M. J. Glimcher. 1996. *Carbonated Apatite Nanocrystals of Bone. Synthesis and Processing of Nanocrystalline Powder*. The Minerals, Metals and Materials Society, Warrendale, PA.

63. Hadjidakis, D. J. and I. I. Androulakis. 2006. Bone remodeling. *Annals of the New York Academy of Sciences* 1092: 385–396.

64. Barrère, F., T. A. Mahmood, K. de Groot, and C. A. van Blitterswijk. 2008. Advanced biomaterials for skeletal tissue regeneration: Instructive and smart functions. *Materials Science and Engineering: R: Reports* 59: 38–71, doi:http://dx.doi.org/10.1016/j.mser.2007.12.001.

65. Koeppen, B. M., B. A. Stanton, R. M. Berne, and M. N. Levy. 2008. *Berne and Levy Physiology*. Mosby, Philadelphia, PA.

66. Raggatt, L. J. and N. C. Partridge. 2010. Cellular and molecular mechanisms of bone remodeling. *Journal of Biological Chemistry* 285: 25103–25108, doi:10.1074/jbc.R109.041087.
67. Jayakumar, P. and L. Di Silvio. 2010. Osteoblasts in bone tissue engineering. *Proceedings of the Institution of Mechanical Engineers, Part H: Journal of Engineering in Medicine* 224: 1415–1440, doi:10.1243/09544119jeim821.
68. Liu H. and T. J. Webster. 2007 Bioinspired nanocomposites for orthopedic applications. In: *Nanotechnology for the Regeneration of Hard and Soft Tissues*, ed. Webster TJ, vol. 3, pp. 1–51. World Scientific, Singapore.
69. Stein, G. S. and J. B. Lian. 1993. Molecular mechanisms mediating proliferation/differentiation interrelationships during progressive development of the osteoblast phenotype. *Endocrine Reviews* 14: 424–442.
70. Long, F. 2012. Building strong bones: Molecular regulation of the osteoblast lineage. *Nature Reviews Molecular Cell Biology* 13: 27–38.
71. Stevens, M. M. 2008. Biomaterials for bone tissue engineering. *Materials Today* 11: 18–25, doi:10.1016/s1369-7021(08)70086-5.
72. Rho, J.-Y., L. Kuhn-Spearing, and P. Zioupos. 1998. Mechanical properties and the hierarchical structure of bone. *Medical Engineering and Physics* 20: 92–102, doi:http://dx.doi.org/10.1016/S1350-4533(98)00007-1.
73. Hench, L. L. and J. Wilson. 1993. Introduction In *An Introduction to Bioceramics*, eds. L. L. Hench and J. Wilson, pp. 1–24. World Scientific, River Edge, NJ.
74. Wynnyckyj, C. et al. 2009. A new tool to assess the mechanical properties of bone due to collagen degradation. *Bone* 44: 840–848.
75. Hutmacher, D. W., J. T. Schantz, C. X. F. Lam, K. C. Tan, and T. C. Lim. 2007. State of the art and future directions of scaffold-based bone engineering from a biomaterials perspective. *Journal of Tissue Engineering and Regenerative Medicine* 1: 245–260, doi:10.1002/term.24.
76. Smith, I. O., X. H. Liu, L. A. Smith, and P. X. Ma. 2009. Nanostructured polymer scaffolds for tissue engineering and regenerative medicine. *Wiley Interdisciplinary Reviews: Nanomedicine and Nanobiotechnology* 1: 226–236, doi:10.1002/wnan.26.
77. Weiner, S. and W. Traub. 1992. Bone structure: from angstroms to microns. *The FASEB Journal* 6: 879–885.
78. Kadler, K. E., D. F. Holmes, J. A. Trotter, and J. A. Chapman. 1996. Collagen fibril formation. *Biochemical Journal* 316: 1–11.
79. Maas, M. et al. 2011. Preparation of mineralized nanofibers: Collagen fibrils containing calcium phosphate. *Nano Letters* 11: 1383–1388, doi:10.1021/nl200116d.
80. Shoulders, M. D. and R. T. Raines. 2009. Collagen structure and stability. *Annual Review of Biochemistry* 78: 929–958, doi:doi:10.1146/annurev.biochem.77.032207.120833.
81. Bonfield, W. 1984. Elasticity and viscoelasticity of cortical bone. In *Natural and Living Biomaterials*, eds. G. W. Hastings and P. Ducheyne, pp. 43–60. CRC Press, Boca Raton, FL.
82. Van Audekercke, R. and M. Martens. 1984. Mechanical properties of cancellous bone. In *Natural and Living Biomaterials*, eds. G. W. Hastings and P. Ducheyne, pp. 89–98. CRC Press, Boca Raton, FL.
83. Black, J. 1988.*Orthopaedic Biomaterials in Research and Practice*. Churchill Livingstone, New York.
84. Discher, D. E., P. Janmey, and Y.-l. Wang. 2005. Tissue cells feel and respond to the stiffness of their substrate. *Science* 310: 1139–1143, doi:10.1126/science.1116995.
85. Daley, W. P., S. B. Peters, and M. Larsen. 2008. Extracellular matrix dynamics in development and regenerative medicine. *Journal of Cell Science* 121: 255–264, doi:10.1242/jcs.006064.
86. El-Amin, S. F. et al. 2003. Extracellular matrix production by human osteoblasts cultured on biodegradable polymers applicable for tissue engineering. *Biomaterials* 24: 1213–1221, doi:10.1016/s0142-9612(02)00451-9.
87. English, A. et al. 2012. Preferential cell response to anisotropic electro-spun fibrous scaffolds under tension-free conditions. *Journal of Materials Science: Materials in Medicine* 23: 137–148, doi:10.1007/s10856-011-4471-8.
88. Badylak, S. F. 2002. The extracellular matrix as a scaffold for tissue reconstruction. *Seminars in Cell and Developmental Biology* 13: 377–383, doi:10.1016/s1084952102000940.

89. Keatch, R. P., A. M. Schor, J. B. Vorstius, and S. L. Schor. Biomaterials in regenerative medicine: Engineering to recapitulate the natural. *Current Opinion in Biotechnology*, doi:10.1016/j.copbio.2012.01.017.

90. Rozario, T. and D. W. DeSimone. 2010. The extracellular matrix in development and morphogenesis: A dynamic view. *Developmental Biology* 341: 126–140, doi:10.1016/j.ydbio.2009.10.026.

91. Adams, J. C. and F. M. Watt. 1993. Regulation of development and differentiation by the extracellular matrix. *Development* 117: 1183–1198.

92. Ramirez, F. and D. B. Rifkin. 2003. Cell signaling events: A view from the matrix. *Matrix Biology* 22: 101–107, doi:10.1016/s0945–053x(03)00002–7.

93. Bettinger, C. J., R. Langer, and J. T. Borenstein. 2009. Engineering substrate topography at the micro- and nanoscale to control cell function. *Angewandte Chemie International Edition* 48: 5406–5415, doi:10.1002/anie.200805179.

94. Cross, M. and T. M. Dexter. 1991. Growth factors in development, transformation, and tumorigenesis. *Cell* 64: 271–280, doi:10.1016/0092–8674(91)90638–f.

95. Zamir, E. and B. Geiger. 2004. Focal Adhesions. In *Encyclopedia of Biological Chemistry*, eds. W. J. Lennarz and M. D. Lane, pp. 128–133. Elsevier, Rehovot, Israel.

96. Cukierman, E., R. Pankov, D. R. Stevens, and K. M. Yamada. 2001. Taking cell-matrix adhesions to the third dimension. *Science* 294: 1708–1712, doi:10.1126/science.1064829.

97. Barczyk, M., S. Carracedo, and D. Gullberg. 2010. Integrins. *Cell and Tissue Research* 339: 269–280, doi:10.1007/s00441–009–0834–6.

98. Shekaran, A. and A. J. Garcia. 2011. Nanoscale engineering of extracellular matrix-mimetic bioadhesive surfaces and implants for tissue engineering. *Biochimica et Biophysica Acta —General Subjects* 1810: 350–360, doi:10.1016/j.bbagen.2010.04.006.

99. Guilak, F. et al. 2009. Control of stem cell fate by physical interactions with the extracellular matrix. *Cell Stem Cell* 5: 17–26, doi:10.1016/j.stem.2009.06.016.

100. George, A. and S. Ravindran. 2010. Protein templates in hard tissue engineering. *Nano Today* 5: 254–266, doi:http://dx.doi.org/10.1016/j.nantod.2010.05.005.

101. Vu, T. H. and Z. Werb. 2000. Matrix metalloproteinases: Effectors of development and normal physiology. *Genes and Development* 14: 2123–2133, doi:10.1101/gad.815400.

102. McCawley, L. J. and L. M. Matrisian. 2001. Matrix metalloproteinases: They're not just for matrix anymore! *Current Opinion in Cell Biology* 13: 534–540, doi:10.1016/s0955–0674(00)00248–9.

103. Engsig, M. T. et al. 2000. Matrix metalloproteinase 9 and vascular endothelial growth factor are essential for osteoclast recruitment into developing long bones. *The Journal of Cell Biology* 151: 879–890, doi:10.1083/jcb.151.4.879.

104. Chan, G. and D. J. Mooney. 2008. New materials for tissue engineering: Towards greater control over the biological response. *Trends in Biotechnology* 26: 382–392, doi:10.1016/j.tibtech.2008.03.011.

105. Lee, S.-H. and H. Shin. 2007. Matrices and scaffolds for delivery of bioactive molecules in bone and cartilage tissue engineering. *Advanced Drug Delivery Reviews* 59: 339–359, doi:10.1016/j.addr.2007.03.016.

106. Jones, J. R. Review of bioactive glass: From Hench to hybrids. *Acta Biomaterialia*, 9(1): 4457–4486, doi:10.1016/j.actbio.2012.08.023.

107. Furth, M. E., A. Atala, and M. E. Van Dyke. 2007. Smart biomaterials design for tissue engineering and regenerative medicine. *Biomaterials* 28: 5068–5073, doi:10.1016/j.biomaterials.2007.07.042.

108. Tai, H. et al. 2010. Studies on the interactions of CO_2 with biodegradable poly(dl-lactic acid) and poly(lactic acid-co-glycolic acid) copolymers using high pressure ATR-IR and high pressure rheology. *Polymer* 51: 1425–1431, doi:10.1016/j.polymer.2010.01.065.

109. Kim, B.-S. and D. J. Mooney. 1998. Development of biocompatible synthetic extracellular matrices for tissue engineering. *Trends in Biotechnology* 16: 224–230, doi:10.1016/s0167–7799(98)01191–3.

110. Stevens, M. M. and G. Mecklenburg. 2012. Bio-inspired materials for biosensing and tissue engineering. *Polymer International* 61: 680–685, doi:10.1002/pi.4183.

111. O'Brien, F. J. 2011. Biomaterials & scaffolds for tissue engineering. *Materials Today* 14: 88–95, doi:10.1016/s1369–7021(11)70058-x.

112. Ma, P. X. 2004. Scaffolds for tissue fabrication. *Materials Today* 7: 30–40, doi:10.1016/s1369-7021(04)00233-0.
113. Mastrogiacomo, M. et al. 2006. Role of scaffold internal structure on in vivo bone formation in macroporous calcium phosphate bioceramics. *Biomaterials* 27: 3230–3237, doi:10.1016/j.biomaterials.2006.01.031.
114. Dennis, J. E., S. E. Haynesworth, R. G. Young, and A. I. Caplan. 1992. Osteogenesis in marrow-derived mesenchymal cell porous ceramic composites transplanted subcutaneously: Effect of fibronectin and laminin on cell retention and rate of osteogenic expression. *Cell Transplantation* 1: 23–32.
115. Jones, J. R., L. M. Ehrenfried, and L. L. Hench. 2006. Optimising bioactive glass scaffolds for bone tissue engineering. *Biomaterials* 27: 964–973, doi:10.1016/j.biomaterials.2005.07.017.
116. Trichet, L. et al. 2012. Evidence of a large-scale mechanosensing mechanism for cellular adaptation to substrate stiffness. *Proceedings of the National Academy of Sciences of the United States of America* 109: 6933–6938, doi:10.1073/pnas.1117810109.
117. Castillo, A. and C. Jacobs. 2010. Mesenchymal stem cell mechanobiology. *Current Osteoporosis Reports* 8: 98–104, doi:10.1007/s11914–010–0015–2.
118. Badylak, S. F., D. O. Freytes, and T. W. Gilbert. 2009. Extracellular matrix as a biological scaffold material: Structure and function. *Acta Biomaterialia* 5: 1–13, doi:10.1016/j.actbio.2008.09.013.
119. Hutmacher, D. W. 2000. Scaffolds in tissue engineering bone and cartilage. *Biomaterials* 21: 2529–2543, doi:10.1016/s0142–9612(00)00121–6.
120. Dalby, M. J. et al. 2006. Osteoprogenitor response to semi-ordered and random nanotopographies. *Biomaterials* 27: 2980–2987, doi:10.1016/j.biomaterials.2006.01.010.
121. Sjöström, T. et al. 2009. Fabrication of pillar-like titania nanostructures on titanium and their interactions with human skeletal stem cells. *Acta Biomaterialia* 5: 1433–1441, doi:10.1016/j.actbio.2009.01.007.
122. Kunzler, T. P., C. Huwiler, T. Drobek, J. Vörös, and N. D. Spencer. 2007. Systematic study of osteoblast response to nanotopography by means of nanoparticle-density gradients. *Biomaterials* 28: 5000–5006, doi:10.1016/j.biomaterials.2007.08.009.
123. Dalby, M. J., D. McCloy, M. Robertson, C. D. W. Wilkinson, and R. O. C. Oreffo. 2006. Osteoprogenitor response to defined topographies with nanoscale depths. *Biomaterials* 27: 1306–1315, doi:10.1016/j.biomaterials.2005.08.028.
124. Oh, S. et al. 2006. Significantly accelerated osteoblast cell growth on aligned TiO2 nanotubes. *Journal of Biomedical Materials Research Part A* 78A: 97–103, doi:10.1002/jbm.a.30722.
125. Dalby, M. J. et al. 2007. The control of human mesenchymal cell differentiation using nanoscale symmetry and disorder. *Nature Materials* 6: 997–1003.
126. Washburn, N. R., K. M. Yamada, C. G. Simon Jr., S. B. Kennedy, and E. J. Amis. 2004. High-throughput investigation of osteoblast response to polymer crystallinity: influence of nanometer-scale roughness on proliferation. *Biomaterials* 25: 1215–1224, doi:10.1016/j.biomaterials.2003.08.043.
127. Albrektsson, T. and C. Johansson. 2001. Osteoinduction, osteoconduction and osseointegration. *European Spine Journal* 10 (Suppl. 2): S96.
128. Burdick, J. A. and G. Vunjak-Novakovic. 2009. Engineered microenvironments for controlled stem cell differentiation. *Tissue Engineering Part A* 15: 205–219, doi:10.1089/ten.tea.2008.0131.
129. Griffith, L. G. and M. A. Swartz. 2006. Capturing complex 3D tissue physiology in vitro. *Nature Reviews Molecular Cell Biology* 7: 211–224, doi:http://www.nature.com/nrm/journal/v7/n3/suppinfo/nrm1858_S1.html.
130. Fisher, O. Z., A. Khademhosseini, R. Langer, and N. A. Peppas. 2009. Bioinspired materials for controlling stem cell fate. *Accounts of Chemical Research* 43: 419–428, doi:10.1021/ar900226q.
131. Holzwarth, J. M. and P. X. Ma. 2011. 3D nanofibrous scaffolds for tissue engineering. *Journal of Materials Chemistry* 21: 10243–10251.
132. Anselme, K. 2000. Osteoblast adhesion on biomaterials. *Biomaterials* 21: 667–681, doi:10.1016/s0142–9612(99)00242–2.

133. Cuenca-López, M. D. et al. 2010. Action of recombinant human BMP-2 on fracture healing in rabbits is dependent on the mechanical environment. *Journal of Tissue Engineering and Regenerative Medicine* 4: 543–552, doi:10.1002/term.271.

134. Diab, T. et al. 2012. A silk hydrogel-based delivery system of bone morphogenetic protein for the treatment of large bone defects. *Journal of the Mechanical Behavior of Biomedical Materials* 11: 123–131, doi:10.1016/j.jmbbm.2011.11.007.

135. Valente, J. F. A. et al. 2012. Alginate based scaffolds for bone tissue engineering. *Materials Science and Engineering: C* 32: 2596–2603, doi:10.1016/j.msec.2012.08.001.

136. Lee, Y.-M. et al. 2000. The bone regenerative effect of platelet-derived growth factor-BB delivered with a chitosan/tricalcium phosphate sponge carrier. *Journal of Periodontology* 71: 418–424, doi:10.1902/jop.2000.71.3.418.

137. Radomsky, M. L. et al. 1999. Novel formulation of fibroblast growth factor-2 in a hyaluronan gel accelerates fracture healing in nonhuman primates. *Journal of Orthopaedic Research* 17: 607–614, doi:10.1002/jor.1100170422.

138. Badylak, S. F. 2007. The extracellular matrix as a biologic scaffold material. *Biomaterials* 28: 3587–3593, doi:10.1016/j.biomaterials.2007.04.043.

139. Parenteau-Bareil, R., R. Gauvin, and F. Berthod. 2010. Collagen-based biomaterials for tissue engineering applications. *Materials* 3: 1863–1887.

140. Chevallay, B. and D. Herbage. 2000. Collagen-based biomaterials as 3D scaffold for cell cultures: Applications for tissue engineering and gene therapy. *Medical and Biological Engineering and Computing* 38: 211–218, doi:10.1007/bf02344779.

141. Murphy, C. M., M. G. Haugh, and F. J. O'Brien. 2010. The effect of mean pore size on cell attachment, proliferation and migration in collagen–glycosaminoglycan scaffolds for bone tissue engineering. *Biomaterials* 31: 461–466, doi:10.1016/j.biomaterials.2009.09.063.

142. Murphy, C. M., A. Matsiko, M. G. Haugh, J. P. Gleeson, and F. J. O'Brien. 2012. Mesenchymal stem cell fate is regulated by the composition and mechanical properties of collagen–glycosaminoglycan scaffolds. *Journal of the Mechanical Behavior of Biomedical Materials* 11: 53–62, doi:10.1016/j.jmbbm.2011.11.009.

143. Wray, L. S. et al. 2012. A silk-based scaffold platform with tunable architecture for engineering critically-sized tissue constructs. *Biomaterials* 33: 9214–9224, doi:10.1016/j.biomaterials.2012.09.017.

144. Mano, J. F. et al. 2007. Natural origin biodegradable systems in tissue engineering and regenerative medicine: Present status and some moving trends. *Journal of the Royal Society Interface* 4: 999–1030, doi:10.1098/rsif.2007.0220.

145. Ryan, K. B., S. Maher, D. Brayden, and C. O' Driscoll. 2012. Nanostructures overcoming the intestinal barrier: Drug delivery strategies. In *Nanostructured Biomaterials for Overcoming Biological Barriers*, eds. M. J. Alonso and N. S. Csaba. RSC Publishing, Cambridge, U.K.

146. Lim, J. Y. et al. 2008. Surface energy effects on osteoblast spatial growth and mineralization. *Biomaterials* 29: 1776–1784, doi:10.1016/j.biomaterials.2007.12.026.

147. Panyam, J. and V. Labhasetwar. 2003. Biodegradable nanoparticles for drug and gene delivery to cells and tissue. *Advanced Drug Delivery Reviews* 55: 329–347.

148. Dhandayuthapani, B., Y. Yoshida, T. Maekawa, and D. S. Kumar. 2011. Polymeric scaffolds in tissue engineering application: A review. *International Journal of Polymer Science* 2011: 19, Article ID 290602, doi:10.1155/2011/290602.

149. Pan, Z. and J. Ding. 2012. Poly(lactide-co-glycolide) porous scaffolds for tissue engineering and regenerative medicine. *Interface Focus* 2(3): 366–377, doi:10.1098/rsfs.2011.0123.

150. Wang, J., B. M. Wang, and S. P. Schwendeman. 2002. Characterization of the initial burst release of a model peptide from poly(-lactide-co-glycolide) microspheres. *Journal of Controlled Release* 82: 289–307, doi:Doi: 10.1016/s0168–3659(02)00137–2.

151. Lutolf, M. P. et al. 2003. Synthetic matrix metalloproteinase-sensitive hydrogels for the conduction of tissue regeneration: Engineering cell-invasion characteristics. *Proceedings of the National Academy of Sciences of the United States of America* 100: 5413–5418, doi:10.1073/pnas.0737381100.

152. Yang, X. B. et al. 2001. Human osteoprogenitor growth and differentiation on synthetic biodegradable structures after surface modification. *Bone* 29: 523–531, doi:10.1016/s8756-3282(01)00617-2.

153. Woo, K. M., V. J. Chen, and P. X. Ma. 2003. Nano-fibrous scaffolding architecture selectively enhances protein adsorption contributing to cell attachment. *Journal of Biomedical Materials Research Part A* 67A: 531–537, doi:10.1002/jbm.a.10098.

154. Woo, K. M. et al. 2007. Nano-fibrous scaffolding promotes osteoblast differentiation and biomineralization. *Biomaterials* 28: 335–343, doi:10.1016/j.biomaterials.2006.06.013.

155. Woodruff, M. A. and D. W. Hutmacher. 2010. The return of a forgotten polymer—Polycaprolactone in the 21st century. *Progress in Polymer Science* 35: 1217–1256, doi:10.1016/j.progpolymsci.2010.04.002.

156. Kim, J. H. et al. 2009. Electrospun nanofibers composed of poly(ε-caprolactone) and polyethylenimine for tissue engineering applications. *Materials Science and Engineering: C* 29: 1725–1731, doi:10.1016/j.msec.2009.01.023.

157. Yoshimoto, H., Y. M. Shin, H. Terai, and J. P. Vacanti. 2003. A biodegradable nanofiber scaffold by electrospinning and its potential for bone tissue engineering. *Biomaterials* 24: 2077–2082, doi:10.1016/s0142-9612(02)00635-x.

158. Li, W.-J., R. Tuli, X. Huang, P. Laquerriere, and R. S. Tuan. 2005. Multilineage differentiation of human mesenchymal stem cells in a three-dimensional nanofibrous scaffold. *Biomaterials* 26: 5158–5166, doi:10.1016/j.biomaterials.2005.01.002.

159. Lutolf, M. P. 2009. Biomaterials: Spotlight on hydrogels. *Nature Materials* 8: 451–453.

160. Drury, J. L. and D. J. Mooney. 2003. Hydrogels for tissue engineering: Scaffold design variables and applications. *Biomaterials* 24: 4337–4351.

161. Alsberg, E., K. W. Anderson, A. Albeiruti, R. T. Franceschi, and D. J. Mooney. 2001. Cell-interactive alginate hydrogels for bone tissue engineering. *Journal of Dental Research* 80: 2025–2029, doi:10.1177/00220345010800111501.

162. Yamamoto, M., Y. Takahashi, and Y. Tabata. 2003. Controlled release by biodegradable hydrogels enhances the ectopic bone formation of bone morphogenetic protein. *Biomaterials* 24: 4375–4383, doi:10.1016/s0142-9612(03)00337-5.

163. Patterson, J. et al. 2010. Hyaluronic acid hydrogels with controlled degradation properties for oriented bone regeneration. *Biomaterials* 31: 6772–6781, doi:10.1016/j.biomaterials.2010.05.047.

164. Zhu, J. 2010. Bioactive modification of poly(ethylene glycol) hydrogels for tissue engineering. *Biomaterials* 31: 4639–4656, doi:10.1016/j.biomaterials.2010.02.044.

165. Nuttelman, C. R., M. C. Tripodi, and K. S. Anseth. 2005. Synthetic hydrogel niches that promote hMSC viability. *Matrix Biology* 24: 208–218, doi:10.1016/j.matbio.2005.03.004.

166. Seliktar, D. 2012. Designing cell-compatible hydrogels for biomedical applications. *Science* 336: 1124–1128, doi:10.1126/science.1214804.

167. Keselowsky, B. G., D. M. Collard, and A. J. García. 2005. Integrin binding specificity regulates biomaterial surface chemistry effects on cell differentiation. *Proceedings of the National Academy of Sciences of the United States of America* 102: 5953–5957, doi:10.1073/pnas.0407356102.

168. Alsberg, E., K. W. Anderson, A. Albeiruti, J. A. Rowley, and D. J. Mooney. 2002. Engineering growing tissues. *Proceedings of the National Academy of Sciences of the United States of America* 99: 12025–12030, doi:10.1073/pnas.192291499.

169. Benoit, D. S. W., M. P. Schwartz, A. R. Durney, and K. S. Anseth. 2008. Small functional groups for controlled differentiation of hydrogel-encapsulated human mesenchymal stem cells. *Nature Materials* 7(10): 816–823, doi:http://www.nature.com/nmat/journal/v7/n10/suppinfo/nmat2269_S1.html.

170. Elisseeff, J. 2008. Hydrogels: Structure starts to gel. *Nature Materials* 7(4): 271–273.

171. Ladet, S., L. David, and A. Domard. 2008. Multi-membrane hydrogels. *Nature* 452(7183): 76–79, doi:http://www.nature.com/nature/journal/v452/n7183/suppinfo/nature06619_S1.html.

172. Leach, J. K. 2006. Multifunctional cell-instructive materials for tissue regeneration. *Regenerative Medicine* 1: 447–455, doi:10.2217/17460751.1.4.447.

173. Lutolf, M. P. et al. 2003. Repair of bone defects using synthetic mimetics of collagenous extracellular matrices. *Nature Biotechnology* 21: 513–518.

174. Pratt, A. B., F. E. Weber, H. G. Schmoekel, R. Müller, and J. A. Hubbell. 2004. Synthetic extracellular matrices for in situ tissue engineering. *Biotechnology and Bioengineering* 86: 27–36, doi:10.1002/bit.10897.

175. Place, E. S., N. D. Evans, and M. M. Stevens. 2009. Complexity in biomaterials for tissue engineering. *Nature Materials* 8: 457–470, doi:http://www.nature.com/nmat/journal/v8/n6/suppinfo/nmat2441_S1.html.

176. Boccaccini, A. R. and J. J. Blaker. 2005. Bioactive composite materials for tissue engineering scaffolds. *Expert Review of Medical Devices* 2: 303–317, doi:10.1586/17434440.2.3.303.

177. Wang, H. et al. 2007. Biocompatibility and osteogenesis of biomimetic nano-hydroxyapatite/polyamide composite scaffolds for bone tissue engineering. *Biomaterials* 28: 3338–3348, doi:10.1016/j.biomaterials.2007.04.014.

178. Kurashina, K., H. Kurita, Q. Wu, A. Ohtsuka, and H. Kobayashi. 2002. Ectopic osteogenesis with biphasic ceramics of hydroxyapatite and tricalcium phosphate in rabbits. *Biomaterials* 23: 407–412.

179. Dorozhkin, S. V. and M. Epple. 2002. Biological and medical significance of calcium phosphates. *Angewandte Chemie International Edition* 41: 3130–3146, doi:10.1002/1521-3773 (20020902)41:17<3130::aid-anie3130>3.0.co;2-1.

180. Woodard, J. R. et al. 2007. The mechanical properties and osteoconductivity of hydroxyapatite bone scaffolds with multi-scale porosity. *Biomaterials* 28: 45–54.

181. Habraken, W. J. E. M., J. G. C. Wolke, and J. A. Jansen. 2007. Ceramic composites as matrices and scaffolds for drug delivery in tissue engineering. *Advanced Drug Delivery Reviews* 59: 234–248, doi:10.1016/j.addr.2007.03.011.

182. Tampieri, A., G. Celotti, and E. Landi. 2005. From biomimetic apatites to biologically inspired composites. *Analytical and Bioanalytical Chemistry* 381: 568–576, doi:10.1007/s00216-004-2943-0.

183. Valliant, E. M. and J. R. Jones. 2011. Softening bioactive glass for bone regeneration: Sol-gel hybrid materials. *Soft Matter* 7: 5083–5095.

184. Hench, L. L. 1991. Bioceramics: From concept to clinic. *Journal of the American Ceramic Society* 74: 1487–1510, doi:10.1111/j.1151–2916.1991.tb07132.x.

185. Lobel, K. D. and L. L. Hench. 1998. In vitro adsorption and activity of enzymes on reaction layers of bioactive glass substrates. *Journal of Biomedical Materials Research* 39: 575–579, doi:10.1002/ (sici)1097–4636(19980315)39:4<575::aid-jbm11>3.0.co;2–6.

186. Kent Leach, J., D. Kaigler, Z. Wang, P. H. Krebsbach, and D. J. Mooney. 2006. Coating of VEGF-releasing scaffolds with bioactive glass for angiogenesis and bone regeneration. *Biomaterials* 27: 3249–3255, doi:10.1016/j.biomaterials.2006.01.033.

187. Day, R. M. et al. 2004. Assessment of polyglycolic acid mesh and bioactive glass for soft-tissue engineering scaffolds. *Biomaterials* 25: 5857–5866, doi:10.1016/j.biomaterials.2004.01.043.

188. O'Sullivan, C. et al. 2010. Deposition of substituted apatites with anticolonizing properties onto titanium surfaces using a novel blasting process. *Journal of Biomedical Materials Research Part B: Applied Biomaterials* 95B: 141–149, doi:10.1002/jbm.b.31694.

189. Goldberg, M., R. Langer, and X. Jia. 2007. Nanostructured materials for applications in drug delivery and tissue engineering. *Journal of Biomaterials Science. Polymer Edition* 18: 241, doi:10.1163/156856207779996931.

190. Boccaccini, A. R. et al. 2010. Polymer/bioactive glass nanocomposites for biomedical applications: A review. *Composites Science and Technology* 70: 1764–1776, doi:10.1016/j.compscitech.2010.06.002.

191. Xynos, I. D., A. J. Edgar, L. D. K. Buttery, L. L. Hench, and J. M. Polak. 2000. Ionic products of bioactive glass dissolution increase proliferation of human osteoblasts and induce insulin-like growth factor II mRNA expression and protein synthesis. *Biochemical and Biophysical Research Communications* 276: 461–465, doi:10.1006/bbrc.2000.3503.

192. Xynos, I. D., A. J. Edgar, L. D. K. Buttery, L. L. Hench, and J. M. Polak. 2001. Gene-expression profiling of human osteoblasts following treatment with the ionic products of Bioglass® 45S5 dissolution. *Journal of Biomedical Materials Research* 55: 151–157, doi:10.1002/1097-4636 (200105)55:2<151::aid-jbm1001>3.0.co;2-d.
193. Rahaman, M. N. et al. 2011. Bioactive glass in tissue engineering. *Acta Biomaterialia* 7: 2355–2373, doi:10.1016/j.actbio.2011.03.016.
194. Reginster, J. Y. et al. 2005. Strontium ranelate reduces the risk of nonvertebral fractures in post-menopausal women with osteoporosis: Treatment of peripheral osteoporosis (TROPOS) Study. *Journal of Clinical Endocrinology and Metabolism* 90: 2816–2822, doi:10.1210/jc.2004-1774.
195. Borgström, F., B. Jönsson, O. Ström, and J. Kanis. 2006. An economic evaluation of strontium ranelate in the treatment of osteoporosis in a Swedish setting. *Osteoporosis International* 17: 1781–1793, doi:10.1007/s00198-006-0193-z.
196. Pors Nielsen, S. 2004. The biological role of strontium. *Bone* 35: 583–588, doi:10.1016/j.bone.2004.04.026.
197. Bonnelye, E., A. Chabadel, F. Saltel, and P. Jurdic. 2008. Dual effect of strontium ranelate: Stimulation of osteoblast differentiation and inhibition of osteoclast formation and resorption in vitro. *Bone* 42: 129–138, doi:10.1016/j.bone.2007.08.043.
198. Marie, P. J. et al. 1993. An uncoupling agent containing strontium prevents bone loss by depressing bone resorption and maintaining bone formation in estrogen-deficient rats. *Journal of Bone and Mineral Research* 8: 607–615.
199. Gentleman, E. et al. 2010. The effects of strontium-substituted bioactive glasses on osteoblasts and osteoclasts in vitro. *Biomaterials* 31: 3949–3956, doi:10.1016/j.biomaterials.2010.01.121.
200. Peter, M. et al. 2010. Nanocomposite scaffolds of bioactive glass ceramic nanoparticles disseminated chitosan matrix for tissue engineering applications. *Carbohydrate Polymers* 79: 284–289, doi:10.1016/j.carbpol.2009.08.001.
201. Arcos, D. and M. Vallet-Regí. 2010. Sol–gel silica-based biomaterials and bone tissue regeneration. *Acta Biomaterialia* 6: 2874–2888, doi:10.1016/j.actbio.2010.02.012.
202. Valliant, E. M., C. A. Turdean-Ionescu, J. V. Hanna, M. E. Smith, and J. R. Jones. 2012. Role of pH and temperature on silica network formation and calcium incorporation into sol-gel derived bioactive glasses. *Journal of Materials Chemistry* 22: 1613–1619.
203. Wan, Y. et al. 2005. Adhesion and proliferation of OCT-1 osteoblast-like cells on micro- and nano-scale topography structured poly(l-lactide). *Biomaterials* 26: 4453–4459, doi:10.1016/j.biomaterials.2004.11.016.
204. Bellis, S. L. 2011. Advantages of RGD peptides for directing cell association with biomaterials. *Biomaterials* 32: 4205–4210, doi:10.1016/j.biomaterials.2011.02.029.
205. Biondi, M., F. Ungaro, F. Quaglia, and P. A. Netti. 2008. Controlled drug delivery in tissue engineering. *Advanced Drug Delivery Reviews* 60: 229–242, doi:10.1016/j.addr.2007.08.038.
206. Leong, N. L., J. Jie, and H. H. Lu. In *Engineering in Medicine and Biology Society, 2006. 28th Annual International Conference of the IEEE.* pp. 2651–2654, New York.
207. Verrier, S., J. J. Blaker, V. Maquet, L. L. Hench, and A. R. Boccaccini. 2004. PDLLA/Bioglass® composites for soft-tissue and hard-tissue engineering: An in vitro cell biology assessment. *Biomaterials* 25: 3013–3021, doi:10.1016/j.biomaterials.2003.09.081.
208. Hench, L. L. 1998. Bioceramics. *Journal of the American Ceramic Society* 81: 1705–1728.
209. Holzwarth, J. M. and P. X. Ma. 2011. Biomimetic nanofibrous scaffolds for bone tissue engineering. *Biomaterials* 32: 9622–9629.
210. Kim, S.-S., M. Sun Park, O. Jeon, C. Yong Choi, and B.-S. Kim. 2006. Poly(lactide-co-glycolide)/hydroxyapatite composite scaffolds for bone tissue engineering. *Biomaterials* 27: 1399–1409, doi:10.1016/j.biomaterials.2005.08.016.
211. Novak, B. M. 1993. Hybrid Nanocomposite Materials—Between inorganic glasses and organic polymers. *Advanced Materials* 5: 422–433, doi:10.1002/adma.19930050603.
212. Kim, H.-W., J. C. Knowles, and H.-E. Kim. 2004. Hydroxyapatite/poly(ε-caprolactone) composite coatings on hydroxyapatite porous bone scaffold for drug delivery. *Biomaterials* 25: 1279–1287, doi:10.1016/j.biomaterials.2003.07.003.

213. Choong, C., J. T. Triffitt, and Z. F. Cui. 2004. Polycaprolactone scaffolds for bone tissue engineering: Effects of a calcium phosphate coating layer on osteogenic cells. *Food and Bioproducts Processing* 82: 117–125, doi:10.1205/0960308041614864.
214. Ural, E., K. Kesenci, L. Fambri, C. Migliaresi, and E. Piskin. 2000. Poly(d,l-cactide/ε-caprolactone)/hydroxyapatite composites. *Biomaterials* 21: 2147–2154, doi:10.1016/s0142–9612(00)00098–3.
215. Lickorish, D., L. Guan, and J. E. Davies. 2007. A three-phase, fully resorbable, polyester/calcium phosphate scaffold for bone tissue engineering: Evolution of scaffold design. *Biomaterials* 28: 1495–1502, doi:10.1016/j.biomaterials.2006.11.025.
216. Yang, F., J. G. C. Wolke, and J. A. Jansen. 2008. Biomimetic calcium phosphate coating on electrospun poly(ε-caprolactone) scaffolds for bone tissue engineering. *Chemical Engineering Journal* 137: 154–161, doi:10.1016/j.cej.2007.07.076.
217. Al-Munajjed, A. A. et al. 2009. Development of a biomimetic collagen-hydroxyapatite scaffold for bone tissue engineering using a SBF immersion technique. *Journal of Biomedical Materials Research Part B: Applied Biomaterials* 90B: 584–591, doi:10.1002/jbm.b.31320.
218. Al-Munajjed, A. A. and F. J. O'Brien. 2009. Influence of a novel calcium-phosphate coating on the mechanical properties of highly porous collagen scaffolds for bone repair. *Journal of the Mechanical Behavior of Biomedical Materials* 2: 138–146, doi:10.1016/j.jmbbm.2008.05.001.
219. Lyons, F. G. et al. 2010. The healing of bony defects by cell-free collagen-based scaffolds compared to stem cell-seeded tissue engineered constructs. *Biomaterials* 31: 9232–9243, doi:10.1016/j.biomaterials.2010.08.056.
220. Chatterjee, K., L. Sun, L. C. Chow, M. F. Young, and C. G. Simon Jr. 2011. Combinatorial screening of osteoblast response to 3D calcium phosphate/poly(ε-caprolactone) scaffolds using gradients and arrays. *Biomaterials* 32: 1361–1369, doi:10.1016/j.biomaterials.2010.10.043.
221. Pramanik, N., P. Bhargava, S. Alam, and P. Pramanik. 2006. Processing and properties of nano- and macro-hydroxyapatite/poly(ethylene-co-acrylic acid) composites. *Polymer Composites* 27: 633–641, doi:10.1002/pc.20246.
222. Lu, H. H., S. F. El-Amin, K. D. Scott, and C. T. Laurencin. 2003. Three-dimensional, bioactive, biodegradable, polymer–bioactive glass composite scaffolds with improved mechanical properties support collagen synthesis and mineralization of human osteoblast-like cells in vitro. *Journal of Biomedical Materials Research Part A* 64A: 465–474, doi:10.1002/jbm.a.10399.
223. Leong, N. L., J. Jiang, and H. H. Lu. 2006. Polymer-ceramic composite scaffold induces osteogenic differentiation of human mesenchymal stem cells. *Proceedings of the 28th IEEE EMBS Annual International Conference*, New York, August 30–Sept 3, 2006, pp. 2651–2654.
224. Jo, J.-H. et al. 2009. In vitro/in vivo biocompatibility and mechanical properties of bioactive glass nanofiber and poly(ε-caprolactone) composite materials. *Journal of Biomedical Materials Research Part B: Applied Biomaterials* 91B: 213–220, doi:10.1002/jbm.b.31392.
225. Marelli, B. et al. 2011. Accelerated mineralization of dense collagen-nano bioactive glass hybrid gels increases scaffold stiffness and regulates osteoblastic function. *Biomaterials* 32: 8915–8926, doi:10.1016/j.biomaterials.2011.08.016.
226. Poologasundarampillai, G. et al. 2010. Synthesis of bioactive class II poly([gamma]-glutamic acid)/silica hybrids for bone regeneration. *Journal of Materials Chemistry* 20: 8952–8961.
227. Poologasundarampillai, G. et al. 2012. Bioactive silica-poly([gamma]-glutamic acid) hybrids for bone regeneration: Effect of covalent coupling on dissolution and mechanical properties and fabrication of porous scaffolds. *Soft Matter* 8: 4822–4832.
228. Obst, M. and A. Steinbüchel. 2004. Microbial Degradation of Poly(amino acid)s. *Biomacromolecules* 5: 1166–1176, doi:10.1021/bm049949u.
229. Hunter, G. K. and H. A. Goldberg. 1994. Modulation of crystal formation by bone phosphoproteins: Role of glutamic acid-rich sequences in the nucleation of hydroxyapatite by bone sialoprotein. *Biochemical Journal* 302(Pt 1): 175–179.
230. Arvidson, K. et al. 2011. Bone regeneration and stem cells. *Journal of Cellular and Molecular Medicine* 15: 718–746, doi:10.1111/j.1582–4934.2010.01224.x.

231. Tsiridis, E., N. Upadhyay, and P. Giannoudis. 2007. Molecular aspects of fracture healing: Which are the important molecules? *Injury* 38: S11–S25, doi:10.1016/j.injury.2007.02.006.

232. Pohl, T. L. M., J. H. Boergermann, G. K. Schwaerzer, P. Knaus, and E. A. Cavalcanti-Adam. 2012. Surface immobilization of bone morphogenetic protein 2 via a self-assembled monolayer formation induces cell differentiation. *Acta Biomaterialia* 8: 772–780, doi:10.1016/j.actbio.2011.10.019.

233. Patel, Z. S. et al. 2008. Dual delivery of an angiogenic and an osteogenic growth factor for bone regeneration in a critical size defect model. *Bone* 43: 931–940, doi:10.1016/j.bone.2008.06.019.

234. Luvizuto, E. R. et al. 2011. The effect of BMP-2 on the osteoconductive properties of β-tricalcium phosphate in rat calvaria defects. *Biomaterials* 32: 3855–3861, doi:10.1016/j.biomaterials.2011.01.076.

235. Kempen, D. H. R. et al. 2009. Effect of local sequential VEGF and BMP-2 delivery on ectopic and orthotopic bone regeneration. *Biomaterials* 30: 2816–2825, doi:10.1016/j.biomaterials.2009.01.031.

236. Cowan, C. M., C. Soo, K. Ting, and B. Wu.2005. Evolving concepts in bone tissue engineering. In *Current Topics in Developmental Biology*, ed. Gerald Schatten, Vol. 66, pp. 239–285. Academic Press, New York.

237. Wikesjö, U. M. E., Y.-H. Huang, G. Polimeni, and M. Qahash. 2007. Bone morphogenetic proteins: A realistic alternative to bone grafting for alveolar reconstruction. *Oral and Maxillofacial Surgery Clinics of North America* 19: 535–551, doi:10.1016/j.coms.2007.07.004.

238. Gersbach, C. A. 2011. Engineered bioactive molecules. In *Comprehensive Biomaterials*, ed. P. Ducheyne, Editor-in-Chief, pp. 131–145. Elsevier, Durham, NC.

239. McKay, W., S. Peckham, and J. Badura. 2007. A comprehensive clinical review of recombinant human bone morphogenetic protein-2 (INFUSE ®Bone Graft). *International Orthopaedics* 31: 729–734, doi:10.1007/s00264–007–0418–6.

240. Chen, F.-M. et al. 2010. A review on endogenous regenerative technology in periodontal regenerative medicine. *Biomaterials* 31: 7892–7927, doi:10.1016/j.biomaterials.2010.07.019.

241. Madonna, R. and R. De Caterina. 2011. Stem cells and growth factor delivery systems for cardiovascular disease. *Journal of Biotechnology* 154: 291–297, doi:10.1016/j.jbiotec.2011.05.014.

242. Krishnamurthy, R. and M. C. Manning. 2002. The stability factor: Importance in formulation development. *Current Pharmaceutical Biotechnology* 3: 361–371, doi:10.2174/1389201023378229.

243. Flaumenhaft, R. and D. B. Rifkin. 1992. The extracellular regulation of growth factor action. *Molecular Biology of the Cell* 3: 1057–1065.

244. Hynes, R. O. 2009. The extracellular matrix: Not just pretty fibrils. *Science* 326: 1216–1219, doi:10.1126/science.1176009.

245. Santos, E., R. M. Hernández, J. L. Pedraz, and G. Orive. 2012. Novel advances in the design of three-dimensional bio-scaffolds to control cell fate: Translation from 2D to 3D. *Trends in Biotechnology* 30: 331–341, doi:10.1016/j.tibtech.2012.03.005.

246. Jeon, O., S. J. Song, S.-W. Kang, A. J. Putnam, and B.-S. Kim. 2007. Enhancement of ectopic bone formation by bone morphogenetic protein-2 released from a heparin-conjugated poly(l-lactic-co-glycolic acid) scaffold. *Biomaterials* 28: 2763–2771, doi:10.1016/j.biomaterials.2007.02.023.

247. Chen, Y. and T. J. Webster. 2009. Increased osteoblast functions in the presence of BMP-7 short peptides for nanostructured biomaterial applications. *Journal of Biomedical Materials Research–Part A* 91: 296–304.

248. Suzuki, Y. et al. 2000. Alginate hydrogel linked with synthetic oligopeptide derived from BMP-2 allows ectopic osteoinduction in vivo. *Journal of Biomedical Materials Research* 50: 405–409, doi:10.1002/(sici)1097–4636(20000605)50:3<405::aid-jbm15>3.0.co;2-z.

249. Saito, A. et al. 2006. Repair of 20-mm long rabbit radial bone defects using BMP-derived peptide combined with an α-tricalcium phosphate scaffold. *Journal of Biomedical Materials Research–Part A* 77A: 700–706, doi:10.1002/jbm.a.30662.

250. Chen, F.-M., M. Zhang, and Z.-F. Wu. 2010. Toward delivery of multiple growth factors in tissue engineering. *Biomaterials* 31: 6279–6308, doi:10.1016/j.biomaterials.2010.04.053.

251. Richardson, T. P., M. C. Peters, A. B. Ennett, and D. J. Mooney. 2001. Polymeric system for dual growth factor delivery. *Nature Biotechnology* 19: 1029–1034.

252. Moghe, A. K. and B. S. Gupta. 2008. Co-axial electrospinning for nanofiber structures: Preparation and applications. *Polymer Reviews* 48: 353–377, doi:10.1080/15583720802022257.
253. Davis, H. and J. Leach. 2011. Designing bioactive delivery systems for tissue regeneration. *Annals of Biomedical Engineering* 39: 1–13, doi:10.1007/s10439–010–0135-y.
254. Schofer, M. D. et al. 2011. Functionalisation of PLLA nanofiber scaffolds using a possible cooperative effect between collagen type I and BMP-2: Impact on growth and osteogenic differentiation of human mesenchymal stem cells. *Journal of Materials Science. Materials in Medicine* 22: 1753–1762.
255. McLean, G. W. et al. 2005. The role of focal-adhesion kinase in cancer—a new therapeutic opportunity. *Nature Reviews Cancer* 5: 505–515.
256. Li, C., C. Vepari, H.-J. Jin, H. J. Kim, and D. L. Kaplan. 2006. Electrospun silk-BMP-2 scaffolds for bone tissue engineering. *Biomaterials* 27: 3115–3124, doi:10.1016/j.biomaterials.2006.01.022.
257. Kim, H. J. et al. 2008. Bone tissue engineering with premineralized silk scaffolds. *Bone* 42: 1226–1234, doi:10.1016/j.bone.2008.02.007.
258. Hosseinkhani, H., M. Hosseinkhani, A. Khademhosseini, and H. Kobayashi. 2007. Bone regeneration through controlled release of bone morphogenetic protein-2 from 3-D tissue engineered nano-scaffold. *Journal of Controlled Release* 117: 380–386, doi:10.1016/j.jconrel.2006.11.018.
259. Kolambkar, Y. M. et al. 2011. An alginate-based hybrid system for growth factor delivery in the functional repair of large bone defects. *Biomaterials* 32: 65–74, doi:10.1016/j.biomaterials.2010.08.074.
260. Kolambkar, Y. M. et al. 2011. Spatiotemporal delivery of bone morphogenetic protein enhances functional repair of segmental bone defects. *Bone* 49: 485–492, doi:10.1016/j.bone.2011.05.010.
261. Boerckel, J. D. et al. 2011. Effects of protein dose and delivery system on BMP-mediated bone regeneration. *Biomaterials* 32: 5241–5251, doi:10.1016/j.biomaterials.2011.03.063.
262. Yilgor, P., N. Hasirci, and V. Hasirci. 2010. Sequential BMP-2/BMP-7 delivery from polyester nanocapsules. *Journal of Biomedical Materials Research Part A* 93: 528–536.
263. Yilgor, P., K. Tuzlakoglu, R.L. Reis, N. Hasirci, and V. Hasirci. 2009. Incorporation of a sequential BMP-2/BMP-7 delivery system into chitosan-based scaffolds for bone tissue engineering. *Biomaterials* 30: 3551–3559.
264. Evans, C. H. 2012. Gene delivery to bone. *Advanced Drug Delivery Reviews* 64: 1331–1340, doi:10.1016/j.addr.2012.03.013.
265. Park, J. et al. 2003. Bone regeneration in critical size defects by cell-mediated BMP-2 gene transfer: A comparison of adenoviral vectors and liposomes. *Gene Therapy* 10: 1089–1098.
266. Rundle, C. H. et al. 2003. In vivo bone formation in fracture repair induced by direct retroviral-based gene therapy with bone morphogenetic protein-4. *Bone* 32: 591–601, doi:10.1016/s8756–3282(03)00096–6.
267. Santos, J. L., E. Oramas, A. P. Pêgo, P. L. Granja, and H. Tomás. 2009. Osteogenic differentiation of mesenchymal stem cells using PAMAM dendrimers as gene delivery vectors. *Journal of Controlled Release* 134: 141–148, doi:10.1016/j.jconrel.2008.11.007.
268. Curtin, C. M. et al. 2012. Innovative collagen nano-hydroxyapatite scaffolds offer a highly efficient non-viral gene delivery platform for stem cell-mediated bone formation. *Advanced Materials* 24: 749–754, doi:10.1002/adma.201103828.
269. Huang, Y. C., C. Simmons, D. Kaigler, K. G. Rice, and D. J. Mooney. 2005. Bone regeneration in a rat cranial defect with delivery of PEI-condensed plasmid DNA encoding for bone morphogenetic protein-4 (BMP-4). *Gene Therapy* 12: 418–426.
270. Tierney, E. G., G. P. Duffy, A. J. Hibbitts, S.-A. Cryan, and F. J. O'Brien. 2012. The development of non-viral gene-activated matrices for bone regeneration using polyethyleneimine (PEI) and collagen-based scaffolds. *Journal of Controlled Release* 158: 304–311, doi:10.1016/j.jconrel.2011.11.026.
271. Daamen, W. F. et al. 2003. Preparation and evaluation of molecularly-defined collagen–elastin–glycosaminoglycan scaffolds for tissue engineering. *Biomaterials* 24: 4001–4009, doi:10.1016/s0142–9612(03)00273–4.

272. Prasad, C. K. and L. K. Krishnan. 2008. Regulation of endothelial cell phenotype by biomimetic matrix coated on biomaterials for cardiovascular tissue engineering. *Acta Biomaterialia* 4: 182–191, doi:10.1016/j.actbio.2007.05.012.

273. Shin, H., S. Jo, and A. G. Mikos. 2003. Biomimetic materials for tissue engineering. *Biomaterials* 24: 4353–4364, doi:10.1016/s0142–9612(03)00339–9.

274. Williams, D. F. 2011. The role of short synthetic adhesion peptides in regenerative medicine; The debate. *Biomaterials* 32: 4195–4197, doi:10.1016/j.biomaterials.2011.02.025.

275. Barker, T. H. 2011. The role of ECM proteins and protein fragments in guiding cell behavior in regenerative medicine. *Biomaterials* 32: 4211–4214, doi:10.1016/j.biomaterials.2011.02.027.

276. Anderson, D. G., S. Levenberg, and R. Langer. 2004. Nanoliter-scale synthesis of arrayed biomaterials and application to human embryonic stem cells. *Nature Biotechnology* 22: 863–866.

277. Susan L, B. 2011. Advantages of RGD peptides for directing cell association with biomaterials. *Biomaterials* 32: 4205–4210, doi:10.1016/j.biomaterials.2011.02.029.

278. Hersel, U., C. Dahmen, and H. Kessler. 2003. RGD modified polymers: Biomaterials for stimulated cell adhesion and beyond. *Biomaterials* 24: 4385–4415, doi:10.1016/s0142–9612(03)00343–0.

279. Anderson, D. G., J. A. Burdick, and R. Langer. 2004. Smart Biomaterials. *Science* 305: 1923–1924, doi:10.1126/science.1099987.

280. Collier, J. H., and T. Segura. 2011. Evolving the use of peptides as components of biomaterials. *Biomaterials* 32: 4198–4204, doi:10.1016/j.biomaterials.2011.02.030.

281. Martino, M. M. et al. 2009. Controlling integrin specificity and stem cell differentiation in 2D and 3D environments through regulation of fibronectin domain stability. *Biomaterials* 30: 1089–1097, doi:10.1016/j.biomaterials.2008.10.047.

282. Mizuno, M. and Y. Kuboki. 2001. Osteoblast-related gene expression of bone marrow cells during the osteoblastic differentiation induced by type I collagen. *Journal of Biochemistry* 129: 133–138.

283. Yoo, S. Y., M. Kobayashi, P. P. Lee, and S.-W. Lee. 2011. Early osteogenic differentiation of mouse preosteoblasts induced by collagen-derived DGEA-peptide on nanofibrous phage tissue matrices. *Biomacromolecules* 12: 987–996, doi:10.1021/bm1013475.

284. Lindley, E. M. et al. 2010. Small peptide (P-15) bone substitute efficacy in a rabbit cancellous bone model. *Journal of Biomedical Materials Research Part B: Applied Biomaterials* 94B: 463–468, doi:10.1002/jbm.b.31676.

285. Yang, X. B., R. S. Bhatnagar, S. Li, and R. O. C. Oreffo. 2004. Biomimetic collagen scaffolds for human bone cell growth and differentiation. *Tissue Engineering* 10: 1148–1159.

286. Schenk, S. and V. Quaranta. 2003. Tales from the crypt[ic] sites of the extracellular matrix. *Trends in Cell Biology* 13: 366–375, doi:10.1016/s0962–8924(03)00129–6.

287. Comisar, W. A., D. J. Mooney, and J. J. Linderman. 2011. Integrin organization: Linking adhesion ligand nanopatterns with altered cell responses. *Journal of Theoretical Biology* 274: 120–130, doi:10.1016/j.jtbi.2011.01.007.

288. Comisar, W. A., N. H. Kazmers, D. J. Mooney, and J. J. Linderman. 2007. Engineering RGD nanopatterned hydrogels to control preosteoblast behavior: A combined computational and experimental approach. *Biomaterials* 28: 4409–4417, doi:10.1016/j.biomaterials.2007.06.018.

289. Mooney, D. J. and H. Vandenburgh. 2008. Cell delivery mechanisms for tissue repair. *Cell Stem Cell* 2: 205–213, doi:10.1016/j.stem.2008.02.005.

290. Jorgensen, C., J. Gordeladze, and D. Noel. 2004. Tissue engineering through autologous mesenchymal stem cells. *Current Opinion in Biotechnology* 15: 406–410, doi:10.1016/j.copbio.2004.08.003.

291. Simón-Yarza, T., E. Garbayo, E. Tamayo, F. Prósper, and J. Blanco-Prieto. 2012. Drug delivery in tissue engineering: General concepts. In *Nanostructured Biomaterials for Overcoming Biological Barriers*, eds M. J. Alonso and N. S. Csaba, pp. 501–525. RSC Publishing, Cambridge, U.K.

292. Pountos, I., D. Corscadden, P. Emery, and P. V. Giannoudis. 2007. Mesenchymal stem cell tissue engineering: Techniques for isolation, expansion and application. *Injury* 38, Supplement 4: S23–S33, doi:10.1016/s0020–1383(08)70006–8.

293. Caplan, A. I. 2007. Adult mesenchymal stem cells for tissue engineering versus regenerative medicine. *Journal of Cellular Physiology* 213: 341–347, doi:10.1002/jcp.21200.

294. Schipani, E. and H. M. Kronenberg. 2008. Adult mesenchymal stem cells. In *StemBook [Internet]*. Harvard Stem Cell Institute, Cambridge, MA.

295. Howard, D., L. D. Buttery, K. M. Shakesheff, and S. J. Roberts. 2008. Tissue engineering: Strategies, stem cells and scaffolds. *Journal of Anatomy* 213: 66–72, doi:10.1111/j.1469-7580.2008.00878.x.

296. Edalat, F., H. Bae, S. Manoucheri, J. Cha, and A. Khademhosseini. 2012. Engineering approaches toward deconstructing and controlling the stem cell environment. *Annals of Biomedical Engineering* 40: 1301–1315, doi:10.1007/s10439-011-0452-9.

297. Scadden, D. T. 2006. The stem-cell niche as an entity of action. *Nature* 441(7097): 1075–1075.

298. Eshghi, S. and D. V. Shaffer. 2008. Engineering microenvironments to control stem cell fate and function. In *StemBook [Internet]*. Harvard Stem Cell Institute, Cambridge, MA.

299. Discher, D. E., D. J. Mooney, and P. W. Zandstra. 2009. Growth factors, matrices, and forces combine and control stem cells. *Science* 324: 1673–1677, doi:10.1126/science.1171643.

300. Gentleman, E. et al. 2009. Comparative materials differences revealed in engineered bone as a function of cell-specific differentiation. *Nature Materials* 8: 763–770, doi:http://www.nature.com/nmat/journal/v8/n9/suppinfo/nmat2505_S1.html.

301. Even-Ram, S., V. Artym, and K. M. Yamada. 2006. Matrix control of stem cell fate. *Cell* 126: 645–647, doi:10.1016/j.cell.2006.08.008.

302. Engler, A. J., S. Sen, H. L. Sweeney, and D. E. Discher. 2006. Matrix elasticity directs stem cell lineage specification. *Cell* 126: 677–689, doi:10.1016/j.cell.2006.06.044.

303. Bruder, S. P., D. J. Fink, and A. I. Caplan. 1994. Mesenchymal stem cells in bone development, bone repair, and skeletal regeneration therapy. *Journal of Cellular Biochemistry* 56: 283–294, doi:10.1002/jcb.240560303.

304. Papathanasopoulos, A. and P. V. Giannoudis. 2008. Biological considerations of mesenchymal stem cells and endothelial progenitor cells. *Injury* 39, Supplement 2: S21–S32, doi:10.1016/s0020-1383(08)70012-3.

305. Ratanavaraporn, J., H. Furuya, H. Kohara, and Y. Tabata. 2011. Synergistic effects of the dual release of stromal cell-derived factor-1 and bone morphogenetic protein-2 from hydrogels on bone regeneration. *Biomaterials* 32: 2797–2811, doi:10.1016/j.biomaterials.2010.12.052.

306. Stevens, M. M. et al. 2005. In vivo engineering of organs: The bone bioreactor. *Proceedings of the National Academy of Sciences of the United States of America* 102: 11450–11455, doi:10.1073/pnas.0504705102.

307. Francis, L. et al. 2010. Simultaneous electrospin–electrosprayed biocomposite nanofibrous scaffolds for bone tissue regeneration. *Acta Biomaterialia* 6: 4100–4109, doi:10.1016/j.actbio.2010.05.001.

308. Sharaf, B. et al. 2012. Three-dimensionally printed polycaprolactone and β-tricalcium phosphate scaffolds for bone tissue engineering: An in vitro study. *Journal of Oral and Maxillofacial Surgery* 70: 647–656, doi:10.1016/j.joms.2011.07.029.

309. Chung, H. J. and T. G. Park. 2007. Surface engineered and drug releasing pre-fabricated scaffolds for tissue engineering. *Advanced Drug Delivery Reviews* 59: 249–262, doi:10.1016/j.addr.2007.03.015.

310. Wei, G. and P. X. Ma. 2004. Structure and properties of nano-hydroxyapatite/polymer composite scaffolds for bone tissue engineering. *Biomaterials* 25: 4749–4757, doi:10.1016/j.biomaterials.2003.12.005.

311. Huang, Y. X., J. Ren, C. Chen, T. B. Ren, and X. Y. Zhou. 2008. Preparation and properties of poly(lactide-co-glycolide) (PLGA)/nano-hydroxyapatite (NHA) scaffolds by thermally induced phase separation and rabbit MSCs culture on scaffolds. *Journal of Biomaterials Applications* 22: 409–432, doi:10.1177/0885328207077632.

312. Gunn, J. and M. Zhang. 2010. Polyblend nanofibers for biomedical applications: Perspectives and challenges. *Trends in Biotechnology* 28: 189–197, doi:10.1016/j.tibtech.2009.12.006.

313. Lanza, R. P., R. S. Langer, and J. Vacanti. 2007. *Principles of Tissue Engineering*, 3rd edn., Academic Press, San Diego, CA.

314. Dzenis, Y. 2004. Spinning continuous fibers for nanotechnology. *Science* 304: 1917–1919, doi:10.1126/science.1099074.

315. Sill, T. J. and H. A. von Recum. 2008. Electrospinning: Applications in drug delivery and tissue engineering. *Biomaterials* 29: 1989–2006.
316. Lao, L., Y. Wang, Y. Zhu, Y. Zhang, and C. Gao. 2011. Poly(lactide-co-glycolide)/hydroxyapatite nanofibrous scaffolds fabricated by electrospinning for bone tissue engineering. *Journal of Materials Science: Materials in Medicine* 22: 1873–1884, doi:10.1007/s10856–011–4374–8.
317. Maretschek, S., A. Greiner, and T. Kissel. 2008. Electrospun biodegradable nanofiber nonwovens for controlled release of proteins. *Journal of Controlled Release* 127: 180–187, doi:10.1016/j.jconrel.2008.01.011.
318. Nam, J., Y. Huang, S. Agarwal, and J. Lannutti. 2008. Materials selection and residual solvent retention in biodegradable electrospun fibers. *Journal of Applied Polymer Science* 107: 1547–1554, doi:10.1002/app.27063.
319. Phipps, M. C., W. C. Clem, J. M. Grunda, G. A. Clines, and S. L. Bellis. 2012. Increasing the pore sizes of bone-mimetic electrospun scaffolds comprised of polycaprolactone, collagen I and hydroxyapatite to enhance cell infiltration. *Biomaterials* 33: 524–534, doi:10.1016/j.biomaterials.2011.09.080.
320. Schneider, O. D. et al. 2008. Cotton wool-like nanocomposite biomaterials prepared by electrospinning: In vitro bioactivity and osteogenic differentiation of human mesenchymal stem cells. *Journal of Biomedical Materials Research Part B: Applied Biomaterials* 84B: 350–362, doi:10.1002/jbm.b.30878.
321. Nie, H. and C.-H. Wang. 2007. Fabrication and characterization of PLGA/HAp composite scaffolds for delivery of BMP-2 plasmid DNA. *Journal of Controlled Release* 120: 111–121, doi:10.1016/j.jconrel.2007.03.018.
322. Service, R. F. 2001. Coated nanofibers copy what's bred in the bone. *Science* 294: 1635–1637, doi:10.1126/science.294.5547.1635a.
323. Zhang, S. 2002. Emerging biological materials through molecular self-assembly. *Biotechnology Advances* 20(5–6): 321–339.
324. Woolfson, D. N. and M. G. Ryadnov. 2006. Peptide-based fibrous biomaterials: Some things old, new and borrowed. *Current Opinion in Chemical Biology* 10: 559–567, doi:10.1016/j.cbpa.2006.09.019.
325. Hartgerink, J. D., E. Beniash, and S. I. Stupp. 2001. Self-assembly and mineralization of peptide-amphiphile nanofibers. *Science* 294: 1684–1688, doi:10.1126/science.1063187.
326. Beniash, E., J. D. Hartgerink, H. Storrie, J. C. Stendahl, and S. I. Stupp. 2005. Self-assembling peptide amphiphile nanofiber matrices for cell entrapment. *Acta Biomaterialia* 1: 387–397, doi:10.1016/j.actbio.2005.04.002.
327. Lee, J.-Y. et al. 2009. Osteoblastic differentiation of human bone marrow stromal cells in self-assembled BMP-2 receptor-binding peptide-amphiphiles. *Biomaterials* 30: 3532–3541, doi:10.1016/j.biomaterials.2009.03.018.
328. Matthews, J. A., G. E. Wnek, D. G. Simpson, and G. L. Bowlin. 2002. Electrospinning of collagen nanofibers. *Biomacromolecules* 3: 232–238, doi:10.1021/bm015533u.
329. Ma, Z., M. Kotaki, R. Inai, and S. Ramakrishna. 2005. Potential of nanofiber matrix as tissue-engineering scaffolds. *Tissue Engineering* 11: 101–109.
330. Li, W.-J., C. T. Laurencin, E. J. Caterson, R. S. Tuan, and F. K. Ko. 2002. Electrospun nanofibrous structure: A novel scaffold for tissue engineering. *Journal of Biomedical Materials Research* 60: 613–621, doi:10.1002/jbm.10167.
331. Bhattarai, S. R. et al. 2004. Novel biodegradable electrospun membrane: Scaffold for tissue engineering. *Biomaterials* 25: 2595–2602, doi:10.1016/j.biomaterials.2003.09.043.
332. Zhang, S. 2003. Fabrication of novel biomaterials through molecular self-assembly. *Nature Biotechnology* 21: 1171–1178.
333. MacPhee, C. E. and D. N. Woolfson. 2004. Engineered and designed peptide-based fibrous biomaterials. *Current Opinion in Solid State and Materials Science* 8: 141–149, doi:10.1016/j.cossms.2004.01.010.
334. Palmer, L. C., C. J. Newcomb, S. R. Kaltz, E. D. Spoerke, and S. I. Stupp. 2008. Biomimetic systems for hydroxyapatite mineralization inspired by bone and enamel. *Chemical Reviews* 108: 4754–4783, doi:10.1021/cr8004422.

Index

A

Abraxane®, 11
Actuated vesicle release systems, 186–188
Adaptive immune system, 126
Adjuvants
  AF03, 136
  aluminum salts, 132–133
  antigen, 128
  AS03, 135–136
  benefits, 127
  MF59, 134
  types, 127
Adsorption
  advantages, 63–64
  chemical, 66
  disadvantages, 63–64
  disposable H_2O_2 biosensors, 69
  DNA immobilization, 70
  dynamic batch process, 66
  electrode biosensor, 70
  electrodeposition process, 66
  label-free electrochemical impedance
    immunosensor, 70
  physical, 66
  reactor loading process, 66
  S. flexneri immunosensor, 70–71
  static process, 66
AF03 adjuvant, 136
Alginate-enclosed chitosan-coated bioceramic
    nanoparticles, 170–171
Amperometric biosensors, 69, 72, 74, 78, 83
Amphiphilic diblock polymer micelles, 12
Antibody-drug conjugates (ADCs), 3–4
Aptamers
  advantages, 33–34
  biomedical sensing/imaging
    carbon nanotubes, 40–42
    dye-doped silica nanoparticles, 38–39
    gold nanoparticles, 34–37
    hybrid aptamer–dendrimers, 43
    magnetic nanoparticles, 39–40
    platinum nanoparticles, 37–38
    quantum dots, 42–43
    silver nanoparticles, 37
  definition, 30

functionality, 30
in vitro cancer cell detection
  AS1411, 47
  colorimetric assay, 44
  electrochemical determination, 45
  electrochemiluminescence assays, 45
  flow cytometry, 45–46
  fluorescent imaging, 45–46
  G-rich oligonucleotides, 47
  mucin 1, 47
  prostate-specific membrane
    antigen,47
  Ramos cells, 44–45
  strip-based assay, 44
  thiolated aptamer, 44
  two-photon scattering technique, 46
in vivo imaging
  aptamer-adapted silver nanoparticles,
    48–49
  aptamer nanoflares, 48
  cancer targeting using MFR-AS1411
    particles, 50–51
  molecular-specific contrast agents, 48
  molecular recognition principle, 30–31
  nucleic acid antibodies, 31
  properties, 33–34
  smart drug delivery agents, 49–52
  structure, 30–31
  target molecules, 30
AS03, 135–136

B

BBB, *see* Blood–brain barrier (BBB)
Bevacizumab, 3
Bioactive glasses, 165
Bioactive signals
  bioinstructive ligands, 385
  ECM components, 389–390
  growth factors
    encapsulation strategies, 387–388
    gene delivery, 388–389
    GF-based therapeutics, 386
    surface adsorption, 387
Biobattery, 112–113

Printed and bound by CPI Group (UK) Ltd, Croydon, CR0 4YY

18/10/2024

01776253-0011